The Falling Sky

The Falling Sky

WORDS OF A YANOMAMI SHAMAN

Davi Kopenawa

Bruce Albert

Translated by

Nicholas Elliott

and

Alison Dundy

THE BELKNAP PRESS OF
HARVARD UNIVERSITY PRESS

Cambridge, Massachusetts
London, England
2013

First published as *La chute du ciel: Paroles d'un chaman yanomami,*
copyright © 2010 PLON

Library of Congress Cataloging-in-Publication Data

Kopenawa, Davi.
[La chute du ciel. English]
The falling sky : words of a Yanomami shaman / Davi Kopenawa, Bruce Albert ;
translated by Nicholas Elliott and Alison Dundy.
pages cm
Includes bibliographical references and index.
ISBN 978-0-674-72468-6 (alk. paper)
1. Kopenawa, Davi.   2. Shamans—Brazil—Biography.
3. Yanomamo Indians—Brazil—Biography.
4. Yanomamo Indians—History—20th century.
5. Shamanism—Brazil—History—20th century.
I. Albert, Bruce, 1952–   II. Title
F2520.1.Y3K6613   2013
305.898′92—dc23   2013008942

Contents

Foreword

An extraordinary man makes himself heard in this book: Davi Kopenawa. His breadth of vision and the meticulous care with which he describes the Yanomami cosmology and way of life take us on a voyage into an Amerindian spirit world. It is a world that may be imaginary for us, but it is profoundly real for him, as he sees the *xapiri* (images of the mythological animal ancestors), speaks to them, and shares his life with them.

This voice is that of a prophet: "The forest is alive. The white people persist in destroying it. We are dying one after another, and so will they. In the end, all of the shamans will perish and the sky will collapse. Before it is too late," the prophet adds, "I want to talk to you about a time long ago when the animal ancestors transformed. Thanks to my shaman elders, I learned how to call them. I see them, I share life with them, and I listen to them.

"You must hear me—time is short."

A Yanomami shaman and spokesperson, known and admired in the Amazon and beyond, Davi Kopenawa continues to live with great simplicity among his people in his traditional home. In this unique book, which he developed in collaboration with Bruce Albert, an anthropologist at the top of his field, he challenges us to see ourselves as the People of Merchandise and to rebel against the damage wrought by the industrial world on the Amazon rain forest.

Bruce Albert found his calling in working with the Yanomami and Davi Kopenawa. Renouncing conventional academic ambitions—although he has all the necessary credentials—he chose to pursue an intellectual and inner adventure in the Amazon. He met his teachers among the Indians who adopted him. Davi Kopenawa describes the book's purpose, and its origins in their friendship: "A long time ago, you came to live with [the Yanomami] and you spoke like a ghost. Little by little, you learned to imitate our language and to laugh with us. We were young . . . Later I told you: 'If you want to take my words, do not destroy them. They are the words of *Omama* [the Yanomami demiurge] and the *xapiri*. First draw them on image skins, then look at them often . . . Like me, you became wiser as you got older . . . Now I would like [these words] to divide themselves and propagate over long distances so they can truly be heard."

Bruce Albert captured these words from a complex tradition, and undertook the enormous task of translating them into French. Few anthropologists have ever undertaken such a difficult endeavor. The publication of this book is a significant event in the history of great eyewitness accounts.

Eerie and unsettling as they are, Davi Kopenawa's visions and warnings may be hastily dismissed as the phantasmagoria of a rain forest shaman. But if you are skeptical, pause to listen. Remember the enduring power of ancient cosmologies, and the philosophical significance of differences in the long history of human evolution. Our thought is enriched by dialogue with the strange and unfamiliar. For any dialogue to take place, it is necessary to respect cultures and understand the immense variety of long histories of thought in different contexts. A purely technical rationality without conscience or spirituality, dominated by material interests, leads the industrialized world to the destruction of our planet. The supposedly technologically backward peoples may be tomorrow's wise men.

Readers from around the world need to listen to Davi Kopenawa's painful appeal for his people's well-being and our own. With the destruction of Amazon rain forest, mankind may be approaching an ecological collapse—what Davi Kopenawa calls the time of the "falling sky."

"What is the purpose of education?" asked Jean-Jacques Rousseau. "To learn how to live better." Davi Kopenawa, Yanomami philosopher and great advocate of ecology, is one of the teachers we have been waiting for.

—*Jean Malaurie*
Director Emeritus for Research
Centre National de Recherche Scientifique (CNRS)

Maps

The Yanomami Territory in Brazil (Terra Indígena Yanomami).
© F.-M. Le Tourneau/P. Mérienne

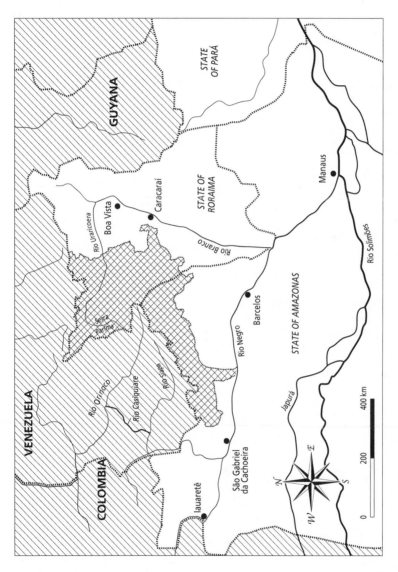

Location of the Terra Indígena Yanomami. © F.-M. Le Tourneau/P. Mérienne

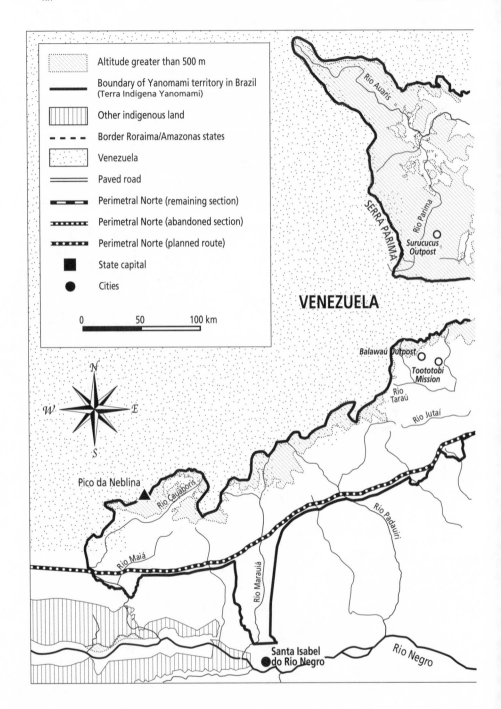

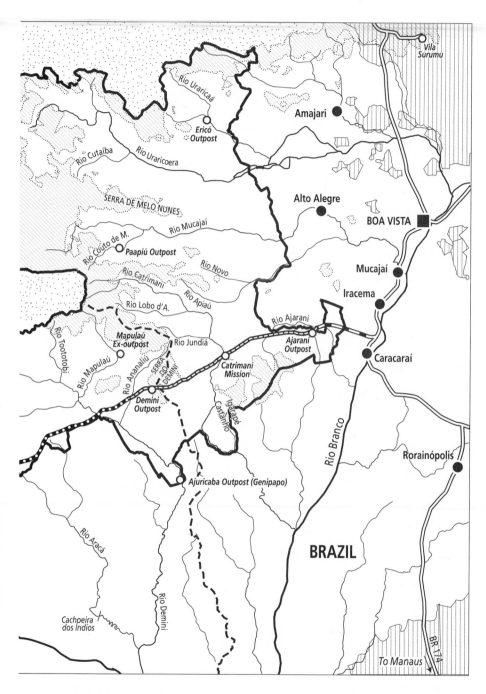

Detailed map of the Terra Indígena Yanomami (cited Portuguese toponyms).
© F.-M. Le Tourneau/P. Mérienne

Niyayopa tʰëri

Amikoapë tʰëri

Xama xi pora

Hero u

Hayowari

Arahai tʰëri

Hʷaxi tʰëri

Amatʰa u

VENEZUELA

Hʷaxima u

Manito u

Mai koxi

Kōana u

Hʷara u

Hayowa tʰëri

Yoyo roopë

Sina tʰa

Mōra mahi araopë

Maima sikɨ u

Ariwaa tʰëri

Warapi u

Paxoto u

Marakana

Wari mahi

Konapuma tʰëri

Weyuku tʰëri

Wanapi u

Weyahana u

Tʰootʰotʰopi

Werihi sihipi u

Parawa u

Maharu u

Iwahikaropë tʰëri

Kapirota u

N

W   E

S

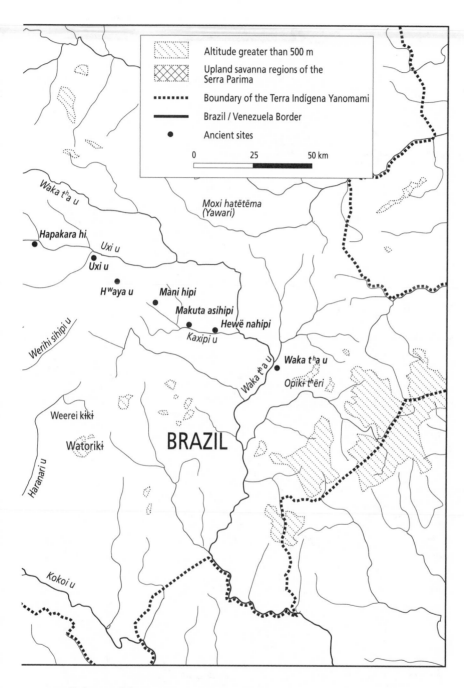

Detailed map of the Terra Indígena Yanomami (cited Yanomami toponyms).
© F.-M. Le Tourneau/P. Mérienne

Location of cited ethnic groups. © F.-M. Le Tourneau/P. Mérienne

THE FOREST IS ALIVE. It can only die if the white people persist in destroying it. If they succeed, the rivers will disappear underground, the soil will crumble, the trees will shrivel up, and the stones will crack in the heat. The dried-up earth will become empty and silent. The *xapiri* spirits who come down from the mountains to play on their mirrors in the forest will escape far away. Their shaman fathers will no longer be able to call them and make them dance to protect us. They will be powerless to repel the epidemic fumes which devour us. They will no longer be able to hold back the evil beings who will turn the forest to chaos. We will die one after the other, the white people as well as us. All the shamans will finally perish. Then, if none of them survive to hold it up, the sky will fall.

—DAVI KOPENAWA

Setting the Scene

THIS BOOK—a life story, autoethnography, and cosmoecological manifesto—is an invitation to travel in the history and mind of Davi Kopenawa, Yanomami shaman. Born in the northern Brazilian Amazon along the upper Rio Toototobi, a region that was at that time very remote from the world of white people, Davi Kopenawa has been confronted in his extraordinary life with a series of representatives from the encroaching frontier: field agents of the Indian Protection Service (SPI),[1] soldiers, missionaries, road workers, gold prospectors, and ranchers. His stories and reflections, which I recorded in his language, transcribed, translated, and then arranged and edited in French (and now in English), present a hitherto unheard version—told with poetic and dramatic intensity, as well as perspicacity and humor—of the historic confrontation between Amerindians and the fringe of our "civilization."

From the time we began working together, Davi Kopenawa wanted his account to reach the largest possible audience. So before the adventure of reading his narrative begins, I offer some essential context here: a brief overview of the Yanomami in Brazil and their history; biographical sketches of Davi Kopenawa, whose spoken words are the origin of this book, and of the author of this chapter, who tried to render the wisdom and flavor of this Yanomami shaman's words in writing; and the story of how we met and produced this book together. All these subjects are addressed in greater detail in the final chapter, "How This Book Was Written," and in the appendixes.

The Yanomami in Brazil

The Yanomami[2] are a society of hunter-gatherers and slash-and-burn farmers who occupy an area of tropical forest comprising approximately 192,000 square kilometers located on both sides of the Serra Parima range, which divides the waters of the upper Orinoco (south of Venezuela) and the tributaries of the right bank of the Rio Branco and of the left bank of the Rio Negro (in northern Brazil).[3] They constitute a vast and isolated cultural and linguistic group, subdivided into several languages and related dialects. Their total population is estimated to be slightly more than 33,000 people,[4] which makes them one of the largest Amerindian groups in the Amazon to have mostly held on to their traditional way of life.

The Yanomami territory in Brazil, legally recognized in 1992 as the Terra Indígena Yanomami, extends over 96,650 square kilometers— an area slightly larger than some European countries, such as Portugal, Hungary, and Ireland. Their population of approximately 16,000 people is distributed among some 230 local groups. These communities are usually formed by what anthropologists call an endogamous set of cognatic kin. They are composed of several families linked through cross-cousin marriages, repeated from one generation to the next, who reside together in one or more ring- or cone-shaped communal houses.[5]

The first sporadic contact the Yanomami of Brazil had with white people—collectors of forest products, foreign explorers, military personnel, and SPI agents—was in the early decades of the twentieth century. Then, from the 1940s through the 1960s, several Catholic and Protestant missions, as well as SPI outposts, were opened on the periphery of Yanomami territory. They provided the first regular points of contact and sources of trade for manufactured goods for the Indians, but such contact also led to deadly epidemics among them. In the early 1970s, these initial incursions by white people suddenly intensified, first with the opening of the northern section of the Trans-Amazonian highway (the Perimetral Norte) on the southern end of Yanomami territory, and then, after a ten-year respite, with an unprecedented gold rush into its heart. Highway construction was abandoned in 1976, and the invasion by gold prospectors had been reined in somewhat by the mid-1990s. Gold prospecting has been recently revived, however, by a surge in gold prices on international markets, and the Terra Indígena Yanomami is also threatened by new interests reaching into the western part of Roraima state, including mining companies, agricultural colonization, and cattle ranching.

Davi Kopenawa, Shaman and Yanomami Spokesperson

Davi Kopenawa was born around 1956 in *Marakana*, a large communal house of approximately two hundred people in the tropical forest foothills along the upper Rio Toototobi, in the far northeast part of Amazonas state in Brazil, near the Venezuelan border. Since the late 1970s, he has lived with his in-laws' community at the foot of the "Wind Mountain" *(Watoriki)*, on the left bank of the Rio Demini, less than one hundred kilometers southeast of the Rio Toototobi.

As a child, Davi Kopenawa saw his origin group decimated by two successive epidemics of infectious illnesses, first one introduced by SPI agents in 1959 (or 1960), and then, later on, one brought by members of the New Tribes Mission. For several years, he was subjected to proselytizing by these missionaries who had settled along the Rio Toototobi in 1963. He owes them his biblical forename, the skill of writing, and a less-than-enchanted view of Christianity. Despite his initial curiosity, he was quickly repelled by the missionaries' fanaticism and obsession with sin. He rebelled against their influence after he lost most of his relatives to a measles epidemic transmitted by the daughter of one of the pastors in 1967.

Orphaned and outraged by the repeated loss of loved ones, yet intrigued by the material power of white people, as an adolescent Davi Kopenawa left the region where he was born to take a job with the National Indian Foundation (FUNAI,[6] which had recently succeeded the SPI) along the lower Rio Demini, at the Ajuricaba Outpost. He then tried, as he says in his own words, to "become a white man," and ended up with a case of tuberculosis. This misadventure cost him a long hospitalization, which he used to learn basic Portuguese. When he recovered, he returned for a period of time to his communal house in Toototobi before being employed in 1976 as an interpreter for FUNAI after the opening of the Perimetral Norte highway. In that capacity, he spent several years traveling across the greater part of Yanomami territory, gaining knowledge of its breadth and its cultural cohesion in spite of local differences. This experience also gave him a more precise understanding of the economic greed animating those he calls "the People of Merchandise," and the threats they represent to the existence of the forest and the survival of his people.

Finally, weary of his travels, Davi Kopenawa settled for good in *Watoriki* in the early 1980s after marrying the daughter of the community's "great man" *(pata t^{h}ë)*, a renowned shaman and firm traditionalist, who became his mentor in shamanic journeying. This initiation enabled Davi Kopenawa to pick up the thread of his shamanic calling, which began in childhood but had been interrupted by the arrival of white people. Shamanism later provided him with the basis for his own cosmological reflection on commodity fetishism, the destruction of the rain forest, and climate change.[7]

In the late 1980s, more than a thousand Yanomami perished in Brazil from illnesses and violence resulting from the invasion of their territory by some 40,000 gold prospectors. This tragedy rekindled Davi Kopenawa's childhood memories of the decimation of his kin, leaving him distraught. Having struggled for several years in Brazil to obtain legal recognition for the Yanomami territory, he began an international campaign to defend his people and the Amazon. His unique experience with white people, his extraordinary firmness of character, and the legitimacy that came with his initiation as a shaman made him the most influential spokesperson for the Yanomami cause in Brazil and abroad. Over the course of the 1980s and 1990s, he visited several European countries as well as the United States. In 1988, the United Nations awarded him the Global 500 Award for his contribution to defense of the environment. In 1989, the nongovernmental organization Survival International invited Davi Kopenawa to stand as its representative to receive the Right Livelihood Award to raise international awareness about the dramatic situation of the Yanomami and their struggle to protect their lands. In May 1992, during the United Nations conference on the environment and development, which took place in Rio de Janeiro (the "Earth Summit," or Eco '92), he finally obtained legal recognition from the Brazilian government for a vast area of tropical forest reserved for the exclusive use of his people: the Terra Indígena Yanomami. He was decorated in 1999 with the Order of the Rio Branco by the president of Brazil "for exceptional merit." In 2004, he became the founding president of the *Hutukara* association, which represents the majority of Yanomami in Brazil.[8] In December 2008 he received special honorary mention for the prestigious Bartolomé de las Casas Award, granted by the Spanish government for defense of the rights of Native American peoples. In 2009, he was decorated in Brazil with the Order of Cultural Merit.

Davi Kopenawa is a complex man, alternately tense or welcoming, introverted or charismatic. Every episode in his personal trajectory attests to his remarkable intellectual curiosity, his unfailing determination, and his great personal courage. He has six children, including a recently adopted little girl, and four grandchildren whom he and his wife, Fatima, lovingly care for. He lives with his wife and his youngest children in a section of the vast communal dwelling of *Watoriki*. Their family hearth is indistinguishable from any other in the house. Despite his fame, he remains utterly detached from material things, and he takes pride only in challenging the arrogant deafness of white people. In the forest, his passion is to respond to the songs of shamanic spirits; in the city, to advocate for his people. A tireless defender of Yanomami territory and rights, he remains a zealous partisan of the tradition of his elders and especially their shamanic knowledge.

Bruce Albert, Anthropologist

I was born in 1952 in Morocco, earned a doctorate in Anthropology from the University of Paris X Nanterre, and am now director of research at the Institute of Research for Development in Paris. I began my long-term fieldwork with the Yanomami in Brazil in March 1975. Just twenty-three years old, I was freshly graduated from a Paris that was, at that time, effervescent in social sciences debates. Exhilarated by intense ethnographic readings, I suddenly found myself plunged into chaos on the Amazonian frontier during the construction of the Perimetral Norte highway near the Venezuelan border, along the upper Rio Catrimani. The Yanomami charmed me immediately with their elegance and mocking pride as they wove their way among the giant bulldozers opening the road, or humorously outsmarted the intrusive good intentions of a local Italian priest. I was revolted by the spectacle of the omnivorous roadwork, blindly gutting the tropical forest, and the ensuing illnesses and social degradation of its inhabitants. Given my temperament—more disposed to the quest for real-life knowledge and social engagement than the pursuit of academic ambitions—I came to understand that the only acceptable ethnographic research for me would require a lasting commitment alongside the people with whom I had decided to work. Anthropology thus became for me an intellectual adventure and a way of life, more than a profession whose institutional aspects were far less seductive. Since then my existence has been guided by the political consequences of that first encounter with the Yanomami. This personal lifelong adventure as an "engaged observer" has not been, however, incompatible with an appetite for anthropological analysis—far from that.

While pursuing ethnographic research on different aspects of Yanomami society and culture, I co-founded a nongovernmental organization in Brazil in 1978, the Comissão Pró-Yanomami (CCPY),[9] which waged a fourteen-year campaign alongside Davi Kopenawa and won legal recognition for the Yanomami lands in Brazil in 1992. For twenty-five years the CCPY led healthcare programs, created bilingual schools, and sustained environmental projects; I had a direct hand in all of them.[10] Along the way, I gained an acceptable knowledge of one of the Yanomami languages—the one that is spoken in the region where Davi Kopenawa was born and where he presently lives. I have visited the Yanomami practically every year, sometimes several times a year, for the last thirty-eight years, and, as one might imagine, I am tied to Davi Kopenawa through a long history of friendship and shared struggle.

The Meeting

I met Davi Kopenawa for the first time in 1978 under odd and amusing circumstances (see the final chapter, "How This Book Was Written"). We were both in our twenties. I had just begun my second phase of ethnographic fieldwork in Yanomami territory (having already spent a year on the upper Rio Catrimani from 1975 to 1976). Davi Kopenawa was an interpreter in the FUNAI outposts set up along the Perimetral Norte highway, the construction of which had been abandoned a year and a half earlier. Then, in 1981, I stayed for several months in the area where he was born along the Rio Toototobi, where we met again while he was visiting his relatives. I had the opportunity there for firsthand encounters with the people and places that were important to him throughout his childhood and adolescence. Then, from 1985 on, the village where he had married, *Watoriki*, became my most frequent destination in Yanomami territory. I knew his father-in-law and shamanic mentor as well as most inhabitants of this community from my first journey in 1975 on the upper Rio Catrimani, where they lived at the time. My friendship with Davi Kopenawa grew increasingly close through lengthy visits to *Watoriki*, and through the bond of our common political involvement against the gold rush then ravaging Yanomami territory. This book originated in Davi Kopenawa's anguish and outrage over the decimation of his people by gold prospectors in the late 1980s. It would not have been possible without our longstanding trust and connection. We began the recording sessions that served as the basis for successive versions of the manuscript in December 1989 and continued through the early 2000s during stays in the forest or political events in the city. These sessions consisted mostly of free-ranging discussions, conducted in fits and starts over a time period spanning more than ten years. This is, therefore, a collection of narratives, thoughts, and conversations recorded in Yanomami about Davi Kopenawa's life, culture, and experience with the world of white people. As one might surmise, recomposing this prolific archipelago of words into French (and then into English) to write this book was no simple undertaking. The challenges of this complex process of editing and writing are also explained in detail in the last chapter, "How This Book Was Written."

The Book

Davi Kopenawa's narrative is the first inside account of Yanomami society and culture published in English since the extraordinary biography of Helena Valero edited by Ettore Biocca in 1970, *Yanoama: The Narrative of a Young Woman Kidnapped by Amazonian Indians*. These two books deal with experiences in two successive periods. Helena Valero escaped from captivity in 1956, the year Davi Kopenawa was born. One takes place in Venezuela, the other in Brazil, and the identity and trajectory of their narratives are the reverse of each other.

Yanoama narrates the tribulations of a young Brazilian girl captured by the Yanomami in 1932 when she was thirteen years old, at a time when the Yanomami warriors of the highland region between the Rio Negro and the Rio Casiquiare tried to repel the rubber tappers and other harvesters of forest products who were invading their territory.[11] Davi Kopenawa narrates a personal history and shares reflections about white people from the point of view of a contemporary Yanomami shaman and spokesperson. His narrative covers a period that begins in his early childhood, before the founding of the first missionary outpost in his native region in the early 1960s, and continues with his unique odyssey, starting in the late 1970s, towards the world of white people.

Yet this book is not a usual Amerindian biography built, like the twentieth-century North American classics, as a documentary life story ghostwritten by an anthropologist.[12] Nor is it an ethnobiography pertaining to a traditional narrative genre, merely transcribed and translated by an ethnographer whose role is reduced to that of a secretary.

Davi Kopenawa's narrative goes far beyond prevailing canons of autobiography—our own or those of the Yanomami.[13] His accounts of key episodes in his life inseparably intertwine personal events and collective history. Moreover, he always expresses himself through complex overlapping genres and styles: myths and dream stories, shamanic visions and prophecies, autoethnography and cross-cultural comparison, reported speech and exhortations. In addition, the book is the result of a written and oral process that was continually shaped by the intersecting projects of the two authors—a Yanomami shaman very wise to the world of white people, and an ethnographer quite familiar with the world of his long-time hosts. In short, this extensive book is the result of a complex collaborative endeavor at the fragile juncture of our two cultural universes.

At a critical time in his life and in the history of his people, Davi Kopenawa decided to entrust me with his words because of my close involvement with the Yanomami, and to put his words in writing so that they would find a path to an audience far beyond the forest where he was born. In doing so he hoped not only to denounce the direct threats affecting the Yanomami and the Amazon rain forest, but to launch an appeal, in his role as shaman, against the widespread damage caused by "the People of Merchandise" and the danger it represents for the future of humanity.[14] Davi Kopenawa's words thus constitute a multidimensional cosmological and ethnopolitical account based on an extraordinary effort at self-objectification and conviction. The text is an unprecedented narrative endeavor rooted in a life story and personal commitment that give him radical singularity, including within the Yanomami universe.

For my part, I did my best to render the poetic sensibility and conceptual richness of his way of thinking and speaking in a translation that stayed as close to possible to his words, yet with a writing style and form of composition that made them accessible to a nonspecialist public. I chose to illuminate his text with this brief introduction, a concluding chapter, and explanatory notes as well as supplementary material at the end of the book to avoid any intrusion of a patronizing interpretive authority that might threaten to overshadow Davi Kopenawa's words. I wished to avoid breaking up his narrative with extraneous reminders of my own presence or feelings. In presenting his account in this way to the reader, in all its singular power and otherness, I hope I have done justice to my mandate to make Davi Kopenawa's words heard and their strength felt. May they resonate and make an impact in our world.

This book is composed of three parts. The first, "Becoming Other," recounts the premises of Davi Kopenawa's shamanic calling and his initiation under his father-in-law's guidance. It also describes Yanomami shamanic cosmology and the multiple tasks of a Yanomami shaman, disclosing the knowledge he acquired by learning from his elders. The second part, "Metal Smoke," deals with different kinds of encounters with white people—initially Davi Kopenawa's own and that of his community, and then the overall experience of the Yanomami in Brazil. It opens with the shamanic rumors about distant strangers that preceded the first actual contacts, then passes through the arrival of missionaries and the opening on the Perimetral Norte highway, and concludes with the deadly encroachment on Yanomami lands by the gold prospectors *(garimpeiros)*. The third part, "The Falling Sky," traces, in reverse order, Davi Kopenawa's journeys—in Brazil, then Europe, and later the United States—to denounce the attacks on his people and the destruction of the forest. This account, told in the form of a succession of shamanic journeys, is intertwined with comparative cultural reflections and critique of certain aspects of our society, and unfolds into a cosmoecological prophecy about the death of shamans and the end of humanity.

Orthography, Pronunciation, and Glossaries

To give an idea of the pronunciation of Yanomami words and expressions cited in this book, the reader need only grasp some basic information, remembering that sounds not specifically mentioned here correspond approximately to similar sounds in English. In terms of vowels: *e* is pronounced like the vowel sound in *fate* in English, *u* is pronounced like the vowel sound in *food*, *ë* is the equivalent of the vowel sound in *but*, and *i* (barred i) is a sound between *i* and *u*. As for consonants: h^w is pronounced like an aspirated *h* with rounded lips (as in *which*), t^h is pronounced like a *t* followed by a light exhalation ("aspiration") as in *top,* and *x* is pronounced like *sh* in *ship.* For more information on the Yanomami language spoken by Davi Kopenawa and its written form, the interested reader can refer to Appendix A at the end of the book.

All Yanomami terms and expressions cited in this text appear in italics while the occasional Portuguese words used by Davi Kopenawa in the recordings we worked from are translated and displayed in boldface the first time they appear. The transcriptions of onomatopoeic forms, which are so delightful and subtly codified in Yanomami speech, were kept to a minimum to lighten the text. On the other hand, several interjections that are frequently used to introduce topics have been maintained. For example: *asi!* indicates anger; *awe!* signals approval; *haixopë!* denotes a positive response to some information; *ha!* indicates surprise (satisfied or ironic); *hou!* shows irritation; *ma!* expresses disapproval; and, finally, *oae!* indicates sudden recollection.

The numeric notations applied to the thirty-five myths (M 4 to M 362) cited in the endnotes refer to my contributions to the collection of Yanomami narratives by J. Wilbert and K. Simoneau, 1990 (see References). Curious readers can, of course, consult this compendium if they would like to deepen their knowledge of Yanomami cosmology. Plant and animal species mentioned in the text are identified in glossaries included at the end of the book, as are details concerning ethnonyms and toponyms. All of the illustrations in this book, except for the maps, were drawn by Davi Kopenawa.

—*B. A.*

Words Given

L ONG AGO you came to live among us and you spoke like a ghost.[1] Little by little you learned to imitate our language and to laugh with us. We were young and at the beginning you did not know me. Our ways of thinking and our lives were different because you are the son of those other people we call *napë pë*.[2] Your **professors** had not taught you to dream like we do. Yet you came to me and you became my friend. You put yourself by my side and later you wanted to know the words of the *xapiri,* whom you call **spirits**[3] in your language. So I entrusted you with my words and I asked you to carry them far away to let them be heard by the white people, who know nothing about us. We stayed sitting and talking in my house a long time, despite the horse fly and black fly bites. Few are the white people who have listened to our words in such a way. I gave you my **story**[4] so that you would answer those who ask themselves what the inhabitants of the forest think. In the past our elders[5] told them nothing of these things, for they knew that the white people did not understand their language. This is why my words will be new to those willing to hear them.

Later I told you: "If you want to take my words, do not destroy them. They are the words of *Omama*[6] and the *xapiri.* First draw them on image skins,[7] then look at them often. Then you will think: '*Haixopë!* This truly is the story of the spirits!' And later you will tell your children: 'These drawn words are those of a Yanomami who once told me how he became spirit and learned to speak to defend his forest.' When these **tapes** that hold the shadow of my words[8] are no longer working, do not throw them out. Do not burn them until they are very old and my stories have long since become drawings that white people can look at. *ɬnaha tʰa?* All right?"

Like me, you became wiser as you got older. You drew these words and stuck them on **paper skins** like I asked you to. They went far away from me. Now I would like them to divide themselves and propagate over long distances so they can truly be heard. I taught you these things so you would teach them to your people; to your elders, to your fathers and fathers-in-law, to your brothers and brothers-in-law, to the women you call "wives," to the young people who will call you "father-in-law." If they ask you: "How did you learn these things?" you will answer them: "I lived in the Yanomami's houses and ate their food for a long time. Little by little their language took hold in me. They entrusted me with their words because they are sad that white people are so ignorant about them."

White people don't think very far ahead. They are always too preoccupied with the things of the moment. This is why I would like them to be able to hear my words through the drawings you made. I would like these words to penetrate their minds. After they have understood my account, I would like the white people to tell themselves: "The Yanomami are other people than us, yet their words are right and clear. Now we understand what they think. These are words of truth! Their forest is beautiful and silent. They were created there and have lived in it without worry since the beginning of time. Their thought follows other paths than that of merchandise. They want to live their way. Their **custom** is different. They do not have image skins but they know the *xapiri* spirits and their songs. They want to defend their land because they want to continue to live there like they did before. Let it be so! If they do not protect it, their children will have no place to live happily. Then they will tell themselves that their fathers must truly have lacked wisdom to have left them nothing but bare and scorched land, permeated with epidemic smoke and crisscrossed by streams of dirty water!"

I would like white people to stop thinking that our forest is dead and placed here without reason. I would like to make them listen to the voice of the *xapiri* who play here incessantly, dancing on their glittering mirrors. Maybe they will want to defend it with us? I would also like their sons and daughters to understand our words. I would like them to make friendship with our sons and daughters in order not to grow up in ignorance. For if this forest is entirely devastated, no other forest will ever be born. I am a child of the inhabitants of this land from which the rivers flow, of these people who are the children, sons-in-law and daughters-in-law of *Omama*. I wish to offer white people these words and those of the *xapiri,* which appear in the time of dream. Our ancestors had them since the beginning of time. Then when it was my turn to become a shaman, *Omama*'s image placed them in my chest. Since then my thought moves from one of these words to the next in every direction and they increase within me endlessly. It is so. I had no other professor than *Omama*. It was his words, which came from my elders, that made me wiser. My words have no other origin. Those of the white people are so different. They are probably clever, but they badly lack wisdom.

Unlike them, I do not possess **old books** in which my ancestors' words have been drawn.[9] The *xapiri*'s words are set in my thought, in the deepest part of me. They are the words of *Omama*. They are very old, yet the shamans constantly renew them. They have always protected the forest and its inhabitants. Today it is my turn to possess them. Later they will penetrate the minds of my children and sons-in-law, then the minds of their children and sons-in-law. It will be up to them to make them new. Then it will continue this way throughout time, again and again. This way these words will never disappear. They will always remain in our thought, even if the white people throw away the paper skins of this book in which they are drawn and even if the **missionaries,** who we call the people of *Teosi,*[10] always call them lies. They can neither be watered down nor burned. They will not get old like those that stay stuck to image skins made from dead trees. When I am long gone, they will still be as new and strong as they are now. I asked you to set them on this paper in order to give them to the white people who will be willing to know their lines. Maybe then they will finally lend an ear to the inhabitants of the forest's words and start thinking about them in a more upright manner?

I

Becoming Other

CHAPTER 1

Drawn Words

Body paintings

WITHOUT OUR knowledge, outsiders decided to travel up the rivers and penetrated our forest. We didn't know anything about them. We did not even know why they wanted to approach us. Yet one day they came all the way to our big *Marakana* house on the upper Rio Toototobi. I was a tiny little child at the time. They wanted to give me a name, "Yosi."[1] But I found this word very ugly and I did not want it. It sounded like *Yoasi, Omama*'s bad brother. I told myself that with a name like that my people would make fun of me. *Omama* had a lot of wisdom. He knew how to create the forest, the mountains, and the rivers, the sky and the sun, the night, the moon and the stars. It was he who made us exist and established our customs in the beginning of time. He was also very beautiful. But his brother *Yoasi*'s body was covered in white spots. *Yoasi* only did bad things.[2] This is why I was angry. But these first outsiders left quickly and their ugly name was lost with them. Then time passed and other white people came. Those ones stayed. They built houses to live among us. They mentioned the name of the one who created them at every opportunity. This is why we came to know them as the people of *Teosi*. It was they who named me "Davi," before my own people had even given me a nickname to follow the custom of our elders. These white people told me my name came from image skins on which *Teosi*'s words are drawn. It was a clear name, which cannot be misused.[3] I have kept it since that time.

BEFORE THE white people appeared in the forest and distributed their names to us without restraint,[4] we bore the names our people gave us. Mothers and fathers don't name the children here. They address them by using the term *"õse!"* ("son/daughter!"). Their small children call them both *"napa!"* ("mother!"). Later, once they have grown, they call their father a different name: *"hʷapa!"* ("father!").[5] Close kin[6] such as uncles, aunts, and grandparents give children a nickname. The people of their house hear the nickname and start to use it. The children grow up with this name and it spreads from house to house. It remains attached to them once they are adults.[7] One of my wife's brothers was called *Wari* because he planted a *wari mahi* tree behind his house for fun when he was a child. As for my wife, she was nicknamed *Rããsi*, "Sickly," because she was always ill as a child. Others among us are called *Mioti*, "Sleeper," *Mamoki prei*, "Big Eyes," and *Nakitao*, "Talk Loud."[8]

Yet once we are adults, malicious people from afar sometimes add other names to these childhood nicknames.[9] These are very ugly words. They do it to mistreat the one they refer to, because for us it is an insult to pronounce someone's name in his presence or that of his people.[10] It is so. We do not like to hear our names, even if it is a child's nickname. It really makes us angry. And if someone should come to say it out loud, we instantly avenge ourselves by doing the same thing. This is how we insult each other, by exposing our names for all to hear. We do not mind being named, so long as our name remains far away from us. It is for others to use, without our knowledge. Yet it often happens that children's nicknames are uttered in their presence. But this must cease as soon as they start to grow up. They don't want to hear them anymore once they are teenagers. It makes them furious if their names are pronounced in front of them. They want revenge and become very aggressive.

WHEN I BECAME a man, other white people decided to give me a name again. This time they were people from FUNAI. They started calling me Davi "Xiriana." But I didn't like this new name. "Xiriana," this is what they call the Yanomami who live on the Rio Uraricaá, a long way from where I was born.[11] I am not a "Xiriana." My language is different from the one of the people who live on that river. Yet I had to keep this new name. I even had to learn to draw it when I went to work for the white people, for they had already put it on a paper skin.[12]

My last name, Kopenawa, came to me much later, when I truly became an adult. This time it was a real Yanomami name. Yet it is not a child's name or a nickname that the others gave me. It is a name I acquired alone.[13] At the time, the gold prospectors had started to invade our forest. They had just killed four great Yanomami men where the highlands start, upstream from the *Hero u* River.[14] FUNAI had sent me there to find their bodies, which were hidden in the forest in the middle of all those gold prospectors who could easily have wanted to kill me too. No one was there to help me. I was scared, but my anger was stronger. It is from this moment that I took this new name.

Only the *xapiri* spirits were with me at that moment. They were the ones who wanted to name me. They gave me this name, Kopenawa, because of the rage inside me to face the white people. My wife's father, the great man of our *Watoriki* house at the foot of the Mountain of the Wind, had made me drink the powder that the shamans extract from the *yãkoana hi* tree.[15] Under the effect of its power, I saw the spirits of the *kopena* wasps come down to me. They told me: "We are by your side and will protect you. This is why you will take this name, Kopenawa!" It is so. This name comes from the wasp spirits who absorbed the blood spilled by *Arowë*, a great warrior of the beginning of time. My father-in-law made their images come down and gave them to me with his breath of life.[16] Then I was able to see them dance for the first time.[17] And when I contemplated the image of *Arowë*, of whom I had only heard the name pronounced, I told myself: "*Haixopë!* So this is the ancestor who put the warrior courage in us! Here is the mark of the one who taught us bravery!"[18]

AROWË WAS BORN in the highlands, in the forest of those we call the War People.[19] He was very aggressive and valiant.[20] He always attacked the houses neighboring his own. But each time, his victims' people surrounded him and sought revenge by arrowing him one after another. When his breath seemed to have stopped and he truly appeared to be dead, they abandoned his bloodstained body on the forest floor. At that moment, the warriors[21] told themselves: "That's good, he will rot here and our anger will subside!" and they turned back, happy to have avenged themselves. Exhausted, they made a stop in the forest and heedlessly bathed in a stream. Yet once abandoned, Arowë's body always came back to life. He was so resilient that no one could really overcome him. He regained consciousness and went in pursuit of his aggressors, caught up with them, and arrowed every last one of them. It always happened the same way. No one could kill Arowë. He was really very aggressive and tough.

In the long run, his perplexed enemies asked themselves, "What can we do? How can we make him perish for good?" Someone proposed: "Let's decapitate him!" All agreed and instantly set out again to try and put an end to it. Once again they shot Arowë's body until it was full of arrows, but this time they did not just leave him for dead on the forest floor. They severed his head, and this is how Arowë, despite all his efforts, was no longer able to escape his enemies' vengeance. A breath of life returned to him and he did attempt to put his head back on his neck several times, but he failed. Finally he really died. Then his ghost divided and spread far off and in every direction. This was how he taught us warrior courage. White people should not think that the Yanomami are courageous without reason. We owe our bravery to Arowë.[22]

Arowë's decapitated body lay on the dried leaves covering the ground. Slowly his blood spread all over. Then the wasps of the forest gathered on this bloody litter and ate their fill. The xiho and kaxi ants did too. It was by eating Arowë's blood that they got so aggressive and that their bites became so painful. When you see a wasp nest under a tree, you don't dare approach it! There are so many wasps in the forest and just as many wasp images. This is why we make them come down as xapiri spirits to attack evil beings[23] or to arrow the warrior spirits of distant shamans. I took the name Kopenawa because it is close to the name of the wasp spirits who fed on the blood of the great warrior *Arowë* and whose images I saw with the *yãkoana* powder. I bear this name to defend my people and protect our land, for it was *Arowë* who taught our ancestors bravery in the beginning of time.

If the white people hadn't appeared in our forest when I was a child, I would probably also have become a warrior and would have arrowed other Yanomami in anger when I wanted revenge. I have thought to do it. Yet I never killed anyone. I have always contained my evil thoughts above me and stayed quiet by thinking of the white people. I would tell myself: "If I arrow one of us, those who covet our forest will say I am evil and devoid of wisdom. I won't do it, for they are the ones who kill us with their diseases and shotguns. And it is against them that I must direct my anger today!"

And so, little by little, my name became longer and longer. First there was Davi, the name the white people gave me in my childhood, then Kopenawa, the one the wasp spirits gave me later. Finally, I added Yanomami, which is a solid word that cannot disappear, for it is the name of my people. I was not born on a land without trees. My flesh does not come from the sperm of a white man.[24] I am a son of the inhabitants of the forest highlands and I fell on the ground from the vagina of a Yanomami woman. I am a son of the people *Omama* brought into existence in the beginning of time. I was born in this forest and have always lived here. Today it is my children and grandchildren's turn to grow up here. This is why my words are those of a real Yanomami. These are words that have stayed with me in my solitude, after the death of my elders. These are words the spirits gave me in dream but also words that came to me by hearing the evil words white people spread about us. They are solidly rooted deep in my chest. These are the words I want to make heard in this book, with the help of a white man who will make those who do not have our language hear them.

You don't know me and you have never seen me. You live on a distant land. This is why I want to let you know what the elders taught me. When I was younger, I did not know anything. Then little by little, I started to think by myself. Today all the words the ancestors possessed before me became clear to me. They are words unknown to white people and which we have kept from the beginning. I want to tell you about the very ancient time when the animal ancestors went through their metamorphosis; of the time when *Omama* created us and when the white people were still very far away from us. In this beginning of time, the day never ended. The night did not exist. Our ancestors had to hide in the smoke from their wood fires to copulate without being seen. Finally they arrowed the *Titi kiki* birds of night, when they cried while naming our rivers, so that darkness would descend upon the forest.[25] At that time, our ancestors also constantly turned into game. Then, after they had all become animals and the sky had fallen, *Omama* created us as we are today.[26]

Our language is the one with which he taught us to name things. He is the one who introduced us to bananas, manioc, and all the food in our gardens,[27] as well as all the fruits of the trees in the forest. This is why we want to safeguard the land where we live. *Omama* created it and gave it to us for us to live on. Yet white people do their utmost to devastate it and if we don't defend it, we will die with it.

Our ancestors were created in this forest long, long ago. I still don't know much about this beginning of time. This is why I often muse on it. And when I am alone my thoughts are never calm. I look deep inside myself for the words of that very distant time during which my people came into existence. I still ask myself what the forest was like when it was still young and how our people lived before the white people's epidemic smoke.[28] I only know that when those diseases did not yet exist, our elders' thought was very strong. They lived in the friendship of their own people and warred to get revenge on their enemies. They were the way *Omama* created them.

Today white people think we should imitate them in every way. Yet this is not what we want. I learned their ways from childhood and I speak a little of their language. Yet I do not want to be one of them. I think that we will only be able to become white people the day white people transform themselves into Yanomami. I also know that if we live in their **cities**, we will be unhappy. Then they will put an end to the forest and never leave us a place where we can live far from them. We will no longer be able to hunt, or even to cultivate anything. Our children will be hungry. When I think about all this, I am filled with sadness and anger.

White people say they are **intelligent**. But we are not any less intelligent. Our thoughts unfurl in every direction and our words are ancient and numerous. They are the words of our ancestors. Yet unlike white people we do not need image skins to prevent them from escaping. We do not need to draw them, like the white people do with theirs. They will not disappear, for they remain fixed inside us. So our **memory** is long and strong. The same is true of our *xapiri* spirits' words. They are also very ancient. Yet they become new again each time they return to dance for a young shaman. It has been this way for a long time, endless. The elders tell us: "It is your turn to answer the spirits' call. If you stop answering, you will become ignorant. Your thought will get lost and no matter how you try to summon *Teosi*'s image to tear your children away from the evil beings, it will be in vain!"

The words of *Omama* and the *xapiri* are the ones I like best. They are truly mine. I will never want to reject them. The white people's way of thinking is other. Their memory is clever but entangled in smoky and obscure words. The path of their thought is often twisted and thorny. They do not truly know the things of the forest. They contemplate paper skins on which they have drawn their own words for hours. If they do not follow their lines, their thought gets lost. Our elders did not have image skins and did not write **laws** on them. Their only words were those pronounced by their mouths and they did not draw them. So their words never went far away from them and this is why the white people have never known them.

I did not learn to think about the things of the forest by setting my eyes on paper skins. I saw them for real by drinking my elders' breath of life with the *yãkoana* powder they gave me. This was also how they gave me the breath of the spirits, which now multiplies my words and extends my thought in every direction. I am not an elder and I still don't know much. Yet I had my account drawn in the white people's language so it could be heard far from the forest. Maybe they will finally understand my words and after them their children and later yet the children of their children. Then their thoughts about us will cease being so dark and twisted and maybe they will even wind up losing the will to destroy us. If so, our people will stop dying in silence, unbeknownst to all, like turtles hidden on the forest floor.

Omama's image told our shaman elders: "You live in this forest I created. Eat the fruit of its trees and hunt its game. Open your gardens to plant banana plants, sugarcane, and manioc. Hold big *reahu* feasts![29] Invite each other from one house to another, sing and offer each other food in abundance!" He did not tell them: "Abandon the forest and give it to white people so they can clear it, dig into its soil, and foul its rivers!" This is why I want to send my words far away. They come from the spirits that stand by my side and are not copied from image skins I may have looked at. They are deep inside me. It was very long ago that *Omama* and our ancestors left them in our thought and we have kept them there ever since. They can have no end. By lending an ear to them, white people may stop believing we are stupid. Maybe they will understand that it is their own minds that are confused and darkened, for in the city they only listen to the sound of their planes, their **cars**, their radios, their **televisions**, and their machines. So their thought is most often obstructed and full of smoke. They sleep without dreams, like axes abandoned on a house's floor. Meanwhile, in the silence of the forest, we shamans drink the powder of the *yãkoana hi* trees, which is the *xapiri* spirits' food. Then they take our image into the time of dream. This is why we can hear their songs and contemplate their presentation dances during our sleep. This is our **school** to really know things.

Omama did not give us any books in which *Teosi*'s words are drawn like the ones white people have. He fixed his words inside our bodies. But for the white people to hear them they must be drawn like their own, otherwise their thought remains empty. If these ancient words only come out of our mouths, they don't understand them and they instantly forget them. But once stuck to paper, they will remain as present for them as *Teosi*'s words can be, which they constantly look at.[30] And so perhaps they will tell themselves: "It's true, the Yanomami do not exist without a reason. They did not fall out of the sky. It was *Omama* who created them to live in the forest." But in the meantime they continue to lie about us by saying: "The Yanomami are fierce. All they think about is warring and stealing women. They are dangerous!" Such words are our enemies and we detest them. If we were so fierce, no outsider would ever have stayed with us.[31] On the contrary, we treated those who came into the forest and visited us with friendship. This lying talk is that of bad guests. When they returned home, they could have said: "The Yanomami set up my hammock in their home, they offered me their food with generosity! Let them live in the forest like their elders did before them! Let their children be many and in good health! Let them continue to hunt, to hold *reahu* feasts, and to make their spirits dance!"

Instead our words were tangled up in ghost talk whose twisted drawings were propagated everywhere among white people. We don't want to hear that old talk about us. It belongs to white people's evil thoughts. I also want them to stop repeating: "What the Yanomami say to defend their forest is lies. It will soon be empty. There are only a few of them and soon they will all be white people!" This is why I want to make white people forget all this bad talk and replace it with mine, which is new and right. When they hear it, they will no longer be able to think that we are evil beings or game in the forest.

When your eyes follow the tracks of my words, you will know that we are still alive, for *Omama*'s image protects us. Then you will be able to think: "These are beautiful words. The Yanomami continue to live in the forest like their ancestors. They live there in big houses where they sleep in their hammocks beside wood fires. They eat the bananas and manioc from their gardens. They arrow the forest game and catch the river fish. They prefer their food to white people's moldy supplies, locked up in metal **boxes** and **plastic** sleeves. They invite each other from house to house to dance at their *reahu* feasts. They make the spirits come down. They speak their own language. Their hair and eyes are still like *Omama*'s. They did not become white people. They still live on that land, which seems empty and silent but only from the heights of our **airplanes**. Our fathers have already made many of their elders die. We must not go on following that evil path."[32]

FAR FROM THE forest, there are many other **peoples** than ours. Yet none of them possess a name similar to ours. This is why we must continue to live on the land that *Omama* left us in the beginning of time. We are his sons and his sons-in-law. We keep the name he gave us. Since they first met us, the white people have always asked us: "Who are you? Where do you come from? What are you called?" They want to know what our name Yanomami means. Why do they insist so? They claim that it is to think straight. On the contrary, we think that it is bad for us. What should we answer them?[33] We want to protect our name. We don't like to repeat it all the time. This would be to mistreat *Omama*'s image. This is not how we talk. So no one wants to answer their questions.

We are inhabitants of the forest. Our ancestors inhabited the sources of these rivers long before the birth of my fathers and even long before the white people's ancestors were born. In the past, we were really very

numerous and our houses were vast. Then many of us died after the arrival of these outsiders with their epidemic fumes and shotguns. We have
been sad and known the anger of mourning too often. Sometimes we are
scared that the white people will finish us off. Yet despite all that, after
having cried so much and put the ashes of our dead in oblivion,[34] we live
happily. We know that the dead go to rejoin the ghosts of our elders on
the sky's back, where game is abundant and feasts are endless. This is
why our thoughts return to calm despite all this mourning and these
tears. We become able to hunt and work in our gardens again. We can
travel through the forest and make friendship with the people of other
houses. Once again we laugh with our children, sing during our *reahu*
feasts, and make the *xapiri* spirits dance. We know that they remain by
our side in the forest and that they still hold the sky in place.

CHAPTER 2

The First Shaman

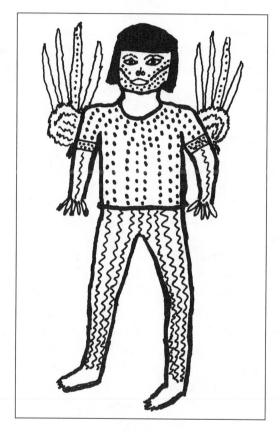

Omama's son

I T WAS *Omama* who created the land and the forest, the wind that shakes its leaves, and the rivers whose water we drink. It is he who gave us life and made us many. Our elders have made us hear his name from the beginning. *Omama* and his brother *Yoasi* first came to existence alone. They did not have a father or a mother. Before them, in the beginning of time, only the people we call *yarori*[1] existed. These ancestors were human beings with animal names. They constantly metamorphosed. Gradually, they became the game we arrow and eat today. Then it was *Omama*'s turn to come into being and to recreate the forest, for the one that existed before was fragile. It constantly became other until finally the sky fell on it. Its inhabitants were pushed underground and became the meat-hungry ancestors we call the *aõpatari*.[2]

This is why *Omama* had to create a new, more solid forest, whose name is *Hutukara*. This is also the name of the ancient sky, which fell long ago. *Omama* set the image of this new land and carefully extended it little by little, like when one spreads clay to make a plate to bake a *mahe* cassava bread. Then he covered it with tight lines traced with annatto dye,[3] like word drawings. He planted immense pieces of metal in its depths so it wouldn't collapse. He also used them to root the sky's feet.[4] Without this, the land would have remained sandy and friable and the sky would not have stayed in place. Later *Omama* turned the remaining metal harmless and used it to make our ancestors' first metal tools.[5] Finally, he put the mountains down on the surface of the land so it would not shake in the storm winds and frighten human beings. He also drew a first sun to give us light. But it burned too hot and he had to get rid of it by destroying its image. Finally, he created the sun we still see in the sky, along with the clouds and the rain, so he could interpose them when it gets too hot. This is what I heard my elders say.

Omama also created the trees and the plants by sowing the pits of their fruit all over the ground. These seeds sprouted in the soil and gave rise to all the vegetation of the forest we have lived in ever since. And so the *hoko si, maima si,* and *rioko si* palms, the *apia hi, komatima hi, makina hi,* and *oruxi hi* trees, and all the other plants from which we get our food grew. At first their branches were bare. Then fruit formed on them. Finally, *Omama* created the bees that came to live on them and drink the nectar of the flowers with which they produce their honey.

At first, there weren't any rivers either; the waters ran deep underground. All you could hear of them was a distant roar, like that of powerful rapids. They formed a great waterway the shamans call *Motu uri u.* One day *Omama* was working in his garden with his son when the boy started to cry because he was thirsty. To quench his son's thirst, *Omama* made a hole in the ground with a metal bar.[6] When he pulled it out, water leaped up to the sky. It pushed back his child, who had come to drink his fill, and shot all the fish, skates, and caimans into the sky. The stream rose so high that another river formed on the sky's back, where the ghosts of our dead live. Then the waters accumulated on the earth and ran off in every direction to form the rivers, streams, and lakes of the forest.

In the beginning, no human beings lived there yet. *Omama* and his brother *Yoasi* lived there alone. There were no women yet. The two brothers met the first woman much later, when *Omama* fished *Tëpërësiki*'s daughter out of a big river.[7] In the beginning, *Omama* copulated in the fold of his brother *Yoasi*'s knee. With time, the latter's calf became pregnant and this is how *Omama* first had a son.[8] Yet we, the inhabitants of the forest, were not born this way. We came later, from the vagina of *Omama*'s wife, *Tʰuëyoma*,[9] the woman he pulled out of the water. The shamans have brought her image down since the beginning of time. They also refer to her as *Paonakare*. This was a fish being that let itself be captured in the appearance of a woman. It is so. If *Omama* hadn't fished her out of the river, perhaps human beings would still copulate behind their knees!

Later *Omama* got angry at his brother *Yoasi*, for *Yoasi* had furtively made appear the evil beings of disease we call *në wāri*,[10] as well as those of the *xawara* epidemic, who are also eaters of human flesh. *Yoasi* was bad and his thought full of oblivion. As for *Omama*, he had created the sun being who never dies and whom the shamans call *Motʰokari*.[11] I am not speaking here of the sun whose warmth settles on the forest and which ordinary people see, but of the image of the sun.[12] It is so. The sun and the moon possess images that only the shamans can bring down and make dance. They have a human appearance, like us, but the white people cannot know them.

Omama wanted us to be as immortal as the sun. He wanted to do things handsomely and put a truly solid breath of life in us. So he searched the forest to cut a piece of hardwood to stand up and imitate his wife's shape with. He chose a *pore hi* ghost tree whose skin is constantly renewed. He wanted to introduce the image of this tree into our breath of life so it remained long and resistant.[13] This way, when we became old, our skin could have sloughed off and remained smooth and new forever. We could have constantly become young again and never died. This is what *Omama* wanted. But *Yoasi* took advantage of his absence and placed the bark of a tree we call *kotopori usihi* in *Omama*'s wife's hammock. This soft bark folded back on the side and hung from the hammock to the ground. The toucan spirits immediately broke into pained wails of mourning.[14] *Omama* heard them and got angry at his brother. But it was too late, the damage was done. *Yoasi* had taught us how to die once and for all. He had introduced death, that evil being, into our mind and our breath,[15] making them so fragile. Since then, human beings have always been close to death. This is why we also sometimes refer to white people as *Yoasi tʰëri*, People of *Yoasi*. We see their merchandise, machines, and the epidemics that constantly bring us death as traces of *Omama*'s bad brother.

It was also *Yoasi* who created the *Poriporiri* moon being. This is why *Poriporiri* also always dies. *Poriporiri* is a man who travels through the immensity of the sky every night, sitting in his pirogue like in a kind of plane. At first he is a young man, but he gets older and older day after day. By the end of his journey, he has become emaciated and his hair has turned white. Then finally he dies. Then his daughters and the toucan spirits cry for him relentlessly. Their tears turn into heavy rain, which falls on the forest for a long time. Once their father's body has decomposed, they carefully gather his bones. Then they bloom again and *Poriporiri* comes back to life. It is so. The moon being is also a thing of death. *Yoasi* wanted it to be so because he lacked wisdom. Unlike him, *Omama* truly wanted us to be eternal. If he had been alone, we would never perish and our breath of life would always be as vigorous. But it was not so and, alas, *Yoasi* made our ancestors become other.

In the end, *Omama* created the *xapiri* so we could take revenge on disease[16] and protect ourselves from the death with which his evil brother afflicted us. Then he created the *urihinari* spirits of the forest, the *mãu unari* spirits of the waters, and the *yarori* animal spirits. He hid them on the mountain peaks and in the deepest part of the woods until his son became a shaman. At first, I thought the *xapiri* came into being by themselves, but I was wrong. I only fully understood who they were later, once I was able to see them and hear their singing. My wife's father also tells how it was *Omama*'s wife, the woman of the waters, who was the first to ask for the *xapiri* to come into being. We are her children and our elders became many starting from her. This is why after procreating she asked her husband: "What will we do to cure our children if they are sick?" This is what worried her. Her husband *Omama*'s thought remained in oblivion. No matter how hard his mind searched, he could not think what else he could create. Then the woman of the waters told him: "Come out of your perplexity. Create the *xapiri* who will cure our children!" *Omama* approved: "*Awe!* These are wise words. The spirits will chase the evil beings away. They will tear the image of the sick from them and bring it back into their body!" This is how he made the *xapiri* appear, as innumerable and powerful as we know them today.

LATER *OMAMA'S* son became a young man and his father wanted him to know how to call the *xapiri* to heal his people. He found a *yãkoana hi* tree in the forest and told his son: "With this tree, you will prepare the *yãkoana* powder! You will mix it with the odorous *maxara hana* leaves and the bark of the *ama hi* and *amat^ha hi* trees and drink it! The *yãkoana*'s power reveals the *xapiri*'s voice. By drinking it, you will hear their clamor and you too will become a spirit!" He blew the *yãkoana* into his son's nostrils with a tube made from a *horoma* palm.[17] Then *Omama* called the *xapiri* for the first time and added to his son: "Now it is up to you to make them come down. If you behave well and the spirits really want you, they will come to you to do their presentation dance and remain by your side. You will be their father. When your children are sick, you will follow the path of the evil beings and fight them to bring back their image! You will also bring down the *ayokora* cacique bird spirit to help you to regurgitate their dangerous objects,[18] which you will tear out from inside the sick people. This way you will truly be able to cure human beings!" This was how *Omama* revealed the use of the *yãkoana* to his son—the first shaman— and taught him how to see the spirits he had just created. Our elders have continued to follow the trail of his words to this day. This is why we continue to drink the *yãkoana* to make the *xapiri* dance. We do not do that without reason. We do it because we are inhabitants of the forest, sons and sons-in-law of *Omama*.

Omama's son listened to his father's words carefully and set his mind on the *xapiri*. He entered a ghost state and became other.[19] He began to see the beauty of their presentation dance. He quickly became a shaman, for he was able to be friendly with all the spirits. The *xapiri* had kept their eyes set on him since he was a little child and his father had often spoken to him about them. Now he was grown up and they had finally come to him in great numbers. He could see them come down, dazzling with light, and hear their melodious songs. He exclaimed: "Father! Now I know the spirits and they have come to stand by my side. From now on, human beings will be able to multiply and fight off disease." *Omama* was the only one to know the *xapiri*. He gave them to his son, for if he had died without teaching their words, there would never have been any shamans in the forest. He did not want human beings to be left helpless and pitiful. This is why he made his son into the first shaman. He left him the *xapiri*'s path before he fled far away. This is what he wanted.

He told him: "With these spirits, you will protect human beings and their children, no matter how many there are. Don't let the evil beings and jaguars come devour them. Prevent the snakes from biting them and the scorpions from stinging them. Divert the *xawara* epidemic smoke from them. Also protect the forest. Prevent the river waters from flooding it and the rains from mercilessly drenching it. Repel the cloudy weather and darkness. Hold up the sky so it does not fall apart. Don't let the lightning strike the earth and soothe the thunder's vociferation. Prevent the giant *Wakari* armadillo spirit from cutting up the roots of the trees and the *Yariporari* storm wind from arrowing them and making them fall!" These were *Omama*'s words to his son. This is why the shamans continue to defend the Yanomami and the forest today. They also protect the white people, even if they are a different people, as well as all lands, however vast and distant.

Omama's son first took the *yãkoana* with his father. Then he continued to drink it alone, again and again, to call the spirits in increasing numbers so he could know their songs. He was magnificent to see when he made their images dance. He was a very beautiful young man. His skin was coated with vermilion annatto and covered in shiny black drawings. His curassow crest armbands held a profusion of scarlet macaw tails, toucan tail pendants, and bunches of *paixi* feathers.[20] His eyes were dark and his hair covered in dazzling white *hõromae* down feathers.[21] A thick black saki tail was wrapped around his forehead.[22] He danced slowly with his back arched, filled with joy by contemplating the beauty of the *xapiri*. He called them and brought them down endlessly because he truly carried them in his thought. This was because he came from the sperm of *Omama*, who is their creator.

TODAY I THINK that *Omama*'s son is dead. Yet his image still exists, very far from here, downstream of the rivers, to the east, or maybe in the sky. I saw him during the time of dream. His image accompanied the image of our forest in tears. Becoming ghost under the effect of epidemic fumes,

the sick forest was asking the *xapiri* to heal the suffering the white people's rage had inflicted upon her. She implored them to clean her trees
and to make their leaves shiny; to make their flowers grow and to make
her fertility return. She told them: "You are mine, you must avenge me!"
I see all that in dreams when the inner part of my body becomes other,
after the *yãkoana* has turned me into a ghost by day.[23] Otherwise I would
not be able to speak about all this like I do.

Omama's son was the first to become spirit, before anyone else. He
was the first to **study** and to see things with the *yãkoana*. Following on
from him, many of our elders became shamans. He taught them to make
the spirits dance. He told them what *Omama* had taught him: "When the
evil beings of the forest capture your children's image to devour it,[24] the
xapiri will take it back and avenge you!" It was by following these words
that the elders started drinking the *yãkoana* and contemplating the splen-
dor of the spirits. We continue to do this to the present day. This is why
shamans can so often be seen working in our houses.[25] Without their
movement, they would be empty and silent. It is so. These words are very
old but they will not disappear, for they are very beautiful and their value
is very high.

CHAPTER 3

The *Xapiri*'s Gaze

Dancing *xapiri*

WHEN I WAS a very young child, my thought was still in oblivion. Yet in dreams I often saw strange and frightening beings we call *yai t^hë*.[1] This is why I was often heard talking and crying in the night. At the time we lived in *Marakana*, an old house on the upper Rio Toototobi.[2] Only a few children in our house dreamed this way. We did not yet know what troubled our sleep, but already the *xapiri* came to us. This is why later, once we were adults, we wanted to drink the *yãkoana* powder to become shamans. The other children grew up without ever knowing what scared us so much.

It was during this period that I saw the spirits for the very first time. At nightfall, the warmth of the fire eased me to sleep in my mother's hammock. After a moment their images would start to come down to me. They made me become ghost and sent me the dream.[3] A path of light opened before my eyes and unknown beings came towards me. They seemed to appear from very far away, but I could distinguish them clearly. They had the appearance of tiny human beings, with black saki tails around their foreheads and their hair covered in white down feathers.

They approached with slow movements, in a blinding light, waving young *hoko si* palm leaves. With their arms decorated with scarlet macaw tail feathers and a profusion of bright, colorful bunches of *paixi* feathers, coated with vermilion annatto dye, they roared at the top of their voices, like a group of guests arriving at a *reahu* feast. There were so many of them, and they kept their eyes set on me. It was beautiful but terrifying because I had never seen *xapiri* spirits before.

When they finally got up close to me, my stomach dropped with fear. I did not understand what was happening to me. I would start to cry and yell and call for my mother. Then suddenly I would wake up and hear her voice softly telling me: "Don't cry! You won't dream anymore, don't be afraid. Sleep without crying now. Be calm!" Much later, once I had become a shaman, I understood that the frightening beings I had seen in my childhood dreams truly were spirits. Then I thought: "It really was the *xapiri* who came to me then! Why didn't I answer them earlier?"[4]

At that time, the spirits visited me constantly. They really wanted to dance for me, yet I was scared of them. These dreams lasted throughout the time of my childhood, until I became a teenager. First I would see the glimmering light of the *xapiri* approaching, then they would take hold of me and take me into the sky's chest. It is true, I often flew over the forest in my dreams. My arms were suddenly transformed into wings, like those of a big scarlet macaw. I could contemplate the tops of the trees below me, like you can from an **airplane.** Sometimes I started to fall into the void and was overcome with fear. Then suddenly my dream was interrupted and I would wake up in tears.

There was a reason I dreamed I was flying so often. The *xapiri* constantly carried my image into the heights of the sky with them. This is what happens when they fondly set their eyes upon a sleeping child to make him become a shaman. They tell themselves: "Later, when he is grown up, we will dance by his side!" and they continue to watch him with interest. So they never stop making him dream, frightening him the whole while. This is why he always becomes a ghost when he sleeps. He does not get sick, but he flails around in his hammock, crying and yelling to the point that some adults in his house get sick of being woken by his wailing. But these are not childish tantrums. Only the children who see the *xapiri* in their dreams yell during the night. If this was not the case, they would sleep peacefully, like the others.

In my dreams, the spirits tied my hammock's ropes high up in the sky. They looked like **radio** antennas extended at my side. They became paths that led the *xapiri* and their songs to me, just like the white people's **telephone**'s talk path. I was lying calmly but I could feel my hammock getting bigger and bigger. Then I had the impression I was growing bigger with it. I was still only a small child, but I felt myself getting huge. I would look around and see nothing but a large void. It made me dizzy. The sky's chest seemed so close, within reach. A sound rose up from it, like the one from groups of dancers yelling loudly when they arrive at a *reahu* feast: *"Aõ! Aõ! Aõ!"* This was the clamor of the *xapiri* dancing as they came towards me, but I couldn't see them very well. Then after a while everything stopped. As I struggled to wake up, I still felt huge. But realizing I was still my normal size, I worried and asked myself: "I'm still so small! But how could I have become so enormous?" and wound up falling back asleep.

Other times, I contemplated the forest from the sky's chest again. But now a mountain of stone suddenly appeared in it, as tall as the one that stands over our *Watoriki* house. It rose in silence, close to me. In reality it was very far away, but its image nearly touched me. My eyes remained riveted on its slopes. I was scared and asked myself: "What is that? What is happening to me?" Much later, I understood why I often saw this great rocky peak in dreams. *Omama* had created the mountains to hide the path by which he escaped from our forest. They are not set down on the ground without reason.[5] Though they may appear impenetrable to the eyes of he who is not a shaman, they are actually spirit houses. But at that time I was very little and I knew nothing of all that. I did not yet know who the *xapiri* were or even that they existed.

I also often dreamed that animals came after me in the forest. The first one I saw was a big tapir. It seemed very threatening to me. It began to chase me. I feared it would trample me. So I quickly climbed up a tree to get away from it. But the tapir started to grow bigger and bigger, until it got to my height. Frozen in place as I crouched on a branch, I watched it with terror. Then, just as it was about to reach me, I screamed and suddenly woke up. Later I understood that it was the image of Tapir, the ancestor *Xamari*, who wanted to dance with me.[6]

I was also often frightened by an enormous jaguar in my dreams. It followed my tracks in the forest and got closer and closer. I ran from it with all my strength, without ever succeeding to put it off the track. Eventually I would trip in the tangled forest and fall before the fierce jaguar. Then it leapt on me. But just as it was about to devour me, I would suddenly come to my senses, crying. Other times I tried to escape it by climbing up a tree. But it chased me, scaling the tree trunk with its sharp claws. Horrified, I hurried to the tree's highest branches. I did not know where to run anymore. The only escape was to throw myself into the void from the top of the tree where I had sought refuge. I started to flap my arms in desperation, as if they were wings, and suddenly I could fly. I glided in circles, high above the forest like a vulture. In the end I would find myself standing in another forest, on another shore, and the jaguar could no longer reach me.

Sometimes I was chased through my dreams by a herd of white-lipped peccaries. They would catch up to me and prepare to trample and bite me. I could hear their menacing tusks loudly crashing through the forest behind me. Yet I managed to escape them by climbing a tree, and once I was at its top I could fly through the sky's chest again. In other dreams I found myself near a watering hole with an enormous anaconda wrapped around me, trying to crush and swallow me in the mud. Or I was fishing on the banks of a river when a giant black caiman suddenly came crawling out of it towards me. I started to run right away but it chased after me and I couldn't outrun it, despite how heavy it was as it raced through the underbrush.

Sometimes I also dreamed that enemies attacked our house. These were people of the highlands, inhabitants of the place known as $H^w axi$ $t^h a$, at the sources of the Orinoco and Parima rivers. These warriors covered in black dye[7] burst into our *Marakana* house's central plaza and started shooting their arrows off all around. I was terrified. The strings of their bows snapped ceaselessly and one by one my elders fell under their arrows. So I tried to escape by slipping out of the house. But a group of warriors set off on my trail. I ran frantically through the forest trying to escape them. I scrambled up a hill, then climbed a steep mountain. When I reached its peak, I threw myself into the void and once again I was able to take flight. The warriors remained frozen on a great rock and followed me with their eyes, powerless. Then suddenly I came out of my sleep.

Sometimes I also dreamed that I was climbing a tall *rapa hi* tree with yellow flowers. I climbed slowly, clinging to its trunk. Then I passed its main branches and finally moved to its top. From there I could contemplate the forest in the distance, seeing in every direction. I could see other houses, a great river, mountains, and hills. I observed spider monkeys jumping from tree to tree, macaw couples, flights of parrots, and herds of peccaries. It was very beautiful. After a while I wanted to go back down. So I looked beneath me. But suddenly all the branches I had used to climb so high seemed inaccessible to me. Alarmed, I asked myself: "How will I get back to the ground? What will I hold on to?" I did not know what to do. I tried to squeeze the tree trunk between my arms but its bark became increasingly slippery. Suddenly my hands lost their grip. I hurtled towards the ground. But at that very instant I woke with a start. Terrified, I would ask myself again: "What happened to me?"

Other times, I answered the call of the water-being women we call *Māuyoma*.[8] These are the daughters of *Tëpërësiki*, *Omama*'s father-in-law; the sisters of the wife he fished out of the river in the beginning of time. I dived in to join them in the deep of a great river. But to my surprise, I came into the inside of a vast house, without getting wet at all. Everything here was dry and you could see as clearly as outside. The house's central plaza was lit by the sun reflecting on the water's surface. I stayed standing without moving, calmly looking all around me. Many doors led to paths cut through the forest. I watched the coming and going of *Tëpërësiki*'s daughters and daughters-in-law, as they went in and out of the home with their children. I found them really beautiful. Though their father terrified me, I couldn't stop admiring them. Yet as soon as I tried to follow them, I woke with a start. Sometimes all it took was for me to turn back to the door through which I came in for my dream to end there and then. I was sorry that I could not stay in the water beings' house!

The next day I asked my stepfather:[9] "Who owns that house under the river I saw when I was sleeping? It was so beautiful; I would have liked to contemplate it longer." He kindly explained to me: "You went to the house where *Omama*'s father-in-law lives with the fish spirits, the caiman spirits, and the anaconda spirits. The *xapiri* are starting to want you. Later, when you are a teenager, if you want to acquire the power of the *yãkoana,* I will truly open their paths to you." This dream recurred often because as a child I spent a lot of my time fishing along the rivers. This is why the water beings constantly captured my image to make me dream.

Sometimes the images of other unknown beings like that of the *ayokora* cacique bird presented themselves to me in my sleep. Its feather adornments were magnificent, and their colors shone brightly in the light. Its presentation dance and songs were outstanding. This spirit did not scare me like the others. I felt happy to be able to admire it. Yet sometimes I also saw the moon spirit, which looks like a human being surrounded by a halo of intense light. It would fly in my direction and come very close to me before loudly bursting into laughter. It showed its prominent canines while its beard and luminous hair quivered in the dark. Then suddenly it disappeared downstream of the sky, where the sun rises.[10] I can still remember it: its image truly horrified me! The unknown beings who appeared in my child dreams were *xapiri* spirits who watched me and were interested in me. At that time I did not know it yet. I was very worried by all these images seen in dream during my childhood. But much later, when the elders gave me the power of the *yãkoana* to drink, I understood that they had come to meet me so I would become a shaman.

THE PEOPLE of our house were often annoyed when I sobbed or yelled during the night. My stepfather would patiently explain to them: "The spirits are watching this child and he is behaving like a ghost. This is why he moans and talks in his sleep." He took a lot of care of me, just like my mother. He was a man of wisdom, a great shaman. When I woke in tears during the night, he would reassure me by saying: "Abandon that dream, come back from that ghost state! Don't be scared! It is the animal ancestors that you are seeing. If you want, when you grow up I will let you drink the *yãkoana* and they will build their house near you.[11] Then you will be able to call them in your turn." Then he passed his two hands over me and blew. After a while I grew calm. Yet a few days later it would all start again. Countless *xapiri* would come back to me, resuming their presentation dance in blinding light, then disappearing as soon as I woke up. My stepfather comforted me again: "Don't be afraid! You will get older and once you are an adult you will become a great shaman. You will really know how to make the spirits dance. You will protect your children and the people of your house from the evil beings and cure them when they are sick." When I heard these words, I calmed down and fell back asleep.

Just like me, my eldest son was worried for much of his childhood. He never slept peacefully. The spirits had also set their eyes on him. He dreamed that he hunted and that he traveled. He often saw the spirits dancing in the night. I told myself that later I too would have him drink the *yãkoana*. Yet now that he is an adult I do not know whether he still sees the *xapiri* in his sleep. He has become a teacher, and he is often busy with the white people's words. Maybe he will be afraid to set his thought on the spirits and forget the word drawings he has learned?[12] Or maybe he has already spoiled himself by thinking about women too much? I do not know.

When I was a child, my stepfather always kept me apart from women. He took care of me so that I could really become a shaman. My mother, my stepfather, my sister, and I most often lived apart from the others in a small house at *T^hoot^hot^hopi*, far from the people of the big *Marakana* house.[13] So I did not live in the company of their daughters and sisters. This is why I feared women when I was a child. When I did mix with girls, I used to tell them: "Don't get near me! I don't want to feel the *puu hana* honey leaves on your arms! That would make my head spin and make me sick." It is true, the smell of those leaves makes the spirits flee. They fear the women who wear them as if they were unknown beings, and if the youngsters start to copulate too soon the *xapiri* no longer want to dance for them. They are disgusted by their penis odor and find them dirty. They no longer come to visit their dream. In the same way, they hate the young hunters who eat their own prey. They don't have dreams either.[14] It is so. The spirits prefer children who grow up without looking at women.

When you are young, it is good to spend your time in the forest. It is bad to always have your thought set on women and to constantly dream of eating their vulvas.[15] It is wrong to spend your nights desiring them and to try to cross the house on all fours to secretly join them in their hammocks.[16] Better to concern yourself with being a good hunter and to always remain attentive to the game in the forest. This is how a young man will please the spirits and they will gladly come to him. Then they will consider that he belongs to them and will be prompt to dance and make him into a shaman.

This is what happened to me when I was a child. I grew up spending my time in the forest, and this is how little by little I started to see the *xapiri*. My attention was always focused on game and during the night the images of the animal ancestors presented themselves to me. Their adornments and paintings shone brighter and brighter in my dreams. I could also hear them speak and sing loudly. Then they suddenly disappeared when I woke up. This often happened to the elders' children, in the time when the white people were still far away from our forest. But since they have gotten close to us, the children and the youngsters are not the way we used to be. Today, the power of the *yãkoana* often scares them. They are afraid that they will die from it, and sometimes they even lie to themselves to the point of thinking that one day they could turn into white people.[17]

When I was a child, I also got sick very often. I was very vulnerable. The evil beings of the forest and those of the epidemics were constantly after me. After a while, the shamans were tired of working to make me better! So they placed my image in a *yaremaxi* carrying sling[18] and hid it in the bat spirit's house. Then it was safe in the darkness, out of its predators' reach. They searched for it everywhere, but could no longer find it. This is how the shamans of old worked. Sometimes they also hid the little children's images in the tapir spirit's pirogue to protect them from sickness.[19] The tapir spirit's own daughter took care of them. She washed them, rocked them, and played with them while sailing on the waters, far from the evil beings starved for human flesh. This is how I finally stopped getting sick so often.

By passing their hands over me to chase the diseases out of my body, the elders of our house also gradually placed the images of the *xapiri*'s precious adornments on me.[20] They tied curassow crest armbands to my arms and put scarlet macaw tail feathers in them. They put parrot feathers through my earlobes. They covered my hair in white down feathers and wrapped a band of black saki tail around my forehead. All these ornaments were invisible to ordinary people's ghost eyes. Yet their images really were there, attached to me, and they protected the little boy I was. They alerted the spirits to the approach of evil beings. Then they could warn their fathers, the shamans, to repel them before it was too late.

The elders also dressed me with the tapir spirit adornments so I would become a good hunter.[21] When a young man wears such precious things, the tapirs fall in love with him. They like him better than all the others.

When they see him walking in the forest, they tell themselves: "How magnificent this hunter is! He is looking for me, I must go towards him!" Failing that, no tapir would let itself be arrowed so easily, just to calm the elders' meat hunger. I think the shamans put these adornments on children's arms so they won't lack game in their old age!

It was thanks to all this finery that the *xapiri* looked upon me with affection and that I constantly saw their images in dreams. These ornaments plunge children like me into a ghost state during their sleep. This also used to happen to the eldest of my three daughters. The spirits' bunches of *paixi* feathers were placed on her when she was still a baby. She dreamed a lot and often screamed in fear during the night. She easily became ghost at that time. She could have turned into a shaman.[22] The spirits watched her with interest, as they had done with me. When she was still a little girl, before her first menstruation, she sometimes told me: "Father! Later when I am more solid, I would like to truly see the beauty of the spirits like you do. You will make me drink the *yãkoana!*" But now she is an adult and she is married. She may still dream of the spirits, but she no longer talks about it. Her thought is obscured by many other things.

THE *XAPIRI* sometimes also simply set their gaze upon children who drink too much honey. We prepare it by diluting it in water and the *xapiri* are very fond of it.[23] One of my brothers-in-law, who was also a great shaman, often fed it to me when I was little. He would tell me: "Drink this honey I just prepared for you! When you grow up, you will be able to make the spirits dance just like me." It was very sweet, I liked that, and I really drank a lot of it. Then I would fall asleep, having drunk my fill. I would instantly enter a ghost state and start to dream. Suddenly everything appeared to me as bright as if it were in full daylight. I heard cries, murmurs, and strident whistling. I saw animals run in the forest, and in the distance I gazed at the *xapiri* dancing joyously. The bee spirits drew close to me, wanting to play. I was then surrounded with such an intense brightness that it frightened me and I wound up bursting into tears. It was so. Honey is the spirits' favorite food and if children drink a lot of it, the *xapiri* easily appear in their dreams, even if they do not yet know how to recognize them.

Once I got older, my mother's brother, my stepfather, and other shamans of our house sometimes offered me a little *yãkoana* powder.[24] They would call me when they gathered to repel the forest evil beings and I was playing nearby: "Come here! Try the power of the *yãkoana!* Enter the ghost state and later you will become a shaman." I was a little intimidated but still accepted a few pinches, which I went to snort by myself; or I got close to them so they would blow a little into my nostrils. I was very curious about what I would see. I remained lying on my hammock for a long time, not moving. I became a ghost and when night fell I dreamed constantly. I could contemplate the magnificent images of the animal ancestors and the sky and river spirits. This happened to me often, for as a little boy I liked to try the *yãkoana*. This is how I was made to grow up.

Sometimes the elders also gave it to me at the end of *reahu* feasts, when all the men take it together at the center of the house before they begin their *yãimuu* dialogues.[25] They had me sniff a little, two or three times over. Then the power of the *yãkoana* took me over and instantly made me die.[26] I rolled and thrashed on the ground like a ghost. I could no longer see anything around me, neither my house nor its inhabitants.[27] I whimpered and called for my mother: "*Napaaa! Napaaa!*" My skin remained sprawled on the ground but the *xapiri* took hold of my image. They sped away with it, far into the distance. It flew with them onto the sky's back, where the ghosts live, or in the *aõpatari* ancestors' underground world. In the end they brought me back to where my skin was lying and I came to my senses. I was getting older at the time and the *yãkoana*'s power no longer scared me at all. Without it, I would not have seen all those things in my dreams. It wasn't having too much plantain soup and peach palm juice that made me dream during my childhood![28] And it certainly wasn't the heady fragrance of the honey leaves that women wear in their armbands!

IF THE ELDERS had not made me drink *yãkoana*, I would never have been able to kill my first tapir when I was still very young and I would never have become a good hunter once I was an adult. Yes, it's true, I killed a tapir all by myself when I was barely a teenager![29] All this happened because I had already seen this animal ancestor's image in dreams. This is how. I had gone to hunt alone, and my stepfather had lent me his shotgun, which he had recently acquired from white people downstream.[30] I had been walking in the forest for quite some time when suddenly I noticed a dark figure on the edge of the path. Frightened, I worried to myself: "What could that be lying in the underbrush?" Then I recognized the tapir's outline. I saw its eyes staring at me in the half-light. That really scared me. I spun around to run away. My heart was beating fast in my chest and I thought: "Maybe it's going to attack me! Tapirs are dangerous. If I fire on it, it's going to turn around and bite me or trample me!" I had already dreamed of tapirs and other animals—peccaries, deer, and caimans—that chased after me to do me harm. That is why I was scared and ran off!

Yet I did not go very far. I stopped my flight and waited for my thought to calm down. Then I slowly returned without a sound to where the tapir was lying. It was still in the same place. It looked at me again. This time I remained impassive. I looked around and located a young tree on which I could hoist myself if it decided to attack me. I quickly made a vine loop to slip over my feet.[31] Then I took aim at the animal and fired. As soon as the **cartridge**'s sound rang out, I threw my gun to the ground and climbed up my tree. Yet the wounded tapir did not come after me as I had feared. On the contrary, after rolling around and moaning on the ground, it tried to escape in the opposite direction. Seeing that, I lost all fear. I came back down from my refuge and slipped another cartridge into my shotgun. The tapir was still lying out in the open and was trying to get back up. I aimed at it again as I drew close and fired. This time it died right away.

I ran back to our house. As soon as I got there I rushed to my stepfather to tell him the news: "*Xoape!*[32] I just killed a tapir with your gun!" He seemed really surprised and did not believe me at first: "You're not lying? Is that true? Where is it?" I answered with pride: "It's true! It isn't far from here, downstream, where the trunk of a *rapa hi* tree is suspended." He still didn't seem convinced: "Is it really dead?" I loudly insisted: "*Awe!* It's lying on the edge of the path. It's true!" Finally, he decided to round up our household: "Let's go cut up the tapir my son-in-law just killed!" Then we set off to bring back the heavy animal's meat.

My stepfather told me I had done well to leave my prey in the forest. He taught me that when you kill a tapir, it is preferable not to touch it and even to avoid breathing in its odor. You have to leave it where it fell and return with your kin to bring back its meat. If not, the hunter who killed it runs the risk of forever coming back empty-handed. Later, I killed many other tapirs. But this one was the very first. I became a good hunter because I dreamed so much at the time. Now I am not as good. I worked with the white people in the forest a lot and they often made me eat my own prey. That made me clumsy with game.

As CHILDREN, we gradually start to think straight. We realize that the *xapiri* really exist and that the elders' words are true. Little by little, we understand that the shamans do not behave as ghosts without a reason. Our thought fixes itself on the spirits' words, and then we really want to see them. We take hold of the idea that later we will be able to ask the elders to blow the *yãkoana* into our nostrils and give us the *xapiri*'s songs.[33] This is how it happened for me a long time ago. The spirits often came to visit me in dreams. This is how they started to know me well. They would tell me: "Since you answer our call, we will dance for you and put up our hammocks in your spirit house!" I never stopped hearing their call throughout my whole childhood. Next I became a teenager, then a young adult, and it continued. I never slept without seeing them come down towards me. They did not frighten me anymore, and I had stopped crying during the night. Yet I kept talking and yelling out in my sleep. In the morning, my relatives would ask me: "What is happening? Are you becoming a shaman?" I merely answered that I had no idea.

This is how it is with us. The *xapiri* first look upon you fondly when you are a child. Then you know they are interested in you and that they will wait for you to become an adult before they truly reveal themselves. They continue to observe you and put you to the test as you grow up. Finally, you can ask the elders to make you drink the *yãkoana* if you want. Then the old shamans will open the paths by which the spirits will come dance for you and build their houses. During childhood, we just become ghosts once in a while, and that's all. But you can only really know the *xapiri* after drinking the *yãkoana* for a long time. Once you have reached that point, they no longer leave your dream. This is how one truly becomes a spirit being in one's own right! During the time of dream, shamans only see the *xapiri*'s presentation dance. They no longer dream of their children, their garden, visitors to their home, or their wife's vulva, as ordinary men do.

For shamans' sons, things go differently. They are born of the spirits' sperm.[34] They become other before the elders even have them drink the *yãkoana*. The *xapiri* owned by their fathers also coupled with their mothers to make them be born. This is why they do not really come from their human father's sperm. The shaman did eat his wife's vulva but it was the *xapiri* who made her pregnant through him. It is so. Shamans' sons are born and become spirits by themselves. They follow their fathers' path. The daughters of the *yawarioma* water beings snatch them as soon as they are teenagers to carry them off to their homes deep in the rivers. Yet it only happens this way if they truly carry the forest in their thought and if they spend most of their time there hunting, paying no attention to women. The spirits look upon good hunters benevolently. They know that they like game, that they track it without respite, and arrow it with agility. Thus, by constantly walking in the forest, young people wind up becoming other during their sleep. They start constantly dreaming of the *xapiri*. The *xapiri* watch them and fall for them. They tell themselves: "We want to come down and set our house up by him. He likes game, let's show him our presentation dance. Maybe he will want us too?"

The water beings are very great hunters. This is why they become fond of the young people whose thought is set on game. They consider them true inhabitants of the forest.[35] So their sisters like to take hold of their image to make them become spirits. Once they are taken this way, the young people enter a ghost state. They run through the forest elatedly yelling *"Aë! Aë! Aë!"* This is how the water beings' daughters attract them far from their own people. Once they have fallen in love, they stay with them underwater a long time. When they finally let them go home, they regain consciousness and suddenly find themselves alone, lost in an unknown forest. Then they tell themselves: *"Oae!* My real house is very far from here!" and they return among their people.

The *yawarioma* water beings are the sons, sons-in-law, daughters, and daughters-in-law of *Tëpërësiki*, *Omama*'s father-in-law, the one who brought him the plants that we grow in our gardens. These are the masters of the forest and the watercourses. They look like human beings, have wives and children, but live deep in the bottom of the rivers, where they are innumerable. They are really very good hunters! They cover the forest paths relentlessly, arrowing macaws, toucans, parrots, *hëima si* birds, and all sorts of other game.[36] Yet they never eat their own prey. Like us, they find that this would be a frightening thing. Instead, they give them to their sisters, who are many and very beautiful. These water beings share their homes with their father *Tëpërësiki*, but also with the electric eel, anaconda, and caiman spirits. Their hammocks are set up side by side underwater, where it is dry, just as they are in our houses. These are the beings that ordinary people's ghost eyes see as fish. Yet their images become *xapiri* made to dance by the shamans.

Omama grabbed one of these women of the waters by the arm. She was one of *Tëpërësiki*'s daughters, who we call *Tʰuëyoma*. But he did not really catch her like a fish. My wife's father told me about it.[37] *Omama* went to the river by attaching a love charm to the end of a liana. Once he reached the riverbank, he cast his line and bait. The woman of the water saw him draw close and found him handsome. Then she took hold of the liana and let herself be pulled out of the water. *Omama* smelled good, he took her arm and hoisted her onto the riverbank. Then he married her and it is from her that we were all born.

Today the same daughters of *Tëpërësiki* make young people inhale *xõa* love charms to capture their image and make them become other. You can hear their murmurs when you are hunting deep in the forest in the afternoon. And if a young hunter comes to meet them, they will seize him. Yet before they appear to him, they will first ask themselves: "Is he handsome and well-groomed?" They will sniff his skin without his noticing. They will inspect his tongue, his chest, and his penis. They will examine his nails. They will ask themselves questions about him: "Is this a good hunter? Does he eat his own prey?" They will only decide to bring him back to their home if they truly find him to their taste. Finally, if he really appeals to them, they will carry him away to their house under the waters.

This is how it goes. The young people start by losing consciousness because they have been tracking game in the forest for so long. They feel very weak and little by little they become ghosts. The animals they approach stare at them and start laughing like human beings.[38] Those they arrow whimper in pain. The trees call to them and the leaves touch them like hands. Then the women of the waters, taking advantage of their weakness, call them and take their image away to their home and keep them there a long time. It is during this stay that they begin to become other. These girls keep them stretched out in their hammocks. They wrap their arms around their shoulders to make them oblivious. They laugh at them when they ask them questions and never answer. Later, when the young men finally succeed in leaving them to go home, the girls will follow them to their own houses. They will hide at the back of their hearth and remain behind them a little longer. Then, after a while, the young men will ask their elders to make them drink the *yãkoana*.

As I SAID, shamans' sons are also spirits' sons. This is why the *yawari-oma* water beings recognize them as sons-in-law and their daughters take hold of them so promptly. As for me, I am only the son of a human being. My father was not a shaman, he did not know the *xapiri*. So I did not know anything of all this when I was a teenager. The women of the waters never took me to their homes and laid me down in their hammocks! They prefer shamans' sons. It is so. But since I was a little boy, I simply never stopped seeing the *xapiri* in dreams, even without knowing who they were. It is only much later, once I had become an adult, that I presented my nose to the elders so they could give me their spirits. I came to want that all by myself. I thought it would be beautiful to really see things as shamans do and so, little by little, I fell in love with the *xapiri*.

The only thing that happened to me in the forest when I was a teenager was that I was attacked by the spirits of the white-lipped peccaries.[39] At the time, I was constantly hunting with the men from my house. Once, we had been tracking a herd of these wild pigs for a long time. This was in the late afternoon. We had just managed to surround them. They had slowed down and were now within our reach. We were preparing to arrow them, each from his own side. Like the others, I chose a prey and calmly prepared my bow. But the peccaries abruptly scattered everywhere. Part of the herd turned back to escape in my direction. Suddenly I found myself face to face with animals furiously charging towards me. Terrified, I tried to get away from them by climbing a young tree, but I tripped and fell. I hit the ground violently and lost consciousness for a moment. All this was very fast. Yet the peccaries had time to jump over me as if I were just a broken tree trunk lying on the ground. They crossed over my chest one after the other at high speed, without ever touching me. There were really so many of them and they smelled very bad. The snapping of their tusks was terrifying. I think it was then that their images attacked me. Yet I did not notice anything at the time.

After they had passed, I stood up, still trembling with fright, and I joined the other hunters, who had managed to arrow several prey. I said nothing about my misadventure. We cut up the downed prey and gathered the pieces in *maima si* and *kõanari si* palm woven bags. Night was starting to fall, we were far from our house. We decided to camp in the forest and to cook the peccary innards in packs of leaves in order to calm our meat hunger.[40] Once I had eaten my fill, I peacefully fell asleep. Yet in the middle of the night I started to feel very bad. I woke with a start and suddenly saw everything that surrounded me with ghost eyes. I started to loudly vomit. Then I thought: "The peccaries are real ancestors![41] Their images attacked me and are making me sick!" The next day we came home. I was very weak, I could not carry anything. The following night, I was still sick. I slept in a ghost state again. That is when I saw the peccary spirits appear in my dream. Countless numbers of them were coming out of a huge hole in the ground, from which a strong wind blew. Wearing their feathery finery, they were dancing slowly on a mirror, which reflected a dazzling light. This lasted a long time, then they abruptly disappeared. I woke up and thought: "What is happening to me? How will I get better?"

Some time later, my mother's sister's husband, who was also a great shaman, began to chase away the sickness that was in me. But he had barely started his cure when I fell deeply unconscious. I was inert, half lying on my hammock. Then my stepfather's mother, who was a very old woman, took a pot full of water and poured large amounts of it on me. I eventually regained consciousness. My ghost returned to my skin and I came to. When I opened my eyes, I saw my mother, her sister, my older sister, and my grandmother crying by my hammock as if I were already dead![42] Then the shaman's work continued for a long time and I was finally able to get better.

THIS IS ONLY what happened to me when I was barely a teenager. I was never taken away by the women of the waters! Yet in the ancients' time, they frequently seized young men's images, as I said. These young people disappeared, running into the forest, and that is how many of them became shamans. My stepfather, who raised me in *Marakana,* often told me about it. He himself had lived through it long ago. Now I want to repeat his words so that the white people can also hear them. This is what he told me.[43]

When I was a teenager, my thought became other and this is how I started to become a shaman. One day I was hunting parrots in the

forest. I heard the tumult of their games in the trees above me.
Suddenly I saw a water being heading towards me. He was imposing. A profusion of scarlet macaw tail feathers, toucan tails, and
colorful *wisawisama si* feathered hides were attached to his curassow crest armbands. Judging by his adornments, he was a great
hunter. Slowly, he came close to me and declared: "Try to arrow
those parrots from here!" Surprised and frightened, I asked him:
"Who are you?" He simply answered: "Me? I want to eat the parrots you will arrow. Move over there and try! But don't arrow their
bodies, aim for the throat, just below the beak." I followed his
words. I arrowed a first parrot, then a second, exactly as he had told
me. Then he took my arm and said: "Brother-in-law! That's good,
that's enough! I'm going to send my sister to take your catch."[44] I
was very hot and sweating abundantly. Little by little, my thought
was losing itself. I stayed in place without saying a word, without
moving, standing up next to the dead parrots lying on the ground.
Then after a while a woman of the waters made her way through
the forest to me.

   The leaves on the trees started to quiver in the wind and the forest filled with a shimmering light. She came close to me with small
steps. Her lips were smiling, for she wanted to make her love
magic take hold in me. She was very beautiful. Her eyes were magnificent and she had a nicely short vulva, with no pubic hair. I
started arrowing parrots again for her. But as soon as they came
spinning down, their cries instantly turned into the songs of *xapiri*
spirits: "*Arerererere!*" The woman of the waters picked up their
carcasses one by one, joyfully approving: "*Awe!* This is very good!
You are an excellent hunter! Keep arrowing these parrots." One after the other, the parrots kept falling under my strokes: "*Arerererere! Arererererere! Arererererere!*" But as soon as they reached the
ground, the arrows sticking in their bodies transformed into
snakes! When I tried to pick them up, these reptiles threatened to
bite me. My eyesight got increasingly blurry and I had trouble dis-
tinguishing things around me. I could feel myself losing consciousness.

   Each time, the woman of the waters came close to me, laughing
in a gentle little voice: "*He he he he!*" Then she picked up my arrows and handed them to me: "Here, take them, this is what you

are looking for." Yet as soon as I tried to grab them, they instantly flew away, producing the same spirit song: *"Arererererere!"* After a while, I truly became other and now I also felt my bow fly away: *"Arererererere!"* I was more and more worried and I couldn't stop wondering what would possibly happen to me. I was now completely under the spell of this daughter of *Tëpërësiki*'s love magic. Then suddenly the spirits of the forest started flocking towards me. The images of leaves and tree roots came down first, making a joyful clamor and whistling with their *purunama usi* bamboo flutes.[45] Their hair was covered in white down feathers, black saki tails were wrapped around their foreheads, and their curassow crest armbands were decorated with a profusion of scarlet macaw tail feathers. They were followed by the images of termites' nests, which ran every which way carrying me on their backs. Then came the turn of the images of stones, which nearly knocked me over and crushed me, then that of the sky, which came to tear out my tongue. Finally other *xapiri* carried my eyes off into the distance and this is how I too started to become a spirit.[46]

After all that, the water being woman grabbed my wrist and pulled me with her into the forest. I started to run at her side, smashing the branches in the underbrush as I passed. I was very excited and kept yelling: *"Aë! Aë! Aë!* A *yawarioma* girl is taking me on her path! The light is blinding me! I'm scared! *Aë! Aë! Aë!"* No one other than me could see her, yet I was really running with her. Her path was very hot and I was dripping with sweat. I could no longer see anything around me. I could not have recognized my people or my own house. I had become other. I ran like this for a long time, crossing unknown forests. At the end, I stopped in a clearing far from my home, exhausted. The woman reassured me in a soft voice: "Don't be afraid! We aren't very far now. Now we are close to my father's house." After this brief moment of respite, we started running down her winding path through the woods faster than ever.

Suddenly I heard the growl of a jaguar accompanied by its cub. Frightened, I instantly warned my companion: "Let's get away from this path, it is going to devour us!" This did not seem to worry her and once again she tried to calm me: "Don't be afraid! This jaguar belongs to me. It will not attack us." Yet I was not reassured at all and insisted: "I'm really very scared! Let's take a detour anyway." Once again, she answered me gently: "No, it will not devour you. It is a domestic animal. Don't be afraid!" I would not relent, so we took a little distance. Yet no matter how I tried to avoid the animal, we always found it on our path. It is so. The water beings consider the jaguar their hunting dog!

Finally, we reached a vast expanse of dark water in the middle of the forest. I remained standing on the shore without moving. I was still just as worried. Then the woman of the waters motioned to the lake's surface with her lips and said: "We have arrived at my father's! Let's go! Let's go in!" I protested forcefully: "No! I don't want to dive into this lake! It is much too deep. Black caimans will devour me in there! I will drown!" She answered with a smile: "Don't be afraid! You won't drown and there are no caimans here. This water is just the outside of our home. The entrance is close by." Despite these words, I continued to resist. Then she dived and came back to the surface holding a handful of earth out to me: "Look! It's dry. It comes from the floor of our house. The door is here, close by. Cross it and you will see the interior with your own eyes. This is the truth!" Then, while I was still hesitating, she grabbed my wrist and pulled me underwater.

I was terrified that I would sink straight to the bottom of the lake. Yet I instantly found myself in a dry place, in an imposing house surrounded by vast banana, manioc, yam, taro, sweet potato, and sugarcane plantations. It looked like one of our dwellings, but much bigger. *Tëpërësiki*, the young woman's father, was lying on one side of the house in his hammock, while all his children were settled on the other. From a distance, I looked in his direction but his daughter warned me: "Don't even dream of getting near my father, or he will instantly swallow you up!" But her many sisters greeted us joyfully. They surrounded me as soon as I arrived and showed me a great deal of friendship. The young woman who had dragged me into the forest was their older sister. There were only two young men with them; they were their brothers. One of them said to the young girls: "Don't make so much noise! Father could wake up!" Then *Tëpërësiki*'s wife, whose hammock was beneath her husband's, said in a small voice: "Daughter! Are you back?" And without setting her eyes on me, she added: "Give the one squatting next to you some of these yams to eat! Have him drink some plantain soup! Also offer him some sweet potatoes! Don't let him stay starving!"[47] The *yawarioma* beings do the *turahamuu* bride-service and we follow in their footsteps.[48] This is why a young man who becomes a shaman calls the parents of the water woman who kidnapped him "father-in-law" and "mother-in-law." It is so.

Once I had eaten my fill, the young girls took turns to come laughing into my hammock and play with me. One of their brothers repeated his injunction that they should not speak too loudly. But their father had finally woken up and we could already hear his deep voice resonating throughout the house. Yet his daughters did not seem to be concerned about it. They continued to follow each other into my hammock to amuse themselves and copulate with me. I was seduced by their love magic. This is why I remained with them for such a long time. Little by little, I metamorphosed to become a shaman. Meanwhile, *Tëpërësiki* had begun to sing his songs so I could get to know them. At certain moments as he chanted, he spit the objects he had just named onto the ground: bamboo arrow points, large fruits from the *aro kohi* tree, and even peccaries and tapirs, for his mouth was truly enormous.[49] This is how I learned the words that allow one to regurgitate the sorcery substances, the spirits' weapons, and the burning cotton of the evil spirits that are in the bodies of the sick. This is how *Tëpërësiki* gave me the mouth of the *ayokora* cacique bird spirits.

But after a while he started to get tired. He stopped singing and spitting things out. Exhausted, he sighed deeply. Then he exclaimed: "Have the visitor squat next to me! I am truly very hungry!" He wanted to swallow me! His sons, who had remained in the house to make arrow points, prevented him from attacking me. Trying to thwart his desires, they answered him: "He can't come now. He is still busy making friendship with our sisters." Despite this, *Tëpërësiki* had me called several more times. But each time the young people lied in the same way. Growing weary, he finally took up his songs again. Then his sons whispered to one of my girl companions: "Sister! Return to the forest with our brother-in-law! Take him back to his house!"

This is how I finally returned home. The woman of the waters who had accompanied me home slept against me in my hammock all night. Then, as soon as the sun rose, she took me back among her people. Then everything started again. Her mother gave me food to eat, her sisters played with me, and her father let me hear his songs. Later another young girl took me back to my house and at dawn I left with her, running through the forest and yelling. All this was repeated day after day. Each time a different woman of the waters dragged me far away and brought me home. I was truly seduced by their love magic and this is how I became a shaman. This is how it usually happens. When a young man's image is captured by *Tëpërësiki*'s daughters, he runs away from his house every day, only to return at nightfall. But he no longer recognizes anyone there. He has become other and leaves at dawn to race through the forest. No matter how hard his people try to force him to stay in his hammock, they do not succeed. He cannot resist the call of these underwater women. No one else sees them, but they always remain by his side. His flights through the forest take him very far from his home. He can even enter and cross unknown people's houses and come back out without noticing, for he is blinded by the intense brightness of the water beings' path through the forest. These *yawarioma* beings can keep him in their power for a long time. The shamans of his house will then have to bring his image back, so he eventually regains consciousness.

This is how my stepfather became a shaman long ago, when he was a young man. At the time, he often arrowed tapirs; he was a very good hunter. This is why the sisters of the water beings took hold of him. He did not just ask the elders to make him drink the *yãkoana* to become other. He did not become a shaman without a reason. It is said that his father was a great shaman in his own right, whose mouth could regurgitate evil objects from sick people.[50] He followed in his footsteps. As for me, I was not seduced by the women of the waters. I only dreamed of them sometimes. As I said, I am not issued from the spirits' sperm, like the shamans' sons. The *xapiri* merely danced in my childhood dreams, long before my wife's father opened their paths to me, and at first I did not recognize them. It was he who weakened me with the *yãkoana* and the *paara* powder so that they would agree to set up their house by my side.[51] Until then they must have found me so ugly and dirty. They must have hesitated to get really close to me! But once my stepfather had me drink the *yãkoana*, I was finally able to truly contemplate their beauty.

CHAPTER 4

The Animal Ancestors

Presentation dance of the *xapiri*

THE *XAPIRI* are the images of the *yarori* ancestors who turned into animals in the beginning of time. This is their real name. You call them "spirits," but they are other.[1] They came into existence when the forest was still young. The shaman elders have always made them dance and we continue to do like them to this day. When the sun rises in the sky's chest, the *xapiri* sleep. When it comes down again in the afternoon, dawn begins to break for them and they wake up. Our night is their day. While we sleep, the spirits are awake, playing and dancing in the forest. It is so. There are so very many of them there because they never die. This is why they call us "the small ghosts"—*pore tʰë pë wei!*—and tell us: "You are outsiders and ghosts because you are mortal!"[2] In their eyes, we are already ghosts because unlike them we are weak and die easily.

NONETHELESS, THE *xapiri* look like human beings. Yet their penises are very small and their hands only have a few fingers. They are tiny, like luminous flecks of dust, and invisible to ordinary people who have only ghost eyes. Only shamans can truly see them. The shiny mirrors they dance on are huge. Their songs are magnificent and powerful. Their thought is right and they work forcefully to protect us. Yet if we treat them badly, they can also be very aggressive and kill us. This is why we sometimes fear them. They are also capable of devastating the trees in the forest as they pass by and even of cutting up the sky, no matter how vast it may be.[3] The real *xapiri* are very brave! Only a few of them prove to be weak and cowardly. These ones fear evil beings and the *xawara* epidemic.

The spirits travel all over the forest, just like we do when we hunt. But they do not walk on rotten leaves and in the mud. They also bathe in the rivers—just like we do when we are too hot—but they do it in pure waters known only to them. They also have children but so many of them that they consider that white people have very few! And even if they become old and blind, the *xapiri* remain immortal. This is why the number of them in the forest is always growing! Those who dance for the shamans only represent a small part of them!

To really see them, we must drink the *yãkoana* for a long time and have the elders open their paths for us. This takes a lot of time. As much time as it takes for your children to learn the drawings of your words. It is very difficult. Yet when I make my *xapiri* dance, the white people sometimes tell me: "We can't see anything! We can only see you singing all by yourself. Where are your spirits?" Those are the words of the ignorant! The powder of the *yãkoana hi* tree did not make their eyes die, like those of the shamans. They cannot contemplate the *xapiri*; that is why their thought remains closed. It is so. The *xapiri* let their voice be heard if their father, the shaman, dies with the *yãkoana* powder. They only come down to their mirrors when they are hungry to drink the *yãkoana* through him. Like their father they die with it and this is how they start to sing and dance for him. Failing that, they could not be seen.

The *xapiri*'s image is very bright. They are always clean, for they do not live in the smoke from houses and eat game like we do. Their bodies never remain gray, without paint or finery, like ours do. They are always covered in fresh vermilion annatto dye and decorated with shiny black waves, lines, and spots. They are heavily scented. When they play with the wind beings' wives, the forest smells of annatto and the hunting charms they wear around their necks. The breeze from their flight spreads fragrances through the forest as strong as those of the white people's **perfumes**. But the *xapiri* dye is one of their own precious goods. It comes from the mixed odors of things of the forest and does not have the acrid, dangerous scent of **alcohol**.

Their arms are decorated with a profusion of bunches of parrot feathers and macaw tail feathers stuck in armbands made of beautifully bright, smooth beads.[4] A multitude of toucan tails and colorful *wisawisama si* feathered skins hang from them too. They really cut a fine figure! *Omama* taught them to dress like this. He wanted them to be magnificent when they come to do their presentation dance for us. Yet there are also very old *xapiri* who danced for our ancestors before us. They have beards and white hair. Sometimes their skulls are entirely bald and even the evil beings are scared of them. They are true elders. All the other younger ones have smooth black hair and wear headbands of saki tails to enhance its thickness. Their eyes are neither reddish nor too pale. Black and limpid, they see far into the distance. Their heads are covered in white down feathers, which cast a dazzling light, preceding them wherever they go. Only the *xapiri* are adorned this way. This is why they shine brightly, like stars moving through the forest.

They stick parrot tail feathers and *hëima si* feathered hide pendants through their earlobes. Their teeth are immaculate and shiny as pieces of **glass**. When they are too small or come to be missing, they replace them with tiny mirrors, for which they ask *Omama* to embellish themselves. Some even decorate their teeth with multicolored *sei si* bird feathers, like white people with their **gold** teeth. Others have long incisors, sharp and terrifying, with which they tear the evil beings to shreds. Others yet have eyes behind their head! They are spirits of distant forests. They are truly other! It is so. Do not think that all the spirits are beautiful!

During their presentation dance, the *xapiri* wave the frayed leaves of young *hoko si* palms, which shine a vivid yellow. They move in rhythm, floating gently in place, above the ground, like a flight of hummingbirds and bees. They whistle through *purunama usi* bamboo stalks, roaring joyously and singing in a powerful voice. Their melodious songs are countless. They sing them without respite, taking turns. Some have teeth that produce a modulated sound: *"Arerererere!"* And others have long nails they use as whistles with a high-pitched screech: *"Kriii! Kriii! Kriii!"* They are so happy to do their presentation dance for us! Their movements are magnificent. They dance eagerly, like young guests entering their hosts' house.[5] But they are even more beautiful!

THE *XAPIRI'S* songs follow each other endlessly. They go gather them from the distant song trees we call *amoa hi*. *Omama* created these wise-tongued trees in the beginning of time so the shaman's spirits could fly there to acquire their words. Since then, the *xapiri* have stopped by them to collect the heart of their melodies before doing their presentation dance for the shamans. The spirits of the *yōrixiama* thrush and the *ayo-kora* cacique[6]—but also those of the *sitipari si* and *taritari axi* birds—are the first to accumulate these songs in big *sakosi* baskets.[7] They gather them one by one with invisible things similar to the white people's **tape recorders**. Yet there are so many that they can never come to the end of them!

Among these bird spirits, those of the *yōrixiama* thrush are really the songs' fathers-in-law, their true **masters**. These *xapiri* are the images of the thrushes whose harmonious call we hear in the forest in the morning and at night. It is so. All the *xapiri* have their own songs: the toucan and aracari spirits, the parrot spirits, the little *wete mo* macaw spirits, the *xoto-koma* and *yōriama* bird spirits, and all the others! There are as many *xa-piri* songs as there are *paa hana* palm leaves on the roofs of our houses and even more than all the white people put together! This is why their words are inexhaustible!

OMAMA PLANTED the *amoa hi* song trees at the edges of the forest, where the earth comes to an end and the sky's feet are rooted, held in place by the giant armadillo spirits and the turtle spirits. Here, these trees tire-lessly distribute their chanting to the *xapiri* who rush to them. These are very tall, decorated with shiny down feathers of blinding white. Their trunks are covered in constantly moving lips, ranged one above the other. These innumerable mouths let out splendorous songs, which follow each other as countless as the stars in the sky's chest. Their words are never repeated. As soon as one song finishes, the next one has already started.

They are constantly proliferating. This is why the *xapiri* can acquire all the songs they want without ever running out, no matter how numerous they may be. They always listen to these *amoa hi* trees with great attention. The sound of their words penetrates them and fixes itself in their thought. They capture them like the white people's tape recorders, in which *Omama* also placed an image of the song trees.[8] This is how they can learn them. Without them, they could not do their presentation dance.

All the spirits' songs come from these very old trees since the beginning of time. Their fathers, the shamans, merely imitate them, in order to let ordinary people hear their beauty. Do not believe that shamans sing at their own initiative, without a reason! They reproduce the *xapiri*'s songs, which follow each other into their ears like into **microphones**. It is so. Even the *heri* songs we strike up when the food at *reahu* feasts is abundant are images of the *amoa hi* trees' melodies.[9] The guests who appreciate them keep them in their chests so they can sing them later, during the feasts they hold in their own homes. This is how they spread from house to house.

These song trees exist in every direction, beyond our land, but also beyond that of the *Xamat^hari* and the mountains where the *Horepë t^hëri* live.[10] But they are other. There are as many *amoa hi* trees as there are Yanomami ways of speaking.[11] So the *xapiri* who come down into our forest possess a multitude of different songs. This is why we hear unknown songs when the shamans of distant houses come to visit. There are also many *amoa hi* trees at the edges of the white people's land, beyond the downstream of the rivers.[12] Without them, their **musicians'** voices would be ungraceful and too narrow. The foreign thrush spirits bring them leaves fallen from these song trees and covered in drawings. This is how beautiful words are introduced into the memory of their language, and into ours. The white people's machines make them into image skins which **singers** look at, not knowing that by doing so they imitate things from the *xapiri*. This is why the white people listen to so many radios and tape recorders! But we shamans have no need for these song papers. We prefer to keep the spirits' voice in our thought.[13] It is so. I relay these words because I myself have followed our elders in seeing the innumerable moving lips of the song trees and the multitude of *xapiri* approaching them. I saw them when I was in a ghost state after my father-in-law made me drink the *yãkoana*. I could really hear their infinite melodies interweaving like they were right next to me!

THE *XAPIRI* never travel through the forest like we do. They only come down to us on paths of brilliant light, which are covered in white down and are as thin as the *warea koxikɨ* spider webs floating in the air. These trails branch off in every direction, like the ones from our houses. They form a mesh covering the entire forest. They split, intersect, and run together far beyond—over the whole vast land we call *urihi a pree* or *urihi a pata* and which the white people call **entire world**. They were opened by the ancient shamans who made the spirits dance long before us, since the beginning of time.

The *xapiri* slowly follow each other down these paths suspended in the heights. We see them sparkling in a kind of moonlight, their feathery adornments swaying, floating gently to the rhythm of their steps. Their images are so beautiful! Some of their trails are very wide, like your roads at night, crisscrossed by cars' headlights, and the most dazzling are those of the oldest spirits. There is no end to the *xapiri* as they come towards us in countless lines. Their images are those of all the inhabitants of the forest, following each other down from the sky's chest with their young. Think of all the scarlet and blue-and-yellow macaws, all the toucans, parrots, gray-winged trumpeters, curassows, guans, *herama*, *wakoa*, and *kopari* falcons, the vultures and all other birds in the forest! And the tortoises, armadillos, tapirs, deer, ocelots, jaguars, pumas, agoutis, peccaries, spider monkeys and howler monkeys, sloths and giant anteaters! And what about all the little fish in the rivers, the electric eels, piranhas, *kurito* catfish, and *yamara aka* stingrays?

Every being of the forest has an *utupë* image. These are the images the shamans call and bring down. These are the images that become *xapiri* and do their presentation dance for us. They are like **photographs**.[14] But only the shamans can see them. Ordinary people cannot. They are the real center, the real heart of the animals we hunt. These images are the real game! White people would say with their words that the animals of the forest we eat are only their **representatives**.[15] And so the *iro* howler monkey we arrow in the trees is other than its image *Irori*, the howler monkey spirit which the shaman calls. These images of animals turned *xapiri* are really so beautiful when they do their presentation dances for us, like the guests at the beginning of a *reahu* feast. The animals of the forest are ugly compared to them. The animals exist, nothing more. They are food for humans. They are merely imitating their images.

Yet when we say one of these images' name, we do not speak about a single *xapiri,* but a multitude of similar images of this spirit. Each name is unique, but the *xapiri* to which it refers are countless. They are like the images in the mirrors I saw in one of your **hotels.** I was alone before them, but at the same time they showed a lot of identical images of me. So there is only one name for the tapir ancestor image we call *Xamari,* but there are an endless number of tapir spirits, *xamari pë.* The same is true of all *xapiri.* We think they are one of a kind, but their images are always very numerous. Only their names are not. They are like me standing before those hotel mirrors. They seem one of a kind, yet their images are juxtaposed far into the distance, without end.

These images of game that the shamans make dance are not those of the animals we hunt. They are those of their fathers, who came into being in the beginning of time. As I said, they are the images of the animal ancestors we call *yarori.*[16] A very long time ago, when the forest was still young, our ancestors—who were humans with animal names—metamorphosed into game. The human peccaries became peccaries. The human deer became deer. The human agoutis became agoutis. These *yarori* first people's skins became those of the peccaries, the deer, and the agoutis that live in the forest.[17] So it is ancestors turned other that we hunt and eat today. On the other hand, the images that we bring down and make dance as *xapiri* are their form of ghosts.[18] These are their real hearts and true inner parts. And so these animal ancestors from the beginning of time have not disappeared. They have become the game that lives in the forest, but their ghosts also continue to exist. They still have their animal names, but are now invisible beings. They have transformed themselves into *xapiri,* who are immortal. Even when the *xawara* epidemic tries to burn or devour them, their mirrors always dawn again. They are true elders. They can never disappear.

This is the truth. When the *yarori* animal ancestors were metamorphosed in the beginning of time, their skins became game and their images became *xapiri* spirits. This is why the *xapiri* always consider animals to be ancestors, like them, and this is how they refer to them! But though we eat animals, we also know they are ancestors turned game! They are inhabitants of the forest as much as we are! They took the appearance of game and live in the forest simply because this is where they became other. Yet in the beginning of time they were as human as we are. They are not different. Today we give ourselves the name of "humans," but we are the same as they are. This is why in their eyes we still belong to their kind.

No MATTER how many *xapiri* there are, they all live on tall hills and mountains. These are their homes. Do not think that the forest is empty. Even if the white people don't see them, the spirits live there in great numbers, like game. This is why their homes are so big. Also, do not think that the mountains have simply been dropped in the forest without reason. These are spirit houses; ancestor houses. *Omama* created them for that. We give them great value. It is from their peaks that the *xapiri* come down to the lowlands where they travel and feed themselves like the animals we hunt. They also descend from there to come to us when we drink the *yãkoana* and call them to dance.

My wife's father's house sits below a small rocky massif we call *Watoriki*, the Wind Mountain. It is also the house of a great number of ancient *xapiri* spirits: *Yariporari* storm wind spirits, scarlet macaw spirits, *ayokora* cacique bird spirits, cock-of-the-rock spirits, spider monkey and capuchin spirits, tapir spirits, deer spirits, and puma and jaguar spirits. Thanks to these *xapiri*, the wind and rain spread from the mountain's heights through the entire forest, making it cool and humid. Those who are not shamans do not perceive any of this. They only see the game they feed on. Only the shamans are able to contemplate the *xapiri* because the *yãkoana* makes them able to see them with spirit eyes themselves.[19]

It was *Omama* who created mountains like *Watoriki*. He planted them in the forest's soil so it would stay in place and would not shake. This is how it happened. One morning, his son was arrowing small birds in the gardens near their house with his child's bow. Suddenly he heard a resounding call ring out in the forest: *"Si ekeke! Si ekeke!"* Frightened, he thought he heard the voice of an evil being who had come to boast of flaying humans by loudly singing: "Tear the skin! Tear the skin!"[20] He instantly ran to alert *Omama*: "Father! Someone is approaching and saying he will skin us alive!" Worried, *Omama* asked him: "What is this evil being really saying?" His son imitated the song he had just heard: *"Si ekeke! Si ekeke! Si ekeke!"* This was actually only the call of a little *si ekekema* bird.

But *Omama*, misled by his son's words, got frightened and cried out: "*Aaaa!* It's true! An evil being is approaching to skin us alive!" He feared the return of *Xinarumari*, the master of cotton who had once skinned a hunter he found on his path.[21] He was overcome by fright and instantly set off running in the direction of the rising sun. Fearing that he would be followed, he carefully covered his tracks by planting large *hoko si* palm leaves behind him. One by one, these palms were transformed into rocky peaks disseminated across our land and the white people's land, where it is very cold. *Omama* placed these mountains on the land to strengthen it and to make a place for the *xapiri* to live.[22] This is how he went away from our forest and left our ancestors alone there. All this because of a little bird's call! He went so far away that he reached the lands where the white people now live. He even went beyond **Europe** and **Japan**, where the sun's path comes out from underground. Then, after having created the white people ancestors, he died and only his image, his ghost form, still exists now. It is his image that the great shamans bring down by drinking the *yãkoana*.

As I said, the *xapiri* never travel on the ground. They find it too dirty, covered in garbage and soiled with excrement. The surface on which they dance looks like glass and shines with a dazzling light. It consists of what our elders call *mireko* or *mirexi*. These are precious objects which belong only to them. They are brilliant and transparent but very solid. White people would say they are **mirrors**. But they are not mirrors to look at oneself, they are mirrors that shine.[23] *Omama* placed them above the earth in the beginning of time so the spirits could dance on them. He decorated their sparkling surface with drawings, like jaguar skins. He used the *xapiri*'s annatto dye to draw tight rows of dots and little lines and wavy lines and circles on them.[24] Then he covered them in white down feathers. Since the beginning of time, the entire expanse of the forest has been covered in these mirrors and the spirits are constantly busying themselves, playing, dancing, and waging war on them. It was on these mirrors that they came into existence and it is from them that they come down to us. It is also where they lay down our image when they make us become shamans. Such huge mirrors are placed where *Omama*'s son and our ancestors became shamans for the first time. They are located in the center of our land, in the savannas that extend beyond the Rio Parima highlands.[25] It was in this place that the *xapiri* were created. Here we find the mirrors of the *xapiri* who imitate the highland people's talk and of those who speak the *Xamat^hari* language. Farther along, we find the mirrors of the spirits who imitate our elders' *Waika* talk.[26] Large father mirrors stand in the middle, surrounded by other smaller ones, scattered like clearings where the *xapiri* stop to adorn themselves on their way to do their presentation dance.

There are many mirrors on the spirits' trails through the forest, for they belong to the *xapiri* of the leaves, the lianas, the trees, and all the animal ancestors! Like guests do, they stop on these mirrors all the time, to rest, eat, and especially to adorn themselves. They coat themselves in annatto dye, put bunches of *paixi* feathers and macaw tail feathers in their curassow crest armbands, cover their hair with white down feathers, make *purunama usi* bamboo flutes, and fray the young *hoko si* palm leaves they will wave during their dance. Once they are ready, they form long lines and come towards us with a joyous roar.

When we drink the *yãkoana*, its great power comes upon us by striking the nape of our neck. So we die and soon become ghosts. Meanwhile the spirits feed on the *yãkoana* through us, who are their fathers. Then they slowly come down on their mirrors from their houses fixed to the sky's chest, singing all the while.[27] They dance on their shiny surfaces without ever touching the ground, covered in feathery adornments and brandishing their machetes, axes, and arrows, ready to do battle with the evil beings. From these heights they see the entire forest in the distance and warn us of the evil threatening us: "Here comes the *xawara* epidemic! A *në wãri* evil being is approaching to devour you! The thunderbolts and the storm wind are angry!" Finally, when their father shaman no longer wants to imitate them, they fly back up to their homes with their mirrors and take their songs back into the sky's chest. Then he returns to his ghost talk.

As I have said, *Watoriki*, the Wind Mountain near which we live, is a spirit house. The *xapiri* who live there are the true masters of the neighboring forest, which is their home's outdoor space. They move around and frolic and rest from their games here. A great many mirrors surround this rocky massif and existed long before our arrival! This is why our ancient shamans had to carefully move them aside when they built our house, after telling the spirits their intentions in a friendly manner. The *Watoriki* site is surrounded by paths belonging to animal, tree, and water spirits. Ordinary people do not see their mirrors, but the *xapiri* see them as clearly as we see the central plaza of our own house. They cover the entire forest, as far as it stretches, and we humans live among them. The spirits constantly twirl and chase each other joyfully through the forest, making a cool breeze that we feel without seeing them. It is so. The wind does not rise up alone in the forest, as those who do not know of the *xapiri*'s existence believe. It is the movement from the invisible flight of the spirits who live there!

Wherever human beings live, the forest is populated with animal spirits. These are the images of all the beings who walk on the ground, climb on branches, or have wings, the images of all the tapirs, the deer, the jaguars, the ocelots, the spider monkeys and howler monkeys, the coatis, the toucans, the macaws, the guans, and the agamis! The animals we hunt only move through the parts of the forest where the mirrors and paths of their ancestors' images that became *xapiri* are. White people never think of that when they look at the forest. Even when they fly over it in their planes, they don't see anything. They must think the soil and its mountains are placed there without reason and that the forest is just a great quantity of trees. But the shamans know it belongs to the *xapiri* and that it is made of their countless mirrors! There are far more *xapiri* than humans in the forest, and all its other inhabitants know them!

OMAMA CREATED the *xapiri* in large numbers and scattered them in every direction from our land and far beyond, to the other side of the waters, all the way to where the white people live.[28] Those who come from those distant lands are truly splendid! They followed *Omama* in his flight long ago and he has kept them near him ever since. He keeps them hidden, for they are the most beautiful and most powerful *xapiri*. This is the case with the magnificent images of the *ayokora* cacique birds, whose mouths skillfully regurgitate the evil spirits' objects and the sorcery plants that they extract from sick people's bodies. Our forest's *xapiri* are the ones *Omama* left here. There are a great many of them, and he thought they would be enough for us. Yet they are weaker and not as wise as the ones he took with him towards the white people's land, where there live as many *xapiri* as in our forest. Yet the white people do not see them. Perhaps their ancestors knew them? But today their children and grandchildren have forgotten them. It is true that *Omama* is sparing with his spirits! He is their true father. He is their master, as the white people say, and he does not want them to be mistreated. If he were to freely send them to youngsters with malodorous penises, who eat too much salt and answer them with a crooked tongue, they would instantly escape, furious and disgusted. *Omama* does not want that! So he keeps them near him and only sends them one by one, solely when experienced shamans call them. He does not surrender his most beautiful *xapiri* so easily! He only lets them go to shamans whom he recognizes and whose bearing he appreciates. First he identifies their adornments and tells himself: "*Haixopë!* They are truly my own!" Then he lets a few spirits go to them: "This is good! You can take them and make them dance far away!"

This is how we must ask *Omama's* image for our most powerful spirits, and only experienced shamans can do that. If a young, ineptly adorned initiate tried it, *Omama* would get furious and instantly rebuff him, stating: "You are truly ugly! Where are your macaw tails? Your arms are bare! Where is your black saki tail headband? Your hair is thin! Where are your ear ornaments made of parrot feathers and *hëima si* feathered hides? Don't you want them? Then you are not one of us! You only know how to stuff yourself into white people's clothes! You are empty! Do not ask me anything!" It is so. If *Omama* did not send us his most beautiful *xapiri*, they would not come to us on their own! In the beginning, when we are still ignorant, only the spirits of leaves, termite nests, logs, firebrands, and dust reach us! These are ghost-tongued *xapiri* who only approach to test the initiate, to prepare his mouth and sweep the clearing where the real spirits will come to settle later. Only once we become experienced shamans will *Omama* send us spirits truly capable of facing disease and epidemic fumes. Once we are elders and have a more robust chest, he finally sends us the powerful spirits of the *ayokora* cacique birds.

The few *xapiri* initially granted us by *Omama* come from very far away, hailing others to join them along their path from house to house. There are very few of them, but little by little their voices join together, steadily increasing as they come in our direction. Decorated with bright ornaments, they gather in a vast cohort that makes a powerful roar. When their troop passes in front of other spirits' houses, these *xapiri* are carried away by their elation and ask: "What are you off to do so eagerly?" They are invited to join the joyful colony as it grows and grows: "We are going to dance for the ghosts; join us! Let's all go together!" This is how it happens. When we diligently answer the songs of the *xapiri* coming towards us, their number keeps increasing. Their joy builds steadily, and finally a great multitude of them arrives to do their presentation dance.

I talked to my wife about all this and one day she asked me: "But if you say *Omama* is sparing with his most beautiful spirits, does that mean that the *xapiri* you usually make dance are weak and ugly?" I instantly protested: "No, that's not it! It is the humans who are hideous compared to the spirits! On the contrary, the *xapiri,* who are our children, are very beautiful! However, the most splendid of them only come little by little, and they do not come easily! It is so!" She answered me: "*Awe!* I understand. They are exactly as you say! If I were a shaman I would see them too!" It is true. Women sometimes also become shamans. This happens when their father was a shaman before them and they were born of the sperm of his spirits—for as I said, when a shaman copulates with his wife, his spirits do so too. Thus as soon as these young girls reach puberty, the *xapiri* show their desire to dance for them. If the girls are not afraid to answer them, the *xapiri* settle by their side.

This is how it used to happen with the daughters of our elders. They did not become shamans without a reason! They followed in the footsteps of their fathers and, like them, cured the sick and chased away the evil beings. At the beginning, they were not supposed to let themselves be sullied by men. But later, once their spirits were well established, they could take a husband. Today there are still some woman shamans, but mostly among the people of the highlands. When these young girls are wise, they do not try to give themselves to boys too early. They grow up without men, and this way the spirits continue to dance for them a long time. Their fathers call the *xapiri* and have their houses built near their daughters. This also happens here in the lowlands, but it does not last. Youngsters wind up copulating with these girls too soon and so they quickly stop answering the spirits. This was the case with the older daughter my stepfather had with a *Xamat^hari* woman of the *Parawa u* River. Since he was a very great shaman, she began to see the *xapiri* and make them dance like he used to do. But she was very beautiful, so the young men lusted after her and their penis odor made the *xapiri* run away. Otherwise she would really have become a shaman!

THOUGH THE images of the animal ancestors are numerous in the forest, they are not alone. The shamans also call down as *xapiri* the images of all its other inhabitants: those of the trees, the leaves, the lianas, and those of the honey, the soil, the stones, the waters, the rapids, the wind, and the rain.[29] They are no less numerous and are truly magnificent to behold when they arrive together to do their presentation dance! But the shamans can also make dance the images of the në wãri evil beings, who devour us in the forest like game.[30] It happens with the dry season being Omoari, who attacks humans when they are fish poisoning during the dry season,[31] or with the evening being Weyaweyari, who steals the image of children playing late outside their houses. This is also the case with the anaconda spirit Õkarimari, who kills women by making them miscarry, and the spirit of the ancient ghost Porepatari, who pierces us with his curare arrowheads.[32] These are fierce spirits who become irascible when they are starving or lacking tobacco.

Yet the xapiri are not all inhabitants of the forest. Some are images of beings who live on the sky's back and even beyond. There are also dangerous spirits, like that of the bird of prey Koimari, who cuts up children with his sharp machete,[33] of the Yãpimari butterfly, who carries away their image, and of the lightning bolt Yãpirari, who we bring down in anger with a powerful crash of light to frighten our enemies. There is also the bloody-mouthed sun spirit Mot^hokari who gives children fever with the burning cotton his wife spins and devours them. There are also all the images of the creatures who live on the new sky we call tukurima mosi.[34] This fragile, transparent sky lies far beyond the one we can see with our own eyes. It is inhabited by prõõri fly beings, warusinari insect beings, as well as watupari and h^wakoh^wakori vulture beings.[35] In the underworld, where darkness and endless rain prevail, everything is putrefied. Yet many other xapiri come from there too! These are the images of the aõpatari ancestors, who devour the sorcerer substances and evil beings thrown away by shamans during their curing sessions. There is also the chaos being Xiwãripo[36] and his peccary spirits, as well as Titiri, the night spirit, Ruëri, the cloudy weather spirit, and Motu uri, the underworld water spirit.

The xapiri are most often magnificent to see, like Yariporari, the storm wind spirit who dances softly, surrounded by whirls of white down feathers, waving large frayed hoko si palm leaves, which sway in his powerful breeze. Yet the images of the në wãri evil beings of the forest can be terrifying![37] This is also true of the image of the jaguar spirit †ramari who brandishes his sharp machete while projecting sparks all around, or the image of the cotton spirit Xinarumari, with his hooked nails, his burning ornaments, and his long venomous tail! There are also the horrible images of Poreporeri, the ghost of dead shamans, with his bald head and emaciated face, and of the moon spirit *Poriporiri,* with his thin beard and sharp canines! And still the images of the flood spirit *Riori* with his stinking hairy body, of the anaconda spirit *Õkarimari,* who dances on a path of embers with his enormous erect penis, and of the great glassy-eyed bird of prey spirit *Ara poko,* who swings an incandescent piece of cotton before him to tie up his prey! When you become a shaman and see these evil *xapiri* dance for the first time, they are really very frightening! Yet once they have tied their hammocks in your spirit house, you eventually get used to them, even if they remain very fierce and aggressive.

It is so. The shamans call innumerable images to dance for them and their words are truly without end! There are many more *xapiri* about whom I have said nothing. There are the *hutukarari* sky spirits who come and go in a blinding light, their heads covered in immaculate down feathers. There are the *waikayoma* women spirits who arrow glass beads[38] and the *amoa hiri* song tree spirits. There is the image of the avenging child *Õeõeri* who taught us war in the beginning of time and that of *Remori,* the bee spirit who gave the white people their tangled talk. There are the *xapiri* of the white people's ancestors created by *Omama* and called *napĕnapĕri.* There are also the old warrior spirit *Aiamori* and *Wixiari,* the death spirit who swallows enemies' breath of life. There are even spirits of dogs, *hiimari,* pots, *hapakari,* and fire, *wakĕri!* These words about the beings whose images we make dance never end! A tape recorder will never exhaust the multitude of their words!

A SHAMAN'S spirits call him "father" because they live by his side and he feeds them with the *yãkoana* powder. They do not use other names in addressing him. If their father does not bother them by getting close to the fragrant honey leaves of women's armbands, if he correctly imitates their song and often drinks the *yãkoana* to make them dance, the spirits will be satisfied and will stay by his side. Having eaten their fill, they joyfully exclaim: "Our father treats us well! He knows how to answer our words!" But if they are starving and exasperated, they feel mistreated and eventually go back to where they came from, never to return. The *yãkoana* is their real food. When their father drinks it, they revel in it through him. They die from its power just as he does. Then they are truly happy and their songs become splendid!

When I was younger, I wondered if the *xapiri* could die like human be-

ings. I know now that though they are tiny, they are powerful and immortal. The spirits that were made to dance by our ancestors are still alive,
long after the death of the shamans who possessed them! It is true. After
the death of the one they called "father," the *xapiri* recognize his son and
son-in-law and become attached to them. Once these are also gone, the
xapiri come down beside their children, who will drink the *yãkoana* in
their turn. It has always been this way. We call these *xapiri* of former shamans who come back to dance for the living *xapiri hapara pë*, the orphan
spirits.[39] The father who once made them dance is no longer. But despite
his death their houses and mirrors still exist. Their eyes, feathery adornments, and annatto body paint remain beautiful. They continue to be
in love with human beings and persist in coming down to us. When a
shaman elder has introduced a young man to his spirits in his lifetime,
these *xapiri* will recognize him and come down to him after their father's
death. As for me, I do not have many of these orphan spirits, for I did not
drink the *yãkoana* in the time when our elders were still alive. They were
not able to give me their spirits before they died and so their *xapiri* do not
remember me. Only one great shaman who died here in *Watoriki* some
years ago introduced me to his spirits in his lifetime. Now they recognize
their dead father's ornaments on me: his armbands' bunches of *paixi*
feathers, his black saki tail headband, and the traces of his annatto dye.
This is why they continue to come down to me. These *hapara pë* spirits
look a lot like their dead fathers. When they come to dance in ghost form,
we see through them the shaman elders who possessed them and their
memory comes back to us with much nostalgia.

ALSO, DO NOT think that the *xapiri* are only male spirits. Many women
spirits also do their presentation dance for the shamans. We call them
yaroriyoma, the animal ancestors spirit women, and also the women spirits *tʰuëyoma*.[40] These are the daughters, sisters, daughters-in-law, and
wives of the *xapiri*. There are very beautiful young coati spirit women
among them, but especially *kumi* vine spirit women, who are good at preparing love charms.[41] Men spirits only undertake their presentation
dance when these female spirits precede and attract them. Their magic
makes the men joyful and so the women spirits succeed in attracting
them in their dance, even when they are lazy or grumpy.

Our women and even our young girls seem so ungraceful compared to these women who can seduce all the *xapiri* and make them jealous! They are truly magnificent! Their long eyes are splendid and their black hair very fine. Their bangs are set off with a line of luminous white down feathers. The little sticks at the corners of their mouths are decorated with small black feathers from curassow crests.[42] Their earlobes hold white flowers from the *weri nahi* tree or red flowers from the *ata hi* tree, green *werehe* parrot tails, and colorful *wisawisama si* bird feathers. Their soft skin is covered in fresh vermilion annatto dye. They dance gracefully, sometimes with their babies sleeping in a carrying sling on their backs.

The male *xapiri* easily fall in love with them! This is why these spirit women always precede them. The male *xapiri* swiftly gather behind them, coming from all parts in constantly growing numbers. They never dance when they are only on their own! The spirit women's great beauty draws their gaze and fills them with emotion! They move beautifully with a joyful roar and encourage each other to dance. So the men spirits only like to do their presentation when they mix with the *xapiri* women. This is why the female spirits always dance first, just like our women at *reahu* feasts. The men spirits then respond to their call and the *xapiri* women lead them into their dance. At first, they pretend to push them away, but the men spirits cannot be stopped in their desire to join them. They are truly in love with them! If it were not so, the *xapiri* would not be so eager to dance!

THE *XAPIRI* are not like animals or humans. They are other. They do not drink river water or eat game. They hate everything salted or grilled and only eat sweet food. Bee spirits eat nectar from the flowers of trees such as the *pahi hi*, *hotorea kosihi*, *xitopari hi*, and *masihanari kohi*. Wasp spirits prefer ripe banana juice. As for spider monkey, toucan, curassow, and agami spirits, they drink fruit juice from *hoko si* and *maima si* palms and *hayi hi*, *xaraka ahi*, and *apia hi* trees. Tapir spirits acquire the image of their fat from the fruits of the *oruxi hi* tree. Do not think that the animal spirits' food is the same as ours! They eat images of what we call *në rope*, the **richness** of the forest.[43] This is real food, both tasty and free of any filth. They only drink flavored water from the high mountains. This is why even their excrement is fragrant. Ours stinks because the game we eat decomposes inside us. But the *xapiri*'s body does not contain any tainted flesh and so even their farts give off a pleasant smell! In fact they are in the habit of inhaling them from the hollow of their hand. They consider them an **energy** they don't want to lose. The emanations from our food and the smoke from our homes seem dirty and malodorous to them. Even the fragrance of the honey leaves on our women's arms are repugnant to them! The jaguar spirits are the only *xapiri* to devour game, while spirits of evil beings such as the *Koimari* bird of prey are also man-eaters. This is also true of vulture spirits who come from beyond the sky and have an insatiable appetite for human fat. These *xapiri* are danger-ous and can fly great distances to devour children of unknown houses. They can attack adults and sometimes even shamans![44] They are very cruel and certainly don't live on flowers!

The *xapiri* like tobacco as much as we do. Yet their plugs aren't any-thing like ours.[45] They are tiny and dazzling white. They are made with the celestial tobacco leaves of the *Yoropori* caterpillar spirit.[46] The kinka-jou, howler monkey, bee, butterfly, and lizard spirits all chew this tobacco. The same is true of the *Poriporiri* moon spirit and the *Yãrimari* thunder spirit. Yet the *warama aka* snail spirit always has the thickest, most hu-mid plug.[47] It is so. When old *xapiri* lack tobacco, the weather gets cloudy. They become irascible and no longer work to hold back the rain and the wind, which become too strong. Nonetheless, once they are satiated and calmed by a fat plug of tobacco under their lower lip, they relax and the weather clears.

THE *XAPIRI* are also valiant warriors and their **weapons** are truly formi-dable. They carry heavy clubs and huge iron blades, which we call *sipa-rari*,[48] like those the *karihirima kɨkɨ* snake spirits and caiman spirits bran-dish during their presentation dance.[49] They are like sabers of **power**. Yet they are not at all like the swords the white people know. They are as high as the sky and as bright and shiny as mirrors. They are made from a dif-ferent steel, slender and sharp, which is the father of metal. This is why they so mortally wound the evil beings. Other spirits such as those of the scorpion and the wasp also kill them by shooting curare-tipped arrows— isn't these insects' bite painful? Some *xapiri* such as the sloth spirit pos-sess a rifle acquired from the spirits of the white people's ancestors. He uses it to intimidate the thunders into becoming quiet and to fire at the evil beings and their hunting dogs. Others spirits fight with spears, like the spirit of the *yamara aka* stingray—isn't this fish's sting dangerous? Others such as the bat spirits use blowpipes to shoot sorcery plants at their enemies. And others such as the *maika* beetle spirit project *mai koko* balls of burning pitch[50] at them or crush them under their weight like the *Maamari* stone spirit.

These are all the weapons that the *xapiri* use to avenge and cure us. It is so. The peccary spirits tear apart the evil beings who take away children's images with their sharpened tusks while the spider monkey spirits untie the ropes that keep it captive. Similarly, the spirits of the little *yaraka asi* fish use their mandibles to tear marks of sickness[51] to pieces, like small fry in a stream fighting over the remains of game. Then the bee and ant spirits gradually devour them, in the same way that insects cluster together on the blood of game being carved apart.[52] The electric eel spirits know how to strike the *xawara* epidemic with their lightning bolts while the moon lacerates it with its sharp fangs.

But the *xapiri* often also wage war to protect us from other hostile spirits sent by distant enemy shamans. This is how it happens. The spirits of the *Xamat^hari* shamans live in the direction of the setting sun while those of the *Parahori* live in the direction of the highlands. For these *xapiri*, our spirits are *Waika* spirits.[53] All are most valorous and quick to make war to avenge themselves. Compared to them, we are all cowards! We often insult and threaten people we dislike, but we rarely actually shoot arrows at each other! On the contrary, the *xapiri* are never satisfied with mere words! They wage war ferociously, truly aiming to wipe each other out. This is why the spirits of *witiwitima namo* kites, small *teateama* birds of prey, and *xiroxiro* swallows fight each other by throwing blocks of stone they tear off the mountains. They are extremely fast and no one can follow their tracks. They attack suddenly, then disappear into the air to instantly reappear someplace else, striking and disappearing again.

Xapiri warriors put sky splinter points on their arrows, which shine with a blinding light as from a luminous metal.[54] They go to find them at the edges of the forest, where the celestial level comes close to the earth and the sun disappears. With these powerful points, they never miss their target, even from a great distance! They can also seize their enemies and shut them in large metal **cases** similar to **prisons** or stick them to the sky's chest with pitch until it kills them. They sometimes also dance while brandishing blazes from distant lands, which they call *mõruxi* *wakë*. These huge fires resemble what white people call **volcanoes.** They burn ragingly and devastate everything along their path. The spirits use them to terrify their enemies and burn their houses. It is so. When *xapiri* sent by enemy shamans are headed towards us, our own spirits fight them with pitiless ardor!

CHAPTER 5

The Initiation

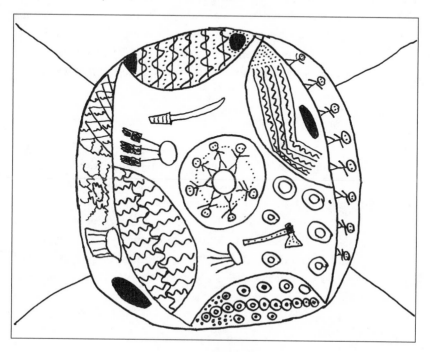

Mirrors of the *xapiri*

O NCE I WAS an adult, the *xapiri* continued to frighten me in my
sleep, just like they had during my childhood in *Marakana*. But I
had not drunk the *yãkoana* yet, and I did not really know them. I was still
an ordinary person; my chest was empty. In my dreams, I only caught
sight of them as dazzling white down feathers, in the form of a distant
swarm of light. I did not have any idea of what they really were! I continuously became ghost at night and never slept peacefully. This is why my
stepfather always wanted to make me into a shaman. When I was a child,
he often told me: "As soon as you grow up, I will give you my most beautiful spirits! I will open their path! I will call them and open a clearing so
they come to you!"

At the time, this frightened me and I would answer him: "I'm still too little, I don't want to!" Yet I continued to become other as I slept and the *xapiri* persisted in visiting my dreams. They kept their eyes focused on me all the time. When it is not so, you cannot dream as if you yourself were a spirit. You are only able to dream of things seen during the day, like ordinary people do. Some young people become other because the *xapiri* show themselves to them when they are hunting. This was not my case. They always revealed themselves to me during the time of dream. They looked fondly at me and wanted to visit me because they recognized the marks of their ornaments on me, for I had worn them since I was a little child.

WHEN I STARTED working for the white people at the FUNAI outpost in Demini, at the foot of the Wind Mountain, my disturbing dreams had not stopped.[1] Then, several moons after I moved there, the father of my future wife decided to settle in the area with his people.[2] They built a new house there. At the time, I often left the FUNAI sheds once my work was done to go sleep in this new house. It was much smaller than our current dwelling in *Watoriki* and situated farther from the white people than we are today. My sleep was very restless there. My old nightmares came back with even more strength, and I became other nearly every night. In the morning, when I woke up, the people of the house would often tell me: "You always behave like a ghost when you sleep!" And even when I went into town with the people from FUNAI, it continued. They also often told me that I talked and thrashed around when I was asleep!

Finally I talked about all this with my wife's father, who is a great shaman. I asked him: "Why do I sleep so badly? What are these visions that frighten me so much during my sleep?" He listened to me carefully, then explained it to me: "You constantly talk and yell when you dream? You thrash about like a ghost in the night? The *xapiri* are making you become other and frightening you when you sleep. Don't worry! They only want to show you their presentation dance so they can settle by your side. To do so, they make you become a spirit like them. Long ago, when your elders treated your ills, they put spirit adornments on you. This is why they recognize you and come to you so easily now! You do not become a ghost without reason!" As I listened to him, my thought hesitated and I did not know what to say about all this. In the end, I merely responded: "I don't know!" Then he asked me: "Does this also happen to you when you are awake?"[3] Of that, I was certain: "*Ma!* I only see the spirits come to me when I'm dreaming!" Then he added: "That's good! Stop crying out in vain during the night! Don't act as a ghost without reason anymore! Drink the *yãkoana* with me and answer the spirits who want you! Then you will be able to heal your people! If you want, bring your nose over here so that I can blow through it their breath of life! I will truly make you become a spirit!"

Worried and wavering, I asked him questions about the *xapiri:* "What are they like? Are they truly very beautiful? Are they powerful? Can they kill us? If we are unable to answer them, can they become hostile?" He simply declared: "If you do not become a shaman, you will be helpless when you have children and they get sick!" Then I told myself: "*Haixopë!* I understand! It is my turn to emulate our elders who have become spirits since the beginning of time! I did not know our grandfathers, but I know they were great shamans. I must follow in their footsteps and make dance the spirits they possessed before me!" Since my childhood, I had often caught sight of the *xapiri* in my dreams and had thought that it would be good to become a shaman to know how to cure people. But without really knowing them yet, I felt at a loss. I told myself that if my people got sick, I would be unable to do anything to avenge them from the evil beings and epidemic fumes.

So I decided and I finally answered: "*Awe!* I am willing to try and drink the *yãkoana!* I am unaware of all these things, but I want to truly know the beauty and strength of the *xapiri!* I want to become a spirit!" My father-in-law looked at me with a smile and replied: "Is that really true? You won't be afraid?" I retorted: "*Ma!* I truly want to follow the path of the elders! I want to be able to continue to make the spirits come down when they are no longer! I want to drink the *yãkoana* so my eyes will die in their turn!" And so he began to give me his spirits by blowing the *yãkoana* in my nostrils for the first time. He is an elder, a great shaman. His *xapiri* are very numerous and powerful. His thought travels very far and his spirits' house is very high.

He was generous in giving me his own *xapiri*'s breath of life because he really wanted to make a shaman of me! I was initiated in his house, the people of *Watoriki*'s house.[4] This was their first dwelling in the Wind Mountain's forest. At the time, I was still working as an interpreter for FUNAI. Yet the white man who was the head of the Demini Outpost did not try to prevent me from taking the *yãkoana* and becoming a shaman.

He did not like me and kept me at a distance. He did not concern himself with what I could do. So, most of the time, he ignored me.

THIS IS HOW it happened. I started to drink the *yãkoana* one day during the dry season. The house was nearly empty. It was not a *reahu* feast period, for the *xapiri* prefer silence. They do not like to come down when the home of the one who calls them is crowded, noisy, and smoky. The previous day, my father-in-law had gone into the forest and cut out and heated strips of bark from the *yãkoana hi* tree. He had collected and cooked its red resin in a clay pot. The next morning he carefully began crushing it into a powder. Once he had finished, he called me and had me squat before him. The sun was already quite high in the sky. The freshly prepared *yãkoana* powder had a very strong smell.[5] Then he started to blow large quantities in each of my nostrils with a tube made of the stem of a small *horoma* palm. He blew forcefully, several times over. It was the first time I inhaled so much *yãkoana* this way!

I was very anxious, because I hardly knew the full extent of its power. Then suddenly its image, *Yãkoanari*, violently struck me on the back of the neck and sent me backward onto the ground. I instantly lost consciousness and remained sprawled on the house's central plaza in a ghost state. This lasted a very long time. The *yãkoana* had really made me die! Finally, I came to a little bit and started to moan. My stomach was dropping in fear and I remained frozen, prostrate in the dust. I must really have been a sorry sight! My skull was hurting a lot. I truly believed I would not survive! My terror kept growing. Yet despite my fear, I squatted again and continued to draw my nostrils close, moaning with each new blast of *yãkoana*: "*Aaaa!* I am becoming other! *Aaaa!*"

WE DO NOT become shamans by eating game or the food from our gardens, but through the trees of the forest. It is the *yãkoana* powder, the sap exuded by *yãkoana hi* trees, that reveals the spirits' words and spreads them far and wide. Ordinary people are deaf to them, but by becoming shamans we can hear them clearly. As I said, the *yãkoana* is the *xapiri*'s food. They call it *raxa yawari u*, the juice of the water beings' peach palm fruit. They never tire of drinking it down. As soon as its power increases, they absorb it through their father, the shaman, when it penetrates him through his nose, the entrance to their house inside his chest.[6] There are a great many of them eating it. This is why the shaman who has called them does not collapse to the ground. By drinking the *yãkoana*, he simply enters a ghost state, and once they are sated his *xapiri* come down onto their mirrors, spreading the sweet odor of their annatto body paint all around.

The *yãkoana*'s power is strong and lasts a long time. Yet it is less luminous and violent than that of the powder drawn from the seeds of the *paara hi* tree, which the *Xamat^hari* use. There are several *yãkoana*. Among them, the *yãkoana haare a* powder has the most intense power.[7] If you inhale it without being prepared, its spirit will strike your skull with an axe and violently throw you to the ground. You instantly lose consciousness and do not come back to your senses anytime soon, especially if you mix it with *paara* powder! As soon as they have drunk the *yãkoana,* the *xapiri* seize their father's image and take it on their distant flights while his skin remains sprawled on the ground. Though the distances may appear considerable to our ghost eyes, they are nothing to the spirits, who are extremely fast. When they come down to us, we barely have time to hear a faint humming before they have seized our image and lost it very far away.

Yãkoanari is the name of *yãkoana*'s father. His image still lives in the place where long ago *Omama* had his son, the first shaman, drink this powder. *Yãkoanari* is a true elder, a very powerful spirit. According to the white people's words, he is the master of *yãkoana*. The power of his powder is so high that he makes a blinding light explode inside us. When you do not know him, he violently knocks you senseless and you just collapse to the ground. You thrash about in every direction, your stomach gripped with terror. Then you remain unconscious on the dirt floor for a long time. This is what happened to me the first time. But later, once you have become familiar with using *yãkoana*, that is over. You no longer fall and roll around and moan in the dust! Despite its sudden and powerful blast, you stay standing and then you can truly become spirit by relentlessly singing and dancing. The *yãkoana* spirits, which we call *Yãkoanari* and *Ayukunari*,[8] are by our side. They help us to think right and make our words constantly grow in numbers and spread. It is the *yãkoana* that allows us to see the paths of the spirits and evil beings under the elders' guidance. Without it, we would be ignorant.

Becoming ghosts during the day or the time of dream, we study with the *yãkoana*. As I have said, if we do not take the *yãkoana*, we do not really dream. But when we sleep under its power, we continue to see the spirits sing and dance in our slumber. Our body remains prone in its hammock but the *xapiri* fly off with our image and show us unknown things. They carry our memory with them, in every direction of the forest, of the sky and of the underworld. If it were not so, we would only see human beings like us in our dreams. We would only see those who are close to us, people hunting or working in their gardens. It is so. Do not think that the *xapiri* only show themselves during the day, when we drink the *yãkoana!* On the contrary, they continue to sing for us during the night. They constantly exhort their father to listen to them: "Don't sleep! Answer us, don't be lazy! Otherwise we will abandon you!" If he stayed with his nose stuck to the ashes of his hearth, snoring away, his *xapiri* would be most unhappy. They would leave the house without his knowledge, one after the other, and never come back. This is why we so often hear the shamans sing in our houses during the night!

THE WHOLE time my father-in-law blew the *yãkoana* into my nostrils, he did not let anyone else get near me. I was lying on a hammock made of bark. Even my wife had to keep away. She only came from time to time to carefully leave a few logs to fuel my fire. Everything around me had to stay quiet. You cannot walk loudly or drop a load of wood near someone taking *yãkoana* for the first time. Otherwise the *xapiri* might instantly run away. They are very fearful and quickly vanish when humans make too much noise. They are not used to that. Their houses are very quiet. So the shamans are careful not to frighten them.

I also had to avoid moving around too much. If you are constantly moving, the spirits will also refuse to come down and dance for you. They only approach with great caution, once the shaman elders have carefully cleaned the ground and covered it in white down feathers. My father-in-law also warned me: "The *xapiri* hate cold water! Only bathe in lukewarm water. Do not go into the forest to bathe in the river. The spirit's mirrors will break! Their paths will be destroyed!" It is true. The *xapiri*'s paths, as fine and transparent as spider threads, are very fragile. He also told me: "When people broil meat on their fire, let them eat alone, do not ask them for anything! The spirits hate the smoke and odor of char-grilled food! They are not starved for meat like we humans are. They only eat sweet food. Do not drink river water either. Don't worry, your appetite and thirst will soon disappear."

At first, I really suffered from the hunger, to the point of crying! But that's the way it is, you cannot see the *xapiri* and become a shaman by dozing with your stomach full of game and manioc. I was also intensely thirsty. My tongue was completely dry. Yet after a few days, my hunger and thirst vanished. The spirits pushed them far away. I did not feel anything. I would see a calabash full of water, but have no desire to drink. The people around me would eat peccary, but I had no taste for eating game either. I was happy inhaling the *yãkoana* powder, blast after blast, again and again. The *xapiri* were constantly dancing around me, and it was they who were feeding me. By becoming other, I was starting to eat an invisible food that they placed in my mouth while I was asleep. In my dream, they repeated: "Eat, this is our food! Refuse game and do not use tobacco! Do not bathe either! Do not go near women! The odor of their honey leaves is dangerous! If you truly want us, listen to our voice and repeat the words of our songs!" Then I could smell the scent of their annatto body paint and magic plants spreading around me. I was very weak but in my sleep I happily ate what they brought to me.

This lasted a long time, maybe **five days** or more. During all this time, my father-in-law never stopped blowing the *yãkoana* into my nose. I got thinner and thinner and my ribs started to stick out. I was very dirty and my eyes were gaunt with hunger. I practically didn't eat or drink at all during this period, only sweet things: a little banana soup and sugarcane juice. I did not eat game, braised plantains, manioc, sweet potatoes, or anything else. I did not use any tobacco wad either. Otherwise I would have uttered ghost words instead of answering the spirits' songs! I only drank the *yãkoana* powder, continuously. Bit by bit, the wasp spirits and the *xaki* bee spirits devoured all the fat in my body. There was nearly nothing left of my flesh. I was a sorry sight and could only speak in an almost inaudible faint voice. I was very weak and pitiful. I no longer had any breath of life. All traces of food and rotten game had disappeared from my insides. The *xapiri* had weakened me with hunger and thirst. They had made me much thinner. I had become clean and sweet-scented like I was supposed to be. It is so. The spirits observe and smell us from afar before they come close. If they find us greasy and nauseating, they instantly run away! The smoky stench of hunters who eat their own game also makes them vomit. In this case, they spit on the apprentice shaman and exclaim: "His chest is the chest of someone who devours his own prey![9] How filthy he is! His flesh is acrid and fetid! It tastes like burned game! His chest smells like the frightening odor of the women's honey leaves!" This is why the elders who give us their spirits begin by trying to clean us. They must rid us of any leftover food, of any whiff of charbroiled and decomposed game that remains in us. They must also wash us of any penis smell. Then they can make us become spirits as they themselves once did. As long as we remain dirty and malodorous, the *xapiri* refuse to come dance for us.

During all the time I drank the *yãkoana*, my wife was worried and a little unhappy with me. She wondered why I wanted to do this and see the *xapiri* if it made me suffer so much. Finally, when she saw me so emaciated and weak, she cried. Then she told me: "Before father had you inhale the *yãkoana*, I was angry about your decision. But now you really make me feel sorry for you!" Our house's other inhabitants were as concerned as she was to see me in this frightening state. Yet I did not feel badly at all for I truly wanted to become a shaman! It is so. To receive the spirits of the elder who give us the *yãkoana*, we must have an empty stomach. At the beginning, its powder must be our only food. Once our insides are truly cleaned out, the *xapiri* can finally come to us.

Then we can start eating a little again, but only food that has not been broiled and is not salty or acidic. We can only ingest food that is white and tasteless: boiled plantains, small fish cooked in a leaf, but also sugarcane juice, papayas, and especially honey diluted in water. This beverage truly has the power to put us in a ghost state and make us become spirits! The *xapiri* eat forest flowers and fruits, but their favorite of all is honey. As soon as the young shaman swallows honey beverage, his spirits eat their fill through him and are delighted. This is why the *xapiri* tell us: "We will come to you, but you must eat sweet food like us! Do not be impatient to devour meat!" After that, when we see bees in the trees, we no longer think that they are simply bees. We know that they are also *xapiri* who only like sweet and fragrant flavors. As I said, the *xapiri* do not eat manioc and game like we do. They also do not drink the water of forest streams. They are flower nectar drinkers! This is why they are only happy to come down to us when all we eat is the food they like. But later, once the jaguar, puma, and ocelot spirits have come to us, we can eat meat again. Then the elders tell us: "*Awe!* Your jaguar spirit has danced, you can now calm your hunger for game! But if you add hot peppers, you will have to carefully rinse your mouth!"

This is how the elders protect the *xapiri* whom they bring down to give to us. Now it is my turn to warn the young people who want to become shamans: "Do not go to the river to follow women! Do not eat all the time! If you don't restrain yourselves, you won't be able to see the *xapiri*! You will never hear their songs! They will refuse to dance for you!" If the shaman elders were not vigilantly at our side when we drink the *yãkoana* for the first time, we would run the risk of not taking any precautions and mistreating the spirits. Infuriated by this lack of respect, they might strike us with their machetes and kill us. Yet even if we fear their power, our desire to make them dance like our ancestors did is stronger. This is because we are inhabitants of the forest.

IT IS TRUE that sometimes the *xapiri* terrify us. They can leave us for dead, collapsed on the ground and reduced to a ghost state. Yet do not think that they mistreat us for no reason. They simply seek to weaken our awareness, for if we were merely alive like ordinary people, they could not make us think right. If we did not become other and remained vigorous and concerned with our surroundings, it would be impossible for us to see things the way the spirits do! This is why the *xapiri* say of the initiate: "If he remains strong, he will not hear our voice!" Then the bat spirits weaken us and keep us in a ghost state by blowing their sorcery plants on us. The *xapiri* also try to clean us of any traces of food odor. As I said, they are very concerned with cleanliness. If they find the slightest scrap of putrefying game in our innards, they tear it to pieces and throw it far away. They also carefully wash our mouth and chest in order to remove any burnt game odor. They scrub our skin in order to erase the fragrance of women, as well as the scent of meat, the odor of copulation, and the stench of excrement. If it is contaminated by the *xawara* epidemic, they just tear it off like that of a venomous *yoyo* toad and throw it in the river. Then they sprinkle us with mountain water and energetically rub it on us. Finally, they cover us in a new skin, adorned with white down feathers and annatto drawings. If we remain dirty, we only have a ghost tongue and are incapable of answering the *xapiri*. The spirits of the leaves, lianas, and trees are the first to come clean us. They also tear open our chest and make it bigger so new *xapiri* can build their house there.

Other spirits make us be reborn. We return to being newborns, still red with the blood of our birth! Then the spirit women cut our umbilical cord and wash us with clear water. They put us down on a bed of white fluffy feathers where we gesticulate like babies! When we cry, capuchin and *proro* otter spirit women rock us in their arms.[10] They breastfeed and take care of us. Finally, when we leave their breast behind and grow up, they teach us the spirits' songs: *"Arerererere!"* Then it is the *wari mahi* tree and *mohuma* eagle's turn to cover our body and face with a bright, luminous white down.[11] Next the images of *Omama* and other *xapiri* give us their finery. They wrap bands of black saki tails around our forehead and put bunches of parrot feathers and scarlet macaw tail feathers on our arms. They decorate our bodies with black and vermilion annatto patterns.

Once we are adorned like this, they carry us to the sky's back and put us down in the center of a clearing where they will do their presentation dance. The ground on this clearing is a vast mirror strewn with white down feathers, which scintillate in a blinding light. All this is both magnificent and so frightening! It is our image that the spirits carry away in order to set it to rights. First they extract it from inside our body and put it down on their mirrors in the sky. Meanwhile, our skin, which has become so weak, remains stretched out on our house's plaza in the forest. Then the spirits lose our thought and language in order to teach us theirs. Next they make us learn the pattern of the forest so that we can see it like they do and protect it. The *xapiri* are superb and sparkling with light. They seem fragile but are very powerful. From their mirrors, they reveal the approach of epidemic fumes, evil beings, and storm wind spirits to us. White people do not know this. Yet this is how our elders have always become shamans. We are only following in their footsteps.

WHEN MY WIFE'S father made me become other, everything happened just as I have described it. First he drew all the vigor out of me with the *yãkoana*. Little by little the *Yãkoanari* spirit ate all my flesh. I got so frail that I was really a pitiful sight! Then the *xapiri* washed my chest of any bitter and salty flavors. They cleaned my insides of any leftover putrefying game. They weakened me and made me be reborn. After some time my father-in-law called other spirits to settle by my side. He told them: "This young man who I have made to drink the *yãkoana* desires you and wants to become a spirit as his elders did before him! Will you agree to do your presentation dance for him?" They answered him: "*Awe!* He is one of yours. We have always danced for your ancestors. We know you. Since he wants us, we will come dance for him in his turn!"

Encouraged by these words, my father-in-law resolutely continued to make me drink the *yãkoana* so I could think right. This is how we study to become shamans. The elder who calls the spirits for us must blow their food into our nostrils for days. Then little by little, we eventually see them approaching and dancing during the night, and it does not stop. This is what my father-in-law did for me. He made me see the *xapiri*'s path, brought them down, and gave them to me. He did not want anyone to be able to call me a liar! He is a great shaman. He is really very wise. It is so. We follow the words *Omama* gave to his son: "If you really want to see the *xapiri* and answer them, you must often drink the *yãkoana*. You must remain at rest in your hammock and stop eating and copulating without restraint. If so, the *xapiri* will be satisfied. If not, they will find you dirty and run away." This is why my wife's father warned me: "Your thoughts will have to remain calm and you will have to answer the *xapiri* carefully, otherwise they will get angry and could mistreat you!"

UNDER THE effect of the *yãkoana*, I remained stretched out on the ground, unconscious for a long time. Then the jaguar and deer spirits came to me and started licking my body with the tips of their rough tongues. They tasted my flesh to see if it was still bitter and salty. They asked themselves: "How is he? Will we be able to clean him and set him to rights?" This is how the spirits first evaluate us. If they notice that our chest is too smoky, soiled by the leftovers of our own prey, or stinking of penis, they get angry and instantly reject us. They violently hit us. Next the *pirima ãrixi* tick spirits clung to my image with their mouths while the sky spirits carried it into the heights to lay it down on their mirrors. Then I drank more and more *yãkoana*.

Finally it was the turn of the *yawarioma* water-being women's images to frighten me. They lived in the underworld before *Omama* made the rivers flow out of the earth. They are his wife's sisters. Their love charms make young people become shamans. They only come down to us when our body is really empty of any game meat; when we are no longer eating bananas and manioc or even drinking water. They do not come down until the *yãkoana* has consumed all our flesh. They are very beautiful and of the highest value. Only the elders can call them for us. As soon as they arrive, they are anxious to carefully examine us. If they find us acceptable, they take us with them. When this happens to a young man, he abruptly dashes out of his house like a ghost. He runs far into the forest, off the paths, moaning and calling his mother at the top of his lungs: "*Aaa! Napaaa! Aaa! Napaaa!*" He only comes back much later, once an elder has followed his trail to bring him home. This is what happened to me! After I had drunk a lot of *yãkoana*, the spirits of the forest and of the women of the water came to me during the day and took me with them. I started running like a ghost, accompanying their light as it rushed far ahead of me. I followed their paths through the forest for a long time, constantly shouting: "*Aë! Aë! Aë!*" I truly ran to the end of my strength! My wife's father feared I would lose myself forever and he protected me. He intervened to stop these spirit women from taking me to their house underwater. So they left me on the forest floor, unconscious, and my father-in-law sent his own *xapiri* to bring me back to our house.

At the beginning, when you do not know the *yãkoana*'s power, you do not stay standing for long! This is what happened to me. Its power made me die and instantly threw me backward. I rolled on the ground, twisting with fear and moaning: "*Akaaa! Akaaa!*" I had become a ghost, but the *xapiri* remained invisible. This made me very anxious. I kept asking myself: "Why don't I see anything yet?" Several days passed like this, without any spirits appearing to me. I was sweating profusely, and my skin was completely covered in dust. I was tormented and deeply agitated. I drank the *yãkoana* without respite and was scared. Each time its power seemed more terrifying to me, and I was feeling increasingly weak. This is why few young people dare to present their noses to the elders! And when they do, they often quickly give up for fear of dying of it! Yet I wanted to continue because despite my fear I really wanted to know the *xapiri*.

This is why I was so worried that I would not be able to see them at the beginning. It is true! I was taking the *yãkoana* all the time and could not see anything! This is what usually happens, but I did not know. When you start drinking the *yãkoana*, you do not see anything at all. Your head is seized by a strong pain and your thought remains closed. You get weaker and weaker and simply lose consciousness. That is all. The *xapiri* do not instantly reveal themselves to him who drinks the *yãkoana* for the first time and, if he is not well prepared, he will not come out of this state. The spirits only start to do their presentation dance after they have stretched the initiate out on their mirrors. You have to spend several nights in a ghost state and be very worn out before they manifest themselves.

First they contemplate you from the heights of the sky. They see you lying in the open, like a small clear stain on the ground. Then they start to come down in your direction because they really want you. As for you, at first you only hear their voices converging from the distances. Then suddenly they approach you and seize your image before you have had a chance to catch sight of them. It is so. The first day you really do not see anything. The following day you are unable to distinguish between day and night and you can no longer fall asleep. The day after that you become ever more exhausted. The next day the *xapiri* finally start to appear. You no longer experience hunger or thirst. You no longer know pain or sleep. The *yãkoana* spirits have devoured your flesh and your eyes are dead. At that moment, you start to see the dawning of a vast and blinding light. Then you distinguish the cohort of *xapiri* singing as they head in your direction. Called by the elders from the ends of the sky, they approach you little by little as they dance along their luminous path. The first to arrive are still relatively few. They call the others as they pass. They gradually come together until they form a noisy throng.

THIS IS WHAT happened to me, and I was very frightened, for I had never seen anything like it. The dreams I had been having since I was a child were nothing next to this! Seeing the *xapiri* come down to me for the first time, I truly knew what fear was! What I started to see, though I could not clearly distinguish them yet, was truly terrifying. The forest initially became an immense void, which was spinning around me without letting up. Then suddenly everything was immersed in a blinding brightness. The light exploded with a great crash. Now I could see only the ground and sky in great distances strewn with brilliant white fluffy feathers. This luminous down covered everything as it floated gently in the sky. There was no longer any shade anywhere. I was watching over everything from a horrifying height. I understood I was truly starting to become other! I told myself: "Father-in-law knows the spirits! This is why he truly knows the forest! He was not lying to me!"

When the *xapiri* want to put us to the test, they snatch our image and go put it down very far away, on the sky's back. The spirits of the *paara* powder trees, the spirit father of *yãkoana,* and the *urihinari* forest spirits are those who carry your image and your breath to lay them down on their mirrors. This is how one really becomes a shaman! This is what I lived through, and it was truly very painful. My thought was seized with oblivion and my skin was lying on the ground, inert. My people told themselves: "He is a pitiful sight, collapsed like a dead person in the dust!" But that was not what it was. My body was indeed prostrate on the ground, yet the *xapiri* were keeping my image on their mirrors, in the heights of the sky. This is why I got dizzy and was so scared I would fall. I found myself suspended over an enormous chasm, sprawled in a heap of white down feathers. I could no longer distinguish the people of the house around me. I could only hear the sound of their voices, like husky, inarticulate rumbling. They sounded like the voices of evil beings. It was truly frightening!

Suddenly everything around me started to be covered in yellow and white flowers like those of the *masihanari kohi* and *weri nahi* trees. Then several luminous paths unfurled to the ends of the sky. They were undulating in my direction and making an indistinct clamor. Full of apprehension, I asked myself what this could be. I wondered: "Who are these unknown beings coming towards me like this? What are they going to do with me?" I was still so ignorant! I asked my father-in-law: "Are these already things of the spirits?" He confirmed: "*Awe!* The *xapiri* are starting to approach you. They are coming little by little, but you cannot really distinguish them yet. You will only be able to see them clearly once you are very weak and have truly become other!" This is what happens when the initiate begins to become spirit himself, and his thought is still searching!

Then, squatting beside me, my wife's father started to teach me to hear the spirits' songs. He told me: "If you want to become a shaman, you must answer the *xapiri*'s voices by imitating their songs and speaking to them. At first, of course, you won't be able to. But little by little they will truly reveal their words to you. Your mouth must not be scared. Even if you don't sing very well yet, they will still be happy that you answered them. They will think: 'That's good! He really wants us!' But if you don't make an effort and don't behave as they hope you will, they will mistreat you. If you make the spirits feel sorrow, they will kill you and run far away!" Somewhat troubled by these words, I did my best to prick up my ears to listen to their voices and try to answer them.

When you start to drink the *yākoana*, you do not perceive anything of the spirits' song. They must first remove everything obstructing your ears and preventing you from listening to them. Next they manifest themselves by letting their melodies gradually be heard while we sleep. At the very beginning, I did not know anything of the *xapiri*. Though I kept tak-ing the *yãkoana*, I could not see them, and I could not even hear their sound yet! This tormented me and I told myself: "What is happening to me? I am dying and behaving like a ghost, but it's no use! I make for a pitiful sight rolling in the dust, all for nothing! What will I do? If I don't see the *xapiri*, should I pretend?" Yet I did not want to lie. All the residues of food had disappeared from my guts and I was exhausted. My very flesh had become that of a ghost. But before I could see the spirits, I first did my best to hear their words. I tried to imitate their songs little by little, as my father-in-law had recommended.

He was the one who started to teach them to me. He introduced me to the *xapiri* as our ancestors have always done with their sons and sons-in-law. Then, as my attention focused, the spirit's words began to be audi-ble to me. They replaced my tongue and throat with their own. Piece by piece, their songs were revealed to me and each time they became clearer. I began to sing like them. But it came slowly. Whatever you do, you must not be impatient! First, you must try to imitate the upstream of the words of the spirits' song.[12] This is how you can begin to truly hear them and this is what I did. Finally they removed the obstacles that blocked my ears.

They suddenly exploded with a dull sound. Then I began to perceive a faint melody, without seeing anything yet. It sounded like the hum of a swarm of mosquitoes. It was the whistling of the *purunama usi* bam-boo flutes into which the spirits blow as they dance. Their high-pitched sound came from far away and gradually drew closer. Suddenly, a low-pitched sound spread like a wind whirling over the whole expanse of the forest. Then I started to distinguish the calls and songs of the *xapiri* com-ing towards me in the distance, from the edges of the sky. Despite the distance, their voices became increasingly clear. Ordinary people could not hear them, but for those who had become ghosts, they were very sharp.

At the moment the *xapiri* finally reveal their voices, your fear vanishes and you experience an intense bliss, even if you are lying in the dust! Then you must try to answer them so that they feel happy to listen to you and encourage you with their clamor. This is how I began to sing, despite all my fears! I still only perceived very faint sounds. Yet I decided to an-swer the *xapiri*'s voices by echoing them. After a moment I began to hear their joyful exclamations: "*Awe!* This time he is answering us as he is supposed to!" Their voices seemed perfectly discernible to me. Satisfied, I applied myself to imitating their sound and their words, again and again, without stopping. Seeing my efforts, the *xapiri* came to my assistance. They told themselves: "He probably doesn't hear us well! Let's start over! What can we do so our songs become audible to him?" Then they started to sing again, making their voices stronger. This is how I finally truly heard them and began to sing like them. If we try hard to answer the spirits, the images of the *yōrixiama* thrush and of the *reã hi* song tree quickly come down to help us.[13] They lend us their throats and reinforce our tongues. This way, the words of the *xapiri*'s song rapidly increase within us as in a tape recorder. We drink the *yãkoana* with our eyes set on their presentation dance and lose all fear of singing before the people of our house. This is what happened to me!

AFTER ALL this time, I was so skinny it was frightening. My face was covered in mucus and *yãkoana* powder. I had died under its power and my eyes were those of a ghost. The spirits had finished cleaning my inside. Several days passed before I finally started to see them dance. I had become a spirit too. Their voices and dances had become mine. Now they showed that they were truly satisfied with me. It is so. The *xapiri* are happy when you answer them by vibrating your tongue: "*Arerererere!*" As soon as they hear you imitate their songs, they let out clamors of satisfaction and rush in from every direction, making a joyous roar, like guests at a *reahu* feast: "*Aë! Aë! Aë!*" But if the answer of your songs lacks vigor, they soon become irritated that they are not desired. They start insulting you: "*Hou!* Your voice is ugly and quavery! You are filthy! You stink of penis and you are a coward! If you're scared of us, don't call us!" They are furious that you merely thrash about in the dust uttering ghost talk without answering them in the manner they expect. They tell themselves: "*Hou!* Our songs are perfectly clear! Is this ghost truly deaf! Can't he see us? Is he asleep? Doesn't he want us? He claims to bring us from afar so we'll dance for him and now he remains mute!"

If you do not apply yourself to drinking the *yãkoana* and singing for them, the *xapiri* refuse to settle by your side. They never come close to ordinary people who are satisfied with living sprawled out in their hammocks. They find them dirty and think they are incapable of hearing their voices. If you call the spirits in vain, they accuse you of having a bitter taste and start mocking your ghost talk. They call you lazy and reproach you for not making them dance. Exasperated, they spit on you and cover you with ashes, then run away. When this happens to a young apprentice shaman, he begins to waste away. He soon becomes very thin and ugly. Instead of transforming into a spirit, he exposes himself to the risk of dying.

To become a shaman, your eyes must not roll around in every direction contemplating the ground or observing the people who live in the house. This is why I tried to keep my gaze focused on the sky. My eyes were those of a ghost and I no longer saw anything around me. My sight and my thought were concentrated on the *xapiri*. And so in the long run they finally showed themselves. I saw them coming towards me from the sky's heights in a shimmering bright light. They were descending slowly, gathering in ever-growing numbers, like a blinding fall of white fluffy feathers. The powerful vibration of their songs was gradually drawing closer: *"Arererere!"* They started to whirl around in place in the air, like a multitude of hummingbirds. Gradually I started to distinguish their splendid ornaments: their curassow crest and agami neck armbands, their black saki tail headbands, and their hair covered in king vulture down feathers. Their immaculate teeth sparkled and their skin shone with vermilion and black annatto drawings. They swirled around above me as they danced and whooped with elation. From that moment, my sleep escaped. I was stretched out on the floor of our house's central plaza and the forest had disappeared around me. I could not stop contemplating the *xapiri*.

They truly made me become other so I would not lie. They really wanted to make me become a spirit. They made the forest disappear and replaced it with a land covered in white down feathers. They laid my image on the back of the sky at the center of their mirror. It was very frightening, but my fear disappeared rapidly, for everything that I saw was magnificent. Despite the distance, I could see the *xapiri* and their colorful and shiny ornaments perfectly. Their gaze landed upon me. Their troop descended from the edges of the sky, carried by a multitude of glimmering paths that swayed in the air. They were as fast as planes and kicked up a powerful wind. This vast distance was nothing for them. They rushed in without interruption, countless, coming from every direction like television images. Then little by little they gathered in front of me like guests at a *reahu* feast crowding together at their host's door, impatient to do their presentation dance.

Their previously imperceptible paths became increasingly clear and bright. As fine as spider webs, they floated and sparkled in the air and came to be suspended near me one after the other. The *xapiri* are always preceded by the images of their paths. One by one, they stick themselves along the edge of the mirror on which the young shaman is stretched out. They attach themselves to it as if they were pictures like the white people's photographs. Meanwhile you must remain lying perfectly straight so the *xapiri*'s paths do not break and they can reach you. Then the spirits use your arms and legs as trails, and your elbows and knees are clearings on which they stop to rest. Finally, they penetrate your mouth and into your chest, which is the house where they will do their presentation dance.

THE *XAPIRI* arrive pressed tightly together in stunning lines, covered in body paint and colored feather ornaments. Once you can finally distinguish them, their appearance is deeply beautiful. The sound of their clamor is loud and their songs are extremely melodious. They avoid the ground's dirtiness and remain suspended in the air. They are sent by *Omama*, who makes them fly at high speed thanks to his own airplane image. It is very powerful and carries them all in its flight, though they are so numerous. This is how they travel above the forest, beyond the sky, and under the ground. They arrive before us on huge sparkling mirrors, which they suspend in the heights. They dance on them like guests at a *reahu* feast dance on the central plaza of their hosts' home. First, the animal ancestors and the water-being women make their entrance, waving frayed young *hoko si* palm leaves. They move forward and backward slowly, in a perfect line, stomping their feet on the ground in step. They are truly magnificent! Then the male spirits dash in after them and dance, making a large circle and whooping joyfully.

The *xapiri* are skillful and very cheerful dancers. Didn't the *yarori* animal ancestors use their funny dancing to make Caiman laugh so hard that he let the fire fall out of his mouth?[14] This is why we try to follow in their footsteps when we too become spirits. We imitate the anteater, spider monkey, deer, and tapir ancestors. We also imitate the *Poriporiri* moon spirit, the *Yãpirari* lightning spirit, the *Hutukarari* sky spirit, and many others. The *xapiri*'s ways of dancing are as diverse as their songs are different. When you reproduce their movements, their images take you by the arm and teach you to follow their steps with confidence. If you look awkward, with stiff legs, they get impatient and admonish you: "Follow me! Look! This is how I dance! Pay attention!" Then they train you in their movements so that your gestures are as graceful as theirs. They cover the circle of their mirror, coming and going with superb agility. They move slowly while assessing the spirit house in which they are about to settle. "Will it be beautiful enough? Will its floor be smooth and gleaming enough?"

DESPITE ALL its beauty, the *xapiri*'s presentation is also frightening. They dance around your prone body on their mirrors while waving huge blades of shiny metal. They observe you, measure your strength, and judge your appearance. Once they have finished their turn, they come back to their starting point, passing you along the way. Suddenly, one of them turns around and brutally strikes your back with his machete's sharp edge. But the spirit strikes this blow without brandishing his weapon. You are just hit by the blade swinging on his back. The intense pain knocks you down unconscious at once. Then the *xapiri* group slow down, stop, and observe you without moving.

The spirits who wound you like this are the hostile spirits of the *waroma kiki* snake and the *poapoa* caiman. Have I not said that certain *xapiri* can be very dangerous? This is very true of *Ara poko,* the **chief** of the *koimari* bird of prey evil beings. When a shaman makes his image come down, the others must protect the children of the house from the breeze of his poisonous tail. When you make this spirit dance for the first time, he cruelly injures you. It is so. The spirits do not content themselves with dancing for you! When they arrive, they also hurt you and cut up your body. They divide your torso, your lower body, and your head. They sever your tongue and throw it far away, for it only speaks ghost talk. They pull out your teeth, considering them dirty and full of cavities. They get rid of your guts, full of residues of game, which disgusts them. Then they replace all that with the images of their own tongues, teeth, and entrails. This is how they put us to the test!

This is what happened to me and I was truly very scared! These *xapiri* elders are really fearsome! They silently drew close to me at the end of their presentation dance. They did not seem threatening. Yet suddenly I felt their blades violently hitting me. They cut my body in half in a single stroke down the middle of my back! The impact drew a long moan of pain out of me. But that did not stop them! After they had cut me in two, they sliced off my head. Then I staggered and collapsed, crying. My thought had wandered and I had become blind, like a dead dog lying on the ground. I remained prostrate like this for a long time, devoid of any sensation. During this time, the spirits continued to dance around me, without my knowledge.

A little later, I regained consciousness. I stopped drinking the *yãkoana* and my thought became calm again. Then I started feeling the shooting pain from the wounds the *xapiri* had inflicted upon me. I felt it in the nape of my neck and my back, where they had hit me so harshly. I could no longer walk without being bent over, as if I had become a crippled old man. At first, all this is terrifying because you really wonder if the spirits intend to kill you. It is true! Yet in the long run the sharp pain from these wounds gradually diminishes, though we continue to ache. This is what I felt and I was really a pitiful sight! My father-in-law did not go easy on me when he gave me his spirits!

Every time new *xapiri* come to you, they hit you in the same way, with the cutting edge of their metal blades. They start doing it before you can really distinguish their image. Then they start again once you are already stretched out on their mirrors and you see them dancing around you. Yet you must not think that this only takes place when you drink the *yãkoana* for the first time! This happens again later, even when you already own a big spirit house and you have become a great shaman! Each time new spirits come to you, they hurt you as badly. This is why shamans' necks and backs become so painful in the end. These parts of the body are the ones the spirits like to hit, and the suffering they inflict on you is really intense. Do not think that I am lying! It is truly appalling! You feel carved up through and through, torn by a deep and acute pain!

Yet when you are curing your kin, the *xapiri* do not attack you like this. On the contrary, they valiantly come down to attack the evil spirits and the *xawara* epidemic. Also, they do not cut you up when you simply call them to dance for you. The *xapiri* who are already settled in your spirit house are not the ones who wound you. It is only those who have come from afar and are dancing for you for the first time. These are the new spirits who come to you little by little over the course of time. There are many of them, which is why shaman elders have so many scars. When they get old, their spines are increasingly frail and painful.

AFTER THEY cut me up, the *xapiri* quickly escaped with the different parts of my body they had just sliced off and flew far from our forest, far beyond the land of the white people. I had lost consciousness, and it was my image they dismembered while my skin stayed on the ground. They flew off on one side with my torso and on the other with my back and legs. They carried my head off in one direction and my tongue in another. The images of the *yõrixiama* thrush, the *ayokora* cacique, and the *sitipari si* birds, masters of songs, tore out my tongue. They seized it to remake it, to make it beautiful and able to utter wise statements. They washed it, scraped it off, and smoothed it out to fill it with their melodies. The cicada spirits covered it in white down feathers and annatto drawings. The spirits of the *remoremo moxi* bees[15] licked it in small dabs to rid it of its ghost talk. Finally, the thrush and cacique bird spirits put the words of their beautiful songs into it. They gave it the vibration of their call: *"Arer-ererere!"* They made it other, luminous and bright as if it were producing lightning. This is how the *xapiri* prepared my tongue! They made it into a light and slender tongue.[16] They made it supple and agile. They transformed it into a song tree tongue, into a spirit tongue. Then I was finally able to imitate their voices and answer their words with right and clear songs.

Later the *xapiri* came to reassemble the segments of my body, which they had dismembered. They put my skull and torso where the lower part of my body goes, and they put that part where my arms and head go. It is true! They put me back together upside down, placing my rear where my face was and my mouth where my anus was! Then, they put a large belt of colorful *hëima si* and *wisawisama si* bird feathers at the juncture of the two parts of my reconstructed body. They also replaced my entrails with those the spirits have, which are smaller, dazzling white, carefully wound around themselves and covered in luminous down feathers. Then they replaced my tongue with the one they reconstructed and put teeth in my mouth that were as beautiful as theirs, colored like the plumage of the *sei si* birds. They also replaced my throat with a tube, which we call *purunaki*, so that I could continue to deftly learn their songs and speak clearly. This tube is the spirits' larynx. This is where they get their voice's breath. It is a door through which our words can come out beautiful and right.

I HAD JUST taken the *yãkoana* with an elder for the first time, and the spirits had put me to the test while I did not yet know them. It happened the way I just told it. Yet despite the painful wound they inflicted upon me, I was still alive. My blood had not run, and I could not even see the traces of their gashes! As soon as they put the parts of my body back together, my thought gradually started to bloom again. Then I felt overwhelmed by the smell of the annatto dye in which they had covered me and by the fragrance of their *yaro xi* and *aroari* magic plants. The *xapiri* stood beside me, unmoving, magnificently dressed. They had finished their presentation dance. They were now impatient to build a house in which to settle!

CHAPTER 6

Spirits' Houses

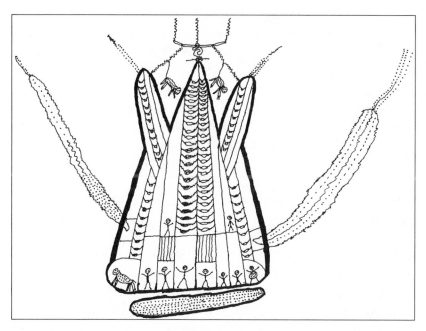

House of the *xapiri,* with hammocks, paths, and mirrors

W HEN YOU die under the effect of the *yãkoana* for the first time, the *xapiri* who come to dance for you do not yet have a home in which to settle. After having sung and danced for a long time, they remain standing or squat down and think: "*Hou!* If this place stays empty, if there is no house for us, we won't stay here!" This is why the first *xapiri* called by our elders are those who will open the clearing where the initiate's spirits' house will be built. The first to come are images of birds who know how to sweep the forest floor to find their food: the spirits of agamis, guans, curassows, great tinamou birds, but also of *pokara* partridges, as well as *makoa hu* and *maka watixima* antbirds. The spirits of the leaves, lianas, trees, and roots follow them to clean the detritus and dust from the new clearing, then those of the *iprokori* wind, the *wahariri* breeze, and the waters. Finally, the *xapiri* of the stones and termite mounds spread white down feathers all over it. All these spirits follow one another, dancing awkwardly in great numbers. They jostle each other, elbow to elbow, in a tangled mess. They do not have real songs and only speak ghost talk. They do not really know the forest's words because they are too close to it. These first *xapiri* only come to prepare the site where the new spirits' house will be built. As soon as their presentation dance is over, they instantly disappear into the sky's heights.

The shaman elders who have us drink the *yãkoana* must then hunt the repulsive spirits of *warama aka* snails off the paths. They must also fend off the spirits of the *yupu uxiri* ashes and the *wakoxori* firebrands, as well as those of the *rio kohiri* cotton hammocks, *wïïri* carrying baskets[1] and *pesimari* pubic aprons. Indeed, if all these household spirits came to dance for the initiate, the real *xapiri* would be loath to approach him. These kinds of spirits would ruin his chances of becoming a shaman, for they are unable to fight evil beings and their hands cannot cure the sick. It is so.

Once the clearing is ready and its surroundings are protected, other *xapiri* instantly begin coming down from afar, carrying the initiate's new spirit house, already built. The spider monkey spirits pull the peak of its roof to hang it in the sky's chest. The *hutukarari* sky spirits carry all its weight while the *yariporari* storm wind spirits push it to further heights. All these *xapiri* work hard, all together, for a spirit house's stakes are made from trees that make those in the forest look puny! Their shafts are huge, carved from one piece, and their weight is tremendous. These are not just wood posts whose bases rot like the ones in our houses! They are like metal bars. These are sky stakes, and they are as heavy as the sky.

Yet the *xapiri,* as tiny as they are, manage to carry them all by themselves. They gather on the clearing where they will put down the new spirit house in growing numbers, dancing slowly from front to back. They let out intense cries of effort and joy accompanied by the high-pitched whistling of their bamboo flutes.[2] Little by little, they point the rooftop poles in the direction of the sky for the spider monkey spirits already perched in its heights to hoist their tips towards them. It is very difficult, for these huge stakes seesaw with the violence of the wind. They swing heavily from one side to the other. The *xapiri* struggle, trying as best they can to slow their movement. Cries of warning ring out in the tumult: "*Aë! Aë!* We're going to flip, we're going to fall, watch out!" It is frightening! Only the spirits can do something like this!

Finally, they succeed in sticking the tip of the roof beams into the sky's chest, pushing them in so vigorously that they pierce it with an enor-

mous cracking sound. At that moment, the spider monkey spirits grab
the beams' ends and bend them to tie them together with ropes covered
in celestial pitch. Sloth spirits fire **nails** into them with their shotguns
while the spirits of the white people's ancestors we call *napënapëri* fix
them in place with long metal pins. Once they have finished all this work,
the new spirit house is solidly fastened to the sky's chest by its rooftop. Its
wood pillars can no longer loudly swing through space. Then the *warea
koxiki* spider spirits quickly cover the house with palm leaves carried for
them in huge loads by the spirits of the great anteater. Finally, it is decorated with patterns drawn by the boa constrictor spirits. This is how it
happens. Only a few kinds of *xapiri* work to build a new spirit house, and
they leave as soon as their work is finished. Later, other *xapiri* will come
from every direction of the forest, sky, and underworld to dance and settle
in the new house.

Spirits' houses are not put down on the ground the way ours are and
they are also not constructed in the same way. They are really other! The
xapiri, sent by *Omama,* carry them from far away. They are already built,
with their pillars and roofs already assembled. But since the *xapiri* fear
dust and dirt, they do not dance on their floors the way we do. Their
home's central plaza looks like a vast surface of smooth and shiny immaculate glass. The white people's ancestors once imitated this spirits'
glass, which is why their sons and sons-in-law continue to use it. They
call it **transparent** in their language. We say that it has a value of bril-
liance, *në mirexi.* The *xapiri*'s mirrors are also very fragile. This is why the
shamans complain when we stamp our feet near them while they are
making the spirits dance! The *xapiri* hate these dull sounds, which make
them think we want to chase them away. They can get irritated and hurt
anyone standing nearby. As I said, they hate a dirty floor, and so they only
travel over these mirrors covered in splendid down feathers and per-
fumed with annatto dye. When we call them to fend off the evil beings or
the *xawara* epidemic, their entire houses do not come down to us. It is
only their mirrors, which remain suspended in the air while they perform their presentation dance on them.

THE *XAPIRI* who will inhabit the new house brought for a young shaman do not come and settle there alone, on their own initiative. The elders who blow the *yãkoana* into his nose must first send their own spirits
to call them. To do this, they dispatch the images of the cock-of-the-rock,

the dove, and the *tārakoma* birds.[3] Only they know how to summon the other *xapiri*, who do not answer any other call. These emissaries travel far, passing from one spirit house to the next, urging their inhabitants to accompany them. They hurry off to call new spirits in every direction they can find them. The *xapiri* who see them pass are intrigued and ask: "What are you doing? Where are you going so joyfully?" The messenger spirits take advantage of the opportunity to encourage them to follow in their footsteps and go down to the young man drinking the *yãkoana*. They initiate *hiimuu* invitation dialogues with their elders to ask them to settle in his new spirit house in large numbers.[4] They vaunt its beauty and exhort them to join them: "Come all! Aren't you eager to travel with us? Come do your presentation dance for our father! He is becoming a shaman too!" The invited *xapiri* answer them with elation: "*Awe!* These are beautiful words! We will follow you! Let's all go together!" and they join in with them in an ever-growing troop.

Just as our wives are often the ones to convince us to go to a *reahu* feast, the *xapiri* women lead the male spirits to dance in the new spirit house. If the women are enthusiastic, the men eventually follow suit, no matter how indolent or sullen they are! The same is true with the *xapiri!* As I said, they only feel exalted when they follow the women spirits. This is why the elder initiating a young shaman begins by bringing down the women spirits with their love charms and intoxicating perfumes. As soon as they pass the male *xapiri*, the latter fall in love with them and go after them in a fervent dance. Other spirits listen to their joyous clamor from a distance, like the guests expected at a *reahu* feast lend an ear to their hosts' uproar from their last forest campsite.[5] Just like the *reahu* guests, the spirits lose sleep, waiting impatiently to make their entrance by dancing on the new home's mirror and setting up their hammocks there. Yet the *xapiri* women only accept to dance once an initiate's spirit house is truly ready to receive them. They are as far-sighted as our guests' wives! They refuse to expose themselves to the rain on a poorly opened clearing or on the muddy floor of an unfinished dwelling! If the new home that must receive them is under construction or they feel cramped in it, the *xapiri* are very unhappy and instantly turn back. Furious at having been misled, they disappear for all time.

HOWEVER, IF the new house is vast and beautiful, they are eager to dance and settle there. They rush in on bright trails descending from all sides

and starting from the place where the sky draws close to the earth. These
are the paths our shaman elders opened for them in the past. The *xapiri*
travel them making a loud noise, impetuously cutting everything up as
they pass. The ground flies into pieces and the trees come crashing down
behind them. Their march's force and violence make our stomachs drop
with fear! Yet despite this terrible din, we begin to hear their approaching
clamor and, more and more clearly, the melodious sound of their voices.
Then we can distinguish the magnificent songs of the *yōrixiama* thrush,
the *ayokora* cacique, and the *sitipari si* birds. Finally, the *xapiri* reveal
themselves to our terrified eyes! They brandish huge sabers projecting
flashes of light in every direction, as if they were waving mirrors around
them. They advance in blinding light, like that of car headlights in the
night. This is why many young people get frightened and forever give up
on becoming shamans!

Finally, the *xapiri* gather around the initiate's new spirits' house and
one by one they enter through the door their path leads to, as the guests
at a *reahu* feast would. Then they begin their presentation dance on the
central plaza's vast mirror, moving very slowly. Each one dances and
sings in his own way. They are adorned like guests, their bodies covered
in reddish orange paint and decorated with black drawings, their armbands heavily loaded with scarlet macaw tail feathers and their hair covered in bright white down feathers. They dance in a radiant light, gracefully waving frayed young palm leaves of bright yellow. They take turns
bursting into melodious songs, never stopping. They eagerly blow into
their thin bamboo flutes and let out cries of joy. The powerful rhythm of
their steps strikes the ground with a dull thud. In the tumult and sparkling light, their annatto dye releases its heady scent. Then suddenly silence returns.

ONCE THEIR presentation dance is finished, the *xapiri* start to settle into
their new home. Some set up their hammocks along its posts, while others lean against them, hang on them, or simply sit on the ground. All
of them keep wearing their feather ornaments and black saki tail headbands. But they put down at their feet the bamboo flutes, machetes, *sakosi* songs baskets, and frayed palm leaves with which they were dancing. When the house is vast, they begin by settling into the base of its
posts, then they pile up in tight rows to their very tops, innumerable. But
later, as new spirits continue to arrive in ever-growing numbers, this first

house will no longer be sufficient. It will constantly need to be extended so the new arrivals can also settle in. Little by little, new dwellings will be adjoined to it, placed on top of it or attached to its sides, piled up layer by layer like wasp nests.[6] This is why the spirits' houses are sometimes so tall and huge, propped up by countless posts and as tall as *komatima hi* and *aro kohi* trees.

In the beginning, when you are a young shaman and your *xapiri* have still recently arrived, their first house is low and narrow. They cannot accumulate in greater numbers there and it is useless to call more of them. Yet that is what I did! At the beginning, I was ignorant and impatient. I wanted to acquire too many *xapiri* at once! Many of those who came to me at the time soon left, declaring: "Wait! You are still young! We will come back to dance later! Do not be so impetuous!" It is so. As you get older, you go on calling new *xapiri* little by little and so your spirits' house constantly grows. Other dwellings are gradually added to it, all bound together. Eventually, the spirit house of an elder, a great shaman, looks like a tall **building** in a big city and can even extend beyond the sky's back.

Different kinds of *xapiri* do not mix in these dwellings. There are only tapir spirits in the tapir spirits' house, joined by their sons-in-law, the spirits of the *herama* falcon and the *xoapema* bird. The spirits of the storm wind and of the thunders also live together. But in the home of the *yoyo* toad spirits, there are only *yoyori* spirits. A big spirit house made of several homes attached to one another has many, many doors. There is the entrance for the *remoremo moxi* bee spirits, the one for the anaconda spirits, and towards the bottom, the one for the *xiwãripo* chaos spirits. The entrances for the *koimari* bird of prey evil spirits and *yãpirari* lightning spirits are near the top. And when the doors are too narrow, the *xapiri* do not think twice about enlarging them with their machetes so more of them can get in!

Each home has a single spirit name, but is occupied by numerous *xapiri*, all of whom are fellows. This name is a house name and a mirror name; it is their ornament.[7] There are as many of these names as there are spirit names. So we name the mirrors of the boa spirit, the jaguar spirit, the puma spirit, the armadillo spirit, and of the agouti spirit. There are also those of the *ixaro* and *napore* cacique bird spirits, of the toucan and bat spirits, of the *koxoro* bee spirit and the cicada spirit, as well as those of the lizard and worm spirits. There is also the mirror of *Titiri*, the night spirit, of the *yawarioma* water beings, of *Poreporeri*, the shamans' ghost spirit, and of *Aiamori,* the warrior spirit. But there are also many spirit houses that the *Xamat^hari* gave us.[8] Long ago, our ancestors knew their *xapiri* but did not make them dance. It was only once our fathers tried the *Xamat^hari*'s *paara* powder, when I was still a child, that they finally heard their spirits' songs and imitated them. This is why these *xapiri* now build their houses next to those of our spirits. The same is true of the spirits of the people of the highlands, whom our ancestors had known since *Omama*'s time.

The homes of a great shaman's evil *xapiri* are hung at his spirit house's highest point, beyond the sky's back, while those of his other spirits are situated in its lower part, in the sky's chest. These fierce *xapiri* starved for human flesh must be isolated, for they are dangerous. They could even strike out at the people of their father's house. Our elders are the only ones to possess such *xapiri,* but they only bring them down to avenge our children devoured by distant enemy shamans.[9] They are the images of the *Koimari* evil being, the anaconda, the *waroma kiki* snake, and the jaguar. They are those of the dry season being *Omoari,* of his sons-in-law, the cicada and butterfly spirits, and his hunting dogs, the caterpillar spirits;[10] but also those of the sun being *Omamari,* of the moon being *Poriporiri,* and of the storm wind and lightning beings. They even include the *poreporeri* dead shaman spirits, and the fly and vulture spirits of the new sky.

Yet not all the spirits at the top of the house are evil. The spirits of the *koxoro* and *õi* bees live there too, as well as that of their leader, the *maihiteriama* long-tailed bird, who travels with them high in the sky to wage war on the sickness beings. But, apart from that, the homes of most of the other animal ancestors who know how to heal people are located at mid-height in the spirit house. As for its lowest level, it is where we find the homes of the *xapiri* who arrived first. These are the images of the toads, the trees, the leaves, and the lianas. Finally, the homes of the spider monkey, black saki, blue macaw, and *kopari* falcon spirits are located a certain distance from the house, overlooking it, for these spirits are responsible for watching over the surroundings and protecting the house from enemy *xapiri* incursions.

As I SAID, spirits' houses' roofs are not thatched with *paa hana* palm leaves the way ours are. They are covered in solid leaves, shiny as mirrors and dotted with luminous down feathers. This is how *Omama* created them in the beginning of time. This is why they are so splendid! Yet their roofs also age. Their leaves wither away, blacken, and tear, like those on our houses. If they are not renovated and the *xapiri* have to live starving and silent in a broken-down home, their father will finally get sick. Then the other shamans will have to treat him and fix his spirit house's roof. Similarly, when a shaman becomes very old, his fellows shamans must replace the rotting posts of his *xapiri*'s home with new poles.

It is never good for a shaman to neglect his spirits' home. He can fall seriously ill if the house deteriorates and becomes black with smoke or if its surroundings are lying fallow, covered in *pora axi* gourd creepers as if the place was abandoned. This happened to some of our elders, and they wasted away. We know this, and we are on our guard. Death easily draws in on a shaman who lets his spirit house get old! This is why he will do his best to maintain it and always take good care of it. To stay in good health, he will often renovate its leaf roof and carefully tend to its clearing. But this is not all. He will also have to regularly give the spirits who live there their *yãkoana* food. Otherwise they will run away, and once abandoned their home will get old, empty, and silent. The *xapiri* will not stay there if their father does not take care of them! It is not enough to call them, make them settle in, and never to think of them again! If they remain famished, if they cannot make their songs be heard, or if they are bothered by loud noises and odors of rot and smoke, they will soon go away, leaving nothing but empty hammocks. It is so. We must also be sure to always call new young spirits and make them dance in the footsteps of the older *xapiri* who were the first to arrive in our spirit house when we were initiated. This is also how a shaman avoids aging too fast!

Like others, at the beginning I thought that the *xapiri* resided in shamans' chests. Yet I was wrong, that is not true. Their houses cannot be so close to the ground, within reach of our smoke and bad odors! They live elsewhere, suspended very high in the sky's chest. This is how the *xapiri* can contemplate the entire forest, however vast it may be. Nothing escapes their gaze from the heights where they live, not even the edges of the earth and sky. In fact, it is their image and that of their mirrors which are present in shamans' chests. A spirit house is far from an ordinary dwelling! Its posts imitate the inside of the chest of the shaman, the *xapiri*'s father.[11] The collarbones in his torso are the beams that support the roof ring. His hips are the stakes' bases in the ground. His mouth and throat are the main entrance. His arms and legs are the paths that lead there. His elbows and knees are mirror-clearings where the spirits stop before they come in.

So if a shaman is too thin and his chest too narrow, his spirits' house will be cramped. The *xapiri* will not be able to grow in numbers there. It will have to be torn open and extended for new spirits to settle in it. An overly small spirit house never leads to anything good. It must be as vast as a mountain.[12] This is why an angry person who wants to insult a shaman will sometimes say: "Your chest is empty! You claim to possess many *xapiri* but it is a lie. You are frail and your spirits' house is skimpy, cluttered, and dark!" This is also why the *xapiri* refuse to settle in a young man's chest when they find it is unclean, sour, salty, and smoky. But if a shaman's chest is broad, his spirit house will be too, and the *xapiri* will come dance there in large numbers. And if he is portly, the house will truly be huge, like the **United Nations building.**[13]

WHEN WE ARE young and want to drink the *yãkoana* powder for the first time, we do not know anything about the *xapiri* yet. The elders just tell us: "Come squat in front of me! The spirits will come to you, they will perform their presentation dance!" Then they blow the *yãkoana* they have prepared into our nostrils. Overcome by fear at having become ghosts, we wonder: "What is going to happen to me?" Then the elders call the *xapiri* for us and while we are knocked senseless by the *yãkoana*'s power, with our eyes fixed on the heights, our thought suddenly opens. We start to hear the spirits' songs, then they reveal their images to our eyes. This is how the great shamans of my village gave me the *xapiri* and had their house built for me. They were really very generous! It is true that sometimes I was very scared. Yet I continued, without ever wanting to quit. I drank the *yãkoana* without respite. My thought was concentrated on its power, for I wanted to see the spirits with all my strength. And when I was finally able to contemplate their presentation dance, I was very happy! I told myself: "So here are the *xapiri* that our ancestors had brought down since the beginning of time! Now I have truly seen them with my own eyes!"

After I had finished drinking the *yãkoana* with my father-in-law, I was washed with hot water and my body was covered in annatto dye. During this time, the spirits continued to visit me with their dances and I spoke to them in silence. I still knew very little and was not sure how to go about it. I told myself: "How am I supposed to sing? Is it really like this?" I was not at all self-confident! As I said, you cannot see the spirits very well at the beginning and it is by drinking more and more *yãkoana* that you finally succeed in seeing them clearly. And so I continued to learn, the same way the white people study, from one **class** to the next, so that my thought would truly become right. This time I took the *yãkoana* alone, in the afternoon, then again the next day and the day after that. I continued like this for entire days. This is how little by little I began to understand the *xapiri*'s words, and my thought started to spread in every direction. Since that time, the spirits of the forest and the sky have never stopped coming to me.

After drinking the *yãkoana* for the first time like I did, you must continue to behave in an upright manner if you want go on seeing the *xapiri*. An imprudent initiate would quickly scare them away if he started to avidly eat game or wanted to copulate too soon. The spirits who had joyfully come to him would turn back, disgusted and furious: "*Hou!* What a filthy one! Does he think we're going to put up with these fetid odors?" You must also feed your *xapiri* with great care. If they are deprived of food, they will protest angrily: "Our father is starving us! He never drinks *yãkoana!* So he must not want us!" Then they disappear, without the knowledge of the one who called them. The young shaman may think they are still in his spirit house, but it is not so. Their hammocks have been left empty. All that remains of them are words in his mouth! He may continue to evoke their names with lies, but doing so he is only invoking rejected hammocks and abandoned ornaments. The real *xapiri* are already far away, back in the mountains from which they came down. The elder who called them for him knows very well they have abandoned his spirit house. So he will reproach him: "*Hou!* You chased my *xapiri* away! What did you do? Don't you realize? Your chest smells of burning and penis, that's why!" In this case, no one can feign ignorance and claim: "Why in the world did my spirits run away?"

After having been put to the test by the *yãkoana*'s power, he who wants to become a shaman must continue to drink it alone, without respite. Otherwise he will not succeed. Even if you have a shaman father who gives you his own spirits, you do not get anywhere if you do not answer them zealously. If I had not continued to take the *yãkoana* with determination, without the support of my elders, it would all be over. I would quickly have come back to my ghost talk. I would no longer have been able to answer the spirits' songs. Then the people of my house would have started to tell themselves: "What a liar! He does not reveal any words of the distant land from which the *xapiri* come down! His mouth speaks without knowing anything! He is just vainly imitating the *yãkoana*'s power! He should shut up and stay drowsing in his hammock!" That is what people think if a young shaman's voice falters. And if he just goes on humming without speaking a word, they will be prompt to mock him: "He only possesses leaf and termite mound spirits! He only knows how to call the *xapiri* of firewood logs and women's carrying baskets! He does not make us hear the animal ancestors' speech!" But if his songs carry distant words, those who listen to him acknowledge it and tell themselves: "It is true! He made the spirits dance, he really knows them! He is repeating talks from other lands unknown to us!"

Ordinary people fear the *yãkoana*'s power and cannot see the *xapiri* work. They only hear the words of their songs. This is why the inhabitants and guests of our houses lend an ear when we become spirits and sing. They seem absorbed by their occupations but do not remain indifferent to what they hear. They tell themselves: "*Haixopë!* It is truly so! If I were a shaman myself, I would work just like them! They see things that we do not know!" Even those who are eating pay attention. Like the others, they want to listen to the words of the spirits who carried our image to the edges of the forest and the sky or to the land of the ancient white people, beyond the big water. These places seem out of reach in the ghost eyes of humans, yet the *xapiri* come down from there in an instant and constantly describe these places in their songs.

SHAMANS REVEAL what they saw when they accompanied their spirits' flight in a ghost state. They give their accounts to those who do not know such things. These innumerable words are of great value. This is why the shamans make them heard at such great length, one after another. By seeing their images, they speak the words of the animal ancestors who turned game in the beginning of time, those of the sky and the underground world, and those of *Omama*, who created the *xapiri* and made his son into the first shaman. These words of the spirits are like the words coming out of the radios that allow people to hear tales from distant cities in **Brazil** and other **countries.** Those who listen to them can think right and tell themselves about the shamans: "It is true! These men truly became spirits! The words of their songs are unknown to us!"

We do not become spirits alone, just for ourselves! Everyone listens to the *xapiri*'s words with attention: adults, young people, even children. Ordinary people know nothing about the places of which they speak. Their mind is too short and their thoughts never stray very far from them. They do not have many thoughts, for they do not see the spirits in dream. Their thoughts remain focused on their hunts, the goods they wish to acquire, and the women they desire. They only know the places they have visited or in which they have lived. When they drink the *yãkoana* during *reahu* feasts, young men who are not shamans roll in the dust in fear and call out for their mothers! Instead of songs, all that can be heard from them are moans and cries: "Mother! Mother! Throw water on my head! I am becoming other, I'm scared!" As for the shamans, they unceasingly answer their *xapiri*. The spirits' songs are always behind them and never leave them mute.

IF YOU WANT your elders to make you drink the *yãkoana* and give you their own spirits, you must really want it and not lie. You must ask them with fervor and great strength of will. At the beginning, the house that shelters an initiate's first *xapiri* is not very imposing. But little by little the elders make others spirits come down, in ever-larger numbers, from every direction of the forest and sky. With time, their grouped dwellings form an increasingly tall, wide spirit house. This is the moment when the young shaman becomes wise and truly able to cure people. It is so.

Everything I have recounted so far is what happened to me once I decided to make the spirits dance and to think right thanks to their words. I was a young adult and I told my elders: "Now I want to make the *xapiri* come down too! Give me their paths!" They accepted to make me drink the *yãkoana* and to bring down their *xapiri* for me. Nobody calls the spirits alone, out of nothing. They never come to us without being called by someone who really knows them. First we have to ask those who made them dance before us to open their paths. When we make this request, the one who offers us his spirits lets us choose those we like best. Then he passes them to us one after another. But if you do not solicit them with determination and show your eagerness, the *xapiri* think you do not truly desire them and refuse to approach you.

This is how you have to study to make the spirits come down. The elders are your professors. They make you drink the *yãkoana* and are always by your side. They give you the first *xapiri* you will have: the spirits of the cock-of-the-rock, the toucans, and the peccaries; of the sloth, the kinkajou, and the butterflies. They simply do it out of generosity. Yet if they should desire to put you to the test, they can make you work a long time before they let you really see the spirits! They give you their *xapiri* by projecting the *yãkoana* powder into your nostrils with their breath of life. The *yãkoana* they make you drink is more than just a tree bark powder. The *xapiri* follow it inside you, like grains of dust would. This is how you will acquire your first spirits. As soon as you drink an elder's breath of life, you are overcome by weakness and teeter under its impact! And when he is a brave warrior, his breath of life makes you valiant too. The same thing happens when he is a good hunter.[14]

The first to make me drink the *yãkoana* when I was a young initiate was my father-in-law, followed by other true elders from our *Watoriki* house and the *Weyahana u* and *Parawa u* river houses. Several of them passed on long ago. To become other, I have always wanted to ask for great shamans' breath of life. I did not want to present my nostrils to young people who spoke of the *xapiri* without really knowing them! Otherwise I could only have joined them in lying. These elders really made me see the spirits for the first time when they blew their *yãkoana* in my nose. They gave me their *xapiri* with their breath of life and since then I have kept it inside me. They never deceived me. This is why the spirits they gave me always come down when I call them. They continue to sing and dance during my sleep, and other *xapiri,* orphans of dead shaman elders, follow them down to me little by little. And so my spirit house continues to grow.

The shamans who initiated me did not ask for anything in return for their *xapiri*. If they had, I would have offered them machetes, hammocks, pots, and a lot of other trade goods. I really yearned to drink the *yãkoana* to be able to become a spirit in my turn. I wanted my thought to become right and extend in the distance along many paths. I wanted to become wise. I did not want to remain alone, lost in my ignorance, after the death of our oldest shamans. If the elders had not blown the *yãkoana* into my nose, if they had not made the breath of their spirits enter my chest, I could never truly have become other. If I had eaten and copulated without restraint, I would not have been able to become a shaman like my elders. My thought would have remained obstructed, and I would never have seen the spirits. I held my nose out to drink the *yãkoana* with our great shamans so my thoughts could travel in every direction of the forest and sky. I really wanted to see the animal ancestors of whom my grandfathers had so often told me. It is so. If I had not wanted it so much, I would be napping in my hammock all the time, like many others today.

Yet by making me drink the *yãkoana*, my father-in-law really scared me. He is not a large man, but his courage is great and his spirits truly numerous. His fathers and grandfathers died long ago, but their *xapiri* did not leave at their death. They came to settle in his own spirit house. This is why it became so huge. Its peak reaches far beyond the sky's back! Even the other shamans are scared of its size and say about him: "He is a real elder, truly a very powerful shaman!" His spirits struck me with terror. They made me cross the sky's chest through-and-through, enveloped in a blinding brightness. Yet I wanted to fly with them even farther! Finally I got so high that I was afraid I would die! I was suddenly terrified that I would not be able to come back to the forest and that I would fall in an unknown place somewhere far away. As he made me drink the *yãkoana*, my father-in-law did not want to let me think the *xapiri* could be a lie. He really put me to the test!

He made them cut up the forest beneath my feet and I felt myself fall endlessly. He also made them carve up the sky, then make it almost completely vanish. It was nothing more than a tiny bright spot in the distance, like a white down feather. I cried in a state of dread! Sometimes the *xapiri* are truly terrifying! Aren't they unknown beings? The woman spirits of the water beings and of the *waikayoma* bead spirits carried my image very far away. They made me run with them with all my strength, stumbling through the woods for days on end! They made me get lost in the forest and only let me come back to our house once evening had come. It was so. My father-in-law did not lie to me and he did not want to make a liar of me. He truly made me know the *xapiri*. At the beginning, when I did not know anything about them, sometimes I told myself: "Maybe he is lying and deceiving us!" Yet after he made the spirits dance for me and I saw them with my own eyes, I lost my doubts. Today I only think with sadness that the elders we have lost were very great shamans!

This is why I still want to imitate them by being able to become spirit and cure people in my turn. As soon as those of our house are sick, I drink the *yãkoana* to chase away their illness. I attack the evil beings who attempt to devour them, I extract the arrow points from their animal double, and push back the *xawara* epidemic fumes that burn them.

SHAMANS ALWAYS want to expand their spirit house. If they remained too narrow and low, it would not be possible for them to heal anybody. Only those who have a very tall spirit house know how to fight the evil beings of sickness, for their *xapiri* are numerous and powerful. Yet you cannot achieve this by presenting your nose to the elders a single time! You have to do it over and over and that takes a long time. Each time, these great shamans must dispatch their cock-of-the-rock and dove spirits again to invite new *xapiri* from far away. Happy to receive this invitation, these spirits will also come to settle in the young initiate's spirit house. Getting to know the spirits well takes us as much time as it takes white people students to learn in their books! After that, when we drink the *yãkoana* and call them, the *xapiri* come down from their houses fixed in the sky's chest. They arrive dancing on their mirrors, as I said, like television images. They take paths invisible to ordinary people, fragile and luminous like what white people call **electricity**. This is why their intense brightness vanishes as soon as they break. These innumerable spirit paths come from very far but get close in an instant, like words in a telephone talk.

When you have often drunk the *yãkoana* with the elders and they have been generous, other *xapiri* come to you easily, in ever-larger numbers. One of them starts to come down alone, in the distance. Then he calls others as he passes, joyously exclaiming: "Our father is calling us! He wants us! We are going to do our presentation dance for him!" Little by little, this is how the *xapiri*'s numbers swell. By drinking the *yãkoana* and becoming other so often, the young shaman's tongue becomes increasingly firm and he stops speaking like a ghost. The spirits' words truly reveal themselves to him then. The *xapiri* constantly sing their songs, one after another, as they hear their father answer their calls. As soon as one of them finishes his melody, he moves aside while another starts to make his heard, without any interruption. Their words come from wise song trees growing at the ends of the earth, which is why they have no end. But as I said, for the initiate to acquire such beautiful songs, the *xapiri* must also gradually replace his throat with their own. Failing that, he would continue to sing as badly as the white people! Learning the *xapiri*'s song is as difficult as trying to learn to draw words on paper skins. At first the hand is stiff and the line crooked. It is truly awful! You must refine your tongue for the spirits' songs as much as you must soften your hand to draw **letters**.

Finally, as you grow older, the *xapiri* continue to flock to your spirit houses in ever-growing numbers. They even come down to the oldest shamans' houses unbidden, while these elders sleep after having drunk the *yãkoana* all day long. They arrive in bands to dance there of their own accord, simply out of homesickness; they do not need to be called anymore. These are unknown spirits, coming from very far away, who were long ago made to dance by great ancient shamans. So the elders welcome these orphan *xapiri* in their spirit houses and make them dance again, as their fathers and grandfathers did before them. This is how the spirits have followed one another to us since the beginning of time.

CHAPTER 7

Image and Skin

Xapiri warriors

W HITE PEOPLE often ask me why one day I decided to ask the elders to give me their spirits. I answer that I became a shaman to be able to heal my people. This is the truth. If the *xapiri* did not avenge us by repelling evil beings and epidemic fumes, we would always be sick. In the beginning of time, *Omama* advised our ancestors: "If you drink the *yãkoana* you will be able to bring back the image of your children captured by evil beings. If you cannot call the *xapiri* to protect them, you will be a pitiful sight before their suffering and you will mourn their death in vain!" Only the spirits know how to tear harmful things from inside our body and throw them far away from us.[1] They are immortal and very good at curing us. This is why we appreciate them so and have continued to make them dance to this day. Very long ago, before the white people's **medicine** had reached us, our shaman could only rely on the spirits to avenge their people, children, women, and elders. They drank the *yãkoana,* made their spirits come down, laid in ambush with them to attack the sickness and make it flee.[2] Of course they did not always succeed and despite their efforts some children were devoured by the evil beings' diseases.[3] This was no different from the white people's **doctors,** who sometimes attempt treatment with medicines that fail. After the shamans' work, our elders' wives, also very wise, used healing plants from the forest.[4] They used them to rub and bathe the bodies of the sick who had just escaped being devoured by the evil beings or the epidemic spirits.[5] Unfortunately, few women today still know how to use these plants. People still think that only the *xapiri* can truly cure them, but now they also count on the help of the white people's medicine.

Before these strangers arrived in the forest, people did not die very often. Once in a while, a very old man or woman would pass away, when their hair had become really white, their eyes blind, their flesh dried up and flaccid. Their chest became other, affected by aches and smoke. They died for no other reason, because they no longer ate and drank. They died the right way, at a very advanced age. Sometimes enemy *oka* sorcerers killed an elder, a woman, or a young man. Now and then an elderly woman who already wanted to depart could die of the wound inflicted on her *rixi* animal double by far-away hunters. At times an angry guest took the earth of one of his host's footprints, rubbed it with *hʷëri kiki* sorcery plants and gave it to a snake to bite or poured poison from burned *paxo uku* spider monkey hair in his food at night.[6] And occasionally enemy warriors arrowed one or two people at dawn. But it was mostly the evil beings of the forest who tried to devour humans by seizing their image. These were the ones the shamans constantly had to fight to cure their people. Though our ancestors mistreated each other from one house to another with their sorcery, they always recovered because their shamans were able to pull its harmful things out of their bodies and throw them into the underworld. In the end there were very few people for whom they mourned. The *xapiri* had been our ancestors' doctors since the beginning of time. This is why I wanted to know them and possess them in my turn.

At the time, the forest did not know all the human-flesh-starved epidemics that came with the white people. Today the *xapiri* can only fight the *xawara* epidemic when it is very young, before it has shattered its victims' bones, torn their lungs, and rotted their chests. If the spirits find it on time and rapidly avenge the sick, they will get better. Yet these new ills that white people call malaria, pneumonia, and **measles** are other. They come from very far away and the shamans do not know anything about them. They attack these evil things with all their strength, but nothing can reach them in return. Their efforts remain in vain and we easily

die of them, one by one, like fish killed in a pond by fish poisoning.[7] The
xapiri only know how to struggle against forest diseases that they have
known forever. When they try to attack the *xawara* epidemic, its evil beings—which we call *xawarari*—finally devour them too.[8] This is why shamans today also need to rely on white people's medicine to keep these
new diseases far from us, even though their *xapiri* know how to cure
people.

A young shaman must continue to drink the *yãkoana* all the time so
that his *xapiri* can feed themselves through him. Without eating *yãkoana*,
starved and angry, they would no longer dance for him. To go on dancing, the *xapiri* need to die and become ghosts like the shamans who became their fathers. They only come to us once they are sated with *yãkoana*. Their mirrors arrive from the sky's chest, slowly preceding them.
They suddenly stop in the air and remain suspended there. The *xapiri*
come down from them one by one, performing their beautiful presentation dance. Then their fathers begin to imitate them one after the other,
singing and dancing while they too become spirits.

The *xapiri* float down through the air from their mirrors to come protect us. When they arrive, their songs name the distant lands they came
from and traveled through. They evoke the places where they drank the
waters of a sweet river, the disease-free forests where they ate unknown
foods, the edges of the sky where, without night, one never sleeps. Once
the parrot spirit has finished his song, the tapir spirit begins his, then
comes the turn of the jaguar spirit, the giant armadillo spirit, and all the
animal ancestors' spirits. One by one, each offers his account, then asks
why their fathers called them and what they should do.

Then the *xapiri* work ceaselessly to cure the sick. The agouti, acouchi,
and paca spirits tear out the harmful things that the evil beings stuck in
their image. The spirits of the *aroaroma koxi* toucanets cut them up and
those of the *kusãrã si* birds rip them into pieces.[9] The spirits of the tad-
poles and *yoyo* toads cool them in their mouths. The spirits of the water-
being women dance while rocking feverish children in their arms and
bathe them with their gentle hands, then the *titiri* night spirits shelter
them in darkness. The *waikayoma* bead-spirit women clean sorcery plant
burns and arrow wounds. The *hokotoyoma* rainbow-spirit women wash
the bodies of the sick with cool water and the tapir spirits lick their
bruises. The *masihanari kohi* tree spirits give them strength.

As soon as a spirit has finished his curing work, he returns to his mirror and waits for the others to finish theirs, one by one. This can last a

very long time, but it is how the sick can really be healed. Once all the *xapiri* have danced and sung in succession and the shaman has imitated all of them, the *Weyaweyari* evening spirit arrives so that their work can be brought to an end and their father can stop being other. Then all the *xapiri* go back to the sky's chest with their mirrors, taking with them all the magnificent songs they guard so jealously.

WHEN THEY encounter us in the forest, the *në wãri* evil beings consider us their game.[10] They see us as spider monkeys and our children as parrots. It is true! This is the name they give us! We could never survive without the protection of the *xapiri,* who are feared by the evil beings as fierce enemies. If one of these evil beings encounters a hunter in the distant forest, when the weather is cloudy and darkness lingers in the morning, he will instantly try to seize his image. Then he will take it to his home, where he will hide it in a wooden box or a **bag** in order to devour it later.

The *në wãri's* houses are overloaded with dangerous incandescent goods tainted with dizziness. There are so many that it is terrifying! These homes look like gold prospectors' **shacks** in the forest or white people's houses in the city. When one of these evil beings attacks us, we instantly start moaning in pain in our hammock. Yet it is not his victim's body that he tears at with his fangs, but his image, which he keeps captive in his distant hideout. If the *xapiri* do not arrive quickly to snatch it away from him, he will devour it whole and the sick person will die. This is why the shamans try to avenge us quickly when we are attacked like this! Under an elder's guidance, they instantly send their *xapiri* to follow the evil being's trail. Then, once these spirits have reached his home, they search everywhere for the victim's image he wants to devour. They turn all the evil being's goods upside down and, as soon as they find the captive image, they free it and escape with it. Upon their return they put this image back inside the sick person who lost it, and, finally, he gets better. This is how the spirits work to cure us!

Yet it is very difficult for them to track down the *në wãri* evil beings, whose paths are so narrow and tangled and often hidden. They must patiently be spotted and tracked one step at a time. Many *xapiri* are good at following evil beings' trails, including the hunting dogs and the *poxe* peccary spirits, who sniff their tracks, and above all the *Koimari* bird of prey spirits, who can follow the most winding trail through the air or under the ground, in the wind and in the night. But in fact the *xapiri* who are really best at chasing the *në wãri* are their own sons-in-law! They know their paths and are the only ones who can easily approach them, for they do not have to fear their hostility. They lead the curing search expeditions. As soon as they catch sight of the evil beings, they pretend to engage them in a *hiimuu* dialogue of invitation in order to elude their suspicion. Then they start to hit them with their machetes and their fellow spirits instantly follow suit.

Thus the sons-in-law of *Omoari*, the dry season evil being, are the first to find him. They are the cicada, butterfly, and lizard spirits, but also the spirits of the *remoremo moxi* bees, the *hãtãkua mo, kõõkata mo,* and *õkra-heama* birds, as well as the giant *wãsikara* lizard.[11] Yet *Omoari*'s path is as burning as a trail of live embers. The spirits of the large *yoyo, hwat^hupa,* and *prooma koko* toads must ceaselessly pour pots of water on it so that the *xapiri* searching for him do not burn their feet. At the same time, the *porari* spirits of the waterfalls and the *proro* and *kana* otter spirits must constantly spray them down to keep them cool. When the trackers are finally within earshot of the evil being's house, they stop and lie in ambush at the edge of the paths leading to it. As soon as his sons-in-law have started speaking to him, they attack him, while protected by the powerful *Porepatari* ghost spirit and his curare arrow points.[12] They hit the evil being's lips and break his teeth so he will release the human image he captured. If he has hidden it, the squirrel monkey and coati spirits turn his house upside down looking for it everywhere. The other *xapiri* hold him back and twist his arms. They squeeze his neck and throw him to the ground. The caiman spirits hit him with their powerful machetes. The capuchin monkey spirits arrow him from every direction. The sloth spirits fire on him with their shotguns. The kinkajou spirits skin him alive. The spirits of the *simotori* giant beetle blind him with a fiery liquid and slit his throat. The *iramari* jaguar spirits burn him and the *yokotori* lake spirits suffocate him. It takes all this to force a *në wãri* being to let go his human prey!

AFTER THAT, the lightning spirits tear the bonds holding the victim's image captive and set it free. The spirits of the small sloth and the *aroaroma koxi* toucanet prop it up and chase *Omoari*'s daughters away as they run over shouting to help their father. They also break his dogs' fangs and scare off his other pets, the *kraya* caterpillars and giant snakes. Then the *xapiri* warriors escape with the sick person's image in a ghost state. Once they have brought it back home, the spirits of the *hraehraema* frog clean it and the spirits of the *hoari* marten bathe it in water mixed with honey. The water-being women spirits dress it in bunches of feathers and the annatto spirits cover it in vermilion dye. Then the deer and jaguar spirits lick its eyes and chest with their rough tongues to bring the sick person back to consciousness. The bee spirits rinse his dry and bitter mouth with their healing water so his saliva returns and he can finally feed himself. This is how the shamans must work to cure children whose image have been captured by the *në wãri!* They must be very valiant and swift in their fight. If their *xapiri* are too slow getting on their way, the evil beings will already have half devoured their victim's image and it will be impossible to bring it back unharmed. The child's state will then become very serious and she will certainly die.

Many *xapiri* bravely come down in answer to our call to attack the evil beings of the forest and avenge us! Aside from all those I mentioned, there are also the bat spirits who have lights to guide themselves in the dark and who blind the *në wãri* by blowing darts in their eyes. The *piri-mari* star spirits bite their kidneys and stomach with their sharp teeth, then cut off their arms. The wasp spirits arrow them, the spirits of the *witiwitima namo* kite lacerate them with their sharp blades, and the coati spirits knock them out with their clubs. The jaguar spirits tear them to shreds and the giant anteater spirits pierce them with their powerful claws.[13] The spirits of the *aro kohi, apuru uhi, komatima hi,* and *oruxi hi* trees bump into them and knock them over. Those of the *wari mahi* tree thrash them. With their skulls split and their bodies covered in wounds, the stunned evil beings eventually stumble. Then the *xapiri* can force them to let go of their prey and give up the fight. It is so and I do not speak of all this in ignorance! I myself have often seen *xapiri* troops dance to battle the *në wãri!* They gather to prepare their attack in the sky's heights, so high and so many that the eye can no longer catch them all! They are warlike and extremely valiant. Only they can bring back the images of children captured by *Omoari,* the dry season evil being, but also extract the hairy, purulent penis of *Riori,* the rainy season evil being, from our feverish wives' vaginas. They are the only ones who can cure those arrowed by the maggot being *Moxari* for eating his leftovers;[14] the only ones, finally, who can send all these beings of disease far away, like the *Watorinari* spirit of the great wind does by sweeping them off with his impetuous breeze.

Yet the evil beings are not the only ones who are hostile to us and make us sick. We can also die when very distant people such as the *Parahori* of the highlands arrow our animal doubles, which we call *rixi*.[15] The *rixi* animal for women is the *hoahoama* forest dog; for men, the *mohuma* harpy eagle. These animals, which are also those of our ancestors, live in the forest close to unknown people on the upper Rio Parima around a great waterfall called *Xama xi pora,* which is protected by innumerable wasps' nests and the gusts of a powerful wind. If unknown hunters kill them, their wound comes to us and can kill us all the way in our house. It is so. Our real inner part is there, a great distance away from our skin, which is here, lying in our hammock! When a person's wounded *rixi* animal tries to escape by running or flying through the highlands forest, she instantly becomes ghost in her home and feels a sharp pain in the place where the bamboo or monkey bone arrow point penetrated her *rixi* animal.[16] This is why we get so sick when distant enemies attack our animal doubles.

WHEN THIS happens, our house's shamans instantly send their *xapiri* to help the wounded *rixi* animal. Their storm wind spirit hurls itself at the enemy hunters so they get lost in the forest, while their *koimari* bird of prey spirits attack them mercilessly. Then their *purupuru namo* monkey spirits dash over to help the arrowed animal and hide it. Once sheltered, they tear the arrow point out of its wound and try to put it back in its den or nest. When the *xapiri* come to the aid of a wounded *rixi* animal, they try to bring it back close to the huge swirling waterfall where its fellow creatures live. But after that, they must still cure the human being who became ghost when his animal double was wounded. The spider monkey and howler monkey spirits are sent to remove the arrow point that hurt his *rixi* animal from his body. Then they give it to the *ayokora* cacique bird spirits who make the shamans able to spit it up in front of everyone. Only then will the sick person truly begin to get better. Yet if his *rixi* was not quickly put out of the enemy hunters' reach, they may find it again. Then, if they finish it off with a club, the sick person will die abruptly and his people will start crying and wailing for him.

SOME OF US, though not many, like to possess sorcery things we call *hʷëri*.[17] These are people whose hand wants to leave a trace of anger.[18]

When invited to a *reahu* feast, they hide these substances in their bamboo quiver. Once they have arrived, they perform their presentation dance, then when night falls, they will probably let their hostility rise up and quarrel with their hosts while talking through the *wayamuu* dialogues.[19] Later they may attack one of the hosts' wives for turning down their advances, making her sterile with *manaka ki* and *xapo kiki* sorcery plants. Spite can also drive them to throw h^w*ëri* evil things on one of their hosts who has refused them a machete or on another who they consider stingy with his food. In the past, the elders often mistreated each other this way. Today it happens less often. Most of us do not really know how to use the h^w*ëri*. We are even scared to touch them, for fear that we will become sick ourselves! They are very dangerous! Yet this does not prevent some people from soothing their resentment with these harmful things.

As soon as a guest secretly projects or rubs a h^w*ëri* on one of his hosts, this person begins to feel ill. Once the *reahu* feast is over, he suffers severe headaches, then is taken with a strong fever. His vision becomes yellowish, while he sees the forest spinning before his eyes. He is overcome with dizziness and his ears start to whistle. Even if he does not quickly consult with one of his house's shamans, they will understand the gravity of his sickness by themselves. So they will start avenging him by trying to destroy the sorcery evil that affected him.[20] In these cases, the sick person himself does not say anything. He remains stretched out on his hammock in a ghost state. His relatives speak for him. When a man's mother or sister declares loudly to everyone at large: "*Osema a* is very ill![21] A h^w*ëri* sorcery thing was thrown on him!" the shamans who hear her are instantly concerned. They soon gather and begin drinking the *yãkoana* together.

When their eyes die under its power, they start to search for the harmful things hidden inside the sick person's body. Once they have discovered them, they seize them in order to cut them out, burn them, and throw their debris far away. This is the only way their patient can get better. These h^w*ëri* diseases are very powerful. They make an intense humming sound. The shamans' eyes see them as bee swarms or mosquito scourges; like yellow and orange clouds attached to the sick person's image. At the same time, they see the sorcery plants from which they come in the form of tufts of grass or leaf-shoots growing from the forest floor. Their *xapiri* must then try to tear them out despite their nauseating odor. The giant armadillo and peccary spirits uproot them and set them on fire. The *repoma* bee spirits dig a hole in the ground[22] through which the capuchin monkey, spider monkey, and black saki spirits get rid of their charred remains by throwing them underground to feed the *aõpatari* ancestors who fell with the sky in the beginning of time. In these meat-hungry underworld people's eyes, the sorcery substances are game, just like the corpses of evil and epidemic beings torn to pieces by the *xapiri*. By devouring these spirits' leftovers, they avenge us of all the evil things tormenting us. So, as soon as they hear the *xapiri* working, they call out to them: "What are you going to send us? It's game! Go ahead! Throw it to us! We're starving! Is it nice and fatty at least?" The spirits throw their pittance underground and the *aõpatari* watch it fall, avidly crying out: "Game! *Aaa!* Look at that meat! *Aaa!*" No sooner has the food landed in their forest than they are carving it up and voraciously feeding in a wild fray. They are truly insatiable and do not share among themselves! You can often hear *Okosiyoma*, one of their old women, crying with hunger because they have not given her any guts to eat! Their teeth are as sharp as metal blades. These are not human beings! Yet they describe us as their distant children who have stayed above them. From the world below, they hear our elders' *hereamuu* speeches as booming rumbles from the sky,[23] much as we hear the harangues of the thunder beings saluting ghosts' arrival on the sky's back as the roar of a storm.

THE GUESTS at a *reahu* feast can also include a resentful man who will attempt to harm one of his hosts, though they offer him food and treat him as a friend. Such a spiteful visitor can collect the soil from one of his host's footprints and rub it with sorcery things to quickly make him die of disease.[24] This happens sometimes when a man is invited to a house inhabited by an elder who once killed his father. As soon as this man sets eyes on the killer, the anger of his mourning returns and he tells himself: "*Asi!* That is the one who ate my father when I was a child!" He will then collect his footprint out of revenge, even after a very long time. But sometimes an ill-intentioned guest can do the same thing simply out of jealousy for a woman or even in reaction to his hosts' stinginess! It is true! Once the hostile visitor has taken the soil of his victim's footstep, he carefully wraps it in leaves and hides it in his bamboo quiver. Once he is back home after the *reahu* feast, he will wait some time then give this footprint away to visitors from a distant forest, who have come to his home as guests. These people will be the ones to rub it with sorcery things, for they are enemies of the man whom he wants to harm, distant people who never visit his house.[25]

These distant enemies begin by dividing the earth from the footprint in several little leaf packets, which we call *mae haro*, footprint packages. They hide most of them in the ground behind their house or in the forest. After that they roll one of them in the palms of their hands, rubbing it with clay, sorcery substances, and *hore kiki* plants that make people cowardly. Then their victim falls ill without delay, burning with fever, while the leg that left the footprint begins to swell up. If the enemy sorcerers want to kill him rapidly, they attach one of his footprint packets to a stick and have a *karihirima kiki* viper bite it several times. This is what our ancestors used to do and some elders still do it today! In this case, as soon as the leaf packet comes apart from being bitten and the snake retreats, the sorcerers exclaim: "Snake! Quick! Run away! Hide in a hole underground or under the leaves of the forest floor! Lie in ambush!" This way, the viper will bite their victim. One day, this person will go staggering into the forest to defecate. Having become a ghost under the rubbed sorcery's effect, he will walk without caution. Before long a snake will bite him and this time he will die quickly. No one will be able to save him!

On the other hand, if the footprint packets are only rubbed with sorcery things and buried, the shamans can recover them and cure the sick person. But to do so they must find all of them and pull them apart one by one! Only then can the victim leave his ghost state and get better. The agouti and acouchi spirits, as well as those of the *paho* rats, search for the footprint packets by sniffing and scratching the ground. Once they have located them, they untie and tear them up with their powerful knives, which cut through the strongest bonds. Then they disperse their contents through the forest. But sometimes they give them to the *ayokora* cacique bird spirits without taking them apart, and these *xapiri* help the shamans regurgitate them in plain view of everyone in the sick person's house.

PEOPLE CAN also be attacked by enemy *oka* sorcerers coming from the highlands or the *Xamat*ʰ*ari* region when they are working alone in their gardens. Traveling from distant houses by night, these sorcerers lie in ambush at the edge of the forest to blow their *h*ʷ*ëri* sorcery on us. They use *horoma* palm wood blowpipes to project tiny arrows attached to a ball of cotton containing dangerous plants.[26] These darts strike the man they are aimed at in the back of the neck and the evil substances instantly spread over his body. He begins to feel weak at once. Rapidly overcome with dizziness, he stops working and squats in his garden, sighing deeply, his head spinning. Then the *oka* sorcerers come out of their hiding place and bolt towards him. Taking advantage of his weakness, they drag him into the nearby forest. They break his limbs, back, and the nape of his neck by twisting them or hitting them on a piece of wood. Then they leave him agonizing on the forest floor. They often also try to hide the signs of their attack with passes over his body to make it look unharmed, so he can return to his house without reporting their attack. They stand him up and say: "Go back home and do not say anything about us! Do not tell that we are here! You will just tell your people: 'I felt ill while working in my garden! It must be *Omoari,* the dry season evil being, who struck me!'"

The *oka* sorcerers' victim then returns to his house. He stokes his fire and, having become ghost, stretches out in his hammock. He keeps repeating the sorcerers' words, attributing his illness to *Omoari.* Then his state quickly gets worse. In this case, nothing will help, even if the shamans instantly tackle his sickness. They will not succeed at all. The *oka* victim will die very quickly, for his bones have all been shattered inside his body. There is nothing to do![27] You can only treat someone who has been attacked by enemy sorcerers if they only harmed him with their sorcery and did not have time to break his bones. But this only happens to wise people who ran away very quickly, as soon as they felt the evil darts' impact on the nape of their neck. In this case, the shamans can still destroy the $h^w\ddot{e}ri$ sorcery inside them and heal them.

ORDINARY PEOPLE do not see sick people's image beyond their skin. Only the *xapiri* can do that. This is why they are able to remove the burning hot cotton and fangs left by the *në wãri* sickness beings, the *Kamakari* evil that devours bones and teeth,[28] or the spider webs that obscure our eyesight. This is why they can extract arrow points that have struck a *rixi* animal, the weapons of enemy *xapiri,* sorcery plants, and footprint packets. They also know how to make all these harmful things that make us sick come out of the mouth of their father, the shaman. They make him able to regurgitate even those most solidly attached to the inside of our body. We do not say that the *xapiri* are powerful in vain! The spirits of the *ayokora* cacique birds and of the tapir are especially gifted with this power,[29] but also the spirits of the toucan, the *wayohoma* nightjar bird, and others yet, who relay each other when the former are exhausted.[30]

Yet of all these, the most skilled at extracting diseases and putting an end to pain are undoubtedly the *ayokora* cacique bird spirits. The *xapiri* of the agouti and paca locate the diseases in our bodies for them so that they can tear them out and throw them far away from us. When you have spirits like these, you do not worry as much when your children get sick! These are our real doctors! The white people's doctors open chests and stomachs with metal blades, without really knowing what they are looking for, and wind up leaving big scars. Our *ayokora* cacique bird spirits treat us from the inside, without spilling blood. I know it because I myself was cured by these *xapiri* when a *Poreporeri* evil being, a ghost of ancient shamans, attacked me. One of my eyes had suddenly become other and could no longer move. It was fixed in place and my eyelid could not close anymore. My mouth had also become a ghost's mouth, numb and twisted on the side. Two shaman elders from *Watoriki*, both of whom have since passed on, cured me with their *ayokora* cacique bird spirits. These *xapiri* extracted the harmful things that the evil ghost being had put inside me. They untied the cotton strings with which he pulled on the side of my face and made their fathers' shamans regurgitate them. Then they washed the strings' evil marks with healing water. This is how I recovered, and I did not have to go into town for that![31]

Only a few elders possess these cacique bird spirits and they only pass them on from time to time. But when they really want to settle in a shaman's spirit house, these *xapiri* generally come of their own accord. If you try to call them, they approach warily and are quick to take flight, the moment they are bothered by noise, smoke, or grilled game odors. They instantly disappear and never come back. This is why shamans take such good care of them. They are protected by their sons-in-law, the *kurira* wasp spirits, and only come towards those who possess such *xapiri*.[32]

These cacique bird spirits have sometimes come down to me in my sleep. I saw the shining images of the *ayokora*, *ixaro*, and *napore* cacique birds dancing in my dream in a noisy troop, followed by those of the *wayohoma* and *taritari axi* birds. They were accompanied by the tapir spirit, who regurgitates the sorcery footprint packets that are too heavy for them, as well as the caiman, macaw, toucan, and peccary spirits. All these *xapiri* were covered in superb feather decorations. Yet the *ayokora* cacique bird spirits were truly the most beautiful. They stood out from all the rest! They live very far away, in a magnificent forest by a great river, which the shamans call the river of the *kurira* wasps, where they are protected by these warrior spirits' huge nests. My father-in-law took me all the way there with his own *xapiri* to visit them. They are countless and never stop succeeding each other in singing their splendid songs. These are the spirits I like best and I always carry their path in my thought. I would really like to know them well and have them settle in my spirit house like the elders did before me!

MOST *XAPIRI* behave in a friendly manner. Yet some can be very hostile and, as I said, only travel through the forest to kill. In this way, distant shamans can travel in the form of evil spirits and steal our children's images in order to devour them.[33] The elders of the distant *Xamat^{h}ari* houses of *Iwahikaropë* and *Konapuma* sometimes attack us like this![34] Sometimes they even send jaguars and snakes near our home to harm us! Their spirits relentlessly wage war on ours and riddle us with their pointed arrows, making us feel their sharp pain. These faraway people do not know us. Yet as soon as one of their children dies from being weaned too early,[35] they mistakenly see the mark of our hands in it and hold us responsible. Furious, they send us their hostile *xapiri* to seek revenge. But they are wrong, our spirits never go to war against their home to eat their children![36] They only attack evil beings and the white people's epidemic fumes. It makes us angry that these unwise people try to assault our children for no reason! We never attack other houses first because we fear there will be no end of reprisals. But if their inhabitants attacked our people, we would not hesitate to seek revenge by sending them our own flesh-starved *xapiri!* It is so.

When a great shaman from a distant village kills one of our children, we respond to his hostility in the same way! Our evil spirits instantly fly to his house and devour a child as if it were a parrot. And when we want to put an end to one of these distant shamans' misdeeds, we send the same fierce *xapiri* to kill him. They take roundabout paths to surprise his own spirits, surround them and rapidly wipe out the most valiant of them. Then they furiously break their house apart, burning it and hurling its charred bones into the water. Finally, they attack the shaman himself, striking him with machetes before throwing his blood in the river to leave no trace of who killed him.

These aggressive *xapiri* are images of *në wãri* evil beings whom we only bring down for revenge. Aside from their threatening weapons, they possess many things of disease.[37] For example, *Hutukarari* the spirit of the sky sticks bright star splinters into his victims' images, which no one can recover from. The spirit of the evil being *Herona*[38] projects urine as dangerous as curare on them, while *Mõeri,* the dizziness spirit, violently strikes the back of their neck and makes the forest spin around them. These evil *xapiri* are really very dangerous when they wage war. The *koimari* bird of prey spirits usually lead the charge, under the command of the most powerful of them, *Ara poko.* They brandish incandescent ropes and keen blades to tie and cut up their victims. Ferocious *iramari* jaguar spirits accompany them with their sharp-edged machetes, as well as spirits of *Xinarumari,* the master of cotton, who grips children with his burning adornments. There are also anaconda spirits among them, who copulate with unknowing pregnant women, making their fetuses rot inside them, or sodomize men whose viscera begin to swell until they explode. That is not all, there are many more *xapiri* of evil beings, such as the *yurikori* fish spirits who slash children's tongues and throats, and the *pirimari* star spirits who tear them apart with their sharpened teeth. But these dangerous *xapiri* only attack people from other houses, located far away. Here at home, they work to cure us, like the rest of the spirits. The dangerous *xapiri* are good at fighting the *në wãri* evil beings because they are these beings' own images turned into spirits. For instance, the *koimari* bird of prey spirits are able to track *në wãri* evil beings to the entrance of their dwellings as tall as mountains and the anacondas to immobilize them by tying them up. The spirits of *Xinarumari,* the master of cotton, can also restore children's skin when it is covered in infected wounds and cannot stop rotting. It is so.

Enemy *xapiri* are always trying to overcome the vigilance of the shamans of the house they will attack. They fly fast, but never travel in a straight line or out in the open. They try to stay hidden and scramble their paths with constant detours to remain undetected. If they want to capture a child in our house, they begin by heading very far in the opposite direction, as far as the white people's land. Then they secretly come back, following a winding path, flying through the depths of the underworld. This way they seem to disappear in the distance, so far away that even the shaman who sent them eventually loses sight of them. But once we have forgotten them, they suddenly spring from the ground inside the house where they have come to find their prey. They blow sorcery things all over it, which daze shamans who could fight back. Then they choose a beautiful, vigorous, and joyful child to kill while he is playing.

These bird of prey and jaguar spirits suddenly descend upon him and lacerate him with their machetes. The child instantly starts wailing in fear and collapses to the ground. The master of cotton's spirits clasp his head, chest, and stomach with their burning disease adornments. The child's eyes roll back, he develops a burning fever. One could think he has been affected by a *waka moxi* sorcery plant.[39] Yet this is not the case. This is certainly the mark of these nefarious enemy *xapiri*! They quickly carry away their victim's wounded image while his empty skin lies on the house's floor. The little one loses consciousness and enters a ghost state. At that moment, if the fierce moon *xapiri* start to cut up the child and feast on his body with the other evil spirits, it is too late to cure him. The child dies instantly, despite his home shamans' efforts to avenge him of the *xapiri* who captured his image. They cannot do anything more. Yet if these shamans are wise enough to drink the *yākoana* as soon as the child begins to lose consciousness, they are still able to track down his aggressors. Then they can send their own evil spirits to take back his image before it is devoured. If the wounded child tied up with burning cotton has not been cut up yet, he will get better. After having brought him back to his people, the shamans will relentlessly wash his body's inner part, until he regains consciousness and really recovers.

The people who devour children's images like this are ancient shamans, powerful and fierce enemies, whose spirit houses are full of very dangerous *xapiri*. Yet a shaman's evil spirits sometimes also go looking for prey on their own initiative, without their father shaman even knowing it! They fly to distant houses to hunt there, driven by their hunger for meat. They go there to devour children, whom they see as game, and only return to their house once they have eaten their fill of fat. Finally, their father notices their misdeed and laments himself: "*Hou!* What have they done? I didn't send them to wage war! I didn't say anything to them!" When a shaman's spirits have killed, we say that their father is *õnokae*, for he is sated with human flesh and fat. His forehead is damp, greasy, and sticky, like that of warriors who have eaten an enemy with their arrows or that of someone who has killed the *rixi* animal of someone living in a distant house.[40] He has to stay stretched out next to his hearth fire, immobile, and fast so that his õnokae state dries out after a while. It is so. We really fear these distant shamans who send their warrior spirits to us, but we get revenge on them in the same way!

THIS IS HOW human beings die in the forest. It is true. Our ancestors' ghosts always want to bring the living to them on the sky's back.[41] The dead are nostalgic for those they left alone on earth. They tell themselves: "My people are so few and so hungry in that forest infested with xawara epidemic and evil beings! They really make me sad! I must quickly go and get them!" This is why we see them in dream displaying the same appearance they had before death. But if they constantly come down to call their living relatives, those are more affected every time. Sometimes they even die of it. In this case, the xapiri must interfere to send the ghosts back to the sky. They tell them: "Ma! Don't come back down like this! Don't come close! Let us live here a while in this forest! Don't be so eager to call us among you!"[42] To which the ghosts answer: "Ma! You should be in a hurry to come back to us!" And the xapiri answer back: "Ma! We are not in torment down here! We will come back to you some day, of course! But we will do so without haste! Go back where you came from!"

This is how the xapiri and the ghosts speak to each other. I heard them after drinking the yãkoana and during the time of my dream. If the xapiri did not interfere like this, the nostalgic ghosts would soon carry their relatives away with them and humans would die constantly, one after another, far too fast. This would not be a good thing! The ghosts, they live a very long time, but they do finally die again. Then they go to live farther still into the heights, metamorphosed into fly and vulture beings on a new translucent sky that is above the one whose chest we see from the forest.[43]

CHAPTER 8

The Sky and the Forest

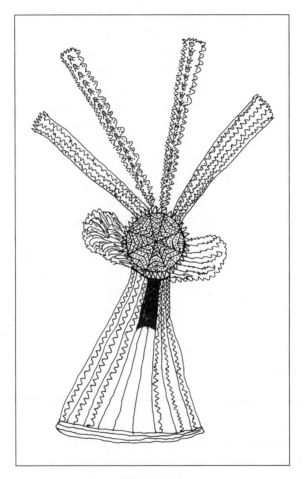

Mirror and paths of the *xapiri*

SOMETIMES, WHEN the sky makes threatening noises, women and children whimper and cry in fear. These are not empty cries! We all fear being crushed by the falling sky, the way our ancestors were in the beginning of time. I still remember an occasion when that nearly happened to us! I was young then.[1] We were camping in the forest, near a small stream that flows into the Rio Mapulaú. I had accompanied a few

elders on the search for a young woman of the *Uxi u* River who had been
taken away by a man from a house in the Rio Toototobi highlands. It was
early in the night. There were no sounds of thunder or lightning in the
sky. Everything was quiet. It was not raining and we could not feel a
breath of wind. Yet suddenly we heard several loud cracks in the sky's
chest. They came in rapid succession, each more violent than the last,
and they seemed very close. It was really alarming!

Everyone in our camp started to yell and weep in fear: "*Aë!* The sky is
starting to collapse! We are all going to perish! *Aë!*" I was also scared! I
had not become a shaman yet and I anxiously asked myself: "What is going to happen to us? Is the sky really going to fall on us? Are we all going
to be hurled into the underworld?" At the time, there were still great shamans among us, for many of our elders were still alive. Several of them
instantly started working together to hold up the sky. Their fathers and
grandfathers had taught them this work long ago, this is how once again
they were able to prevent its fall. Then, after a moment, everything got
quiet. Yet I think that this time the sky nearly did shatter above us again.
I know it has happened before, far away from our forest, where it is closer
to the edges of the world. These distant places' inhabitants were wiped
out because they did not know how to hold it up. But where we live the
sky is very high, and more solid. I think this is because we are at the center of the terrestrial layer.[2] But one day, a long time from now, it may finally come crashing down on us! It will no longer want to stay in place. It
will come apart and crush us all. But this will not happen so long as the
shamans are alive to hold it up. It will lurch and roar but will not break.[3]
This is what I think!

All the beings who live in the forest fear that they could be crushed
and wiped out by the sky's immensity, even the *xapiri!* Thinking of this
makes the people of our houses scared and they begin to cry. They know
very well that the sky has fallen before! I know some of this talk about the
sky's fall. I heard it from my elders' mouths when I was a child. It was so.
At the beginning, the sky was still new and fragile. The forest had barely
come into existence and everything there easily returned to chaos. It was
inhabited by other people, who were created before us and have since
disappeared. It was the beginning of time, during which the ancestors
changed into game one after another. And when the sky's center finally
collapsed, many of them were hurled into the underworld. There they
became the *aõpatari,* those sharp-toothed carnivorous ancestors who de-

vour everything the shamans throw them. They still live underground with *Yariporari,* the wind storm being, and *Xiwãripo,* the chaos being. They also live surrounded by peccary, wasp, and earthworm beings who also became other.

The back of this sky that fell in the beginning of time is now the forest where we live and the ground that we walk on. This is why we call the forest *wãro patarima mosi,* the old sky, and the shamans also call it *hutukara,* which is the name of this ancient celestial layer. Later another sky came down and fixed itself above the earth, replacing the one that had collapsed. It was *Omama* who drew its **project,** to use the white people's word. He asked himself how to consolidate it and put rods of his metal inside it, which he also buried like roots in the ground.[4] This is why this new sky is more solid than the old one and will not come apart so easily. Our shaman elders know all this. As soon as the sky starts to shake and threatens to crack, they instantly send their *xapiri* to reinforce it. Without that, it would have collapsed again long ago!

The people of the beginning of time were not as wise. Yet they worked hard to prevent its fall. But they were overwhelmed by fear, so they cut overly fragile stays from the soft hollow wood of the *tokori* and *kahu usihi* trees to sustain it. Most of these ancestors were crushed or thrown underground, except in one place where the sky finally came to rest on a wild cacao tree, which bent under its weight but did not break. This happened in the center of our forest, where you find the hill region we call *horepë a.*[5] Finally, a *werehe* parrot slowly gnawed at the surface of the sky, lying on the cacao tree canopy, and made a hole through which these first people finally escaped. They went out into the new forest on the back of the old sky and continued to live there. The shamans call them *hutu mosi horiepë t^{h}ëri pë,* the people who came out of the sky. Yet later these ancestors also died. They metamorphosed and were carried away by the waters or were burned when the entire forest went up in flames long ago.[6] This is what I know. We came into existence after them and we too have grown in number. So we are their ghosts.

WHEN A VERY old shaman is sick for a long time and eventually dies of his own accord, his *xapiri* silently leave his spirit house. Once abandoned, it gradually falls apart. Nothing else happens. On the other hand, if a still-young shaman dies suddenly, arrowed by warriors or eaten by enemy sorcerers, his spirits get angry. The sky gets dark and the rain does not stop coming down. The wind storm lashes into the trees in the forest, the thunder beings angrily roar, and the lightning beings crash. The rain falls endlessly, and the spirits of the sky pour snakes all over the forest. The jaguar spirits' mirrors come loose and these fierce animals start prowling. All this happens after the death of a shaman who owned a very tall spirit house.[7] His *xapiri*, furious to be orphans, start cutting up the sky. The spirits of the *ëxama* and *xot^{h}et^{h}ema* woodpeckers, followed by those of the *yõkihima usi* birds, thrash its chest with their axes and sharp machetes. Entire patches of the sky then start to come apart with a tremendous noise, so loud that even the surviving shamans are terrified![8] They must rush to send their own spirits to consolidate it and contain the fury of the orphaned *xapiri*.

The sky moves, it is still unstable. Its center remains solid, but its edges are already heavily damaged and have become fragile. It warps and sways with terrifying cracking sounds. The feet holding it in place at the ends of the earth are so unsteady that even the *xapiri* worry. Yet one of them, *Paxori*, the spider monkey spirit, proves particularly courageous. He travels from far away, but is always first to hold up the patches of sky threatening to come apart and to try and reinforce it. He is not a forest monkey but a celestial being, an ancient and powerful *xapiri* with skillful hands. Yet it would be impossible for him to carry out these repairs if he were working alone. Many other spirits come to help him, including those of the night monkey, the kinkajou, the *hoari* marten, and the *wayapaxi* squirrel. But he also calls the *hutukarari* celestial spirits, the *yãripirari* lightning spirits, and the *yãrimari* thunder spirits to back him up.

All these *xapiri* arrive in great numbers. They tear the axes and machetes out of the angry *xapiri* orphans' hands. They squeeze them in their arms to make them squat and try to calm them down. Then, by joining their efforts, they succeed in preventing the sky from breaking up. The sloth spirits consolidate the cracks with metal rods fired from their shotguns. The *ahõrõma asi* ant spirits pour glue to fill in its gaps. Then the cracking gradually comes to an end. Once silence has returned to the forest, the people of our houses—and even those who often doubt the shamans—tell themselves: "It is not a lie! They really become spirits and know how to contain the sky's fall!" Our ancestors have done this work since the beginning of time. If it had not been so, the sky would have collapsed on us long ago! Yet despite their efforts, it always remains unstable and fragile, at the mercy of the spirits of dead shamans who constantly want to cut it up.

The *xapiri* also constantly work to prevent the forest from turning into chaos. When the rain falls without interruption for days on end and the sky remains full of dark low clouds, we start to get tired of it. We cannot hunt or burn our new gardens to plant banana plants. We feel sad for our women and children, who are hungry for game. We are tired of the dampness and we also long for eating fish.[9] Eventually we turn to our shaman elders for help, for they know the rain being *Maari* well and can ask him to stop. So they drink the *yãkoana* and start working. Their spirits wash the sky's chest and call the sun being *Mot^hokari* and the dry season being *Omoari*. Then they turn the **key** that holds back the rain and bring light back to the sky. In my childhood, I often saw my father-in-law work this way to make the rain retreat and the forest happy. We call this *payëmuu*.

During flood times, the sons and daughters of the rain being *Maari* and the cloudy weather being *Ruëri* dance joyously above the forest waving young *hoko si* palm leaves, like guests performing their presentation dance. When these palm leaves are very damp, the rain never stops! To put an end to it, the spirits of the *rõrõkona*, *kutemo*, *kreemo*, and *tãitãima* cicadas, as well as those of the *kori*, *ixaro*, and *napore* cacique birds, must seize them and lift them towards the sky's heat. They shake the leaves to dry them off, producing a light breeze. This is the summer wind we call *iproko*. All these *xapiri* are the daughters and sons-in-law of the dry season being *Omoari*. This is why they can do this work. But for the showers to end once and for all, the spirits of the *ëxama* woodpecker and the small *roha* iguana must also lift the *Maari* rain being's penis and tie it back around his belt.[10] Other *xapiri* have to make him lie down in his hammock, hand him a tobacco wad to calm his anger, and delicately remove his wet feather headdress to put it somewhere dry. Then daylight and warmth finally return to the forest. The dry season settles there and the waters start to go down. White people do not know the images of the rain being and his children. They probably think the water falls from the sky for no reason! But me, I have often contemplated them in my dreams, like my elders saw them before me. It is so. The people of the forest's words are other.

We also do not return to the low water time so long as the daughters of *Motu uri*, the underground water being, continue to play happily in the rivers. The shamans must send their *xapiri* to interrupt their games and bring them back to dry land. The cicada and butterfly spirits are responsible for this, along with the wife, daughters, and daughters-in-law of the sun being *Mot^hokari*. But *T^horumari*, the celestial fire spirit, must still arrow *Motu uri* himself, then drag him by the arms and burn him.[11] Finally, the *Kõromari* ibis spirit pierces the ground with its metal bar so that the waters run off underground; only then will the water level go down. But the *xapiri* can also tackle the *Maa hi* rain tree to bring the rains and flooding to an end. Our elders know it well, and my father-in-law told me it stands at the borders of the earth and sky. It is gigantic and its leaves are endlessly dripping with dampness. Everything around it is cold and dark. The ground is covered in mud. It is the home of the *titiri* night beings and the *horemari* earthworm beings.

When the *Maa hi* tree blooms, it begins to rain in the forest and the rivers rise. To stop it from dripping, the *napore* cacique and howler monkey spirits must vigorously shake its branches and make its flowers fall. Then the macaw spirits must cut its branches with the help of the tapir spirit, who comes with them in his big pirogue. When this happens, the rain tree wraps itself in heat and the cicadas start to make their voices heard there. The *xapiri* who are the sons-in-law of the *Omoari* dry season being go after their father-in-law and call him back into the forest through a *hiimuu* invitation dialogue. They pick up the dead fish in the dried-up streams to offer them to him. Then he agrees to slowly return from the distances where he had taken refuge. It is so. *Omoari* does not answer to the spirits of the leaves and the trees, nor to those of the animal ancestors. If the *xapiri* who are of his people did not go get him, he would not come of his own accord. Then dampness and darkness would overrun the forest forever and it would finally return to chaos.

WHEN THEY also want to put an end to the thunder beings' anger, the *xapiri* go to their homes on the sky's back. They squat before them and reprimand them: "Your loud vociferations are bothering us! What are you doing? Why won't you stay quiet?" The furious thunders respond by threatening to strike them. To appease them, the spirits stretch out in their hammocks to demonstrate their friendship, as we do with a brother-in-law.[12] Then they offer them food and tobacco. Sometimes they also blow a little *yãkoana* in their nostrils to calm them. Then little by little the thunders finally grow quiet. If it were not so, the storm's roar would have no end, as it was in the beginning of time.

Thunder was an animal then, a kind of big tapir who lived in a river, near a waterfall.[13] At the beginning our ancestors did not know him. They were merely exasperated by constantly hearing the powerful roar of his voice echoing through the forest. They grew weary and decided to make him shut up. Finally they went after him and arrowed him. They cut up his body, careful to avoid spilling his blood on the ground. They cooked his flesh for a long time and ate it with great satisfaction. At the end of this meal, a mocking, well-fed hunter insisted on offering a leftover piece of raw liver to Thunder's son-in-law, the ancestor of the $h^w\bar{a}ih^w\bar{a}iyama$ bird. Furious, the latter struck the tiresome hunter's hand and the piece of the tapir's liver was projected onto the sky's back, where it came back to life and multiplied in every direction, like so many thunders with booming voices. These are the ones we hear above the forest today and who the shamans admonish to be quiet.

As for the lightning beings, they look like big macaws covered in splinters of light. When they flap their wings, they cast blinding glints of light with a loud crash. They are very powerful too, and do not hesitate to express their violent anger when they are famished. Their feet of fire fall into the forest from the sky's back with a terrifying din. This is why the shamans try to contain their fury. To get them under control, they make their own images dance and send them back to them as *xapiri*. These spirits then grab the lightning beings and reason with them: "*Ma!* Do not be so irascible! Do not destroy the forest like this! Other people live there! Humans have children there!" Then they play with them and tickle them, but if the lightning beings do not calm down after that, they will hit them and scold them sharply. Finally, they make them simmer down and become quiet and the storm goes silent in the forest.

The storm wind being *Yariporari* is also very dangerous.[14] He grows innumerable arrow-canes in his vast garden. When he wages war, he travels through the forest, angrily blowing its arrows in every direction. He is so frighteningly strong that even the *xapiri* fear him as he knocks over everything in his path. He turns our houses upside down and pushes the big trees onto our campsites. He smashes the branches, tangles up the underbrush, and slams into the tree trunks. He is accompanied by the giant armadillo being *Wakari,* who cuts out the trees' roots with his enormous machete. *Yariporari* is a formidable wind who fell into the underworld in the beginning of time. He hides in a hole underground, covered by a heavy lid that *xapiri* mourning their dead fathers or shamans angry at their enemies can lift to get revenge. Then *Yariporari* lets loose all his brutal power, ravaging the forest and terrorizing its inhabitants. When this happens, the spirits of the *witiwitima namo, xiroxiro,* and *teateama* birds and the *Koimari* bird of prey spirits attempt to grab him and tie him up. They must then destroy his arrow-cane plantations and lock him back up in the underworld. If not, his violence would eventually wipe out everything in the forest and sweep us far away. Until the elders taught me about the wind spirit *Yariporari,* I did not think there could be such a powerful evil being in the underworld! Yet despite the fact that he is so dangerous, the shamans can also make his image dance as *xapiri.* But in this case it is the image of his ancient form, his father spirit, whom they bring down to repel the epidemic fumes with which the white people fill the forest. It is so. Without the shamans' work, the forest would soon return to chaos, as I said. Rain and darkness, the thunders' anger, lightning and wind would never stop there. Only the *xapiri* can protect it and keep it solid and steady. This is why we follow in our ancestors' footsteps and become spirits with the *yãkoana.* This makes the *xapiri* happy and therefore they continue to take care of us. White people do not know these things! They are satisfied with thinking that we are more ignorant than they are, just because they know how to produce machines, paper, and tape recorders!

PEOPLE ALSO complain to the shamans when the dry season lasts too long, when the banana plants and the sugarcane are parched in the gardens and the streams dry up in the forest. To put an end to drought, the shamans try to bring *Toorori,* the damp weather being and master of rain,[15] back into the forest. They send him their flood, rain, and chaos spirits, who are the images of the evil beings *Riori, Maari,* and *Xiwãripo,* to invite him to return. Then they send him the images of the cloudy weather and night beings, *Ruëri* and *Titiri,* as reinforcements. Then little by little the shriveled rainy season being *Toorori* is able to pull himself out of the belly of the sun being *Mot^hokari,* who had swallowed him. Little by little, he comes back to life by pouring water on his head, then takes revenge and establishes himself in the forest in his turn. Once this happens, the rain starts to fall again.

I tried to make the rains come back myself, without really knowing this work the elders do. It was here, in *Watoriki*, already a long time ago.[16] The drought would not come to an end. The heat was getting more and more intense. The sun being *Mot^hokari* had come down from the sky's chest and had really put his feet down in the forest. *Omoari*, the dry season being, also seemed to want to settle in forever! He had dried up all the watercourses and had eaten his fill of fish and caimans. He had scorched the trees and roasted the ground. The stones had gotten burning hot. Game and humans suffered from thirst. The time had come to burn the plots of land we had cleared in the forest. We did so, but the wind carried sparks into the underbrush, which was too dry and covered in dead leaves. The surrounding forest began to burn. Then little by little the fire spread in every direction. When fire is so powerful, it is no longer friendly. It becomes an unknown and dangerous being who seizes every tree around him to build his house. He even started to ascend the slopes of the *Watoriki* mountain, not far from our house, right where the evil beings of the forest grow their own sorcery plants. We were very worried because we thought that these burning plants could spread a *xawara* epidemic on us. The smoke was constantly increasing. First it rose very high in the sky's chest. Then it came back down, getting lower and lower and thicker and thicker, and covered the entire forest. Our eyes were irritated and our chests dried out. We could not see anything around us and we were coughing all the time. It had become very difficult to breathe. We were afraid everything would burn and that we would die suffocated. We were really worried for our children, our houses, and our gardens.

Then I joined my father-in-law and the shamans of *Watoriki*, as well as a few others from neighboring houses whom I called up by radio,[17] and we drank the *yākoana* and started working to attract the rain. First we made *Omama*'s image dance to strike the fire and squash it. Then we called the thunder spirits and those of their sons-in-law to make the storm waters come and pour them on the blaze. We also brought down the storm wind being's image to push the smoke back into the sky and throw it far away from us. After that, little by little, the huge fire started to diminish. Our spirits drove away the dry weather being *Omoari* while admonishing him: "Go home! Don't try to stay here or the entire forest will burn with all its inhabitants!" Then they started to call back the rainy season being *Toorori* so he would wash the forest.

We worked like this for days, and finally the rain started to fall. If we had not done so, all the trees of the forest would have burned, all the way to the white people's land, for this fire was not just a fire. It was a formidable evil being, a flesh-hungry fire spirit we call *Naikiari wakë*. This was the spirit of the *Mõruxi wakë* blaze who came out of the ground, the same one who had previously consumed the entire forest in the beginning of time. This fire comes from where the sun lives, and in the place where he comes from, the waters never stop boiling. His representative is what white people call a volcano. He is so powerful that he even burns sand and stones. In their night talks, the elders often told us of the fire that ravaged the highlands in *Omama*'s time. They told us that in some places the trees never grew back. So the bare land we call *purusi* at the sources of rivers is the trace of this ancient blaze's path. It did not appear by itself, without a reason![18] Yet the forest grew back in other places because the being of the earth's fertility, whom we call *Huture* or *Në roperi*, worked ceaselessly to replant it. He is an indefatigable worker. He repopulated the scorched earth with all its trees, but also its garden plants, manioc, banana plants, and *rasa si* peach palms so that our ancestors, their children, and grandchildren could feed themselves. If he had not existed, we would forever be famished and we would be sad to see!

IN THE PAST, when our elders used to become shamans under the effect of *hayakoari hana* sorcery leaves,[19] they were also able to call the images of the white-lipped peccaries and use them to attract these animals in the forest around their houses. One of the *Watoriki* house elders, whom I called brother-in-law, knew how to make the peccary spirits dance like that, but he is no longer. When he died, I saw his spirit house collapse and tear apart the *xapiri*'s fragile paths as it fell. He had warned us: "As soon as my ghost leaves for the sky's back, you will no longer see peccaries in the forest. Then you will lament yourselves over your meat hunger!" Yet while he was still alive no one had told him: "*Awe!* I too want to know how to tend the peccary spirits' paths so they won't run away!" I myself said nothing to him. I was still ignorant at the time. If I had done so, maybe this game wouldn't have vanished from the forest for so long?[20] But it is true that no one in that time was wise enough to hold on to these spirits' paths!

Only the shaman elders truly knew how to make the peccaries come out of the ground by calling their image. In the past, people often used the *hayakoari hana* leaves for sorcery. But it is a plant that belongs to the *hutukarari* spirits of the sky. This is why those who were hit by its power became other without delay. They saw the image of the *Hayakoari* tapir-like being in front of them. Then they began to gesticulate elatedly and go yelling out of their house. But it was not really in the forest that they started running. Though their relatives could not see this, their image fled and rode the *Hayakoari* being all the way to his home, very far away. They remained lost in the forest like this for a long time while they had become other. At that moment, they started to see the images of the peccary ancestors dancing for them. Finally, they left the *Hayakoari* being's path and little by little they calmed down. They came back to their homes, guided by the *xapiri* of the shaman elders who came to their rescue. Without them, they would have died of hunger and exhaustion, forgotten on the *Hayakoari* being's mirror.

Later, as well-tried shamans, they were able to open the paths of the *worëri* peccary ancestors and make their images come down again. To call them, they would first send the spirits of the *xotokoma* birds,[21] who are their sons-in-law. These emissaries cut down trees to make an entrance into the forest for their fathers-in-law and hung magnificent bead ornaments there to attract them. Then they let their *tʰora* bamboo flutes' call ring out so the *worëri* spirits came to dance near the shaman who sent them. After that the animal peccaries also drew close to our houses. This is how our elders worked to satisfy their people's hunger for game. Yet the peccary spirits' paths are very fragile. As soon as their father dies, they break and sink deep underground. The other shamans try everything to bring them back, but in vain. The peccary ancestors remain in the underworld until another young man becomes other under the effect of the *hayakoari hana* leaves and learns to call them again.

As for tapirs, they only appear within range of the hunters in the forest when the shamans succeed in bringing the image of the tapir ancestor *Xamari* there. To do so, they must first send their ocelot and hunting dog spirits to find his trail, then the spirits of the *xoapema* birds, *herama* falcons, and *ëxëma* woodpeckers to call him. Otherwise *Xamari* would continue sailing his pirogue on distant rivers. Tapirs like to laze about in the water, don't they? All these birds' spirits are his sons-in-law,[22] which is why he gladly answers their bamboo flutes' call and their invitation: "Father-in-law! Come to us! We are hungry for meat! We want you!" But he barely has time to say a friendly word to them before they tie a rope to his pirogue and pull it to the riverbank with the help of the *kana* giant otter spirit. The tapir ancestor then disembarks to set foot in the forest. Right away, his sons-in-law take great pains to show him where he can find his favorite food: the fruits of the *rio kosi* and *ëri si* palms, as well as those of the *apia hi, oruxi hi, makina hi, hapakara hi,* and *pirima ahit^ho^tho* trees. This is how shamans attract tapirs onto dry land so they can be hunted in the forest.

But even then they can only be spotted by very skillful trackers; those we call *xama xio,* "tapir's bottom."[23] Though they are not shamans, these hunters possess images of the tapir spirits and his sons-in-law inside them. These images come down to them and set up their hammocks in their chests, for their fathers were also great tapir hunters. Without their skill, we would never eat meat from this game! It is true! You never see a tapir when you merely think about your things and hunt without real interest. You only find tortoises on the forest floor like that! But if a hunter is truly in love with the tapir ancestor's image and longs for him,[24] he will easily spot one of these animals who have come from deep in the forest, even very close to his house.

This is how our long-ago shamans drew the peccaries and tapirs close to their homes, but also the spider monkeys, parrots, curassows, and deer. They drank the *yãkoana* and made the *yarori* animal ancestors' images dance. And when they decided to bring the macaw spirits down to them, these birds were seen appearing in the forest. It was really so. Animals are only happy if their *xapiri* let their songs be heard, and the spirits do not like their fathers to laze about in their hammocks, not drinking the *yãkoana.* Game only become easy to hunt if the shamans bring down their ancestors' images. It is so. Our elders from the past were very wise and knew how to do this work. They did not just sing in vain, as the white people often think, for if the shamans do not work relentlessly, game becomes irascible and fierce. It keeps complaining about hunters: "*Ma!* These are other people! They treat us without care. They are filthy, pouring our cooking juice outside their houses! Casually throwing our bones and skin into the forest! It is painful to see! Let's stay far away from them!" Animals are also human beings. This is why they turn away from us when we mistreat them. In the time of dream, I sometimes hear their unhappy and angry talk when they want to refuse themselves to the hunters. If you are really hungry for meat, you have to arrow the game with care and it must die on the spot. If it happens like this, the animals are satisfied to have been rightly killed. Otherwise they flee far away, wounded and furious at humans.

THE KORI and *napore* cacique birds and *piomari namo* jaybird never gather in the forest trees if they are not bearing fruit. Then no other bird will perch on them either. It is so. Parrots, toucans, macaws, curassows, agamis, guans, and *pokara* quails are in the habit of following caciques and jaybirds to their food. They feed on their leftovers, the fruits their noisy flocks peck at on the treetops or drop on the ground. This is why the shamans make cacique and jaybirds' spirits dance when they want winged game to be abundant in the forest again. These spirits' images make the fruit of the trees ripen to feed all the other bird spirits who are in love with them and follow them. People who have never drunk the *yãkoana* do not realize this. They only hear the shamans sing in the night, without really understanding what they are doing. Yet when the forest has the value of hunger, the shamans send their cacique and jaybird *xa-piri* very far in the direction of the setting sun in order to bring back the image of its fruit. Upon their return, the other bird spirits call out with joyful impatience: "*Awe!* We're finally going to eat! Let's ask them for our share of the food they're bringing back! We're ready for it! We are starving and in pain!" Then they all dash for this unhoped-for pittance in a happy troop. This is how winged game starts reappearing in the forest! First it comes back very far from us, then little by little it gets closer to our houses. The hunters pass the news along enthusiastically: "Game is eating near that river, and also that colony of trees, over there, and also in this other place!"

This was the work our great shamans used to do to attract game to their forest. Today we have lost this knowledge and many of our fathers had even forgotten it before us! Only the real long-ago elders were able to do it. They could gather a multitude of parrots and macaws on *hoko si* and *õkarasi si* palms, where they played tamely, staying within range of hunters while they nibbled young leaves. It is true! Long ago, when my grandfathers lived at the sources of the Rio Toototobi, they really had this power! Sometimes they used it so the people of their house could eat their fill of bird flesh and decorate themselves with their feathers. They wanted to make them happy. And when their relatives were too hungry for meat, they even brought game from the ghosts' forest, which lies on the sky's back! To do so, they used to send their *xapiri* to set the game running, so they could drive it and make it fall to earth. Shamans know that the ghosts' forest is covered in ever-abundant fruit trees and that there are far more peccaries, spider monkeys, curassows, and guans there than down here!

FOREST TREES and garden plants do not grow all by themselves, as white people think. Our forest is vast and beautiful. But it is not like this without reason. Its value of growth makes it so. This is what we call *në rope*.[25] Nothing would grow there without it. It comes and goes like a visitor, making plants grow everywhere on its path. When we drink the *yākoana*, we see its image spreading all over the forest and making it damp and cool. In the trees, the leaves are bright green and the branches loaded with food. We see a profusion of *rasa si* peach palms, covered in heavy bunches of ripe fruit, hanging low on their thorny trunks, as well as vast plots of banana plants and sugarcane! The land's value of growth is at work everywhere. It creates the forest's abundance, feeding human beings and game. It makes all the plants and fruit we eat come out of the ground.[26] Its name is that of everything that flourishes in our gardens, as well as in the forest.[27]

In the beginning of time, *Omama* placed this value of growth inside the land where we live. Its image was later disseminated everywhere until it reached the white people's territory. Its real center is our forest home, where *Omama* came into existence. This is true. We live in the place where the father of *në rope* growth resides, the place from which he came into being. This is why his image, which we call *Në roperi*, dances with those of the animal ancestors as soon as we bring them down. When the forest has the value of hunger, shamans can drink the *yākoana* to bring its value of growth's image back. Yet we never need to do this work where we live now, at *Watoriki*. Our land is beautiful and full of richness.[28] The evil being of hunger, *Ohiri*, remains far away and *Në roperi*, the image of growth, has danced by our side since we settled there. After each time of rain, she generously makes the trees' fruit and our garden plants grow. Everything flourishes easily and game eats in abundance from the trees, on the ground, and in the waters.

The image of *Në roperi*, the forest's richness, looks like a human being, but is invisible to ordinary people. She only lets their ghost eyes see the food she makes grow. Only shamans can truly see her presentation dance. She comes to them preceded by a noisy troop of cacique birds and jaybird spirits, followed by a multitude of macaw, parrot, toucan, and curassow spirits. The *xapiri* leading the other birds are the image of growth's **companions**, they are her **collaborators**. She never dances without them. The shamans make them come down when the people of their house are hungry, for no food grows where their resounding call is not heard. These animal ancestors discovered and propagated the earth's fertility in the beginning of time. This is why today's birds, who are their ghosts, continue to eat the fruit of the forest. They are their representatives. This is what the elders say. But the forest's richness is also the images of the *yamanama* bees who make the trees' flowers bloom and spread sugar in their fruit and those of the papaya tree and sugarcane. It is also the images of the banana plant women and the *aro kohi* and *wari mahi* big trees, whose leaves are so lush.[29] In the highlands, the spirits of the *witiwitima namo* kite are responsible for the abundance of the *kaxa* caterpillar, the fruits of the *momo hi* trees and *xoo mosi* palms, and the edible flowers of the *nãi hi* tree.

As soon as the cacique birds and jaybird spirits' strident call rings out from every direction, one can also hear the low-pitched song of *Në roperi*, the spirit of plants' growth. He arrives dancing joyously, carrying all the food of the forest on his back. He looks like a human being, but he is other. He is much more beautiful! His eyes are magnificent and his hair looks like a profusion of yellow and white flowers. His body is covered in luminous down feathers and he wears a stark black saki tail headband. He moves slowly, followed by a company of images of trees, lianas, and leaves. He is surrounded by a noisy cloud of colorful bird spirits: *sei si, hutureama nakasi, ayokora* caciques, and small araçari toucans. He is also followed by a multitude of *yarori* animal ancestor spirits and *urihinari* forest spirits, waving frayed young palm leaves in an inebriating fragrance of flowers. He dances among them, brandishing the forest fruit he carries, which are also covered in a dazzling white down. I saw this image of the forest's richness in a dream after drinking the *yãkoana* all day long. She is truly superb! I even felt the smooth, sweet taste of her ripe fruit in my mouth!

Once he has finished his presentation dance, the *Në roperi* spirit of growth feeds the shaman who called him and sets up his mirror in a separate dwelling inside his father's spirit house, like the other *xapiri* do.

From this moment on, this shaman will know how to bring the forest's value of growth to his people. Without their knowledge, he will be able to make all the plants flourish and to cure the forest of its sterility. As soon as he makes his *Në roperi* spirit dance, the flowers will start blooming on the trees. Then their branches will become fertile and the fruit will develop in abundance. If the spirit of growth did not come down with his cacique birds and jaybird spirits, our land would remain barren and game would not come into our forest. These images of fertility make the animals' food grow, and ours too. And after that, it is *Omoari*, the dry season being, who puts heat down on the ground to help make the forest fruit ripen, for he eats it too!

IN THE PAST our shamans drank the *yãkoana* and exhorted their *xapiri*: "Our women and children are starving! Make the food in the forest grow!" Then they sent them to fetch the image of *në rope* growth very far away, where the *Huture* being who owns her lives, and they brought her back with them. Then the plants grew and the trees flowered along her path. She came all the way to their forest and continued on her way beyond. Today we are not as wise as our grandfathers, but we try to follow in their footsteps. They did not teach us how to bring the forest's growth back before they died. Yet now that we have also become spirits as they did, we too have learned to know the image of forest richness by making her dance in the time of dream. It is so. When the value of growth moves away from our houses, she does not come back by herself. The shamans must really work hard to bring her image back because without her the fruit of the trees and the plants of our gardens stop growing. Then they have to work often to keep her by our side, for she can always run away again, this time never to come back.

When this happens, it means *Ohiri*, the hunger being, has settled in the forest in her place. Having come from very far away, where the white people have nothing to eat, he lies in ambush to mistreat us. No matter how much we plant or how hard we work, nothing grows in our gardens, not bananas, not manioc, not sugarcane! All the cultivated plants shrivel up and the branches of the trees remain empty. Game becomes increasingly scarce. Then we say: "*Urihi a në ohi!* The forest has taken the value of hunger!" *Ohinari* is what white people call **poverty**. He is an evil being who kills little by little, through hunger. Once he has decided to settle in the forest, he can stay in the same place for a very long time. When this happens, people soon have nearly nothing left to eat. Day after day, *Ohiri* blows his *yãkoana* powder in their nostrils and makes them become other. They constantly get weaker. Their limbs have no energy and they get dizzy. Their ears get blocked, their voice is dry, and their empty eyes are sad to see. Little by little they waste away and finally lose consciousness. Then they die, completely emaciated.

To avoid this, the shamans must drink their *yãkoana* again and again and send their *xapiri* after the *në rope* fertility image in distant forests or even on the sky's back. It is true. As I said, there is another value of growth above us. It is that of the ghosts and thunder beings who also eat the fruit of their forest and the plants from their gardens. Their forest abounds with *oruxi hi, mõra mahi, yawara hi* trees and many other fruit trees! Its richness is truly very great and the cacique birds and jaybird spirits can bring it back on earth. Yet the ghosts can also decide to let a little of this abundance fall down to humans on their own. This sometimes happens during their feasts, once they are sated and sing their *heri* songs, when they hear the women of the living complain of their hunger and ask them to give up some of their leftovers. In those places where the ghosts prove generous, the fruit of the forest trees and the *rasa si* peach palms in the gardens become abundant and the happy humans eat their fill.

IN THE beginning of time, *Koyori*, the ant ancestor,[30] discovered the gardens' value of growth when the forest was still transforming and passed it down to our elders of long ago. But he was not the one who made the forest's trees grow. It was *Omama*. *Koyori* worked alone in the forest all day long, until his people were intrigued by his long absences. He eluded their curiosity by pretending to cut down trees on the search for wild honey. But he was lying! In fact, he was secretly spending his time clearing the forest to open a huge garden, which he was constantly expanding. Yet there were no cultivated plants at the time. To make them grow, *Koyori* just tapped his foot on the ground and repeated: "May the roots of these plants grow! The maize will come out here! The banana plants here!" Then the maize and banana plant plantations instantly started to grow before his very eyes. *Koyori*'s mother-in-law was called *Poomari*. She was a quick-tempered old woman who constantly complained about her son-in-law. She was furious that he spent so much time in the forest without bringing back any food for her. One day she finally lost her temper and insulted him by making fun of his prominent bottom. He decided to get revenge. He sent her to get maize farther and farther into his vast plantation so that she would lose her way. Finally, she got lost and in her distress she turned into a *poopoma* bird. You can still hear her resounding call in gardens: *"Pooo! Pooo!"* As for her son-in-law, he changed into a *koyo* ant.

Since then, shamans have known how to bring down the images of *Koyori* and his mother-in-law *Poomari*. I heard their songs when my wife's father made them dance and I often saw them in my dreams after drinking the *yãkoana*. These images also own the earth's value of growth. This is how it first appeared. In the time when *Koyori* came into existence, there were no gardens yet. People only ate fruit from the forest. *Koyori* then called the *Në roperi* being of land richness for cultivated plants. He was the first to grow maize, banana plants, manioc, taro, and yams. He taught us this work. So even if a man is not a shaman, if he possesses *Koyori*'s image, it will help him to work in his garden without getting tired, whether he is sound or sick. You will never see him slumbering in his hammock! It will constantly make him want to open new plots in the forest and plant all sorts of food. It is so. When we work in our gardens, we also imitate the image of the giant *wãsikara* lizard who enables us to work hard in the sun without weakening. These images pass from a father to his son through his sperm, through the blood of his sperm.[31] They are invisible; they are deep inside us, in our thought, in our ghost, in our own image.[32]

In gardens, the spirits of the *horeto* dove look after the banana plants. They plant them along with humans and follow their growth, for these are also *në ropeyoma* fertility spirit women. However, the bat and spider monkey spirits are the ones to play and copulate with the banana plant shoots when they are still young women.[33] These *xapiri* make them pregnant with their value of growth and so they start to carry voluminous bunches of fruit.[34] This is true. Plants do not grow by themselves, without reason! Banana plants are plant-women. Their fruit are born because they are gravid and they give birth. The same is true of every plant that grows in gardens and the forest. First the plant-women are pregnant. Their pregnancy lasts for some time, then they deliver. This is when their fruit appear. They are born like humans and animals. This is why the people of a house also call on shamans when their banana plantations have trouble growing or when they are in a hurry to have enough bananas to hold a *reahu* feast and their gardens are still young. They ask them to make their bat and spider monkey spirits dance so they will impregnate the banana plant–women and their fruit will develop quickly. These *xapiri* put their offspring and the taste of sugar in young banana plant shoots,[35] like human beings with their sperm! This is what they do; I have often seen them copulate like that in the time of dream.

As for the giant *waka* armadillo spirits, they are the owners of manioc tubers and their fertility.[36] They plant them along with humans and are the ones to make them grow. A man who possesses the image of this animal inside himself will have a very beautiful manioc plantation! This image will help him while he is working in his garden and his arms will be full of its value of growth. The man's manioc plant tubers will then become long and firm. It is so. If you ask them, the shamans can also call the spirit of the giant armadillo and make him dance to enlarge the tubers of a manioc plantation that is not yielding much. For *rasa si* peach palms, the shamans can also bring down the spirit of the *marokoaxirioma* bird,[37] who impregnates the *raxayoma* peach palm–women's image by slipping the eggs of their fruit around their necks. Then these fruit start to grow profusely. To stop them from prematurely dropping from their trunks, the *napore* cacique bird spirits also have to give their mothers baby slings in which to carry their heavy bunches like newborns.[38] Finally, the macaw spirits will be in charge of making them ripen.

THE SPIRIT of the small *yõriama* tinamou makes the *ara si* taro plants grow in our gardens. To increase the number of their tubers, shamans can also call his image and make his value of growth dance. Yet it is merely the forest soil that makes yams grow;[39] this soil that *Koyori* made fertile in the beginning of time. This ant ancestor's image also makes the maize plantations thrive, like he did in the past by tapping his foot on the ground. A very long time ago, our ancestors hosted their *reahu* feasts by offering their guests maize.[40] Today we do not grow much maize anymore. But the ant ancestor *Koyori* is the real keeper of the forest soil fertility anyway. He is also responsible for the growth of sugarcane and

sweet potato. We do not need to water the ground the way white people do for the food in our gardens to be abundant! The forest's *ně rope*
value of growth is enough. Without it, plants would remain ugly and
shriveled up.

WHEN WHAT WE have planted in our gardens really does not grow well,
we sometimes think that enemy shamans might have diverted the forest's richness far from our home. Yet a shaman from a friendly house
might also take it with him without meaning any harm. Once he has
eaten his fill, a guest at a *reahu* feast might unwittingly steal the image of
his hosts' forest's value of growth by dreaming. Having become a ghost
under the effect of the plantain soup they had offered him,[41] he can bring
the bat spirits who made these fruits grow back to his home so they can
dance there too. It is so. If we drink a lot of plantain soup or peach palm
fruit juice at a feast, we become other and at night the images of their
fertility come to visit us. This happened to me once during a *reahu* feast
with the *Xamatʰari* of the *Kapirota u* River. I drank so much of their peach
palm fruit juice that I stole the image of the *marokoaxirioma* bird who
made these palm trees grow in their garden! He appeared to me during
my sleep and followed me to make my own plantation in *Watoriki* grow!
My hosts noticed, but did not hold it against me. They simply told me:
"You can keep these fruits' *ně rope* fertility! We will make another one
come to our gardens!" But even when visitors carry away our plantations'
richness, it does not last for long. The value of growth remains abundant
in the forest and if our gardens take the value of hunger, our shamans
drink the *yãkoana* to bring it back home. And if need be we can also borrow the forest's fertility from a friendly house. So we tell the shamans
there: "My people are hungry because my plantations are not growing
well. I too would like to obtain the value of growth you possess! But I do
not know how to do it!" In this case, they will prove generous and will
make its image dance to give it to whoever asked for it.

ANIMALS ARE just like humans. We eat our fill when our gardens are full
of bananas and *rasa si* palm fruit and they eat their fill when the fruit of
the forest trees is abundant.[42] This is their food as it is ours, for the animals we hunt are the ghosts of our ancestors transformed into game in

the beginning of time. Another group of the first human beings were hurled into the underworld, but these remained in the forest in which we too were created. We refer to them as game, but in fact we are all human. It is so. When the forest's richness runs away, the game becomes skinny and scarce, for this richness is what makes game prosper. The animals get fatter, then make young that grow and multiply in their turn because they feed on the forest's sweet, ripe fruit.[43] To live, their images must feed on the image of the forest's value of growth. This is why shamans also bring down the image of the game's fat with that of the forest's fertility. This tapir, peccary, and spider monkey fat comes from beyond the land of the white people's ancestors. It fattens up their **cattle** and also makes some of the white people very big! We call it *yarori pë wite,* the fat of the animal spirits.

Shamans must send the *napore* cacique and *hutuma*[44] birds' *xapiri* far into the distance to bring this game fat back into their forest. It comes from an unknown and ancient being who looks like a giant spider monkey and remains hidden downstream of the sky, where the sun is born.[45] This being is extremely adipose, for he keeps all the game's fat on himself and does not easily give it away. If he is late distributing it, the animals might remain too scrawny and sick to be hunted. Yet when his image is willing to come dancing down into the forest, they start to fatten up again, however many there are of them: monkeys, deer, tapirs, peccaries, curassows, guans, macaws, and parrots, as well as turtles and fish. When we sleep in a ghost state, having eaten our fill of plump game, this fat's image comes to stouten us up too! I myself have only seen this giant spider monkey being once, by drinking the *yãkoana.* When he wants to make the game he owns fleshy, his image travels alone through the forest. He divides his fat up among all the animals along the way. Only the elders, the great shamans, can call him to fatten up the game. I do not know how to do it yet, and I do not want to pretend. I will try when I am certain that I will really know him. I do not want to behave like those shamans who constantly lie and brag about bringing down *xapiri* whom they have barely glimpsed and do not know anything about!

As I SAID, the *xapiri* travel and work in the forest, on the sky's back and under the earth, in every direction, innumerable and powerful, in order to protect us. They relentlessly attack evil beings and the epidemics that try to devour us. They clean the wombs of women made sterile by *xapo kiki* sorcery things and copulate with them so they can have children by their husbands again.[46] They consolidate the forest when it becomes other and wants to transform itself. Without them, the plants of our gardens would not grow, the trees of our forest would not bear fruit, and game would remain skeletal. The forest would never stop having the value of hunger. They hold up the sky when it threatens to collapse, contain the thunders' anger, send the rain being's daughters away, and shut the storm winds in. They admonish the cloudy weather being and delay the nightfall being. They repel the night spirit and call the dew so that dawn breaks faster. They contain *Xiwãripo*, the chaos being, who wants to tangle up the forest when he smells the menstrual blood of girls who have left their seclusion enclosure too soon. They return the snakes and scorpions who fell from the sky's back to the place they came from. They keep the jaguar spirits' mirror closed to keep these animals from coming out of the ground, from the place where our ancestors found the egg that gave birth to them. It is true, the first jaguars were born from a huge egg! In the beginning of time, old women collecting crabs and shrimp in a stream found it floating in the water. Their curiosity piqued, they approached and heard it making a muffled roar. They carried it in a basket to their house where the people, though perplexed, finally cooked and ate it. After that they threw away the pieces of its shell outside their house and the people changed into jaguars who scattered all over the forest!

It is so. The *xapiri* defend us against all the evil things, darkness, hunger, and sickness. They repel them and combat them unceasingly. If they did not do this work, we would be sad to see! Wind, lightning, and rain would leave us no respite; the rivers would constantly rise and flood the forest. It would be infested with snakes, scorpions, and jaguars; invaded by evil beings and epidemics. Night would cover everything. We would have to stay hidden in our houses, starving and terrified. Then we would start to become other and the sky would eventually break apart again. This is why our ancestors started to make the *xapiri* dance in the beginning of time. Their concern has always been to protect their people, as *Omama* taught his son. We are only following in their footsteps. Yanomami shamans do not work for money the way white people's doctors do. They simply work so that the sky and forest remain in place, so that we can hunt, plant our gardens, and live in good health. Our ancients did not know of money. *Omama* did not give them any talk of this kind. Money does not protect us, it does not fill our stomachs, it does not create our joy. For white people, it is different. They do not know how to dream with the spirits the way we do. They prefer to ignore that the shamans' work is to protect the earth, as much for us and our children as for them and theirs.

II

Metal Smoke

CHAPTER 9

Outsider Images

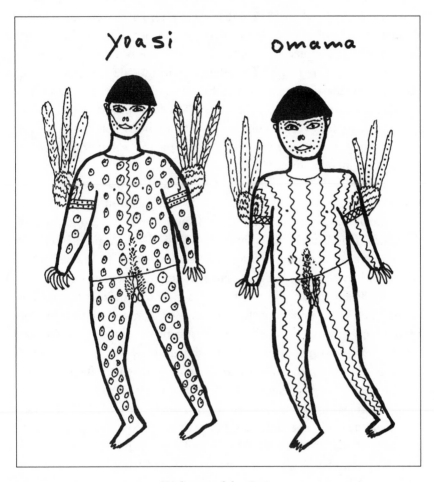

yoasi   omama

Trickster and demiurge

L ONG AGO the white people did not exist.[1] This is what the elders
taught me when I was a child. At the time, *Omama* lived in the forest
with his brother *Yoasi* and his wife *Tʰuëyoma*, whom the shamans also
call *Paonakare*. His father-in-law *Tëpërësiki* lived in a house under the water. There was no one else. It was so. *Omama* had given us life long be-

fore he created the white people and it was also he who possessed metal before they did. The first pieces of iron that our ancestors used were those that *Omama* left in the forest when he ran far away, towards the downstream of all rivers. They did not own real axes, nor real machetes, like we do today.[2] They just bound worn pieces of iron to a handle to make them into hatchets.[3] These tools were rare. There were really few of them in their homes! Only a few elders carefully looked after them. They worked with these pieces of metal and called them *Omama*'s tools because they were very tough.[4] The other men had to take turns borrowing them to clear their gardens. Visitors from friendly houses also came and asked to use them. In that time, it was so. The white people's objects were not yet distributed everywhere, the way they are now! This is why today I think of our elders' hard work, and it makes me think that I don't need to own a lot of merchandise.

Omama was the only one to own metal, and he had always worked with it in his garden. In the beginning of time, he even turned himself into a metal bar out of fear! He had just fished *Tëpërësiki*'s daughter out of the river when his underwater father-in-law decided to come visit him. So his spouse's father set on his way, weighed down with a huge bag of woven palm leaves full of banana plant sprouts, manioc, sugarcane, yam, taro and sweet potato cuttings, tobacco, papaya, and maize seeds. He was coming to teach *Omama* how to use cultivated plants. Yet he made a terrifying racket from a distance, like a hurricane or a big **tractor.** *Omama* was frightened to meet this fearsome father-in-law and instantly metamorphosed into a piece of metal and planted himself in the ground inside his house.[5] His brother *Yoasi* followed suit right away but only transformed himself into a palm-wood digging stick.[6] When *Tëpërësiki* came into their dwelling, he only saw his daughter. He asked her: "Where is your husband?" She gestured to the metal rod with her lips. "Where is your brother-in-law?" She indicated the piece of wood. Then *Tëpërësiki* declared: "You will plant these things I brought and make them multiply. When you have children and you make the Yanomami grow in numbers, they will be able to feed themselves." Then he went back to his home under the waters. These are the foods we eat to this day. But *Omama* did not get the metal from his father-in-law, he already owned it. By drinking the *yãkoana*, I saw him turn into a steel tool. His image still stands there where it happened, in the highlands, at the sources of the rivers. After that, he returned to his human form and he taught our ancestors to work with this metal in their gardens.

OUR ANCESTORS did not use knives. They cut game up with bamboo blades. They broke their catch's bones with pieces of hard wood. They also fished with hooks carved from armadillo bones or the *ërama t^{h}ot^{h}o* vine's hooklike thorns tied to *yãma asi* leaf fibers. Women grated manioc on stones or the rough bark of the *operema axihi* tree.[7] Men made fire by rubbing cacao tree drills between their hands. People cut their hair with sharpened splinters of reed or the teeth on piranha jaws. There were no combs. They combed their hair with the spiny pit of *ruapa hi* fruit. There were no mirrors either. When you wanted to pluck your eyebrows or paint yourself, you had to ask someone else to do it for you. At the end of *reahu* feasts, people traded bows, arrows, bamboo quivers, arrow points, feather ornaments, tobacco, annatto dye, gourds, dogs, cotton hammocks, and clay pots. When our elders lived on the upper Rio Toototobi, they owned a clay deposit there. The women fashioned the pottery and the men traded it with people from other houses. This is how we lived in that time. I often heard my stepfather talk about all that when I was a child. At that time, there were no white people's objects. There were still none of their hammocks, nor aluminum pots, nor cassava bread cooking plates made from the lids of metallic barrels. Men slept in hammocks made from strips of bark[8] or cotton. Women cooked in pottery and baked their cassava bread on clay plates.

In the time of our ancestors, the white people were very far away from us. They had not yet brought measles, coughing disease, and malaria into our forest.[9] Our people were not sick as often as we are today! They were in good health most of the time and when they died their ghosts were not tainted with the fumes of epidemics. Now, when someone dies of white people's diseases, even his ghost gets sick and returns to the sky's back with fever. His breath of life and flesh are soiled all the way there! In the past, people never all got sick at the same time! They did not die as much as now. The *në wãri* evil beings of the forest ate a man or woman's image. A young girl expired when a distant enemy hunter arrowed her *rixi* animal double. A child was devoured by enemy shamans' spirits. Sometimes an elder suddenly died before his time. When this happened, people in-

vited relatives from friendly houses and all cried and wailed for him together. If people thought that enemy *oka* sorcerers had shattered his
bones, a group of warriors instantly left to avenge him. People mourned
for an elder who died this way, then later it could also be an old woman.
Sometimes men could also be arrowed by enemies. Someone could die
from a snakebite or an old person started to cough all the time and eventually expired. It was so. People only fell dead once in a while.

In that old time, the Yanomami truly loved the beauty and coolness of
the forest. The elders died out like the logs in a fire, when their heads
were white and their eyes had become blind. They dried out like dead
trees and snapped. There were many shamans at the time. They often
made their spirits dance to treat the sick. Then the old women rubbed
their bodies with forest remedies. People also drank wild honey when
they felt badly, and that cured them. The ancestors knew all these things
so well. Today it is no longer so. The gold prospectors soiled the forest. It
has been permeated with epidemic fumes, and we were caught in a death
frenzy. There were really a great number of us when I lived on the Rio
Toototobi as a child. There were three big houses close to one another
and many elders. Then the white people came with their fever and measles and a lot of our people died. Today the great shamans are nearly all
gone, our houses have become much smaller, and we die young.

When *Omama* created our ancestors and taught them the things of
this world, their thought was calm. They cleared new gardens in the forest and worked arduously in their plots. They planted banana plants,
manioc, sugarcane, yams, taro, sweet potatoes, maize, and tobacco. They
also grew many *rasa si* peach palm trees. Their concern was to widen
their plantations so that the guests at their *reahu* feasts would be many,
and that once sated with food they would evoke their hosts' generosity
with beautiful words. And as soon as the plants in their gardens began to
grow, they all went hunting together far away in the forest. There, they
killed large numbers of monkeys, tapirs, and peccaries, which they
smoked before they brought them back to their homes. Then throughout
the dry season they invited each other to big feasts from one house to the
next.

Guests decorated their hair with white vulture down, they tied black
saki tail bands around their foreheads. They covered their faces and bodies with fresh annatto dye and drew curved black lines, circles, and dots
on themselves. They wore green parrot feathers and turquoise throats of

hëima si bird feathered hides in their ears. They stuck long scarlet macaw tail feathers and bunches of black and white guan feathers in their cotton armbands. They dangled toucan tails and orangey cock-of-the-rocks' feathered hides from them. They were very beautiful and danced euphorically to show themselves at their best in their hosts' house. Then their hosts offered them abundant portions of banana soup, peach palm fruit juice, and sweet manioc juice. At night, men and women took turns singing *heri* songs and constantly joked and danced. Sometimes some of the male guests took the wrists of partners they chose from their hosts' daughters and wives and sang around the house's central plaza. This is what the elders called *hakimuu*.[10] But impatient fathers or jealous husbands were often quick to get angry and start a fight! When this happened, adversaries took turns pounding each other's chests with their fists to put an end to their anger. But if they were truly furious and that did not work, they started hitting each other's heads with clubs. Failing that, their rage could not truly come to an end!

Our long-ago elders' thought was only truly in sorrow when someone from their house died. And if he had been eaten by an enemy, the anger of mourning could only be appeased once he had been avenged. So people drank his bone ashes mixed with banana soup during a *reahu* feast and went to war. Their shamans' thought was always set on the *xapiri*. Once they became old, they passed what they knew to the youngest by having them drink the *yãkoana* powder. By doing so, they gave them their breath of life and their words of truth. They told them: "These are the spirits *Omama* created to stand by our side. They are powerful and immortal beings!" These were the things that were on our people's minds in the past. Their thought was not yet obscured by the white people's merchandise and their epidemics!

OUR ANCESTORS loved their own words. They were truly happy this way. Their mind was not set elsewhere. The white people's words had not made their way among them. They worked with uprightness and spoke of what they did. They grew their own thoughts, turned to their people. They did not constantly tell each other: "An airplane will land tomorrow! White visitors will arrive! I will go ask for machetes and clothing!" and also: "The gold prospectors are drawing near! Their malaria is dangerous, it will kill us!" Today all these speeches about the white people stand in the way of our own thought. The forest has lost its silence. Far too many words come to us from the cities. Several of us went there when they were sick or to defend our forest.[11] White people often visit our houses. Their words sneak into our thought and darken it. They constantly worry us, even when these outsiders are far away from us.

Our minds get entangled with words about the gold prospectors who eat the forest's floor and foul our rivers, with words about the **settlers** and the **cattle ranchers** who burn its trees to feed their animals, with words about the **government** that wants to open new roads here and tear **minerals** out of the ground.[12] We fear malaria, flu, and **tuberculosis**. Our mind is constantly attracted by white people's merchandise. We are too often thinking about obtaining machetes, axes, fishhooks, pots, hammocks, clothes, guns, and ammunition. Young people play **soccer** all over the house's central plaza while the shamans are working there. They no longer tie their foreskin with a cotton string around their waist like the elders did. They wear **shorts,** want to listen to the radio, and think they can turn into white people. They struggle to babble the white people's ghost talk while sometimes dreaming of leaving the forest. Yet they know nothing of what the white people truly are, their thought is not yet opened. Try as they might to imitate them, it will not lead to anything good. If they continue on this dark path, they will wind up drinking *cachaça*[13] and become as ignorant as the white people can be.

Our elders of the past did not ever think about these white people things. Today our eyes and ears are too often set far from the forest, elsewhere than on our people. The words about the white people stand in the way of our own words and tangle them up with smoke. This makes us worried. Then we try to slow down and quiet our thought. We tell ourselves that the shamans will avenge us of the white people's diseases and that we will not all die. We think that our *reahu* feasts will continue no matter what. But we also know that all the white people's words could only disappear from our mind if they stopped invading and destroying our land. Then everything would be quiet like it used to be and we would live alone in the forest again. Our minds would become as untroubled as our ancestors' in the beginning of time. But this will probably never happen!

LONG BEFORE they met the white people in the forest, our shaman elders already knew how to make their ancestors' images dance. These images came from a very distant land, downstream of all the rivers, where *Omama* had brought the first outsiders into being. The long-ago shamans named these outsider's images *napënapëri*.[14] They had known them since a time when the current white people's grandfathers were not born yet and when their faraway land was nothing but forest without paths. These images come down to us as *xapiri* from the eastern sky. They come from where the sky's feet are set on the terrestrial layer, from a land to which *Omama*'s image escaped after his death. Yet this distance is nothing to those *xapiri* whose flight is so quick. The paths they follow are threads of bright light like the lines of the *t^h oru wakë* fires that cross the sky's chest during the night. When other spirits hail them on their course, they listen to their songs with pleasure and prove impatient to follow them. This is how they reach us. At first there are not many of them, but on their way they build into a large troop.

But the outsiders whose images were first called down by our ancient shamans were called *Watata si*. These were outsiders but not white people, whom our long-ago elders had barely heard of then, and whom they called *napë kraiwa pë*.[15] The *Watata si* lived on an arm of the Rio Parima's middle course.[16] The men of this group had their hair cut like ours. They wore red loincloths around their waists and tight rows of glass beads around their wrists. They also had ear pendants made of mirror fragments and toucan tails. Their women hid their pubes with long aprons of colored beads. They often drank fermented manioc juice. These were the people who gave our forefathers the pieces of worn metal, red cloth, glass beads, and manioc graters they used so long ago. But they also went to get such goods from the *Mait^h a*, other people who were closer to them.[17] In exchange, they brought them big balls of cotton. It was so. Our elders of the past made very long trips to trade, which they often returned from with the coughing sickness. Today all these forest people they used to visit no longer exist.

We have not called down the spirits of these ancient outsiders in a long time. Instead we now call the *napënapëri* spirits of the city white people's ancestors. We know them well and we often call them and make their images dance. They own airplanes and they are fierce warriors. They look like white people, but compared to them they are very beautiful. They are not human beings. These *napënapëri* spirits are very tall. They are also very different from the spirits of the forest and of the animal ancestors. They are wrapped in **uniforms**, like very long white shirts. Their eyes are hidden by shining metal skins. These are **eyeglass**-like

mirrors that allow them to see the evil beings coming from a great distance. Their heads are also covered with burning iron hats that scare
away the epidemic fumes. They have beards as bushy as saki tails and
black hair like the hair of *Omama*, who sends them to us. They carry
heavy metallic blades to pierce their enemies. These iron swords are very
long and made of a very hard metal. They hang all around their arms and
from their belts. When one of them breaks, they instantly replace it with
a brand new one, and when they are attacked, their enemies' blows rico-
chet off these heavy pieces of sparkling steel.[18]

THESE WHITE people's spirits are the images of the *Hayowari $t^h\ddot{e}ri$*, a
group of Yanomami ancestors carried away by a flood and turned into
outsiders by *Omama*.[19] They came into existence in the beginning of
time, on the land where their fathers had been created before them. They
are the ghosts of the first white people. They are white ancestors turned
other who dance for us now as *xapiri* spirits. They are the true holders of
Omama's metal. It was they who taught today's white people to build airplanes, machines to capture songs, and image skins. They are powerful
and able to clean the entire forest by driving the epidemic fumes away.
Only they as *xapiri* truly know the *xawara* epidemic smoke, for it also
comes from the white people. This is why they are so good at making it
lose its grip on its victims' images. Other *xapiri* prove weak and clumsy
when faced with its power. They no longer know how to heal sick people.
It is so. Our shaman elders of the past possessed words about the white
people since very long ago. They knew their distant land and heard their
twisted tongue long before they met them. They were familiar with the
image of their ancestors who boiled metal to make tools and they often
called these images as *xapiri* when they studied by drinking the *yãkoana*.
Then their sons and grandsons made them dance in their turn. And we,
we now follow in their footsteps. But do not think that the *napënapëri* images we make dance are those of the white people who are nearby around
us. Those white people of the present only want our death. They want to
take our place in the forest, and they are our enemies. We do not want to
see their images!

There are a great number of *napënapëri* spirits living on the white people's lands. These spirits tirelessly protect them from the epidemics that
spread there. This is why the white people do not die of the epidemic
fumes as much as we do! The *napënapëri* spirits make their doctors wise.

Our shamans value these spirits' bravery highly and many are those who want to call them down to dance for them. Yet it is not easy. As I have said, *Omama* does not always prove generous with his *xapiri*. He often keeps the most powerful ones close to him and only sends us the weakest. The white people healers in the cities, whom they call *rezadores*,[20] also know how to call down the *napēnapēri*'s images. But they are miserly with them too. Drinking the *yākoana* is not sufficient to make these white people ancestors' spirits come to us on their own initiative. My wife's father, who is an elder, never saw them in his youth, back when he still lived in the highlands. They only came down to him much later, when he was nearly killed by malaria fevers. It was they who cured him and since then he can call them down and make them dance as often as he wants.

The same thing happened to me when I was younger. The epidemic beings, whom we call *xawarari,* had severely struck and wounded me. I was very sick and I truly believed I would die. I slept in a ghost state. My breathing was very short and my chest only made weak rattling sounds. That is when I first saw the spirits of the white people's ancestors come down to me. Suddenly they rushed in to battle to fight the *xawarari* beings who threatened to devour me. They pierced them with their iron blades and cut off their arms and poked out their eyes! This is how I was finally able to escape death. Since then, I have continued to call down these white people's ancestors' *xapiri* who so valiantly avenged me. Their dance mirror is in place in my spirit house and I often answer their songs when I am drinking the *yākoana*. Sometimes they also come to visit me of their own will, during the time of dream. Then I make them dance in silence. At night, I do not sing out loud, for I fear the people of my house would complain: "Shut up! You're frightening our sleep! We want to sleep! You're bothering us with these songs!" These white people ancestors' spirits speak to me in their ghost talk. Yet I understand them because little by little I have learned some of their language since I was a teenager. Then I can share their words with those who listen to me when I become spirit myself. But our elders of long ago, they knew nothing of the white people. When their shamans made these outsider spirits dance, they simply imitated their twisted talk without understanding it!

It was *Omama* who created us, but it was also him who brought the white people into existence. There is only one and the same sky above us. There is only one sun, one moon. We live on the same earth. The white people were not created by their governments. They come from *Oma-ma*'s **factory!** They are his sons and sons-in-law as much as we are. He created them a long time ago from the foam of the blood of our ancestors, the inhabitants of *Hayowari*. *Hayowari* is the name of a hill located between the headwaters of the Rio Parima and the upper Orinoco, which we call *Hʷara u*. This is where the rivers' origins are, where *Omama* pierced the ground in his garden to quench his son's thirst.[21] When I was a child, my stepfather often told me about these people of the distant past. Now that I have also become a shaman, I too sometimes see their images and hear their words. This is why I can talk about it. *Omama* created the Yanomami after fishing the daughter of *Tëpërësiki*, the being of the deep waters. He copulated with her and it is from her womb that we became many. The people of *Hayowari* were among the first inhabitants of the forest in the beginning of time. They were the children of *Omama* and his wife *Tʰuëyoma*. They became outsiders much later, after *Omama* made the waters spring from the ground and fled far away, downstream of all the rivers, in the direction of the white people's land.[22]

These *Hayowari tʰëri* ancestors became other during a *reahu* feast to which they had invited their allies to bury one of their people's bone ashes. This is how it happened. It was the last day, just before the many guests returned to their homes. The man in charge of distributing the smoked game to celebrate the deceased placed a small mound of *yãkoana* powder on a clay plate in the middle of the house.[23] A group of guests and hosts gradually gathered around it while conversing and started snorting large pinches of the powder. It was strong and everyone sniffed with loud exclamations of approval. Then after some time the men joined each other in pairs, squatting face to face, to begin a *yãimuu* dialogue. Seized by the *yãkoana*'s power, they were soon all very excited.[24] They punctuated their words by slapping their sides with the flat of their hands. After a while, their anger increased to the point that they began to take turns pounding each other's chests. Suddenly a group of guests violently attacked one of their hosts, who had remained isolated. From the other side of the house, his mother, an old woman, started furiously insulting them to defend her son. Then she cried out for her daughter's husband to come help his brother-in-law. The young man was still in seclusion in an enclosure of *yipi hi* leaves with his young wife, who had just had her first menstruation.[25] Hearing his mother-in-law's call, he dashed out to avenge his brother-in-law, without thinking of the danger.

The forest was still young and fragile at that time. The young man had barely crossed the threshold of his seclusion enclosure before the *Xiwãripo* chaos being started to soften and break up the ground around him. Then *Motu uri u,* the river of the underworld, violently gushed out, tearing the earth apart. Its foaming torrent quickly covered the neighboring forest and soon destroyed the *Hayowari* people's house. It was terrifying! They were all carried away by the flood while they were still squatting, singing, and pounding their chests. Their clamor could still be heard disappearing in the distance as they were quickly dragged downstream. Some tried to escape into the forest: they turned into deer. Others tried to climb the trees: they became termite nests. But most drowned and were devoured by giant *kana* otters and huge black *poapoa* caimans. This is why to this day shamans need to work to stop *Motu uri u*'s waters from gushing up from beneath the ground. You can still see the big hole from which they sprang in that time in the highlands, though it has since been covered over by the forest. You can see it from an airplane at the headwaters of the Orinoco, Catrimani, and Parima rivers. We also call it *Xiwãripo.*

The waters that came out of the ground at *Hayowari* then made a large bend by coming down the hills and spreading very far into the forest in the direction of the rising sun.[26] Once they reached the place where the land becomes flat and windy, they started rapidly whirling around. Then after some time they gradually lost speed and their motion calmed down. They have remained like this ever since, unmoving, forming a lake as vast as the sky. This is what the white people have named **ocean.** A *Yariporari* storm wind lives in the center of this watery vastness whose depths are inhabited by giant electric eels and *tëpërësiri* vortex beings who swallow humans.[27] It is also the hiding place of enormous sharp-toothed epidemic-fish whose tails throw lightning and giant angry tadpole beings who destroy white people's boats.[28]

All that remained of the Yanomami who drowned in *Motu uri u*'s torrents was a bloody foam drifting on the waters where the rivers become very wide. It slipped slowly to the downstream, towards the place where *Omama* had settled after his escape from our forest highlands. As soon as he saw it, he entered the big river and little by little gathered the reddish foam floating towards him in a small basket. Then he carefully put it down on its bank and began shaping it with his hands. It warmed up and new human beings finally appeared on the land along the edge of the river. At first the foam that passed over the water was barely tinted. *Omama* gathered small mounds of it, which he brought back to life by placing them in a distant land, on the other side of the waters. This is the land of the white people's ancestors that you call **Europe.** He began by creating the people who our ancestors called *napë kraiwa pë* and whose skin was as white as their paper. With the increasingly dark reddish foam carried by the water, he created other outsiders. This time they were people who looked like us. He settled them near us, in the forest. So he brought this foam of our dead ancestors back to where it came from and kept their image on the land of Brazil, which we consider *Omama's* land. These were the people our elders called *napë pë yai*, the "true outsiders," the other **Indians:** the Pauxiana, the *Watata si*, and the people of the Rio Demini who were once close to us,[29] but also the Ye'kuana, the Makuxi, the Tukano, the Wayãpi, the Kayapo, and many others.[30]

REMORI, THE ancestor of the big orange *remoremo moxi* bee, gave white people their twisted tongue. Doesn't their talk resemble the buzzing of the bumblebees? He put a different throat than ours inside them. *Remori* lived by *Omama*, far away downstream of the rivers, where they become very wide and are bordered by vast stretches of sand.[31] Wanting to bring the foam of the people of *Hayowari* back to life, *Omama* urged *Remori* to teach the outsiders he had just created another language. This is why our forbears did not understand anything that the first white people they met said to them. It was truly frightening for them to hear the white people's inarticulate talk! When these strangers spoke to them, they tried to prick up their ears but they stayed confused. They thought: "What could they possibly mean? Is that really all they are able to say? What an appalling way of speaking! Isn't this language a ghost talk? No, it must be another language, the one that *Remori* gave to the outsiders!"

No matter how they tried to imitate this talk, it never led our long-ago elders to anything understandable! They only managed to utter words as ugly as they were twisted! Our words as inhabitants of the forest are very different! They are the ones *Omama* taught us and the white people cannot understand us either. It is so. *Omama* and *Remori* decided that the different people they had created should not possess the same language. They thought that using a single language would provoke endless conflicts between them because one group's evil words could be heard by all the others without impediment. This is why they attributed other modes of talking to the outsiders and then separated them on different lands. While making these languages open inside them, they warned them: "You will not hear the words of others. You will only understand your own and that way you will only quarrel among yourselves. The same will be true of them."

Omama, Remori, and the inhabitants of *Hayowari* disappeared from the highlands of our forest long, long ago. Yet it is only so in the eyes of ordinary people. The shamans, they know that their ghosts are still present there. They constantly make their images dance and their songs heard. When I was younger and I listened to the elders when they became spirits, I used to ask myself: "How do they do it? Where do these words from the beginning of time really come from?" Later, when I began to drink the *yãkoana* myself, they called these images down for me too. Then it was my turn to see the people of *Hayowari* become other and to be carried along by the waters of *Motu uri u* as far as the vast stretches of sand where *Remori* lives downstream. Since then, I often continue to contemplate them in the dream time of my ghost sleep.[32]

CHAPTER 10

First Contacts

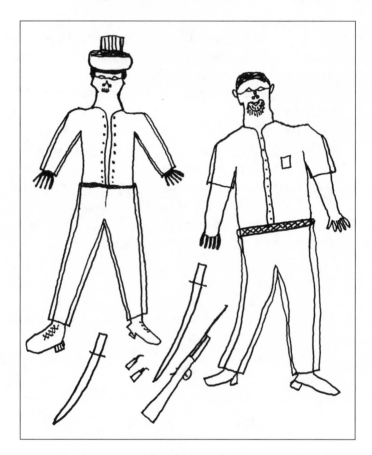

The white people

M Y FATHER died when I was still a very young child. The elders told me that he was killed by enemy *oka* sorcerers. He had been working alone in his garden for a long time when he began to get hungry. He went off into the forest to collect *yoi si* palm fruits. The *oka* took advantage of this to blow a sorcery powder on him with their blowpipes. He started to feel sick and to lose consciousness. Then they took hold of him and shattered his limbs, his neck, and his back. I was told that their group was led by the great man of the *Hero u* River's people, with his allies from the upper Rio Mucajaí, the people called *Amikoapë t*^{h}*ëri*. At the time, there were many of them still and they were our enemies. I learned all this not so long ago, from my wife's father. No one had said anything to me about it before. If I had known about this when I was young, I might have avenged my father by killing this *Hero u* leader still in *õnokae* homicide state.[1] But today a long time has passed, and I am no longer angry. Anyhow this man already died of malaria when the gold prospectors arrived in our forest.

My mother was still carrying me on her breast when my father died. I have no memory of him. I do not know his name. No one revealed it to me, not even my mother. My older sister also never spoke to me of our father. Her mouth was probably afraid. Only the elders who knew him still remember his name. They may occasionally mention it among themselves. I think that my wife's father knows it. But they must all fear my reaction and tell themselves: "If we reveal his father's name to him, Davi will get angry!" And so my thought has remained closed. Among us, when someone dies, we stop speaking his name for all time. If a person inadvertently spoke the deceased's name before his kin, they would be overcome with pain and nostalgia to the point of getting furious. Then they would seek to get revenge, through sorcery or with their arrows. Only distant people can still sometimes bring up a dead person's name, but only when the inhabitants of his house are not present. Otherwise we do not say anything. This is why when a small child's father dies, none of the people who knew him will tell the child his father's name. He will never hear it.

I occasionally call up this time of my childhood to answer white people's questions. I do it without anger, for their thought is ignorant of all these things about our names. They are not afraid to blurt out their names and those of their dead all the time! This is not true with us. Pronouncing a man's name to his face will immediately send him into a rage, and after his death, saying his name will be zealously forbidden among his relatives.[2] We are so. We refuse to reveal the names of the dead, for we give them high value. We have great **respect** for them. We think that white people like mistreating their own deceased. They shut them up underground and insult them by mentioning their names at any opportunity. I wonder how they can shed tears over them after having done that kind of thing! We Yanomami mourn for our dead all together and for a very long time, but we never say their names.

After my father's death, another man took my mother as a wife. I was still a baby and he took me with her. This man protected me and raised me. He fed me with the game he hunted and the wild honey he collected and the bananas and manioc he cultivated to make me grow. Today he is very old and lives far away, in another house than mine. I do not see him often, but I always carry him in my thought with affection. Every so often I go to visit him and bring him trade goods. I also send white **medics** to treat him and in this way I protect him the way he once protected me.[3] He was a great shaman and he liked to give us his words. When I was a child he often spoke to me of the ancestors who turned into game in the beginning of time. He also told me how *Omama* came into being and made of his son the first shaman, then how he created the outsiders. He told me all this at great length, in the night, when I was stretched out in my hammock watching the fire my mother blew on from time to time. He did not want me to grow up in ignorance. To this day, when I see the spirits dancing in my dream, I remember his words. They remain so vivid in my mind. He always joked and smiled, but he was also a fierce warrior. He had inside him the images of *Aiamori*, the war spirit, and of *Õeõeri*, the ancestor who taught us how to arrow our enemies! It was he who avenged my father's death, for he was his friend. My father was younger than him, and they called each other brother-in-law. They often hunted together. At that time, our elders did not hesitate to kill the enemies who had eaten their kin. They were very valiant. They did not avenge them furtively, by blowing sorcery powders on them from a distance. They preferred to go to war and use their arrows.

As a child, I lived in a place that was called *Marakana* on the Rio Toototobi. That is where my father died. When I was born there, the area had only recently been cleared. The elders had opened new gardens but were still living in the forest, under temporary shelters.[4] My stepfather told me about it. I do not remember it myself. When you are so little, you are not really conscious of these things. Adults speak to children, but their minds remain closed. Words do not really settle on it. It is only later, as they grow up, that their thoughts truly start to unite with each other and that their minds begin to dawn. I still have a few memories of the time of the *Marakana* big house. Yet I do not remember seeing the elders plant its posts or cover it in *paa hana* palm leaves. I only remember this home once it was built. It was vast and our elders were numerous in it. At the beginning, two communities were gathered there, for we were at war with the people of the upper Rio Mapulaú and the Rio Catrimani, who lived only a few days away.[5]

Later we separated again because the elders often quarreled. After *Marakana*, there were three houses quite close to one another.[6] Ours was situated upstream, in a place called *Wari mahi*, where a huge silk-cotton tree stood. The others lived a little farther downstream, very close to the banks of the Rio Toototobi. But soon my stepfather began taking his distances from the people of *Wari mahi*. He only lived with them from time to time. He had built a small house by himself and cleared a garden a half-day's walk downstream. I lived there with my mother, my older sister, and another **family**. This place was called *T*ʰ*oot*ʰ*ot*ʰ*opi*, the place of lianas.[7] We spent most of our time here and rarely visited *Wari mahi*. My stepfather did not like living there because he thought there were too many people. I think he found the house too noisy. This is why after *Marakana* I mostly grew up in *T*ʰ*oot*ʰ*ot*ʰ*opi*.[8]

I remember this period of my childhood well. This is when my mind opened, thanks to the game and the food my stepfather gave me. He took me with him on all his travels. We often went to *reahu* feasts in friendly houses. We also left for long expeditions during which we lived in forest camps. The elders bivouacked for long stretches like this far from their houses to hunt and gather wild fruit.[9] In that time, we really spent a lot of time in the forest. It is less so today. Young people are always roaming around the white people's outposts. But I really grew up in the forest, constantly drinking wild honey. This is what made my thinking right and allowed it to grow. When I was very small, I began to observe the elders who went to hunt and work in their gardens. During this time, I also first saw them dance to present themselves at their allies' *reahu* feasts and imitate the vulture spirits to go to war against their enemies.[10]

My mother often took me with her when she went into the forest to find freshwater crabs, to poison fish in small streams or ponds, and to collect fruits of all kinds. I also followed her to our garden when she went there to pick manioc and bananas and chop dead tree trunks for firewood with her axe. As soon as I got a little older, the adults started to call me at dawn to accompany them to hunt. I followed them through the dew-soaked forest and when they arrowed small animals, they gave them to me, telling me: "Carry this game, when you get back you will eat it!" At the time, we were a small group of boys, all the same age. The others were a little older than us. We grew up constantly hunting and fishing together. We also spent our time imitating everything the adults did. This is how little by little we began to think right. We arrowed all sorts of lit-tle birds and lizards in the forest or neighboring gardens. We brought them back, proudly entering our big house like hunters. Then we smoked them and organized small *reahu* feasts with this "game," as we had seen our elders do.[11] They jokingly encouraged us. They added real pieces of game to our catch. Then we joyfully launched into *heri* songs, as people do when the food at a *reahu* is abundant. We also imitated the guests' pre-sentation dances. We even danced in couples, girls and boys, holding each other by the wrist, like we sometimes saw the adults do on feast nights. We really had a lot of fun!

All this took place in the house's central plaza. The adults laughed with pleasure as they watched us. They were amused to see us so boldly parody them! We were truly playful! We pretended to drink the *yãkoana* powder, like the men all do on the last day of the *reahu*. We also mim-icked their anger during the *yãimuu* dialogues. We manhandled each other, squatting two by two and tightly gripping each other's necks. Like them, we sang by yelling into each other's ears and we slapped our sides with the flats of our hands. The only adults we did not dare to imitate were the shamans. The elders had warned us. It is very dangerous to do so because their spirits could be irritated and take revenge. This is how we lived. Our only example was the ways of our elders. We did not follow the white people's ways, as today's children often do when they build little wood airplanes or play with **balls.** We did not listen to the noise of radios or tape recorders. Our ears only paid attention to the words of our people and the voices of the forest.

Our elders invited each other from one house to the next for *reahu* feasts, to drink banana soup and offer each other bundles of smoked game and piles of cassava bread. Often they also had fights. They chal-lenged each other by yelling elatedly and they insulted their adversaries by angrily speaking their names. Then they took turns striking each oth-er's heads with long clubs. They confronted each other this way because food was stolen from their garden, because they were jealous of their wives, or simply because they had called each other cowards. I would watch them from a distance, a little frightened, and tell myself: "*Haixopë!* This is how you must fight to appease your anger!" Sometimes they also launched incursions against their enemies. At the time, they were at war in the direction of the rising sun with the elders of the people of the Rio Catrimani—who lived on the Rio Mapulaú at the time—and in the direction of the setting sun, with the *Xamat^hari* of the upper Rio Demini.[12] As I have said, my stepfather was very valiant and prompt to avenge the dead of his house. In that time, he arrowed a good many of our enemies from the Rio Catrimani and took away the two sisters who are still his wives from the *Xamat^hari*.[13] I was by his side when he launched these raids with other warriors from the Rio Toototobi. I often saw them line up with their bows and arrows on our house's central plaza and imitate the vulture spirits before they set off.[14] I intensively focused my attention on them, and I told myself: "This is how we must seek revenge! When I am older, I will join them!" I was so sorry that I was much too young to go with them! But it was by constantly observing adults this way that my thinking became wiser and that I grew.

BEFORE THEY came to *Marakana*, our elders lived on many other gardens in the highlands. For a long time they lived in the place of the sitting *yoyo* toad, *Yoyo roopë*, at the source of the Rio Toototobi.[15] My stepfather often spoke of this forest because he lived there for a long time as a child. From there, his elders traveled to the *Xamat^hari* who lived on the *Kapirota u* River to get metal tools because the *Watata si* of the Rio Parima were now too far away.[16] The *Xamat^hari* obtained these tools by going down the Rio Demini to the camps of the white people living on the banks of the Rio Aracá. Here, these outsiders fished for freshwater turtles and gathered Brazil nuts, balata, and fibers from the piassava palm.[17] Though they lived far upstream, the people of the *Kapirota u* River knew these white people well. They often visited them and worked for them for several moons during the dry season. This way they acquired all sorts of manufactured objects from them.[18] It was through them that our elders first came to know these white people of the river, very far from their houses. Yet they did not approach these outsiders for the simple pleasure of admiring them. In fact, they were afraid of them, and with good reason. Once, white people they met in the forest had offered them **poisoned**

food and several of them died. This happened near the Rio Aracá waterfall, which the white people call Cachoeira dos Índios. I heard this story
from my stepfather when I was a child. He told it sometimes during the
night when he exhorted the people of our house with his *hereamuu*
speeches about the time long ago.

If our elders traveled so far, it was not only to go after matches, aluminum pots, and salt. They knew how to make fire with cacao tree drills,
their wives cooked in clay pots, and they salted their boiled bananas with
yopo una vine ashes. What they really wanted from the white people of
the river were their brand new metal tools, for they did not possess any.
At the time, it was very difficult to obtain such tools. So they returned
from their long journeys downriver with a few machetes and sometimes
an axe head, but it was always with great difficulty. Then they could clear
new, vaster gardens, and grow the plants that would feed their relatives.
Yet they still needed to take turns lending each other these few tools, as
they once did with the worn pieces of iron they received from the *Watata
si* on the Rio Parima. This way, when a man had finished clearing his
plot, another could take a turn working, and then another and another.
Eventually these iron tools were also borrowed by the people of neighboring houses, as in the past. The elders often told me this when I was
a child.

FOR MY PART, I met the white people for the first time when I was still
very little. I really knew nothing about them. In fact, I did not even think
such beings could exist! These were people from the Inspetoria and **soldiers** from the **Boundary Commission**.[19] One day they arrived close to
our house in *Marakana*. They had traveled upriver towards us for days
and days, packed into big canoes piled with food and large wooden chests
full of merchandise. There were very many of them. Then a group of
these outsiders entered our home to ask our elders to help them. They
wanted to recruit men to accompany them and carry their heavy loads in
the forest. They wanted to reach the sources of the rivers in order to dig
holes in which to plant big straight rocks. Our elders did not understand
their ghost talk at all. So it was a *Xamat^hari* who had taken a wife among
us who spoke with them. He had already met white people while working
on the Rio Aracá, near the Cachoeira dos Índios waterfall, and knew their
language a little. These white people from the Boundary Commission

worked for several moons in the forest highlands before leaving as suddenly as they had arrived.[20]

I do not remember everything that happened at this time, because it was so long ago. Yet I did not forget these outsiders' arrival, for they truly terrified me! As soon as their approach was announced, all the mothers of *Marakana* warned their little children: "The outsiders are getting close! Hide! Otherwise they might take you with them!" Then they immediately hid them behind their hammocks, sheltered by the firewood logs resting against the house's wall.[21] The older ones like my older sister ran off by themselves to hide in the neighboring forest. My mother, she made me crouch by her side, then she covered me with her big carrying basket. I was terrified but she managed to calm me down by speaking to me in a low voice: "Don't be afraid, the strangers won't see you! Keep quiet!" Once I was out of view, I was a little reassured. Then I stayed crouching, silent, observing the group of white visitors entering our house. I found them frighteningly ugly and my heart was beating hard in my chest. I would have liked to run away, like the older children, but I did not want to attract their attention. So I had to wait a long time, not moving at all, holding my breath, before they left and my mother delivered me.

The mothers of our house did not want the white people to see their children. They were truly afraid they would take them away. The elders remembered that the Boundary Commission's soldiers had already asked for Yanomami children in the past when they had traveled up the Rio Mapulaú for the first time.[22] At that time, our people lived in the highlands, at *Yoyo roopë*. Yet the people of the Mapulaú told them that the white people had insisted on taking away several of their children. Of course no one wanted to give them their own children! Yet the elders of that time feared the white people's anger and their epidemics. So finally the great man of the Mapulaú gave the soldiers a little boy and a little girl who were not the children of people of his house. These were captives brought back from a raid on the *Yawari*, who lived on the upper Rio Catrimani at the time.[23] I had often heard my fathers and my grandfathers tell this story. This is why I was so scared of the white people! I was afraid they would want to take me away too! To this day, I wonder what these outsiders wanted to do with these Yanomami children. Maybe they intended to raise them to send them back later to ask our elders' permission to work in their forest? I do not know.

Today, our children no longer fear the white people. But in the past they really scared me a lot! They truly seemed to be other. I watched them from afar and told myself that they must be *në wari* evil beings! The mere sight of them horrified me. They were ugly and hairy. Some were alarmingly white. I wondered what their **shoes**, their **watches**, and their glasses could possibly be. I listened hard to try and understand their words, but it was useless. They sounded like inarticulate sounds! They also feverishly handled all sorts of objects, which were as strange and frightening in appearance as they themselves. Even long after this first visit, I ran off crying whenever one of these white people made to approach me. Truly, they terrorized me! I feared the light that emanated from their **flashlights**. I was even more afraid of the rumble of their motors, the voices of their radios, as well as the explosions of their guns. The smell of **gas** disgusted me. The smoke from their cigarettes made me fear I would get sick. I really thought they must be evil beings starved for human flesh!

In *Marakana,* the adults were not as afraid of these outsiders as we children were. They already knew them. Many of them had already met white people during their trading journeys downstream. Yet what truly terrified everyone was their airplanes, which flew over our houses several times. No one had ever seen any before.[24] The houses emptied out as soon as their roar could be heard! Men, women, and children ran off as fast as they could and scattered in the woods. The elders thought that these unknown flying beings could fall into the forest and burn everything as they crashed. They thought that we were all going to die and sometimes they were so frightened that they even cried as they talked about it! This is how it happened. Our fathers and our grandfathers did not trust the white people and had always feared their epidemic fumes. Yet they never really tried to find out what had brought them to their home. They did not know that they had come to mark the edges of Brazil in the middle of our land. They made themselves available and friendly. They readily gathered to accompany the white soldiers and transport their food and metal tools in big carrying baskets. They merely looked on with curiosity when the white people cleared large paths and erected big stones at the sources of the rivers. They never imagined that later these people's children and grandchildren would come back in large numbers to dig gold from the rivers and make their cattle eat in the forest. They never thought that these outsiders would one day want to chase them from their homes to take their land! On the contrary, once their initial fright passed, our elders were happy about their visit. Day after day, they examined the big wooden boxes full of machetes and axe heads that these white people had brought all the way up the Rio Demini.[25] A single thought was on their minds: "From now on, we will never lack for metal tools again!"

Much later, once I had become an adult, I began to ask myself what these white people had come to do in our forest. I came to understand that they wanted to know it and plot its limits in order to take possession of it. Our elders did not know how to imitate these outsiders' language. This is why they let them approach them without hostility. If they had understood their words as well as they understood ours, they would probably have prevented them from coming into our forest so easily! I also think these strangers duped them by flourishing their merchandise with good words: "Let's be friends! See, we offer you so many of our goods as presents! We do not lie!" This is always how the white people start talking to us! Then the *xawarari* epidemic beings arrive in their footsteps and we immediately start dying one after another! Our elders did not know anything of all this yet. They simply wanted to trade for machetes, axe heads, clothes, rice, salt, and sugar. They spoke to the white people by joyfully repeating a few of their words like parrots. They told themselves: "These outsiders are truly friendly, they are very generous!" But they were wrong! Once they obtained the precious things and food they coveted, they soon fell ill, then perished one by one. It makes me sad to think about it. Our elders were taken in by all this merchandise, and that killed every one of them. This is how my older relatives disappeared, wanting to make friendship with the white people. And after their death, I remained alone with my anger. It has never left me since. It is the anger that makes me **fight** today against those outsiders who think only of burning the forest's trees and soiling its rivers like hordes of peccaries! I always feel sad when I see the emptiness of the forest that my elders traveled, for the *xawara* epidemics never left it. Since that first time, our people have continued to die in the same way.

IN THE BEGINNING, our elders carefully washed the machetes that the white people gave them before bringing them home. They dipped them into the water of the streams, then rubbed them with sand for a long time. Their blades were sticky and gave off an alarming sickly sweet smell. They were covered in grease and wrapped in paper skins.[26] As soon as the white people opened their enormous wood boxes to give out their trade goods, curls of fine sweet-smelling dust came out. This penetrating odor spread everywhere. All the things they distributed smelled strongly of it: the metal tools as well as the cotton cloth and hammocks. Our fathers and grandfathers did not have a white person's nose. They recognized the metal tools' sickening smell from a distance. This scent seemed dangerous to them and they were afraid of it because it made them cough and immediately made them very sick.[27] Old men, women, and children died of it very quickly. This is why they called it *poo pë wakëxi*, the metal tools smoke. They thought this was the origin of the *xawara* epidemics that were devouring them.[28] At that time, our people did not know much about the white people. They did not know their odor or those of their goods. That is why these odors seemed so intense and frightening. It was like when a young hunter is overcome by the odor of a herd of wild pigs in the forest for the first time! They had never smelled anything like it before and it scared them!

At that time, the white people also gave out a large amount of pieces of red fabric. Men used them to make themselves loincloths. But this red cloth was also dangerous. Shortly after they acquired them, the elders would start coughing and their eyes would get infected.[29] This is why they called them *t^hoko kiki*, the cough things. These are evil trade goods woven far away, in the white people ancestors' land, with the cotton of epidemic trees, the *xawara hi*.[30] The shamans who fight their disease see their image appear in the form of strips of scarlet fabric. Today we often wear shorts and sometimes other clothing.[31] Yet we are still wary of these pieces of red cotton.[32] Their illness struck our elders too often. When the white people tore this dangerous cloth, a disgusting smoke rose from it and made everyone sick. Our fathers' chests were too weak to resist it and they quickly died of its cough. This smoke came from the **storehouses** where the white people piled these pieces of fabric; it was the smell of the fumes from the machines that had woven them.

Our elders were equally frightened of the fumes from the debris and trash the white people used to throw in the fire. When they saw them burn pieces of **newspapers,** they told themselves: "The fumes of these image skins with red and black drawings is dangerous! It will slit our throat and bruise our chest. Its cough will make us perish!" They were also frightened by the smoke from the burned tobacco these outsiders constantly swallowed.[33] In fact, all of the white people's objects struck our elders with their power of disease: the machetes, fabrics, papers, cigarettes, soaps, and the plastic things. Their malodorous fumes spread freely among these strangers, and those of our elders who came too close and breathed it in immediately started to cough and vomit.[34] They did not have any of the white people medicine, so it killed them very quickly. Even the song tree things that these outsiders called **harmonicas** made people sick! When they distributed them, all the young people had fun heartily blowing into them as if they were *purunama usi* bamboo flutes. But soon these careless youngsters got sore throats and the cough spirits immediately lacerated their chests.[35] It was so. The white people's objects were very dangerous for our elders. They did not know them and had never seen anything like them. They were born very far from the cities and the **factories**, in the middle of the forest. The insides of their bodies were very vulnerable to the fumes from all this merchandise.

LATER, WE received another visit at *Marakana* from white people of the Inspetoria. They brought our elders several shotguns as presents. They gave a brand new one to my stepfather, who was our house's great man. The elders treated them as friends and so they stayed among us as guests for some time. Then their leader, who was called Oswaldo, started wanting to take one of our young girls with him. He lusted after one of the daughters of the people of *Sina t^ha*, whose house was a little downstream of ours.[36] I called her sister. She had just had her first menstruation. Oswaldo lived in a small hut that the people of *Sina t^ha* had built for him in front of their home. After some time, he began giving game and manioc flour to the girl's parents, as we usually do to obtain a wife. His mind was set on her beauty. He really wanted to eat her vulva. More and more, he insisted on possessing her. In the end, my stepfather would have accepted to give her to him, for he feared his anger, but the elders of *Sina t^ha* did not agree. The girl's parents and grandparents did not want such a thing. They knew that the white man would never stay to live with her in the forest. They feared that he would take her downstream, to the city, and that after some time he would abandon her there.[37] They knew that they would never see her again. And also a young man from their house wanted to take her for his wife.

At the beginning, Oswaldo was trying to show his friendship for everyone. His lips smiled at every occasion. Yet finally he got angry at being refused. He began to let himself express recriminations. Then one day he found the girl lying in her young suitor's hammock. His desire turned into fury. He gathered his belongings and left without saying a word. He went down the river with his anger sticking in his chest. The elders heard no more about it. Yet much later he returned among the people of *Sina tʰa*. Once again, he asked the parents for their young girl. This time he was no longer smiling. His face was tense and hostile. Having been refused again, he began angrily threatening the elders: "I want this woman right away! If you do not give her, I will make you all die!" Our elders were valorous and did not let themselves be impressed by this rage born of his desire to copulate.[38] They had no intention of letting him take the young girl and they did not give in. Yet no one suspected that Oswaldo was speaking the truth and that he had decided to take revenge. The people of *Sina tʰa* did not pay enough attention to his threats!

Some of their elders told me that Oswaldo got increasingly furious and that during the night he buried a box of metal containing powerful epidemic fumes near the house. The next day, the heat of the sun was very strong and this box heated up under the earth for a long time. After a while, the **poison** blew its lid off and let out a thick smoke. But my stepfather told me it did not really happen this way. He told me that to get revenge, Oswaldo called the young girl's suitor to a place where he had hidden a package in the ground. There was a long, straight rope sticking out of it, which he immediately set fire to with dried leaves tied to a stick. As soon as the fire started to spread, Oswaldo ran for shelter. Soon after, the package exploded underground like a huge gun shot. Clods of earth were thrown in every direction and a thick cloud of smoke suddenly enveloped the *Sina tʰa* house.[39] Terrified by the explosion, its inhabitants wondered what would happen to them.

Some time later, Oswaldo ran off, vociferating in his ghost language. No one understood what he was saying! Yet soon after his departure the people of *Sina tʰa* started to die one after another. It all took place during a *reahu* feast. The women were still grating the manioc to prepare the cassava bread to be distributed to the guests with smoked game for their journey back. Abruptly, several elders fell severely ill and an old man died. His people wrapped his body in a bag of palm leaves and tied it to the trunk of a young tree in the forest.[40] They mourned for him, then hurried to finish preparing the *reahu* travel supplies, which they hastily distributed to their guests. Meanwhile, the children began burning with fever. Soon the illness reached all the house's inhabitants. Overcome by panic, everyone who still could escaped into the forest in every direction.

Oswaldo's smoke was not a mere coughing illness. His victims were struck with violent itching and their skin came apart in shreds. People did not stay sick for long, they died without delay, one after another.[41] In very little time there were dead bodies all over the *Sina tʰa* house, sprawled across the floor or doubled over in their hammocks. Many people had died just as suddenly in the gardens, the forest, and on the riverbanks. The *xawarari* epidemic beings voraciously devoured a great number of women, old men, and children, as well as several shamans. The young girl Oswaldo had so intensely desired did not escape them either. This is what I was told by the elders who survived the epidemic by running away. Some time later, they came back to their houses and found these putrefying corpses lying everywhere. They gathered and incinerated their dead relatives' bones and never stopped crying in all the time it took to fill the many funerary gourds with their bone ashes. But the smoke from the pyres of the epidemic dead was also dangerous and now several more of them died. It was frightening! The rare grief-stricken survivors were seized by a deep mourning anger. They wanted to take revenge on Oswaldo, who had taken flight in an *õnokae* homicide state.[42] Wanting to arrow him, they looked for him all the way to the white people's Inspetoria outpost downstream at Ajuricaba.[43] It was in vain. He had probably left for Manaus and, as far I know, he never came back to our forest.

This is how most of our elders were eaten by the *xawara* epidemic for the first time. Before Oswaldo's smoke, there were still many of them. Today barely a few remain.[44] Only the people of *Yoyo roopë* were able to escape this epidemic, under my stepfather's guidance. Oswaldo had taken a liking to him. He always brought him presents. He told me that when the people of *Sina tʰa* began getting sick, Oswaldo warned him as he was about to board a canoe and go back down the river: "Leave this place! Do not go near those people or you will also be contaminated! They will all die! I am very angry with them! Let them perish, do not go back to their home! Warn your people and escape into the forest, otherwise you will also disappear!" As soon as he heard these words, my stepfather exhorted the people of our house: "The *xawara* epidemic is close by! We must abandon everything and leave at dawn! We must not go mourn for the *Sina tʰa* dead or we will also die!" Yet the next day some hesitated to leave. To cut short their indecision, my stepfather immediately set fire to our house. He was a great man, he was truly very valiant and strong! This is how we came to leave the *Marakana* region in the greatest haste. We traveled from camp to camp, very far downstream, along the Rio Demini. We stayed hidden in the forest for several moons, until finally we came back to settle on our *T^hoot^hot^hopi* site, far from *Marakana*. If we had not fled, most of us would also have perished. In the end, only a few of us died, and these were the ones who went against my stepfather's advice during our flight and came back to get manioc in their old gardens by going through *Sina t^ha*.

What did Oswaldo use to make this epidemic explode? I have no idea, but the white people must know! Our ancestors knew nothing of these epidemic fumes' burning fevers. Their bodies were always as cool as the forest in which they had always lived without medicine or **vaccines.** Maybe Oswaldo set fire to the powder they use to make big rocks explode?[45] In any case, our elders only had to inhale this unknown smoke to die of it very quickly, like fish who still do not know the deadly power of *koa axihana* fishing poison leaves. This is how we came to know the power of the white people's *xawara* epidemic near *Marakana*. Then we understood how dangerous they were for us! It was long ago now. Yet all the survivors still remember the smoke Oswaldo spread to take revenge! They still talk about it with their grandchildren to this day. We never want to experience such suffering again! More than enough of us have already died of the *xawara* epidemics the white people propagated in the past. We, who are what remains of our fathers and grandfathers, want to be as numerous as they once were. We do not want to keep dying before our time! We only want to pass away once we have become old men with white heads, emaciated and blind, all doubled over! We only want the death being *Nomasiri* and the night being *Titiri* to make us disappear when the time has really come! Then we will be happy to die because we will have lived long enough, as our elders from the past did before they met the white people. In *Marakana*, our people were many and in good health until suddenly all were decimated—women, children, and old men. This is why their death still makes me angry. These words of mourning have existed in me since my childhood. I also draw the strength to speak harshly to the white people from them.

WHEN THEY saw these outsiders for the first time, our elders thought they were ghosts. They were really very scared and they told themselves:

"The ghosts of our dead have come back among us!"[46] Later, they understood that these could only be people coming from the *Hayowari* ancestors whom *Omama* had turned into outsiders. Then they told themselves that these inhabitants of distant lands had probably returned to the forest out of generosity, to bring trade goods to the Yanomami who lacked them.[47] Today no one thinks that anymore! We saw the white people spread their epidemic and kill us with their guns. We saw them destroy the forest and the rivers. We know that they can be greedy and evil and that their thought is often full of darkness. They have forgotten that *Omama* created them. They have lost their ancestors' words. They have forgotten what they were in the beginning of time, when they also had a **culture.**[48]

Omama put down the foam with which he created the ancient white people very far away from our forest. He gave them another land, a distant one, to protect us from their lack of wisdom. But they copulated relentlessly and had more and more children. Then they were overcome with a feverish joy at making countless goods and machines. Eventually they found their land too small for them. They still had ancient words from their grandfathers about *Hayowari*'s inhabitants and their forest. So they told their children: "Far away from here there exists another land, a very beautiful one, where long ago *Omama* created our people. The inhabitants of the forest from which they came still live there. These people are not other from us!" These words must have spread among the white people of long ago. Finally, these strangers crossed the big lake that separated them from us. They sailed on it for moons on vast boats. They escaped the storm wind and the evil beings that populate the center of these waters. Then they finally succeeded in coming back to this land of Brazil.

Yet *Omama*'s real words had not been in them for a long time. It was his bad brother *Yoasi*, creator of death, who guided them to us, like a father leads his children. The ancestors whom the white people call **Portuguese** were truly *Yoasi*'s sons! As soon as they arrived, they began lying to the people of the forest: "We are generous and we are your friends! We will give you our merchandise and our food! We will live by you and we will occupy this land together!" Then they spoke among themselves and started to flock to the land of Brazil in ever-growing numbers. In the beginning, they were seduced by the forest's beauty and they were friendly to its inhabitants. Then they began to build houses there. They cleared bigger and bigger gardens to cultivate their food, and they planted grass for their cattle everywhere. Their words began to change. They started to chain and whip the people of the forest who did not follow their words. They made them perish of hunger and tiredness by forcing them to work for them. They chased them out of their houses to seize their land. They poisoned their food and contaminated them with their epidemics. They killed them with their guns and flayed their bodies with knives, like prey, to bring their skin to their great men. The shamans know all these ancient words. They heard them by making the image of these first inhabitants of the forest dance.[49]

The white people tell that a **Portuguese** man said that he discovered Brazil long ago.[50] They really think that he was the first to see our land. This is nothing but a thought full of oblivion! *Omama* created us with the sky and the forest, where our ancestors have always lived. Our words have been present on this land since the beginning of time, as are the mountains where the *xapiri* live. I was born in the forest and I have always lived there. Yet I do not say that I discovered it or that I want to own it because I discovered it. Just as I do not say that I discovered the sky or game! They have always been here since before my birth! I am satisfied with looking at the sky and hunting the game of the forest. This is all and this is the only right thought. Our elders of the distant past did not ask themselves: "Do the white people exist?" As I have said, our shamans were already calling down the image of these outsiders' ancestors long before their children reached us. The images of these long-ago white people danced for them and the shamans sang and danced by imitating their twisted words. Ordinary people listened to their ghost talk with curiosity and told themselves: "I would really like to know these outsiders! What are they like? Will I see them one day?"

Our *xapiri* spirits travel very far, to the edges of the earth and sky. This is why our shaman elders have known the big lake the white people crossed for a very long time. They often made its image dance with those of the wind storm and whirlpool beings that populate it. Its waters come from the great river that sprang out of the underworld they called *H^wara u.*[51] *Omama* created these outsiders with this river's foam. Our long-ago shamans were already talking about the white people long before they reached us in the forest! These outsiders' ancestors did not discover this land! They arrived in it as visitors. But they relentlessly devastated it and cut its image up into pieces, then shared them out amongst themselves. They claimed it was empty in order to take control of it. This same lie continues to this day. This land was never empty in the past and it is no more empty today! Our ancestors and those of all the inhabitants of the forest already lived here long before the white people arrived. This has been *Omama*'s land since the beginning of time. Before the epidemic fumes decimated them, our people were numerous here. In those distant times, the factories and the metal machines did not exist. There were no motors, no airplanes, and no cars. There was no **oil** and no gas. Men, the forest, and the sky were not sick from these things yet.

The Mission

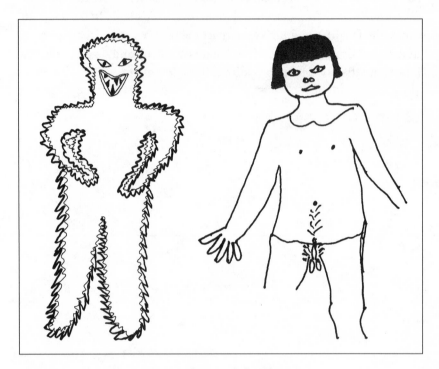

Xawara, the cannibal epidemic

MY ELDERS met the people of *Teosi* for the first time by going to visit the *Xamat^hari* who had settled near the Ajuricaba Outpost, downstream on the Rio Demini.[1] These white people, whom they had never seen before, told them they wanted to know their *Marakana* house and visit it. This was at the beginning of the rainy season. The outsiders invited them to get into a heavy motorboat and traveled up the river with them. After a few days, they reached the mouth of the Rio Toototobi. All our people were gathered in a vast forest camp. There were really many of us. Small *ruru asi* leaf shelters were scattered all over the place. My stepfather told me about this. The elders had just launched an incursion against the people of the upper Rio Catrimani.[2] Fearing retaliation, they had left *Marakana* to take refuge in the forest for a while.[3] Despite this, the white people insisted to go all the way to our home. Finally, a few men accepted to accompany them there so they could get some bunches of plantain bananas from their garden. So the white people finally visited *Marakana* and after several days they returned to our forest camp.[4] Then they traveled back down the river towards the Ajuricaba Outpost without explaining anything about their intentions. Several moons passed. Next it was the soldiers of the Boundary Commission whom we saw appear on the Rio Toototobi. They worked a long time in the highlands planting large stones at the sources of the rivers, then they also went downstream without saying a word to us.[5]

A dry season came to an end, and then another arrived. This is when the people of *Teosi* finally returned to our home.[6] At the beginning, they were only visitors. They had not yet cleared airplane paths or built houses in our forest. The elders simply invited them to tie their hammocks to the posts of our house. Then for the first time, they began to let us hear *Teosi*'s songs on a machine, then to recite his words at great length.[7] It was so. At that time, these people of *Teosi* still lived far away from us. They had settled at the Ajuricaba Outpost, by the people of the Inspetoria and the *Xamat^hari*.[8] But the head of this outpost did not like them.[9] This is why they resolved to abandon the *Xamat^hari* and make friendship with our elders by telling them they wanted to live on our land. Yet since these missionaries' very first visit our *Marakana* house, a great number of our people had been devoured by the epidemic smoke of this SPI agent I talked about, Oswaldo.[10] Our elders had nearly all disappeared. We had become other people. After returning from a *reahu* held by the people of the *Warëpi u* River,[11] an upstream highland group also decimated by the epidemic, my stepfather decided to permanently settle in our new house at *T^hoot^hot^hopi*. All the survivors from *Wari mahi* followed him. But those from *Sina t^ha* stayed a little upstream, near an abandoned Boundary Commission camp. Then, after their second stay with us, the people of *Teosi* returned to Ajuricaba. But this time they were quick to come back up the river. They chose to settle near the *T^hoot^hot^hopi* site opened by my stepfather. They called it "Toototobi" in their language. They had found the forest beautiful. They began building their houses there and planting what they needed to eat.[12] This is how the people of *Teosi* began to live with us!

AT THE VERY beginning, they only knew their ghost language. They did sometimes try to sing and talk like us, but we did not understand much of what they wanted to say to us and it made us laugh![13] Yet little by little they endeavored to draw our words on paper skins so that they could really imitate them. And so after some time they were able to speak with a straighter tongue. This is when they began to frighten us with *Teosi*'s words, and to constantly speak to us in anger: "Don't chew tobacco leaves! It's a **sin,** your mouth will be burned by it! Don't inhale *yãkoana* powder, your chest will become black with sin! Don't laugh and copulate with others' wives, it's filthy! Don't steal what you are refused, it's evil! *Teosi* will only be happy with you if you answer him!"[14] It was truly so. They repeated *Teosi*'s name relentlessly, in everything they said: "Accept *Teosi*'s words! Let us return to *Teosi* together! It is *Teosi* who sent us! *Teosi* wanted us to protect you! Do not refuse him, or you will burn in the great *Xupari*[15] fire after your death! If you follow *Satanasi*[16] and his words, you will burn there with him and you will be a sorry sight! But if on the contrary you all imitate *Teosi* as we do, one day when he wants it, *Sesusi*[17] will come down towards us and we will see him appear in the clouds!"

These words were very different from those of our elders! We had never heard words like these! We knew nothing of *Teosi* or *Satanasi*. We had never even heard their names spoken, nor *Sesusi*'s. We only knew *Omama*'s and *Yoasi*'s words. Yet in that time our elders deeply feared the white people. Many of them had just been devoured by Oswaldo's epidemic smoke. They thought that the people of *Teosi* might be speaking the truth. They were worried to hear these unknown words. This is why everyone began repeating them, including the great men and the shamans. It was a pitiful sight! I often still think of it. The people of *Teosi* focused their anger on those who continued to make the *xapiri* dance despite it all. They kept telling these shamans that they were evil and that their chests were dirty. They called them ignorant. They threatened them: "Stop making your forest spirits dance, it is evil! They are **demons** whom *Teosi* has rejected! Do not call them, they belong to *Satanasi*! If you remain so evil and persist in not loving *Sesusi*, you will be thrown into the great fire of *Xupari*! You will make a pitiful sight! Your tongue will dry up and your skin will crack in the flames! Stop inhaling the *yãkoana* powder! *Teosi* will make you die! He will smash you with his own hand, for he is very powerful!"

These evil words, repeated unceasingly, finally frightened all the shamans, and they no longer dared to drink the *yãkoana* or even sing during the night. They only asked themselves who this *Teosi* could be to mistreat them so! *Omama* had never said anything like this. Our ancestors only knew the beauty and strength of the *xapiri,* and they preferred their songs to anything. They did not understand why the white people started speaking to them so badly. These new words of *Teosi* left them baffled and anxious. Then, one after the other, they rejected their *xapiri* and their spirits ran away. The last great shamans did not even have the courage to bring them down to treat the sick anymore. They also became mute. Seeing this, all our houses' other inhabitants gradually accepted *Teosi*'s words.

As soon as the missionaries finished building their houses in Toototobi, they began living there with their wives and children. From that time, we all attempted to follow *Teosi*'s words exactly as they did. Every day, the people of our house gathered at their call, even the children and old people. It was early in the morning. It was cold and we were sleepy, but we had to go anyhow![18] Everyone told themselves: "If I don't imitate *Teosi* with the others, I will burn in the *Xupari* fire alone!" Despite our sleepiness, we eventually came down from our hammocks. We were very docile at that time! We complied with everything that the people of *Teosi* told us. After having gathered us all together, the white people began to sing: "Who created the sun? It isn't me who created it! It is *Teosi* who created it! Who created the moon? It isn't me! It is *Teosi* who created it! Who created the forest? It isn't me! It is *Teosi* who created it! Who created the game and the fish? It isn't me! It is *Teosi* who created them!" They also sang that *Teosi* made the earth and the sky, the light and the night, the wind and the rain. They told how he had given life to **Adam** and **Eve:** "It was *Teosi* who brought us into the world. He took the earth and kneaded it in his hands, then he gave it his breath of life to create a man. His name was Adam. Later, he made him go to sleep and pulled a rib out of him to create a woman. It is also him who gave children to women. *Teosi* is very powerful! We call him Father! He makes us happy. Accept his words. Later, he will come to get you and take you with him."[19]

We would ask them: "But where does the one you call *Teosi* live?" They would answer us: "He lives beyond the sky. He is building our houses there. This is why he himself isn't coming to get us yet. But he already sent us his son *Sesusi* so he would wash the blackness from our chests with his blood. After our death, we will live with *Teosi* forever. We do not really die!" Hearing this, we would tell ourselves: "That's good! We will imitate *Teosi* like the white people do. That way our chests will stay clean. And when we die we will go to live with him!" The missionaries talked to us about *Teosi* while showing us pictures, about which they said: "These are the **Bible**'s words!"[20] Then we would think: "It may really have happened the way they say. Don't these outsiders tell the truth? *Teosi*'s words may be true!" This was how they managed to deceive us. Their words led our thought astray and left us worried. Once we were gathered, after having sung and listened to the white people, we took turns trying to talk to *Teosi*, like them. Everyone had to do it! The men and the women, the youngsters and the elders. First we closed our eyes and put our hands over our foreheads.[21] Then we spoke out loud, without fear. When we wanted to be successful at hunting, we said: "Father *Teosi*, you are very good. You alone are generous. I want to go hunting today. Protect me from the snakes. Make their fangs harmless. Make them run away when I approach. Protect me from the *xiho* ants. Take the pain out of their sting. You are the one who created game. Send it on my path in the forest. We are all hungry for meat. Make me encounter a tapir. I will arrow it and I will **thank** you. We will all eat him together. Our bellies will be full, and we will be happy. And if I eat too much tapir, protect me from diarrhea. Otherwise, send me howler monkeys and curassows. I will also arrow them. Show me a caiman so I can knock it out. Make it cowardly so it doesn't bite me if I inadvertently step on it. Or at least make me find a turtle on the forest floor. I will thank you! Do that so we can think that you are truly good!"[22]

The elders also talked to *Teosi* about women. They would tell him: "Father *Teosi*, you are good. I am happy because of you. No one else is as great. Chase *Satanasi* far away from me when he makes me look at another man's woman. Prevent me from listening to him when he tells me: 'Look at that young woman, she is so beautiful, eat her vulva!' Make me copulate only with my wife. As soon as we simply want to make friendship with a woman, *Satanasi* makes us lecherous. That is evil! Only you can make him back down. You must make me strong!" The shamans also asked *Teosi* to wash their chests: "Father *Teosi*, my chest is black. Wash it with *Sesusi*'s blood. When the *xapiri* come close to me, chase them away, send them back where they came from. It was *Satanasi* that led them and orders me to make them dance. *Teosi*, I want to make your spirits come down in their place. You who created the **angels**, send them to me! Only they are beautiful and truly powerful!"

Often we also sang: "Father *Teosi*! We love your son *Sesusi*. When he comes down from the sky, we will follow his path. We will go live with him in your forest, where there are no enemy sorcerers, nor snakes, thorns, nor *kaxi* ants. Here below, the forest is hostile. This is why we want to join you. Then we would not suffer from hunger anymore, for in your house bread and **coffee** are abundant. We will be happy, we will eat our fill. Our father *Teosi* is generous. His forest is magnificent. I will go to *Teosi*'s! In his house, I will no longer do wrong. I will no longer eat the vulva of women other than my wife. Near him, I will no longer be sick and I will not die! I am scared to burn in the *Xupari* fire with *Satanasi*. Only those who ignore *Teosi*'s words will perish there. Me, I will go to *Teosi*'s forest! *Teosi* is very powerful. I no longer fear the enemy sorcerers. *Teosi* knows how to make their sorcery substances harmless. They can try to blow them on me with their blowpipes, but they will no longer be able to kill me. I pushed my fears far away from me. I will live with my father *Teosi*. I will follow *Sesusi* there!"

These words of *Teosi* are other people's words. They are not our ancestors'. Yet in that time we tried to repeat them again and again in the company of the white people. Sometimes some of us started to quietly chuckle if someone twisted his tongue when clumsily trying to imitate them. I often made fun of others this way! And I used to tell myself: "We must be a pitiful sight to see! We close our eyes to talk to *Teosi* and we don't see anything. We speak to him without even knowing who he is!" It is true. At the time everyone tried deep in his chest to speak to *Teosi*. But despite how hard we listened, we never truly heard his words. This is why at the time I often asked myself: "What does *Teosi*'s voice sound like? Will he eventually answer us?"

SOME TIME after settling in Toototobi, the people of *Teosi* asked all the adult men to gather together. They simply told them: "You have to open a long clearing, which will be an airplane path. Other white people who possess *Teosi*'s words like we do will soon come down there!" Then our elders obeyed and started working under the guidance of a new mission- ary who had just arrived, a Brazilian called Chico. The others were **Amer- icans.**[23] Our fathers really toiled away to clear this landing strip![24] Though they were hardened to work, it was a sad sight to see them felling tall trees with axes under the blazing sun for days on end. Chico was very ag- gressive. He repeated *Teosi*'s words and only interrupted himself to give orders. As soon as a man stopped to take a little rest, he angrily yelled: "Get back to work! Don't sit there doing nothing! If you don't work, you won't get anything!" It was very difficult. The place the white people had chosen to bring their airplanes down was full of big hard *komatima hi* trees, and the path they had traced through the forest was really very long. A good number of our elders even wondered at the time if this path might be intended to welcome *Teosi!* They were so eager to see him with their own eyes! So they worked without respite and they remained docile. But the missionaries had not really said that, even though they never stopped claiming that one day *Teosi* would come down from the sky's heights. They would tell us: "*Teosi* will soon come get us. When he ar- rives, you will hear the sound of a flute coming from the clouds. At the moment, he is preparing our houses and food to welcome us in the sky. We must wait! He has a lot of work because we, the people of *Teosi*, are numerous!" Our elders thought that these words might be true. They re- mained thoughtful and asked themselves: "Will *Teosi* really come down to us? Will it be soon or in a long time?" Then the day when the people of *Teosi*'s first airplane approached in the sky, they gathered in fear behind the missionaries to watch it come down on the new landing strip. Every- one was very frightened, as in the time of the Boundary Commission air- planes. It is true. Our elders still did not truly know the white people. They had let themselves be taken in by the missionaries' constant talk about the coming of *Teosi*. The missionaries had never really told them what this airplane path would be used for. They had never asked them what they thought of it. They had only promised them gifts so they would stop being scared.[25]

It was Chico, the Brazilian missionary, who started to make us doubt the white people's words. We were curious and we often asked him ques- tions about *Teosi:* "What appearance does he have? What is the sound of his voice? How does he speak?" Yet he invariably answered each of our questions with: "*Teosi* is *Tupã*, the Thunder!"[26] This irritated us because it was obviously a lie. We knew very well that in the beginning of time, Thunder's booming voice had exasperated the people of that time and that they had finally arrowed and devoured him![27] Chico got angry easily and would speak to us very badly. He also sometimes tried to frighten us. Once, he flew into a rage because children had pinched watermelons he had planted along the airplane strip. To deter the little thieves, he planted a stake in front of his plantation with a shotgun attached to it and connected the trigger to a vine. Then he told everyone that his weapon would fire if anyone got near his watermelons. Another time, he ordered us to follow him to his maize plot. He started nervously pouring a white powder on the plants' inflorescences. It must have been a powder to kill mosquitoes and cockroaches. Then he threatened us again: "Now if you steal my ears of maize again, it will kill you!" During this time, he would also sometimes angrily yell at a shaman who refused *Teosi's* words: "I am going to kill you and drink your blood! I like to drink the Yanomami's blood!" Yet this bravado, far from frightening the young man, only provoked his anger and that of his people.[28] His brothers immediately came to his rescue and confronted Chico, vociferating as much as he was. Then they warned him: "If you claim to possess *Teosi's* words, do not address us with such evil words. It is a sin! And if you claim to kill one of our people again, we will not hesitate to arrow you like an enemy!"

One day, a group of hunters came to ask Chico for cartridges. Finally, he reluctantly agreed to give them a few, then hid the rest. This stinginess irritated the men because during their first visit the missionaries had always proved generous to win their friendship. So they decided to wait until Chico turned his back and steal the rest of his munitions. When he realized this, Chico flew into a fury again and started to scream: "You are truly evil! I want to make you die!" Faced with so much rage, my stepfather took it upon himself to retrieve what remained of the cartridges from the hunters. He gave them back to Chico and Chico finally calmed down. Then several moons passed and this incident was nearly forgotten. But suddenly we all got sick, struck with a violent measles epidemic.[29] Then once again many of our people died very quickly. Just after that, Chico suddenly left to work in Surucucus, another one of *Teosi's* people's outposts in the highlands.[30] Rendered desperate and furious by all these deaths so soon after the ones in *Marakana*, the few elders who had sur-

vived wanted to take revenge. They were convinced that Chico had burned
this epidemic smoke to punish them for stealing the cartridges. They
thought he had hastily run away because he was in an õnokae homicide
state and he feared that the survivors would want to arrow him! And that
was the case! But no warrior among them had ever killed a white person.
They only knew how to arrow their enemies in the forest. They hesitated,
then time passed, and they finally gave up on their revenge. This is why
Chico is still alive.

As I HAVE said, we did not know the white people well in that time. We
still feared them a great deal. However, they were not afraid of us! They
probably found us very docile. They must have thought we were cowards!
This is why they treated us roughly. At this time, before the epidemic,
there were two Americans at the mission. The one we called Kixi was
quick to get angry, like Chico.[31] He constantly told us off, repeating: "Satanasi is deceiving you! It is because of him that you are thieves! You belong to him and will all burn in the Xupari big fire!" But all this rage came
to an abrupt end one day when my stepfather nearly killed him. Infuriated by his aggressiveness, my stepfather finally hit him. The missionary
was very afraid and after that he stopped talking to us so badly. This happened at the beginning, when we still accepted Teosi's words. My stepfather's oldest son was still a little child.[32] He would arrow lizards and small
birds around the mission for fun. One of his ruhu masi darts[33] got stuck
in the palm roof of one of the white people's houses. He went to get a pole
and leaned it against the building's wall so he could get his dart back. He
climbed up carefully. Once he reached the top, he tried several times to
use his bow to reach the dart and pull it to him. The missionary saw this
as he was on his way home. He thought the child was trying to get inside
the house by parting the leaves on the roof. He bolted in that direction,
yelling for the boy to come down. The frightened child obeyed him, but
as soon as he was on the ground the missionary started hitting him with
a piece of flat wood he had picked up off the ground.

My stepfather and other men were not far from there, near the river
preparing clay for the adobe of a new house for the people of Teosi. Suddenly, one of his daughters came running towards him and excitedly told
him what had happened: "The white man just hit my little brother! His
mouth is bleeding!" Hearing this, my stepfather dashed to the mission.

He flew into a rage when he saw his young son's blood. He threw himself at the missionary, brandishing his hoe. He was most valiant and *Teosi*'s words had not taken his courage away! Frightened, the white man attempted to calm him: "Wait! Don't get angry! We need to speak to *Teosi* together!" My stepfather did not answer him. He just tried to smash his hoe down on his head! But he was still too far away and he missed him. Then he tried to hit him again, but the panicked missionary managed to dodge the blow, repeating: "Don't hit me! We need to talk to *Teosi* together! Let's talk to *Teosi!* Let's talk to *Teosi!*" Still overcome with fury, my stepfather threw his hoe on the ground and started hitting the white man's face with his right fist. The missionary tried to defend himself. But once he received a violent blow to the nose, he could no longer resist his adversary's ardor. His wife and daughter tried to hold back my stepfather. Even the white man's little boy tried to hit him in the back. In vain. My stepfather repelled them, shoving them away one after another. In the end, they were all crying, frightened and powerless. The missionary was standing there without moving, in a ghost state, panting and gradually sinking down as the blows landed. Finally, my stepfather grabbed a stick to finish him off, but the white man's wife desperately tried to stop him by clinging to it. At that point, Chico arrived. He was returning from a visit upstream, among the *Sina tʰa* people. When he saw the missionary about to collapse and my stepfather brandishing his club, he threw down his backpack and rushed between them. He collared my stepfather, crying: "Don't do that! Stop! Stop hitting! Kixi is your friend!" It was so, finally, that he was able to contain his anger. Kixi was in bad shape, covered in blood and stunned by punches. He had narrowly escaped death! His wife hurried him back to their house to take care of him. They stayed shut up in there all day. The next day, the white man appeared again, with his face swollen and several broken teeth. Later, he left for Manaus to put in brand new ones.

THE EPIDEMIC reached us at the mission some time after Chico's cartridges were pinched and my stepfather hit the missionary. An airplane arrived. Kixi was coming back from Manaus with his family. Without his knowledge, his small daughter had contracted measles there. He only noticed this once he had arrived in our forest.[34] This is what he told us later. But maybe he also wished us to die, like Chico did? He must have been very angry after what my stepfather did to him! Many of us thought he might have brought back a *xawara* epidemic smoke in an iron box and opened it among us to avenge himself. Yet no one saw anything explode like in the time of Oswaldo, at *Marakana*.[35] I don't know! Yet it is true that Kixi alerted us about his little daughter's illness. As soon as he understood that she had fever, he began telling us: "Do not come close to my daughter anymore! Stay far away from her! She is sick, she has measles! She will contaminate you! You will die!" Yet it was already too late. Some of us had carried her in our arms, others had played with her. But Chico did not say a word to us. He never tried to warn us. This is why the epidemic's survivors later wanted to arrow him.

This epidemic started devouring us during a *reahu* feast. Our elders had invited the people of the *Warёpi u* River to our house at Toototobi. These people did not have enough manioc in their gardens for the feast they were thinking of holding, so my stepfather had invited them to use his own gardens. He also had offered them to hunt with us in order to gather the necessary game.[36] As soon as these guests arrived at Toototobi, all the men left to go hunting together on an expedition planned to last several days. But the hunters came back much earlier than intended. They had only arrowed two tapirs. Several of them had started to burn with fever in their forest camp. The same had happened in our house and the disease started to obscure our thoughts.

Despite this, preparations for the feast continued for a few days. A group of women went to the gardens to gather manioc tubers. They peeled them and piled them on one side of the house's central plaza, then covered them with banana leaves. The next day they started to grate them to prepare the flour for the cassava bread to go with the smoked game. The fever had now reached most of the people in the house. The following day, only a few women were still able-bodied enough to bake the cassava bread. Many thought it could simply be a coughing illness and did not worry too much. But they were wrong! It was the measles, which is truly very dangerous for us. We call it *sarapo a wai*.[37] Soon nearly all were contaminated, both among us and among our *Warёpi u* River guests. The disease immediately spread among the people of *Sina t^ha*. Then once again, as at *Marakana* long before, our people started to die in great numbers, one after another, in the house and in the forest; both the children and the adults, the men and the women. Their skin was covered in reddish patches, which they scratched to ease their terrible itching. They lost all their hair and their faces seemed swollen. They were relentlessly rocked by a violent cough and were burning up with fever.

At the beginning of the epidemic, the missionary ordered those who were not affected yet to cut large quantities of firewood so that the sick could keep warm. Along with the other youngsters who were still healthy, I spent my time chopping up dead tree trunks in the garden with an axe. Yet soon after I was also taken by the disease. This *xawara* epidemic was really voracious! It was so hungry for human flesh and nearly killed me too. I was so sick that I finally lost consciousness. I had truly become ghost and the fever was burning my whole body. Then, in my dream time, I began to see the sky's chest collapsing onto the earth.[38] Our house's shamans were working frenetically to try and hold it up. Nothing was working, it swayed and roared and continued to come apart. Vast patches came off with rumbling cracks. Then they slowly fell towards me, shining with a blinding light. All our house's inhabitants were in tears and at the end even the shamans were screaming in horror. I was certain the sky would collapse on the forest and crush all the human beings. But suddenly I came to. Reassured, I exclaimed: "What a fright! I just saw the sky shatter and fall on us!" I was truly very sick during this epidemic! Yet I was finally able to escape death. The people of *Teosi* called their airplane with a doctor and medicine to treat us.[39] This is how my older sister and I were able to get better. My stepfather also survived, though he was really almost dead. All our kin were weeping over him and they had already prepared a bag of palm leaves and stakes in order to hang his corpse in the forest.[40] This is what happened. I did not know the *xapiri* well at this time, but I think they must have protected me.[41] It is probably thanks to them that I am still here to tell this story, and it is also because of this that I later became a shaman.

AN UNCLE[42] whom I loved very much was the first to be affected in Toototobi, before the epidemic spread to our whole house. The missionary had warned him by saying his little girl was sick. He did not listen to him and approached her to speak to her with affection. This is how he was contaminated. Then he died very quickly, before all the others. As soon as he became ghost, the shamans did everything they could to make him recover. Yet their hands had to let go and they were unable to protect his image. I tried to get close to him several times while they worked, for I was very sad about his sickness. But the other adults stopped me. Then I never saw him again. I only learned the news of his death from a distance. From that moment on, I felt really alone. This uncle would show his attachment to me and protect me. He carried me in his arms and often gave me food. His death really made me very sad. I could not stop crying. At first the elders thought that *Amikoapë tʰëri* enemy sorcerers from the upper Rio Mucajaí had blown harmful plants on him, then broken his bones.[43] But it was not that. Just after his ghost went away to the sky's back, other people in the village gradually fell sick and died the same way he did. It was the *xawara* epidemic that killed him. This is why if I had been an adult I think I would have arrowed the missionary to avenge his death. But I was still just a child, and I was very scared of the white people. Later, I never stopped thinking about this uncle as I grew up. He had made me think by saying: "When I die, you will have to go away to the white people's. Don't stay in this house, no one else here will really have friendship for you. They are other people!" I always kept these words in me. It was by remembering them later, as a teenager, that I left my Toototobi village to go away downriver, to work at the Ajuribaca Outpost.

After my uncle, it was my mother who was also devoured by the epidemic. She began burning with fever. She was still young and hardy. Yet she died in a few days. It happened so quickly that I was not able to take care of her. I still remember all that today with a great pain. Spared by their own epidemic, the missionaries put my mother in the ground behind my back, somewhere near the Toototobi Mission. My older sister and our other relatives were also very sick. As I said, my stepfather was nearly dead. None of us were able to stop the outsiders from burying my mother. They did the same thing with many of our people. I learned about all this much later, once I was cured. But I was never able to learn where my mother was buried. The people of *Teosi* never told us, so that we could not gather the bones of our dead. Because of them, I was never able to mourn my mother the way our people usually do. It is a very bad thing.[44] It made me feel a deep sorrow and the anger from her death has persisted in me since that time. It hardened little by little and will only cease with my own end.

After death, our ghost does not go to live with *Teosi*, as the missionaries claim. It tears itself out of our skin and goes to live elsewhere, far from the white people. Our dead live on the sky's back where the forest is beautiful and full of game. Their houses are numerous there and their *reahu* feasts are constant. They live happily, without pain or disease. From up there, we are the ones who are pitiful to them! They feel sad to have abandoned us on earth, alone, starving, and prey to evil beings! This is why my grief eases a little when I think that my mother is living happily in the forest of the ghosts, surrounded by all our dead relatives. It is true. It is us, the few human beings who remained, who suffer in the forest far from our dead who are very numerous up there.

DURING THIS new epidemic, the missionaries did not give up on talking to us about *Teosi*. On the contrary, they prevented the shamans who were still able-bodied from treating us! They would repeat to them: "Do not call down your spirits who belong to *Satanasi!* It is *Teosi* who will heal the sick. And those who will die will go back to live with him. They will be happy there! Do not be worried!" The shamans obeyed out of fear and did not do anything. They did not interfere to battle the *xawarari* epidemic beings. They did not avenge their people who were close to death. Many of us were frightened by this silence and probably died because of it. That is what I think. This time, most of the adults who had escaped the *Marakana* epidemic died. Our elders had wisdom and took care of us. But suddenly they were no longer among us. When I think back to this time, I remain silent and alone in my hammock. All this torments me, and I have never been able to forget it. My thoughts follow one another with melancholy, and I cannot stop them. In order to try and soothe my mind, I tell myself that those who made our elders perish will one day die in their turn, provoking the same sadness among their people.

All these dead, on top of those at *Marakana*, seized us with a great affliction and planted anger in the survivors' chests.[45] They began speaking to the missionaries harshly: "You claim that *Teosi* takes care of us. You gave us his name and yet you are the ones who make us perish! We don't want to listen to his words anymore! He does not chase evil far away from us! On the contrary, he let us be devoured by your epidemic!" We were all distraught and furious. It took a long time before our thoughts were calm again. The white people of the mission did not answer our reproaches. They merely answered: "*Teosi* protected you! He is the one who cured you! We talked to him constantly! He was at your side and he is very powerful! He was the one who chased away the *xawara* epidemic. He brought your dead back to his house. Do not be sad, they are living happily beside him."[46] I remember all that very well! At the time, I was a young boy and the missionaries really wanted to convince me. They constantly told me: "Listen to us! You must accept *Teosi* and his words, for if you die you will go to the sky and he will take care of you."

After all this suffering, faced with the missionaries' insistence, we started thinking they might be telling the truth again. We wound up fearing them again, them and him whose name they evoked at every occasion. We would tell ourselves: "Maybe *Teosi* really wanted our people to join the ancestors' ghosts on the sky's back? Maybe he will soon come down into the forest so that we also die and he can take us to him? Don't we need to accept his words to avoid rousing his anger and burning in the *Xupari* big fire?" Our thought was doubtful and so we started to lend an ear to the people of *Teosi*'s speeches again with fear and docility.[47] Soon after, my stepfather even let them plunge him into the waters of the Rio Toototobi to **baptize** him.[48] Then all followed his example and wanted to become **believers** again.[49]

As I SAID, Chico had left Toototobi just after the epidemic, but now he came back to the mission.[50] He said he was a man of *Teosi*, but he was very different from the other missionaries. He did not have a wife and children. He lived alone and, as time passed, he must have told himself: "Why shouldn't I take a Yanomami wife?" He employed a young girl to take care of his house, to wash his laundry and dishes. She was a *moko*, a girl whose breasts were still hard and pointy. She was very beautiful and he began to desire her. He constantly offered her food and clothing.[51] He had fallen in love with her, and he began to eat her vulva. After a time, he wanted to truly possess her as a wife. Without the other missionaries knowing, he decided to ask my stepfather for her. He told him: "I have lived alone for a long time and I want this girl to be mine! I need a wife too!" I don't know why, but my stepfather finally accepted to give her to him. He probably thought that if he refused, Chico could get angry and he might seek revenge with a new epidemic smoke, like Oswaldo did at *Marakana!* As for me, I was really unhappy about all this. This young girl was a close kin to me and everyone knew that Chico had already made a young married woman pregnant. I was furious that he still claimed to be part of the people of *Teosi!* All this was very bad. Since his arrival at the mission, Chico had constantly told us: "Do not covet other men's wives, do not call them to copulate in the forest! It is a sin!" He had tricked us with all his lies!

So the people of Toototobi got angry again. They began openly taking it out on him: "How can you follow *Teosi*'s words while committing the sins you tell us about yourself! You lied to us!" Chico responded with irritation: "I am not committing a sin! I want to marry her. I do not desire another man's wife. I always obey *Teosi!*" But our elders countered: "Lies! Go ask your people in Manaus for a wife instead. There are plenty of white people's women! If you marry a woman from your home and you imitate *Teosi* with rightness, then we will follow you! But if you want to copulate with our women one after another, it means that you are deceiving us! You are bad! If you were a true son of *Teosi*, you would remain without a wife rather than eating our daughters' and wives' vulvas! You often say that we are wrong, but you imitate us! This means that your words of *Teosi* are lies and that your thought is full of oblivion!"

Our elders believed that if the white people really held *Teosi*'s words, they could not touch our women. Otherwise, it meant that they were liars and that *Teosi* did not exist. After the epidemic, they were still distressed by the memory of their dead and they were tormented by the missionaries' words. Chico's actions truly perplexed and infuriated them. They lost all will to imitate these white people who now appeared to them to be nothing but cheaters. Once again, they had proved to be lazy with *Teosi*'s words. Yes, some of us still listened to them from time to time. But little by little all ceased to be truly interested in them. The missionaries still tried to talk to us about *Sesusi* and sin as much as they could. But our ears had become deaf. Chico continued to repeat his threats: "If *Teosi* is not in your thoughts and you don't love him, he will make you die!" but he had done too many bad things in Toototobi. Even the other white people had finally noticed! The head of the people of *Teosi* sent him back to Manaus, where he finally stopped being a missionary.[52] For our part, we were finished with *Teosi*'s words.[53] Chico's deceptions had made us think and we chased these words of lies and fear far away from us.

DURING THAT time, my father even threatened the people of *Teosi* with his hunting gun! This happened because a great shaman, whom he called brother-in-law, died abruptly during a visit to our home. This was an elder who had come from a house at the sources of the Orinoco called *Maamapi*. This man was his friend. One day, he was working at weeding the mission's airplane path. He started to feel a sharp pain in his stomach. Did the spirits of an enemy shaman arrow him? Did distant hunters wound his animal double? I do not know. His sickness did not last long. His state got worse very quickly and a terrible pain tormented him. Yet none of our shamans tried to tear the arrow points out of his image that made him suffer so badly. Neither my stepfather nor anyone else. They no longer dared to call the *xapiri* to cure people. They had rejected them and no longer drank the *yãkoana* to feed them and make them dance. They feared the white people's ugly words and only addressed *Teosi* now.

My stepfather was still a believer at the time. He tried to cure his friend with the words he had received from the missionaries. He asked *Teosi* to let his friend live: "*Teosi,* I call you Father. I carry you in my thought. You are very good. You are the only one who can heal us. It was you who created the forest and the sky. You are the only one to be so powerful. The *xapiri* are weak. My brother-in-law is in agony. Tear the pain out of his stomach. If he recovers, I will thank you. If he comes back to life like *Sesusi,* I will be happy with you. If he dies I will be very sad. I will think with anger that your words are but lies!" He spent an entire night squatting next to his sick friend who moaned in pain. He kept his head down, with his face in his hands. He doggedly imitated *Teosi*'s words. It was a pitiful sight to see. The sick man kept groaning, repeating: "It hurts so much! I'm going to die!" Then suddenly we could not hear his voice any longer. He stopped breathing. Then all the people of the house approached his hammock to initiate their mourning laments. My stepfather stayed squatting with the dead man's mother, a very old woman. He cried with her a very long time before the pain of his sorrow turned into fury. Then he yelled out before all those crying with him: "Starting from this day, I will no longer vainly imitate the words of *Teosi* who let my brother-in-law die!" It was early in the morning. The dead man was still stretched out on his hammock. My stepfather went into the forest nearby to prepare the frame on which the body would be hung in a tree. Then he came back towards our house by crossing between the houses of the nearby mission. From a distance, he saw the people of *Teosi* absorbed in their **prayers.** One of them called him: "Come with us! We are going to speak to *Teosi* together! Don't be sad. He is protecting you!" My stepfather kept walking, without answering, the anger of his mourning stuck in his chest.

He went to get his shotgun. He came back to the mission house where the white people had gathered with the gun in his hand. They were singing the words of *Teosi* and once again they insisted that he join them. Still not speaking a word, my stepfather crouched among them with his firearm. Their songs stirred up his wrath even more. When the missionaries stopped singing, they exhorted him to imitate *Teosi*'s words in his turn. He remained silent. He was listening to the faraway sound of mourning laments still rising up from our house. Suddenly, he began to yell: "*Ma!* I will no longer sing for *Teosi!* I don't want to lie anymore! He does nothing to heal us! Only our *xapiri* truly worked to defend us! Your *Teosi* is nothing but lazy. I listened to you and I spoke to him because you told me he knew how to cure people. He did nothing for my brother-in-law. Now it is over! I have lost all joy. I have nothing left but my anger!" Surprised by the strength of his words, the missionaries looked at him with frightened eyes. My stepfather continued yelling, standing before them, waving his shotgun: "I have thrown away *Teosi*'s words! I will no longer speak one word of them! I no longer want to sadden my people by uttering such lies! *Teosi* let the one we are mourning for die. I am furious! Now I only have one desire, to kill you!" Then he put a cartridge into the gun and pointed it at the white people and they immediately ran out of the house. Yet the one whom we called Purusi stayed frozen in the entrance, facing my stepfather as he cried out: "You run like cowards, but you will die anyway! You who is standing there, I will kill you first! I am furious! *Asi!*"

Though this American was an adult, he suddenly started to cry in fear. He really thought my stepfather was going to open fire on him. He begged him, sobbing: "Don't kill me! I don't want to die from one of your bullets!"[54] He had slumped onto the ground. My stepfather grabbed him by the shirt with one hand to make him stand up, still yelling: "Stop crying like a child! Get back on your feet! I want to kill you standing up!" He was a fierce and dreaded warrior at that time. Yet he did not kill the missionary. He must have felt sad to see the man in such a state. In the past, this man had treated him with friendship and had given him trade goods.[55] Finally, he lowered the barrel of his shotgun and let him run away to join the other white people, who had locked themselves up in their homes. Then my stepfather went back to our house, where he found the circle of mourners still crying around the deceased's hammock. He made a brief *hereamuu* speech so the body could be wrapped in a bag of palm leaves and carried into the neighboring forest to be hung there. A few men carried the funerary load, followed by a group of tearful women. Once their task was finished, the mourners' lamentations began anew. All were overcome with sadness and anger. My stepfather continued to carry the grievance of his friend's death long after this day. He never again joined the missionaries to sing and he stopped lending an ear to their speeches and reproaches. He began denouncing *Teosi*'s words as a lie of the white people.[56] Later, he even moved away from the Toototobi Mission to go live far away on the upper *Wanapi u* River.

Aerial view of the large collective house at *Watoriki*. Photograph: W. Milliken, 1993.

Watoriki hunters. Photograph: R. Depardon-Palmeraie et desert, 2002.

Davi Kopenawa and *Watoriki* children. Photograph: F. Watson—Survival International, 1990.

Davi Kopenawa during a shamanic session at *Watoriki*. Photograph: B. Albert, 1993.

Davi Kopenawa and Bruce Albert signing the contract with Plon/Terre Humaine in São Paulo, Brazil, March 2009. Photograph: M. W. de Oliveira, 2009.

Davi Kopenawa on top of the Empire State Building in New York.
Photograph: F. Watson—Survival International, 1991.

Davi Kopenawa, president of the *Hutukara* Yanomami Association,
during a meeting of the Indigenous Council of Roraima (CIR).
Photograph: D. Gomes Macario, 2009.

CHAPTER 12

Becoming a White Man?

Napë a, a white man

WHEN I WAS a child, the missionaries really wanted to make me know *Teosi*. I never forget this old-time of the Toototobi Mission. Sometimes I think about it. I tell myself that maybe *Teosi* exists. I really don't know. What I do know is that for a long time now I have not wanted to hear his words anymore. The missionaries deceived us long ago! Too often, I listened to them tell us: "*Sesusi* will come! He will come down to you! He will come soon!" But time has passed and I still haven't seen him! I finally got tired of hearing these lies. Do shamans vainly repeat this kind of thing all the time? No, they drink the *yãkoana* and instantly bring down their spirits' image. That is all. So once I became an adult, I decided to make the *xapiri* dance like my elders did in the time of my childhood. Since then, I have only listened to their voice. Maybe *Teosi* will take revenge on me and make me die? I don't care, I am not a white man! I don't want to know anything more about him. He has no friendship for the inhabitants of the forest. He does not heal our children. Nor does he defend our land against the gold prospectors and cattle ranchers. It is not him who makes us happy. His words know only threat and fear.

It is true. To this day, the people of *Teosi* have not given up on trying to frighten me! When I meet them, they still tell me: "Davi, your thought is black! *Satanasi* possesses you! If you continue listening to his words, you will go burn in the *Xupari* big fire! Stop responding to the spirits so that your thought can bloom anew with the words of *Teosi!* He is the one who will truly protect you!" But I am no longer a child, I am no longer afraid to retort: "I listened to enough of your trickery in the past. Enough! How can you say that your *Teosi* wants to protect us if he is constantly threatening to throw us into an inferno? If we could see him, maybe we would fear his anger to the point of submitting to him. But we only know what you say about him and we have never been able to see him! So if you want to follow his words, do it alone, locked up in your house. As for me, I never want to hear them again!" But all these twisted words no longer bother me. After my death, the people of *Teosi*'s talk and songs will no longer be anything. My ghost will be happy on the sky's back, with those of our dead shaman elders. It is so. There are more of us on the sky's back than here below!

Teosi's words belong to the white people. They used to be unknown in the forest. They appeared among us not very long ago. No one had pronounced them before the missionaries arrived with them. This is why we do not really understand them. We barely know their upstream.[1] Our thought cannot open them out in every direction as we do with those of the spirits. If we go on following them for no reason, we will eventually forget the words of our elders. Then we will be called believers, when in fact our minds will simply have become as forgetful as those of the white people who know nothing of the forest. Yet today the opposite is happen-

ing. Very few of us still imitate *Teosi* and the shamans no longer fear the
missionaries as they once did. The *xapiri* continue to let us hear their
songs, which are our true language.

Even with the *yãkoana*, we have never succeeded in seeing *Teosi*'s image dance! Though we close our eyes and try with tremendous effort, as I
did in the past, it is always in vain! *Teosi* is dead and his ghost disappeared
beyond the sky. You cannot see him or hear him. Yet long ago, when I
became ghost under the effect of the *xawara* epidemic, I glimpsed a big
piece of white cloth that stayed up in the air without feet. It was difficult
for me to really distinguish it, but there were **priests** and **nuns** sitting at a
big table around it.[2] Then I woke up, and later when I went back to sleep
I did not see it again. But maybe *Teosi*'s image is also what the spirits of
the old shamans call *Wãiwãiri*? He is a being with flabby, radiant skin
who merely dances in place when he appears, shaken by limp and frightening tremors.[3] I have never seen him, but my wife's father told me about
him a few times when we drank the *yãkoana* together. He told me that
this image, which he sometimes called down, wore a long piece of cloth
around his neck covered in black drawn words and that perhaps this was
the true image of *Teosi*.

As I have said, my father-in-law is a very great shaman. Our long-ago
elders opened their spirits' paths to him themselves. He nearly died several times and his *xapiri* always brought him back to life. It was by almost
dying in this way that he also saw *Teosi* and *Omama* fight each other. He
told me how both of them sprang up together when the forest began to
exist. Yet very soon, *Teosi* got angry at *Omama*, whom he found too clever.
He was jealous that he could create the things of the forest alone. Finally
he killed him in a rage. Then *Omama*'s ghost took revenge and killed
him in his turn. After this happened, *Teosi*'s ghost went to live beyond the
sky, above the white people's land. *Omama*'s ghost stayed above our forest, close to the *xapiri*. Since then, their images have remained far away
from each other. All this happened after *Omama* ran away from our forest towards the downstream of the rivers, where he created the white
people.[4]

Teosi is dead, as is *Omama*. All that remains of them is their name,
their value of ghost. Perhaps *Teosi*'s image takes care of the white people.
They must know. But we know very well that he does not protect the inhabitants of the forest! In the past the missionaries claimed that *Teosi*
created the earth and the sky, the trees and the mountains. But as far as

we can see these words only brought us the *xawarari* epidemic beings who devoured our elders, and all the other evil beings who have burned us with their fevers and eaten our chests, eyes, and stomachs ever since. This is why to us *Teosi* is more like the name of *Yoasi*, *Omama*'s bad brother, the one who taught us to die.[5] *Omama* was the one who created the *xapiri* to avenge us of diseases and the *yãkoana* so we could make their images dance. He wisely chose to defend the inhabitants of the forest against *Nomasiri*, the spirit of death.

In the beginning, *Omama* was not the only one to possess the *xapiri*. *Teosi* created them with him. These are the ones that the missionaries call angels. Yet *Teosi* finally became hostile to them because they did not obey him. Then he chased them away, accusing them of being dirty and lazy. Seeing this, *Omama* called them back to him and transformed them into *xapiri*. He gave them their gleaming ornaments and magnificent songs. They are much more beautiful than humans, truly like the spirits that white people call angels. The beauty and power of these spirits soon made *Teosi* envious. This is why, as I have said, he finally killed *Omama*, who was their father. *Omama* did not die for no reason! This is why the people of *Teosi* hold so much resentment for the shamans who make his *xapiri* dance today. This is what I think.

THE MISSIONARIES own a book from which they give out *Teosi*'s words. When they looked at it, they often told us that *Sesusi* would whiten our chest and clean our thought. They unceasingly repeated that *Teosi* does not like those who call down the spirits, chew tobacco, steal from others' gardens, copulate with married women, fight with clubs, lead sorcery expeditions, and are brave at war. But this is nothing but twisted talk to us. *Omama*, he always has friendship for us no matter what we do! He does not claim to wash anyone's chest! His image does not constantly tell us: "You are evil! If you refuse my words, I will make you burn or be carried away by the waters! I will make the forest's soil shake under your feet!" He only tells us: "You are like your ancestors were! Continue to follow in their footsteps! Later you will die, so you must not fear anything while you are alive!" It is so. We know nothing about sin, that thing the people of *Teosi* constantly talk about to frighten us. We are not evil! We are simply not white people! We are as our ancestors always were!

To us, all these white people's words about *Teosi* are in vain. If one of my children's image is captured by a *koimari* bird-of-prey evil being, it will be useless for me to hide my face in my hands and talk to *Teosi* instead of calling the *xapiri!* If I merely close my eyelids as if I were asleep and say: "Father *Teosi*, protect him!" no one will answer: "*Awe!* I will take care of him!" My child will die and I will be left with nothing but my sorrow. That is all! You do not see anything by following *Teosi*'s words: not the *në wãri* evil beings, nor the sorcery plants' diseases, nor the spirits of the *xawara* epidemic. *Teosi* must be lazy, for he makes no effort to cure us, even when we are in agony. He is not even concerned when we die! On the contrary, the *xapiri* are always zealous in avenging us. This is why they taunt *Teosi* as we would a lazy shaman: "The white people say you are powerful. You claim to be able to heal, but we never see you working! You never come down from your hammock! You flee from battling evil beings! You only know how to repeat words of fear and death!"

In the beginning, our elders approached the people of *Teosi* to get some merchandise and medicine. It was not much, but at the time there were no others in our forest. Then these outsiders would not stop frightening them with *Satanasi* and the *Xupari* big fire. That is why many of our elders wound up imitating the white people out of fear. Yet these words never succeeded in washing our chest like the missionaries said they would. None of us stopped getting angry or wanting to get revenge. No one stopped lying or having desire for women. Time passed and little by little all returned to their true words. This is what happened to my stepfather. In the beginning, he really made an effort to speak to *Teosi* like the Americans, by repeating after them: "*Sesusi*, whiten my chest! Chase the spirits far from me!" Despite this, the *xapiri* did not stop wanting to come down to him and *Teosi* never managed to repel them. So he did not fear drinking the *yãkoana* again. It is so. As long as we are alive, we will continue to make dance the images of our ancestors turned animals to cure our people. Indeed, we are inhabitants of the forest. We do not remain locked up all the time in our little houses pretending to talk to *Teosi* and eating alone, the way the missionaries do![6]

Yet at first, when I was a young child at Toototobi, I was happy to hear the people of *Teosi*.[7] If they had behaved better towards us, maybe I would have continued to do as they did? I don't know. They taught me how to draw the words of our language, then to recognize the **numbers** the white people use to **count**, as they did with the other children of our house.[8] Then they often showed me image skins about the people of

Israel and about *Sesusi*.[9] They also gave me a book where *Teosi*'s words were drawn. I liked to listen to them speak of these ancient things. I would have liked to speak to *Teosi* and especially to see him. I really thought I would become one of his people, though having heard his name pronounced so often, I feared his anger. To tell the truth, I was more curious about these white people's new words than those of our elders! And at the time my stepfather and brother-in-law had rejected their *xapiri* and become believers.[10] Our thought was focused on *Teosi* and the *Xupari* big fire. Of course, we mixed up the white people's words a little when we imitated them. Yet by dint of being repeated, they began to become solid in us. We would go to visit friendly houses and we would speak to their inhabitants in the manner of the missionaries:[11] "Accept *Teosi* and take his words into you! He created men and women. He created the food in the forest and gardens. He created the fish and game, the monkeys and the tapirs!" The Americans were happy with us. They declared that we were really people of *Teosi*, just like them. Yet we did not understand these words well. They were not those of our elders, who had never told us: "Father *Teosi* exists, he protects us!" This very name was unknown to us before the white people's arrival. But we simply wanted other words than our own! We told ourselves: "These white people are other people, maybe they possess other spirits. Maybe *Teosi* really exists! Maybe he is as powerful as they say?"

So in the beginning I listened to the missionaries a great deal. I wanted to follow their words and I did the best I could to imitate them. I was happy to be considered one of them. They had already plunged my head into the water of the Rio Toototobi while holding my nose, like a **pastor.** I had really made friends with *Teosi!* Yet when I was alone and I wanted to talk to him, I could not even manage to see him in my dreams. And then the white people continued to speak to me harshly, despite my efforts: "Davi, you are in sin, that is bad! Do not chew tobacco! Do not desire married women! Do not drink the *yãkoana! Satanasi* is deceiving you! You sadden us, you will burn in the *Xupari* big fire!" With time, listening to these unceasing reproaches weakened *Teosi*'s words in me. They only seemed to be about sin and recriminations. I was beginning to get tired of them. Then all that finally made me angry. I would tell myself: "I understood *Teosi*'s words well. Now I am one of his sons. My chest has become clean. Yet these white people constantly accuse me of being evil! Why?" Then I began retorting: "Do not speak to me this way! I don't want to hear so much bad talk anymore! It was enough to tell me all this once!

If you try to frighten me by repeating it at every opportunity, I will only think you are trying to lie to me!"

I had not had a father for a long time. My stepfather now had other wives and young children.[12] Those who took care of me the most, my mother and uncle, had recently died. I was distraught at the idea of having to grow up without ever seeing them again. I was tormented by the sorrow of this mourning. I felt alone in our Toototobi house. Of course, I was not truly alone, but I no longer had close kin there who could take care of me and feed me. I was often sad or angry. I thought of nothing other than running away.[13] I constantly told myself: "I don't have anyone here anymore. I want to disappear to very far from here, where the white people live. I want to live with them and become one of them!" I was truly captivated by this idea. I no longer wanted to live in our house or even see the forest. I had decided to abandon them forever. Becoming a white man, that was all I could think of. Yet I had no desire to follow *Teosi* like I used to. The missionaries had deceived me by harassing me with reproaches. I wanted to forget the words they had given me. When I thought about it, the only thing that came back to my mind was that *Teosi* had let my people perish. It revolted me. I told myself: "So be it! Now I don't care about dying. I am not the son of a white man. Let the epidemic devour me too, and let me burn with *Satanasi!*" These were the kind of thoughts that finally led me to leave our house at Toototobi. As soon as I had a chance, I went to work at FUNAI's Ajuricaba Outpost, downstream on the Rio Demini. I started living with other white people who did not speak about *Teosi*. Little by little, the missionaries' speeches were erased from my mind and finally I completely forgot them.

AT THE TIME, the people from FUNAI who had replaced the previous ones from the Inspetoria often came to visit us at Toototobi to barter.[14] We gave them Brazil nuts, as well as ocelot, *kana* otter, deer, and peccary skins.[15] They brought us machetes, knives, and axes, hooks and fishing lines, hammocks and a few clothes, but also shotguns and cartridges. Sometimes they helped us by giving us a little medicine. They also prevented the white people who lived downstream of the river to enter into our forest. Because of all this, I found it good that they came to visit us. I had grown up but I still attended the mission school. I told myself it would be good for me to learn another **custom**.[16] I had become a teenager and I could now leave my people to travel to other lands. I wanted to know other people.[17] This is what I was constantly thinking about at the time!

At first, the FUNAI men who came and went from Toototobi to barter with my elders did not take any interest in me. To them, I was still a child. But one day they asked my stepfather if I could go and work with them at their Ajuricaba Outpost. He instantly refused because he considered that I was far too young to go to these white people on my own. So they called other youngsters, who were older than me. But it seems that they did not much appreciate their work, for they quickly sent them back home. Later, on other visits, a man from FUNAI approached my stepfather again and insisted that I come work with him. He promised that he would bring me back to Toototobi later. This time I had grown up and I was wiser. I had begun to get used to these new outsiders. My stepfather asked me if I really wanted to follow them. I answered him that this really was what I wanted. So this time he finally gave in: "Very well, go work with these white people! But stay vigilant! Be very careful of their diseases and the jaguars in the forest! Don't mess around and don't get yourself into bad situations!" With these words, I finally left with the FUNAI men.[18]

The man who had insisted I accompany him intended to set me up in his own house, downstream from the Ajuricaba Outpost, so I could work there just for him. The head of the FUNAI outpost, Esmeraldino, noticed this, and he did not like it. He took me aside and said: "Don't go with that guy. He will make you work for him without respite. You will be hungry, you will be a sorry sight to see! Come settle at our outpost instead. You will help us in the kitchen, you will take care of the food and the dishes!" So I followed his advice and stayed with him at Ajuricaba. This is how I began working with the FUNAI people for the first time.[19] I assisted the outpost's cook. I cut wood, I lit the fire, and I went to fetch water from the river. I cooked the game. I washed the plates, the cutlery, and the pots. And I also hunted and fished. I really had a lot of work and no time to laze around! But I still liked living with the white people and carrying out the tasks they gave me. I was barely a teenager and with them I was learning a lot of things. I really wanted to know them better and even imitate them.

YET AT THIS time I did not yet know much about them. I knew the missionaries a little but not the white people of Ajuricaba, who were very close but really so different. In fact, I dreaded having to talk to them!

They did not know my language and I understood practically nothing of what they said to me. So once I arrived at the FUNAI outpost, I was content to work without saying a word, trying to follow the orders I was given: "Come here! Go over there! Cut some wood! Go fishing!" I was able not to make too many mistakes because the *Xamat^hari* in that place spoke a little Portuguese and helped me to grasp what I was told. I really wanted to know the white people. This is why I listened to them with so much attention. Yet my mouth was scared of speaking to them. I did not tell myself: "I will learn their language!" Instead I tried to capture their words one by one and fix them inside myself. But it was really not easy! I had a lot of trouble gathering a few of them in my mind. Then little by little those I was able to recognize grew in number. I remained mute, but I was starting to understand what the outpost's people were saying to me. Then my mouth finally lost its fear. So I took a chance on uttering a few of their words with a twisted tongue. But what I said was really ugly to hear! It was really nothing but ghost talk.

The FUNAI people had given me a very big cotton hammock and all sorts of clothing.[20] All of this made me happy. I told myself: "Why not imitate the white people and become one of them?" I only wished for one thing now: to be like them! So I constantly observed them in silence, paying tremendous attention. I wanted to assimilate everything they said and did. I was already used to wearing shorts. The people of *Teosi* had distributed them to us since they lived among us so we would hide our penises. I also knew **flip-flops.** Yet I had never worn long pants or closed shoes or even shirts and certainly not glasses! When I saw the white people slip on their pants, I thought: "I am going to hide my legs just like them!" When they put on their shoes, I thought: "I am also going to shut my feet in to walk!" When they slipped into a shirt, I thought: "I am going to wrap myself in a beautiful cloth like that one!" Their glasses were particularly impressive to me and I started to hope: "One day I will also be able to hide my eyes like the white people do!" I observed their watches, which also made me envious: "It would be so nice to wrap that thing around my wrist and follow the sun, even at night!" These were all my thoughts at the time!

I CONSTANTLY dreamed of the time when I would become an adult and told myself: "Later I will own a motor to run everywhere on the river with a big motorboat like the white people!" My thought was really set on their

merchandise. At the time, I thought they could make it themselves, however they wanted! These goods obscured my thought and made me forget
all the rest. I no longer carried my people or my old house at Toototobi in
my mind. If the outsiders who had taken me with them had been inhabitants of the river, of those white people who lived downstream along the
Rio Demini, I think I would never have come back home. I would have
become a man among the turtle fishermen and the palm fiber collectors.
And if they had consented to give me one of their daughters, I would
have taken a wife among them and stayed there! If I had really wanted to
become a white man, I would have lost myself among these inhabitants
of the river and I would probably still live there.

I am not telling any lies. This happened to one of the youngsters from
our house at *Marakana*. He was a brother-in-law to me. He was older
than I was. He was already an adult when I was still a child. This was very
long ago. After Oswaldo's epidemic, he had left for the Ajuricaba Outpost, as I later did. He worked there for a time, then he followed a white
man who had once worked for the Inspetoria downstream. The white
man had settled on the lower Demini, far from the Ajuricaba Outpost,
near a lake. He had cleared a garden and lived by capturing water turtles
to sell them.[21] He also hunted and sold game skins. He worked alone,
this is why he took away this young Yanomami to help him. And finally
my brother-in-law stayed with him. He did not want to come back to our
home because he could not find a wife among us.[22] When he left *Marakana* to go down the river, he stopped in our little house of $T^hoot^hot^hopi$
and told my stepfather: "*Xoape!*[23] I am taking a pirogue to go downriver
to the white people!" He answered: "That is good. Go, and most importantly, bring us back shotguns!"

Then the young man answered him: "*Xoape!* I will only come back
when you are blind, when your head has whitened, and your lips have
become very thin. I will only come back to mourn your death!" Then
he continued his trip. He never came back to live with us. Yet much later
I eventually saw him again. He sometimes traveled up the river to the
Ajuricaba Outpost where I worked and I also saw him in Manaus. Each
time he saw me, he recommended that I be docile with the white people.
Sometimes he also told me: "Why don't you come settle with me downstream, with the inhabitants of the river! They will really feed you!" Listening to him, I would tell myself that one day I might follow his example. But since I worked for FUNAI's Ajuricaba Outpost, its agents did not

let me lose myself among the white people of the lower Rio Demini as he did. That was better for me. As for my brother-in-law, he started constantly drinking *cachaça* and I think his chest was finally taken by disease. I never saw him again after that time. He finally died among the white people, without ever coming back to our forest. In the beginning, I thought the same way he did. It was only much later, when I understood that the white people could be bad, that my mind turned away from such thoughts.

ONE DAY WHEN I was working at Ajuricaba, the head of the FUNAI outpost, Esmeraldino, took me with him to Manaus. We went down the Rio Demini, then the Rio Negro on a motorboat for days and days.[24] The closer we got, the more impatient I became to get my first sight of the city I had so often heard of! Yet in the end, I was a little disappointed when we arrived. We landed in a place far from any house and remained there during our entire stay. We slept on our boat, in the **port.** At night, I saw all sorts of lights streaming around us in every direction: boats passing each other on the river, big airplanes flying over us,[25] and the cars following each other along the riverbank. I did not feel very safe. I worriedly asked myself what all these fires in the darkness could be. And by day there were so many people and so much noise along the river! A multitude of white people bustled along it, coming and going as they yelled out the names of fish: *"Jaraqui! Curimatā! Tambaqui! Surubim! Tucunaré!"* and palm fruits: *"Açai! Bacaba! Buriti!"* All this to barter them for some old pieces of paper skin. At the time, I did not know what money was and I was still unaware that in the city you could not eat or drink without it. I watched all these white people with a little fear. There were so many of them and they rushed in every direction, like *xirina* ants! I told myself: "Our elders did not know that the white people were so numerous and that their food was so abundant! And all these machines to run everywhere, on water, on land, and in the air! It is really frightening!"

With every **jet**[26] that flew over us, I looked at the sky with apprehension. Of course, I had known the missionaries' small airplanes since my childhood, for they landed at Toototobi from time to time. But I did not know there could be such enormous airplanes and that there were so many of them![27] And, especially, I had never seen an **automobile.** So I was still very worried whenever I had to walk in the city to go to the

Manaus FUNAI house. I constantly remained on alert, watching the cars' movements on one side or the other. I was afraid they would hit me and run me over as they raced ahead. They seemed so heavy to me! I watched them from afar and tried to fix my gaze on their wheels, which intrigued me. I asked myself: "What could that be? Are they made like big iron turtles? Do they have a sort of hands and feet? How can they move so fast?"[28] In the beginning, I did not realize that the cars' wheels turned. I thought they ran! I still knew nothing at all of the things of the city! And I had never seen so many white people! They were really everywhere! I told myself they must never stop copulating to be so numerous and that this was probably why they were coming to live in our forest. But this did not really worry me yet. I merely thought: "The white people are other people, this is why they are so strange. Later, I will know them better and I will feel less anxious in their presence." In fact, I only wanted one thing: to become one of them! I was still young and truly ignorant! At that time, I was still a long way from asking myself: "With all these white people increasing around us, what will happen to us later?"

FINALLY, SOMETHING bad happened to me at the Ajuricaba Outpost. My chest was seized with tuberculosis. This disease was transmitted to me by a young *Xamat^hari* who had himself been contaminated in Manaus. It was the first time he went there, as I had. But he had been working for the white people of the river for a long time. They had even given him a wife. So he had eventually stayed in the city for long periods of time because he really liked living in the white people's company. He had also gotten used to drinking *cachaça* like they do. After a while, he began coughing more and more often. He was already very weakened, and he saw a doctor who recommended that he stop drinking and that he take medicine. This doctor tried to send him to the **hospital.** But there the young man refused to let himself be treated by white people. Over time, he became so sick that he could only think of dying. Then he decided to run away and return to his village. He had become very thin and could not stop coughing and spitting up blood. Yet once he arrived at the Ajuricaba Outpost, coming from the city, the FUNAI people let him move into the same room as me. We ate out of the same pot. We shared the same plates and the same cups. Sometimes he gave me his leftover coffee. At the time I thought his cough was just a kind of flu. I was still unaware that tuberculosis was such a dangerous disease and that it could kill. The young *Xamat^{h}ari* did not know anything either. The white people did not say anything to us. So I lived by his side for a good while, then abruptly he died. His disease had already penetrated my chest long before.

One day, upon returning from a visit to Toototobi, the outpost chief Esmeraldino found me burning with fever in Ajuricaba, prostrate in my hammock. I felt very bad and I could not stop coughing. He was fond of me and was worried to see me in such bad shape. First he tried to treat me right there at the FUNAI outpost. But it did not work. My state got worse and anyhow they were nearly out of medicine at the outpost. Eventually he came to think that it would be wiser to take me to the city. He had really decided to help me. So we took a motorboat down to the Rio Demini's mouth to get to the little town of Barcelos. He instantly took me to the hospital. But I was not able to stay there because the doctor told us he did not have any medicine against tuberculosis. He recommended that we go to Manaus, where it would be easier to get treatment. So we continued our trip, this time by going down the Rio Negro. Other men from FUNAI accompanied us. There was also Yo, a young Japanese man who had come from very far away to visit us in the forest.[29]

As soon as we arrived in Manaus, Esmeraldino took me to a hospital and left me there with another doctor.[30] I found myself alone in this city, worrying about what could happen to me from now on. In that time, I still didn't understand what the white people said to me very well. Luckily, I found someone I knew in this hospital. It was Chico, the former Brazilian missionary whom the Americans had dismissed from Toototobi! Now he worked for FUNAI, and he too had gotten sick. Despite everything that had happened with him in the past, it was good for me that he was there, because he knew my language. So the doctor told him to ask me whether there was blood in my saliva. I answered that there was and that I felt a sharp pain when I breathed. He could also see that I was constantly coughing. Then he understood that the tuberculosis was eating my chest. But he did not explain anything to me. He only notified the FUNAI people. Later, they told me what he said. He also told them that I would have to stay in the hospital a very long time before I was cured. When I got this news, I did not complain and I was not scared. I accepted all this without questioning it because I really wanted to get better. Above all, I would never have wanted to bring this disease back to the forest and contaminate my people!

I believe I stayed in that hospital for one **year.** It was long, very long! If I had wanted to, I could have run away, like many others.[31] Yet that was never my intention because I did not want to die like the young *Xamat^hari* from Ajuricaba who had given me his disease. And also the people of this hospital treated me well and I got used to them. Then I spent all my time lying in a room, without doing anything foolish, taking my medicine every day. I was not impatient. I had decided to calmly await the moment when I would be told I was cured and I could leave the hospital. In the beginning, I continued to observe the white people from a distance like in Ajuricaba, without speaking a word, just to know them. But this time I had to stay locked up with them for a very long time, alone and without anything else to do! During all this time, the other patients, the nurses, and the doctors made an effort to speak to me. So I tried hard to repeat their words, one after the other, like a *werehe* parrot. There was also a school in this hospital. I went there once in a while without learning much. But most importantly I made a friend among the other patients. He taught me words and a little writing. I much preferred to stay free and learn with him. This is how I really lost my fear of speaking to white people. I asked them for water, food, things like that. Gradually, their way of speaking became clearer to me. And little by little I also succeeded in making myself better understood. Yet I was often alone and I constantly thought about my forest with nostalgia. So time passed slowly, very slowly.

Yet one day the doctor must have told the FUNAI people, "Davi is no longer sick. We killed his tuberculosis!" Suddenly they came and told me I was cured. I was not expecting it! I was so happy to be healthy again and to finally be able to leave the hospital! It was Esmeraldino, the chief of the Ajuricaba Outpost, who came to get me and put me up at his home. He kindly took care of me again. If he had not helped me the way he did, I would certainly have died of this ugly disease. Yet once I was cured I did not want to go back and work at the Ajuricaba Outpost. The FUNAI people in Manaus also thought I should go back home to Toototobi. They told me: "Davi, now you know the white people's words. You must go back to your people. That is your place. You will help them. And later, when you have really become an adult, if you want, you can come back and work with us." These words seemed wise to me. So FUNAI brought me back to Toototobi. It was not my people who called me back to them. I decided to come back and live in the forest on my own and little by little the desire to become a white man disappeared from my mind.

Now I sometimes remain awake in the middle of the night and feel alone in the middle of the sleeping people in our big house at *Watoriki*. Then my thoughts escape far away, one after another, and I cannot stop them. I keep moving around in my hammock, unable to sleep. I think about our ancestors who turned into game in the beginning of time. I keep asking myself: "In which place did the night beings really come into existence? What was the sky like in the beginning of time? Who created it? Where did the ghosts of all those who died before us go?" Then my mind begins to calm down and I finally hope to find rest. But thoughts about the white people often also come to torment me. Then I say: "When my mother carried me at her breast, these outsiders were still very far away from us. We knew nothing about them. The elders did not suspect that one day they would make us nearly all perish! Today I have understood that they destroy our forest and mistreat us because we are other people than they are. So if we try to imitate them, things will really go badly for us!"

When I think about all this, sleep escapes far from me. The time of my adolescence is far away. Yet I still remember that long ago I vainly attempted to look like the white people! I hid my eyes behind dark glasses and my feet in shoes. I combed my hair on the side and tied a watch to my arm. I learned to imitate their way of speaking. But that did not lead to anything good. Even wrapped in a beautiful shirt, I still felt like an inhabitant of the forest inside myself! This is why I often tell the young people of our house now: "You think that one day you may become white people? That is nothing but a lie! Do not believe that you merely need to hide in their clothes and display a few of their goods to become one of them! Believing something like that just tangles up your thought. You will wind up liking the *cachaça* better than the forest's words. Your mind will darken and finally you will die of it!" It is true. *Omama*'s words and the *xapiri*'s words are very ancient. Only they can make us happy. Imitating *Teosi* and the white people's words is no good for us. They can only torment us. This is why I think we must follow in the footsteps of our ancestors, like the white people follow in those of their ancestors.

It is true that today I still hide my penis in shorts. It is a habit I picked up with the people of *Teosi* when I was a child. It is also true that I know a little of the white people's language. Yet I only clumsily imitate it when I go to the city or talk to them in the forest. Then, as in the past, I try to be a *werehe* parrot and attempt to make myself understood. But as soon as I am alone with my people, my mouth closes itself to these strange words. They escape far from my mind, and my tongue hardens without being able to pronounce them. The thought of young people who want to become white people is full of smoke! This is why when I became an adult I chose to keep the sayings of our forebears inside me, even if they died a long time ago. It is with the *xapiri*'s songs that my thought can extend to the rivers' sources or towards other distant forests or beyond to the sky's feet. It is with them that I can see what the great shamans of the past saw before me, that I can contemplate the images of the beginning of time as they made them come down long before my birth. It is so! I never want to stop following the path of our long-ago elders, for that is our true way of becoming wise.

CHAPTER 13

The Road

The road and the city

A FTER MY recovery from tuberculosis, I came back among my people and resumed my life in the forest. Then time passed and one day Chico, the former missionary I had met in the hospital, appeared at Toot-otobi again. Now he worked for FUNAI. He had come up the river to our house to recruit people who could help him. He wanted to make friendly contact[1] on the upper Rio Catrimani with Yanomami who had never seen white people, in a distant forest without any trails. He was doing this work for FUNAI because at the time the white people were starting to open a road path on our land.[2] Yet we considered the people Chico was looking for as enemies, and we barely knew them. My wife's father's group had often launched war raids on them in the past. But it was always to avenge the death of elders whom their *oka* sorcerers had killed first. They used to call these people *Moxi hatëtëma,* and we call them that to this day.[3] They had never attacked our fathers and grandfathers in the open, with their arrows, only with their sorcery blowpipes. They had never made friendly contact with the white people and did not possess any of their goods. They only used stone axes to work in their gardens.[4]

Several of us accepted to accompany Chico on this trip: my stepfather and me, three other men from our house, and a *Xamat^hari* who lived downstream on the Rio Toototobi.[5] There was also another white man, whose name I've forgotten. From the Toototobi Mission, we went downstream on a motorboat to the mouth of the Rio Mapulaú. Then we went up this river for a while and we came to a little house inhabited by the elders of the *Watoriki* people, the people of my future father-in-law. At that time they lived on the *Werihi sihipi u* River, a small tributary of the Rio Mapulaú. So we made a stop there to bivouac. But we instantly understood that they had just been struck by an epidemic. They had barely finished the bone cremation *reahu* feast for their dead. Their guests were inhabitants of *Sina t^ha* and *Hero u* who had close kin married among them.[6] The deceased's bones had already been burned and crushed. Their ashes had been kept in gourds sealed with beeswax.[7] But since the smoke from the epidemic dead is dangerous, several more people had died soon after the cremation, just before our arrival. So when we entered their house, all were tormented by mourning and were still in tears, wailing for their recently deceased.

We only slept there a single night. We left the next day at dawn. Chico began by ordering us to hide part of our provisions and the trade goods intended for attracting the *Moxi hatëtëma* in the forest.[8] We were too heavily loaded. Then we went back down the Rio Mapulaú to find the main course of the Rio Demini and eventually we entered another of its tributaries downstream, which we call *Haranari u.* But our boat was still far too heavy for this little watercourse. So we traveled upstream with much difficulty. The stream was more and more obstructed with tree trunks and lianas. Exhausted, we finally gave up on our navigation. We set up camp on the riverbank and unloaded our boat. Next we continued on foot upstream through an unknown forest. It was very difficult to advance through the tangled undergrowth. Yet we did not get discouraged and we were happy, for my stepfather constantly cheered us with his jokes as he cleared the path with his machete. He was a very valiant man who always liked to make people laugh. Finally, after walking for three days, we reached the bottom of a big rocky peak we call *Weerei kiki*. We slept there, and spent a long time over the following days looking for tracks of the *Moxi hatëtëma* in the neighboring forest. But we found nothing at all. It was truly empty of any human being. Finally, Chico gave up and we went back the way we came to return to the Toototobi Mission. All this had been for nothing. I learned later that the *Moxi hatëtëma* lived very far from there, on the upper Rio Apiaú!

During this trip I began to get to know the people of my future wife's father, the ones who lived in the house where we stopped on the way, on the *Werihi sihipi u* River. I had heard about them as a child because they had long been at war with our elders, who called them *Mai koxi*. Yet I had only met them a single time, shortly before I left to go work at the Ajuricaba Outpost. At that time they were living on the upper Rio Catrimani, and the people of *Teosi* wanted to make them live closer to their mission. To do this, they had started by flying over their house, dropping from their plane arrows and trade goods in the forest. Later, they sent us on an expedition from Toototobi to make friendship with them. But we had caught the flu without knowing it, and after a few days of walking we were all feeling sick. We were soon all burning with fever and decided to turn back. In the end, the people of the *Werihi sihipi u* River came to visit us at Toototobi a little later, by themselves. They just arrived one day when we were not expecting them. We listened to their words of friendship, then we cleared a path between their house and ours. This is how we began to visit each other.[9]

I DID NOT stay in Toototobi upon my return from this trip in search of the *Moxi hatëtëma*. Chico invited me to continue working with him and I decided to follow him again.[10] After my tuberculosis, my stepfather did not want me to go back to the white people. Yet I did not listen to him. I had forgotten the city and my desire to become a white man. But in the meantime another of my uncles had also died. Enemy sorcerers from the highlands had blown harmful plants on him and had broken his bones. So I felt the anger of mourning and loneliness again. This is why I left

Toototobi with Chico. Yes, he had acted badly with us in the past, and our elders continued to hold it against him. But I was a child when he worked at the Toototobi Mission. He had left our forest a long time ago. My thought had calmed down and I had forgotten all that. That is how I am. My anger does not last when I no longer see the people who provoked it. And also, Chico had helped me when I was in the hospital. Anyway, he had promised me that I would live with him in the city and he would feed me. He seemed to want to take care of me. So I began to take a liking to him, and when I left Toototobi I moved into his home in Manaus.

He lived in his father's house, a little outside of town, in the forest. We stayed there for a time. But I needed those old paper skins they called money to live among the white people there! So Chico found me work. In the morning, I had to fill buckets at a spring, then go sell them in the surrounding area. This is how I was able to make the money for my food. In the afternoon, I also washed swimming pools in big houses. This was to repay Chico for the few goods he bought for me: shorts, shirts, flip-flops, a hammock, and soap. It was he who truly taught me to work for the white people. He often told me: "In the city, no one will have any friendship for you if you are lazy! White people only like hard workers. Do not think that they give money to layabouts!" Some time later, he found another house, and we went to live there. Then thanks to him, the FUNAI people eventually decided to take me back with them. They knew that I worked well in the past and that I had learned more of the white people's language now. They asked me to become an interpreter. This is how I went back to work for FUNAI in the forest with Chico.[11]

This time we left Manaus on a big boat with two decks headed in the direction of the Rio Branco. It was the heart of the dry season. The waters were very low. We went upriver slowly, then we entered one of the Branco tributaries, the Rio Catrimani, and followed it upstream to the mouth of a small watercourse called Igarapé Castanho. Nearby was a Yanomami house whose inhabitants used to work for the white people of the lower Catrimani.[12] We stopped there for a while. The big boat dropped us on the riverbank, then left for the downstream. We continued to navigate up the Catrimani in a small motorboat. It was very long and difficult because this river is cut through with many rapids. During the whole trip, we only encountered a single white hunter. He was traveling downstream. We stopped and Chico called to him to draw close. When he saw his canoe was full of giant otter and ocelot hides, he started to speak to him with anger. Then he confiscated his load and sent him back home, warning him that it was forbidden for white people to hunt in our forest. After this, we continued navigating upstream, towards the Catrimani Mission where the Fathers lived. We bivouacked there, then left our boat behind because the rapids upstream from the mission are impassable. We continued our trip by foot through the forest along the Rio Catrimani. I traveled with Chico and two other FUNAI men, who were Satéré-Mawé and Tikuna Indians. Only one Yanomami from the house close to the mission accompanied us. We walked for days and days in the direction of the upper Catrimani. First we passed through the home of the people of *Makuta asihipi,* then those of the people of *Mani hipi, H^waya u,* and *Uxi u.* From there, we continued walking along the banks of the Rio Lobo d'Almada to its upper course. Eventually, we reached a last house inhabited by the people of my future wife's father, whom Chico and I had first met during our trip on the Rio Mapulaú. But since that time, they had abandoned the *Werihi sihipi u* River and sought shelter by one of their old gardens of the upper Lobo d'Almada, which they called *Hapakara hi.* They used to live there in the past, before they answered the people of *Teosi's* call and got closer to the Toototobi Mission. Most of them had just been devoured by the epidemic of the *Werihi sihipi u* River, and the survivors had gotten scared. This is why they had decided to turn back and return to live in the highlands, far from the white people.

THIS EPIDEMIC smoke had reached them when they had just finished building a vast house downstream of the little *Werihi sihipi u* River. This is what they told me. My future wife's father lived there with his older brother, who was the great man of the house. He was a true elder, a great shaman. One afternoon, they heard the whirring of a **helicopter** circling above them in the forest. It was the dry season. The waters were low. The river was skirted by sand beaches. After a while, the helicopter landed on one of the beaches, far from the house. Silence returned. Then sometime later it left and disappeared into the sky. The people of the *Werihi sihipi u* River were worried. They asked themselves what these outsiders had come to do in their forest. At nightfall, they heard an explosion. They thought the white people must have left an unknown and dangerous thing of fire in the sand and that it had produced this detonation; something like these **bombs** they had started using to clear their path in the forest.[13] The next day the *Werihi sihipi u* River's great man decided to go there to see what had happened. A group of youngsters joined him, mostly because they thought they would bring back abandoned goods from their visit. They reached the riverbank quite quickly but found nothing but dirty papers, **cans**, rubber boots, and a straw hat on one of its beaches. They also saw tracks of the helicopter's feet and those of its occupants. But most importantly they found several holes dug from one side to the other in the sand. They asked themselves what the white people could possibly have done there. There was nothing else. Finally, weary of their vain search, the people of the *Werihi sihipi u* River went back to their house.

A short time later, their great man fell ill and died very quickly. Then all the house's inhabitants began to burn with fever. They were shaken by trembling and suffered an insatiable thirst. They did not understand what was happening to them. It was not a simple coughing sickness.[14] Very quickly, many other people died too. They fell one after another, more and more numerous, especially women and children. Some tried to escape into the forest, but they died there in the same way. Only a few survived this epidemic smoke. The *Werihi sihipi u* River house was big, yet in a very short time this fierce sickness nearly entirely emptied it of its inhabitants.[15] What did the white people who descended from this helicopter come to do? Did what they burned really contaminate the people of the *Werihi sihipi u* River? I do not know. I would have liked to examine these holes in the sand myself. Chico told me he had also searched there along the river without finding anything else. Maybe these white people blew up an epidemic smoke like the one Oswaldo blew up in *Marakana* when I was a child? Yet they were not angry with the people of this house.[16] They did not even meet them! Maybe they merely wanted to kill them to empty the forest and dig for minerals later? I have never understood what could have happened.

AFTER STOPPING at the *Hapakara hi* house on the upper Rio Lobo d'Almada, we continued our trip with Chico, but this time on our way downstream of the Rio Mapulaú. When we arrived there, the forest was silent. All that remained in the region was the house abandoned by the people of the *Werihi sihipi u* River. Yet Chico decided to build a new FUNAI outpost here. He wanted to attract all the people of the houses we

had just visited along the Rio Lobo d'Almada around it.[17] We began clearing and burning a plot of forest upstream, not far from the mouth of a
small river called *Maima siki u*. Chico wanted to plant a garden there
when we came back to the area at the beginning of the next rainy season.
In the meantime, he was in a hurry to leave. So we had to do all this work
in a few days before returning to Manaus.

In the end, we only stayed in town until a new moon and then we returned to the Rio Mapulaú. But this time we did not go on foot! To get
there, we traveled up the Rio Demini on a motorboat from the Ajuricaba
Outpost.[18] It was a lot easier! But because of the rapids, we had to stop
downstream of the plot of land that we had cleared the previous time. So
Chico had to find a new place to set up his Mapulaú Outpost. He chose
an old patch of forest where the Inspetoria people had settled long before,
when the Boundary Commission had traveled up the Rio Demini for the
first time.[19] We rapidly opened a vast clearing and built two huts made of
manaka si palm wood boards covered in *paa hana* leaves. We were in a
hurry because the rainy season was approaching.[20] Then we went to the
old plantations abandoned by the people of the *Werihi sihipi u* River. We
cleared their tangled vegetation for several days because we wanted to
take banana plant shoots for our new garden.

We had barely finished this work when my future wife's father and his
two brothers-in-law arrived at the *Werihi sihipi u* River accompanied by
their wives and children. They had come from the Rio Lobo d'Almada to
collect taro and bananas in their old gardens. But when they arrived, they
were surprised. They wondered who could have cleaned their abandoned
plantations! They slept in their former house and came to meet us the
next day. Chico asked them if we could pull up the banana plant shoots
we needed, then he invited them to come settle at the new outpost. They
accepted.

At the time, Chico made us work relentlessly! So we hastily planted
our new garden with large plots of banana plants and sugarcane. Then
we prepared everything necessary to really settle in this new place. Notified of our presence, the people who had remained in the *Hapakara hi*
house of the Rio Lobo d'Almada cleared a path from there to the Rio Mapulaú and started to visit us regularly. Then those from the Rio Toototobi
did the same thing and came to get trade goods from us.

This is how I spent my time with Chico at this new Mapulaú Outpost,
but I finally got tired of it. He really made me work too much! He never

stopped giving me orders. I had to clear land, cut posts, cut up palm wood laths, and collect palm leaves for the huts. I had to toil without respite to plant the garden. But he still never seemed satisfied. He lost his temper at me all the time. But it was not because I was lazy! He had asked a young girl from the *Werihi sihipi u* River people to help him and made her his wife, just like during his old missionary time! She had previously been briefly married to my Toototobi stepfather, who had rejected her. So she had come back among her people at the *Werihi sihipi u* River. Chico was very jealous about her. No man could get near her. Yet this girl sometimes passed in front of the place I was working and talked to me. Once, Chico saw us sharing food in the company of other Yanomami. We were joking around and we laughed together. He instantly called me aside and abruptly asked me if I was copulating with her. I denied it. I told him that I merely treated her in a friendly way, nothing more. He did not believe me and that is why he started to hate me. His jealousy really made him enraged! He even threatened me by yelling: "Don't get near her! I want her to myself! Watch out!" These threats made me angry. I answered in the same tone: "You are bad and your thought is empty! You are a white man. Go find yourself a wife in Manaus instead of taking ours and being jealous of them!" In the end, he chased me away from the outpost: "I don't want you here anymore! Go away, go back home!" All this made me furious at him. This time I truly understood why my elders did not want him in Toototobi anymore! So I decided to go back to the city to tell the FUNAI people about all this. I soon left for Manaus with a Xikrin Indian who worked with us.

The FUNAI *delegado*,[21] Porfírio, thought I was still on the Rio Mapulaú with Chico and was really surprised to see me show up in his office by myself. He asked me: "What are you doing here? What happened? Why did Chico let you leave?" I told him everything: "Chico sent me away out of jealousy. He took a young Yanomami girl as a wife and even forbids me to talk to her. Yet she is one of our people and that forest is not his!" Porfírio listened to my words with attention. He seemed upset. Then he answered me: "You are right, Chico did wrong! You are a Yanomami, he should not mistreat you this way!" He was a wise man. Later he called Chico back to send him to work in a different place, in Surucucus, where illegal miners had just invaded the forest in search of **cassiterite**.[22] Once Chico left the Rio Mapulaú, another man from the outpost, a Tukano Indian, took his Yanomami wife for a while. Then he also left and he abandoned her along the way, pregnant and far from her people. Finally, a *Xamat^hari* married her in Ajuricaba. She still lives there. The new outpost we had just opened on the Rio Mapulaú was left abandoned. No white man ever came back there. Later, the people of the *Werihi sihipi u* River burned it down with everything that remained inside, including the radio. They were furious that they had been abandoned there despite Chico's promises. And also, a new epidemic smoke had just struck their people who had stayed on the upper Rio Lobo d'Almada.[23]

THIS IS HOW my first job for FUNAI came to an end. Next, Porfírio, the *delegado* in Manaus, sent me to a new outpost, Iauaretê, which had just been opened on the upper Rio Negro, upstream from São Gabriel da Cachoeira, where the Tukano Indians live.[24] He must have done it to send me away from our forest, because I did not want to work there with Chico anymore. Once I arrived in Iauaretê, the outpost chief who had come with me from Manaus decided I had to go work alone with the Maku. These are inhabitants of the forest who lived very far from the white people, near a mountain called the Serra dos Porcos. He accompanied me to their home, then quickly left, leaving me on my own. I stayed in this place quite a long time. I did not feel comfortable at all there, for the Maku are very distant other people whom I did not know. I did not understand a word of their language, and they knew nothing of the white people's language. I was worried, and I told myself: "How will I live among them? They don't hear any of my words and they speak a ghost language!" But luckily they took a liking to me and generously fed me the whole time I was with them, though they did not understand me.

At the time, I also worked upstream on the Rio Negro, among other inhabitants of the forest near the border with **Venezuela.** I think they were called Warekena. I'm not sure anymore. I only remember that they spoke yet another language. It was very difficult for me to work on the upper Rio Negro. This forest belongs to other people than my own. There are very many of them, and they all have different languages.[25] I never knew how to speak to them. Because of that, I always felt bad doing this work. So I decided not to stay in this region anymore. I asked to leave, and the FUNAI people called me back to Manaus. This time, they decided to make me study so I could become a **health worker.**[26] I started to learn how to make people swallow medicine and how to tie bandages and even give injections. I was very hardworking. I really wanted to know how to heal people in the white people's way. Yet I had difficulty understanding what the teachers explained to me. I was very young, and I still imitated white people's talk so poorly. And all I knew of writing was the little I had been taught as a child at the Toototobi Mission, in my own language. I was not able to read the medicine's paper skins. FUNAI had brought me to study with other Indians who had lived with white people for a very long time. They thought I was like them. But these strange words were not as clear to me as they were to them. I had left the forest too recently.[27]

And so the new FUNAI *delegado* who had replaced Porfírio sent me home to Toototobi as soon as the course ended, without any explanation. It all happened very fast. He only told me one thing: "Go back and work in your village, with your people. You will give them medicine!" He put me in an airplane, and I arrived in Toototobi from one day to the next. That's all. A short time later, one of the missionaries came to me and just told me: "You don't work for FUNAI anymore, they fired you!" This new *delegado* in Manaus did not like me. He was bad and had no friendship for the people of the forest. He must have told himself: "I don't know what to do with that Davi. I don't want to see him anymore; let him go back to his place, in the forest!" Yet in Manaus I had really tried to learn the white people's words to be able to heal people like they do. I had behaved well and never drunk any *cachaça*. I don't know what I did to make this *delegado* throw me out like that, without even saying a word! He was probably a coward and did not dare to speak to me when I looked in his eyes. It is so. I nearly became a health worker! I had started studying, I liked it. Yet since FUNAI threw me out I got angry and abandoned this idea. I told myself: "Too bad! This FUNAI *delegado* is just a know-nothing!" And I decided to quietly live among my people again at Toototobi, just like I used to in the past.

YET ONCE again it did not last. Sometime later, other white people arrived at Toototobi. These were people from the department of **malaria** prevention. We knew them because they sometimes came to spray a medicine to kill mosquitoes in our houses. This time they had come to capture small *ukuxi* gnats to find a disease that causes blindness.[28] They had heard that I spoke the white people's language. So shortly after their arrival, they called for me. They asked me to help them: "We don't know how to make ourselves understood, and we aren't able to work! You who know our words, stay by our side!" So I did it and little by little, they saw that I managed well as an interpreter. Once their work in Toototobi was finished, they asked me to accompany them: "Come with us! We have friendship for you. You will continue to work for us and we will pay you for that!" They still had to go to several places in our forest, on the Rio Mucajaí and Rio Catrimani, then in the highlands, to Surucucus.

Since the malaria fever was burning in me, I decided to follow them, so I could at least be treated in the city. But I was not alone on this trip. I was also accompanying a few Toototobi elders suffering from the disease of the *ukuxi* gnats that the white people were looking for. We were all supposed to be sent to the hospital. A small airplane came to get us to take us to town. This is how I got to know Boa Vista for the first time![29] I had heard about it for a long time, but I had never been there before. When I arrived, I thought it was a beautiful place. It was a small, very quiet city at the time. There were no thieves, and the white people did not kill each other there yet. You could keep your mind calm. No one there knew much about the Yanomami yet. That was good. We could go wherever we wanted without fear. The white people were friendly with us. But it has changed a lot since then. The gold prospectors arrived in large numbers and the **streets** filled with hostile words against us. Now I would be scared to take a walk there alone.

After the malaria people brought me to Boa Vista, I usually stayed at the hospital there, like in the past, when I had arrived in Manaus. I took care of myself and helped my elders who did not speak the white people's language. Then when we felt better we started to visit the Toototobi missionaries who also have houses in this town. But we did not go there to imitate *Teosi* with them! No! What we wanted was to work for them in order to make a little money. We really desired to possess the big cotton hammocks and other brand new goods we had seen in the stores in town! To get them, we had to weed the *Teosi* people's gardens, like we did at the mission. All the time we spent at their place was busy with that!

A little later, once I truly recovered, the malaria people asked me to work with them again. I had liked being a help to them. They had treated me well, and the work they gave me was not difficult. I wanted to follow them on their next trips! Yet one time when I was going from the hospital to work at the missionaries, I met a Yanomami coming out of the nearby

FUNAI house. He was a *Xamat^{h}ari* from the Rio Cauaboris, from the *Wawanawë t^{h}ëri* group. He had left his people at a very young age. The FUNAI people had taken him with them. His name was Ivanildo. I had already met him before, in Manaus, at the time when I was working there for Chico. Now he was an interpreter on the road which the white people were starting to open in our forest and which they called Perimetral Norte.[30] After running into me, Ivanildo spoke about me to an outpost chief I had also known in Manaus in the past: Amâncio. Amâncio was now working in Boa Vista. When he heard of me, he asked Ivanildo to bring me to the FUNAI office so he could meet me. I was curious, so I quickly went to visit him and talk a little.

As soon as I arrived, Amâncio told me we had to meet the *delegado*. Then they started to tell me: "You should not accompany the malaria people anymore! It is us, FUNAI, who really take care of the Indians, you should work for us!" They kept insisting that I should work for FUNAI again! This surprised me and at first I did not want to listen. Not so long ago, the Manaus *delegado* had fired me with no explanation. Now all of a sudden the one in Boa Vista was deciding to take me away from the malaria people, who treated me so well! That really made my chest angry. While they were talking, I told myself: "These white people's thought is really filled with smoke!" So I instantly retorted: "No! I don't want to work for FUNAI anymore. I did it before, at the Ajuricaba Outpost and on the Rio Mapulaú, then I studied to be a health worker in Manaus. In the end, I was dismissed back into the forest without even a word! Your elders have no wisdom and are hostile to me. I am willing to work with the white people, but I do not want to let myself be mistreated this way! Now I would rather help the malaria people!"

Despite these words of refusal, the Boa Vista *delegado* really insisted, speaking to me harshly. He told me that because I was an Indian the malaria people could not make me work without FUNAI's permission.[31] Amâncio added: "The *delegado* who sent you away from Manaus was bad. This is another FUNAI, and this other *delegado* is now in charge.[32] He is a man of good and he really wants you to work for us. You cannot refuse him in this way!" They repeated all this several times and Amâncio seemed very committed to hiring me to work with him personally.[33] Finally I spoke of all this to the malaria people, who told me: "FUNAI will not let us hire you because you are one of theirs. All right. So go back with them, if they want you so much!" This is how I came back to FUNAI once again! Amâncio finally convinced me by promising me I would go

to work among my people. He told me: "We will go settle at the Ajarani
Outpost, along the road, it is in your forest.[34] We will help the Yanomami
who live in this area. We will defend them together, for the road has just
arrived near their home!" Without these words, I would never have accepted. At that time, I knew next to nothing. I basically grasped the white
people's words, but I did not understand their thought. My mind was still
uncertain. I had listened to Amâncio, and told myself that he might be
a white man who thought straight. When he gave his orders at the Boa
Vista FUNAI, he told everyone he would defend the Yanomami. I believed him. I knew nothing about him except what he did in my own
sight and what he told me of his intentions.

He repeated that he would not let our forest be invaded by the white
people. And it is true that often he really acted to defend us and our land.
At the time, he had a lot of money to work, given by FUNAI. When the
miners invaded Surucucus the first time, it was he who had them expelled.[35] He also sent airplanes to bring doctors to our forest. He helped
us that way. And he also traveled with me to know our land better. We
went up the headwaters of the Rio Demini together, very far upstream,
near the border with Venezuela, to the Rio Taraú. We continued walking
in the forest until we reached the houses of *Xamat^hari* who had never
seen white people before, like the *Xihoma t^hëri*.[36] Amâncio appreciated
my work and really had friendship for me. I am sure of that. He helped
me and often supported me within FUNAI. Otherwise I would have
stopped working there long ago! Yet later when I learned that he had
helped the military in Brasília to divide our forest in small parcels, like
pens for cattle,[37] I really did not like it. Despite his friendship, I think he
deceived me in hiding these words of cutting up our land. This made me
very upset.

As soon as I accepted to follow him, Amâncio sent me to get my **identity card**, which I had forgotten in Toototobi.[38] Upon my return, he quickly
made me new FUNAI papers. Then we left to work on the outpost he
had told me about, on the edge of the Perimetral Norte.[39] At the time, it
was just a hut near the Rio Ajarani, where the Yanomami whom we call
Yawari live.[40] It was them who first saw the white people tearing out
the forest's ground with their enormous tractors and **trucks** to open the
road's path. When these machines came onto our land, I was still far
away.[41] I was accompanying Chico on his vain search for the *Moxi*

hatëtëma, then I was in the area of Iauaretê, lost among the Maku! I only came to know the road once it had entered very deep into our forest, nearly to the Rio Demini. Yet Chico had already told me a little about it, when we were on the Rio Mapulaú. He had explained: "The white people are clearing a big path in the forest. They are coming in our direction from their little town of Caracaraí. Then they will cross the Demini and continue very far, all the way to the Tukano!" He sometimes talked about it on the radio too, with the other FUNAI men. I did not understand everything they said, but what I understood was enough to worry me.

When I was a child, the white people came up the rivers in big canoes and made a great number of our elders perish. Then they came back by airplane and helicopter. Once again their epidemic fumes made many more people die. Now they had decided to open one of their roads all the way into the heart of the forest. I knew that their diseases would probably devour those of us who had survived until then. I thought about all this when I was alone in FUNAI's Mapulaú Outpost. It tormented me and made me sad. I told myself: "The white people tear up the forest's floor. They cut down the trees and blow up the hills. They make the game run away. Now will the epidemic fumes from their machines and bombs finally kill all of us?" I already knew that this road would only bring us bad things. No one had warned us before the construction began. Chico had only said a few words about it to the people of the *Werihi sihipi u* River when we opened the Mapulaú Outpost. I myself had warned them about the diseases which would once again spread through the forest. But soon after, I left for Manaus because of my argument with Chico. On my way, I only saw the cleared forest path where the road would be laid out. At the time, there were only small groups of white people in torn clothes here and there, working alone with their axes.[42] The big machines had not arrived yet.

THE WORDS about the road that I was able to understand at that time also frightened me for a reason other than diseases. I had sometimes heard the FUNAI people tell that to open the route from Manaus to Boa Vista, the soldiers had opened fire on the Waimiri-Atroari Indians and had thrown bombs in their forest.[43] These people were valiant warriors. They did not want the road to cross their land. They attacked the FUNAI outposts to prevent the white people from coming into their home. This is what made the military angry. Hearing this talk, I started to fear that the soldiers might also want to strike out at us that way! But luckily that never happened.[44] Yet many of our women, children, and elders died because of this road.[45] They were not killed by soldiers, no, that is true. But they were devoured by epidemic fumes brought by **workers** and visitors. And once again it revolted me to see my people die this way. All this had just repeated itself over and over since my childhood. The pain of my people's death long ago at Toototobi strongly came back to me. Once again anger invaded my thought: "This white people's path is truly evil! The *xawarari* epidemic beings follow their machines and trucks on it. Will their hunger for human flesh really make all the rest of us die, one after another? Did they open this road to make the forest silent of our presence? To build their houses on the traces of our own after we disappear? Are these outsiders really evil beings to continue to mistreat us this way?"

Our elders did not have these worries because, once again, they knew nothing of the road. The government's great men did not gather them together to make their voice heard about its opening. They did not ask them: "Can we clear this path on your land? What do you think? Won't it frighten you?" The few white people who had mentioned it incidentally had said next to nothing. Neither the FUNAI people nor the people of *Teosi* had really warned anybody about what would happen. As for me, who speaks the white people's language, FUNAI had sent me to work very far away from there, in Iauaretê. So one day the machines arrived in our forest without having been preceded by any word. Our elders, kept in the dark, did not show any hostility to the white people who worked on the road. Neither did those of Ajarani, nor those of Catrimani, Mapulaú, or Aracá. None of them said anything, despite their concerns.[46] They thought that no matter what happened, at the end the forest would never disappear and that they would continue to live there as they always had. They just told themselves that they would be able to obtain food and goods from the new outsiders in abundance. They knew that these white people of the road threw them off their airplanes and distributed them generously.[47] They knew nothing of the white people's real intentions. As for me on the Rio Mapulaú, I was too young to convince them of what was really threatening them. So I went down the river to Manaus alone, keeping my worry and my sadness in my chest.

LATER THE men whom I had seen cutting down trees with axes on the road's path went away. Others arrived in far greater numbers to replace

them. They started tearing up the forest's soil with enormous yellow tractors. This time the people of the *Werihi sihipi u* River really understood
that the road might get close to them. They had been invited to a *reahu*
feast in the house of the inhabitants of *Hewë nahipi,* on the Rio Jundiá. At
the time, the white people were working with their machines less than a
day's walk downstream, on the banks of the Rio Catrimani.[48] My future
wife's father and his people heard their incessant rumble for the first
time then and were surprised by it. Their hosts explained it to them: "It is
the white people clearing a path downstream by tearing up the ground in
the forest with huge machines!" They were perplexed by this news but
they did not speak about it much. They came back home with these words
in their thought.

Then a short time later they also started to hear the voice of the tractors digging in the ground all the way from their house on the Rio Mapulaú. They had never listened to such a strange noise in the forest. At first,
it still seemed distant to them. But day after day it got closer and became
more and more distinct. Their concern grew, and they asked themselves
what could possibly be heading towards them. They had still never seen
the giant machines and trucks the white people used to clear the road's
path. Their dull roar never stopped and seemed to them to be the sound
of evil beings devastating everything on their way. Now they heard it day
and night, without respite, and they anxiously began asking themselves:[49]
"Will the white people also destroy our house as they tear up the ground?
Or will they make it blow up and burn us all with it?" Their fears no
longer let them rest, and the detonation of the bombs shattering the
rocks of the hills scared them more than anything. The epidemic smoke
from the helicopter had devoured most of them the previous summer.
They feared it would happen again: "Will the white people of the road
bring us sickness and make us perish again? If it happens this time there
will be no one left to gather our bones and mourn us!" There were so few
of them left that they asked themselves if the fumes from the machines
would now finish them off once and for all. All were very scared, the elders as well as the rest of the people. Yet some youngsters were still curious to get a closer look. A few brave ones sometimes exclaimed: "Let's go
to the white people's path! We'll ask them for shotguns and cartridges!"
Despite their fears, they were seized by the desire for merchandise. So a
small group would gather and set out, guided by the distant roar of the
tractors. But they always turned back before they reached the road. Fear

was stronger and made them change their minds. At the last moment, they always thought: "If we go there, we will die!" and they never dared to venture to the place where the white people's gigantic machines were tearing up the forest's ground.

Finally the rainy season arrived and the work on the road suddenly stopped. All the white people and their tractors and trucks went away. Reassured, my future wife's father and his people stayed near the Mapu-laú Outpost. The rest of the group went back to its former *Hapakara hi* house on the upper Rio Lobo d'Almada. The forest was silent again. Yet it was then that the *xawara* epidemic suddenly returned. All the groups of the Rio Lobo d'Almada were gathered for a *reahu* feast in one of their houses at H^w*aya u*. There were also some people from *Hero u*, who had come from the upper Rio Mucajaí, where there were no white people at the time. These guests had almost none of these outsiders' goods. So during the feast, they decided to go downstream to the Fathers of the Catrimani Mission to acquire machetes, glass beads, and pots from them. After a few days of work there, they came back to H^w*aya u* with the objects they coveted. Yet like the people of *Teosi* in Toototobi before them, the Catrimani Fathers had unknowingly brought a child infected with measles from the city to their mission.[50] Then the people of *Hero u* brought this *xawara* epidemic back to the feast's hosts along with the merchandise they had obtained from the missionaries. This time, no one saw any smoke explode. Yet the elders remembered that a Father from the mission had visited them on the Rio Lobo d'Almada recently and that one of their people had stolen some trade goods from him. They thought he could have made this epidemic burn out of revenge. That's why they called this disease the "Father's epidemic smoke," *patere xawara a wakëxi*.

As for me, I do not really know what happened. This is what I have heard said. In any case, towards the end of the feast, the sick guests ran away from H^w*aya u* to try and escape from the *xawarari* epidemic's beings. But it was in vain! These cannibal beings chased them to their home and devoured nearly all of them. And so, in a short time, the houses of the Rio Lobo d'Almada were emptied of most of their inhabitants.[51] The *Hero u* guests lived several days away from their hosts' house by foot. So they did not bring the disease to their own people. Several of them died on the way back, in the forest. The others barely survived. One of their elders remained sprawled on the ground, unconscious, for several days. The ants ate his eyes and made him blind, but he did not die. This epi-

demic was what we call *sarapo a wai,* measles. It was very dangerous for
these elders whose flesh had never known such a disease. It is the same
sickness that devoured all my people at Toototobi when I was a child. Today the few elders remaining at the Rio Lobo d'Almada houses and my
wife's father's house, *Watoriki*—where I live now—are survivors of these
epidemics of the time of the road. Since then, the white people have left
their gravel path to silence. It has been nearly covered in tangled vegetation. Yet our forest has been soiled by diseases that will no longer leave.

The white people of the road did not burn epidemic fumes like Oswaldo's at *Marakana* or the white people helicopter's at the *Werihi sihipi u*
River. This time the *xawarari* epidemic beings simply accompanied the
outsiders' machines and trucks into the forest. These evil beings are in
the habit of following the white people wherever they go, for the white
people are their fathers-in-law.[52] They keep their eyes set on them and
move in their footsteps. These are evil beings greedy for human fat. Only
the shamans can see them and send their *xapiri* to try and push them
back as soon as they approach us. Yet if they do not succeed in chasing
them away, these *xawarari* epidemic beings set up their hammocks in
our houses and devour us as they please, taking their time. They do not
kill everyone at once. First they eat a group of people, then they come
back to devour some of the survivors. If a few men and women still manage to escape them, they will attack their children later. This is how these
evil beings empty the forest of its inhabitants little by little.

The machines' and motors' fumes are dangerous for the forest's inhabitants. These are also metal smoke, epidemic smoke. We had never
inhaled such a thing until the white people came. We are other. Our flesh
does not bear vaccination marks, and we do not have medicine for the
xawara epidemics. Our elders had always been protected from it by the
forest's coolness. We are of another blood. Unlike the white people, we
have never lived on blazing lands without trees where machines run in
every direction. From the beginning of time, our ancestors lived alone in
the forest, far from merchandise and motors. These epidemic fumes
have a frightening smell which cut our elders' breath of life. As soon as
they inhaled them, they made them all perish, one after another. And to
this day the people of the highlands continue to die of them in the same
way.[53] At the time of the road, I would have wanted to tell the white people: "Do not come back to our forest! Your *xawara* epidemics have already
devoured too many of our elders here! We do not want to know such sad-

ness again! Open your trucks' paths far from our land!" But I did not dare to speak to them. I was still too young and not wise enough. I did not know what it was to defend the forest. I did not know how to make my voice heard in their cities. It was only later, after the road tore up the forest, that I started to think more firmly. I started to dream of the forest *Omama* created for us more and more often and little by little his words grew inside me.

CHAPTER 14

Dreaming the Forest

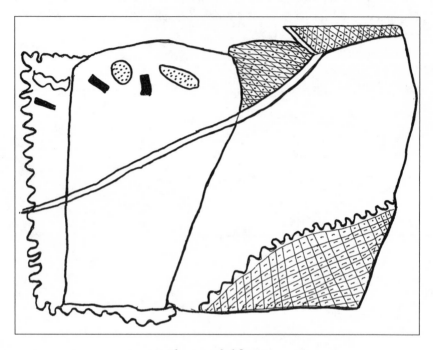

The wounded forest

O NE DAY I accompanied Amâncio in his FUNAI pickup from the Ajarani Outpost we worked at to the end of the new road. For the first time, we came to the foot of the rocky peaks that we call *Watoriki*, the Wind Mountain, and that the white people call the Serra do Demini. There, we found the barracks from an old roadwork site. Everything had been abandoned since the beginning of the last rainy season. Amâncio liked this place very much, because the forest is very beautiful there, and full of enthusiasm he instantly told me: "We're going to open another outpost here and leave Ajarani!" Yet at that time there were no Yanomami houses in the Wind Mountain region. The forest was silent. The only human traces in the surrounding area were of an old garden, abandoned very long ago, and recently destroyed by the roadwork's progress. Despite this forest's emptiness, Amâncio decided to set up a new FUNAI outpost

there. He named it the Demini Outpost[1] and promised to attract Yanomami from other regions to settle nearby. So we began to clean everything so we could occupy the abandoned barracks. Since it was the dry
season, we started to clear a very big forest patch to open a new garden to
feed our future visitors. Then we planted it with vast expanses of banana
plant and sugarcane shoots we brought from the Ajarani Outpost. Many
other FUNAI men came to help us, and everyone worked under Amâncio's orders without respite.[2]

After some time, people from the Catrimani, Ajarani, and Toototobi
rivers started making regular visits to this new outpost. But these occasional visitors were not enough for Amâncio. He wanted some Yanomami villages to settle around his new FUNAI outpost for good. So he
asked me to invite a very faraway group there, the *Opiki t^hëri*, who at the
time lived close to the Catrimani Mission.[3] But despite their distant location, they accepted Amâncio's invitation and came to build a new house
near the Demini Outpost. They promised to settle there permanently,
though in the end that did not happen at all! In fact, they never stopped
coming and going between Demini and their old home at Catrimani.
They did not feel safe in the new forest where they had just settled because it was so remote from their own. Above all, they feared the people
of the Toototobi and Mapulaú rivers, who were hostile to them and lived
only a few days on foot from there. Their wariness was not in vain: some
time after they settled there, a party of Toototobi warriors came to the Demini Outpost to seek revenge on them for the death of a man of their village. This man had gotten sick upon returning from a *reahu* feast in the
Hewë nahipi house on the Rio Jundiá where some *Opiki t^hëri* had also
been invited. The people of Toototobi instantly accused them of having
killed their relative with a *paxo uku* sorcery substance poured in a calabash of banana soup. Their warriors were lying in ambush in the forest
and nearly attacked the *Opiki t^hëri* in the very middle of the FUNAI outpost! They only abandoned the idea because white people were too close
to those they wanted to arrow.[4]

But there was something else that was not right. The *Opiki t^hëri*'s great
man was very old and had several younger wives. He was constantly jealous and often got angry because of the youngest one.[5] Each time this happened, he would become furious and leave the outpost for some time
with his people. Once, in one of these fits of jealousy, several of his sons
even came to me at daybreak with bad talk. They were very excited and

abruptly accused me of eating the vulva of their father's young wife, just because she had moved into the FUNAI outpost the previous evening. They shouted at me: "You work with the white people but your thought is full of forgetting! You copulate with that woman, that is why you are hiding her! You are bad!" But all of this was wrong. The truth was that she wanted to leave her husband and that she had simply sought shelter with her two brothers who worked at the outpost. Yet her husband and his sons thought it was me who had attracted her! Angered by these twisted words, I retorted: "These are nothing but lies! Her brothers are taking care of her. Do not think that she came for me!" But they were seized with such fury that soon after they left for their old house on the Rio Catrimani with their father and all their relatives. They never came back.

AFTER THE *Opiki t*ʰ*ëri*'s departure, the inhabitants of the *Werihi sihipi u* River were the next to get close to the Demini Outpost. These were my future father-in-law's people. They knew me well, from the time when I had worked with Chico on the Rio Mapulaú. The white people had stopped working on the road for two rainy seasons. Visitors from Toototobi on their way back from the Demini Outpost had told the people of the *Werihi sihipi u* River that they had seen me working there. My future father-in-law and his people had never gotten close to this former road-work site. But, now that the forest had returned to silence, they decided to come visit me at the new outpost. My future wife's youngest brother was their first emissary. He saw that the *Opiki t*ʰ*ëri* had left and that the Demini Outpost was only inhabited by FUNAI people and a few Yanomami who stayed to worked there. I told him that Amâncio would really like his people to come settle in the area. He went back to his elders with these words.

A short time later, his father came to Demini. He was accompanied by two young men. It was his very first visit there.[6] At that time, he was robust and still traveled a lot. He had cleared a new path in the forest from his house of the *Werihi sihipi u* River to the new road. Then he followed its gravel path all the way to the Wind Mountain. Since Chico had abandoned the Mapulaú Outpost, there were no more white people there and the people of the *Werihi sihipi u* River felt the lack of trade goods. So my future father-in-law, their headman, had come to Demini to get salt, fishhooks, and metal tools from the new FUNAI outpost. He was also very

concerned because visitors from other villages often brought white people's diseases to his home, and there was no help available on the Rio
Mapulaú. He told me: "We do not want to live alone in the forest anymore, lacking everything. We are constantly struck by the *xawara* epidemics. Now we want to settle within reach of the white people's medicine." I answered him: "These are wise words! Come settle near here and
we will be able to treat your people and help you!"

This is how it happened. Amâncio had told me to invite my future
father-in-law to settle closer to the Demini Outpost. But the old man only
accepted because he already wanted to. Now he was the great man of his
house. His elder brother and nearly all his people had been devoured by
the helicopter epidemic, then the one they attributed to the Father of the
Catrimani Mission. He thought that all these deaths needed to end. He
did not want his forest to age alone, empty and silent, littered with lost
bones. This is why he decided to leave his garden on the Rio Mapulaú
and to get closer to FUNAI, near the Wind Mountain. But he did not settle next to the new outpost right away. First he and his people settled a
day's walk away, on the banks of the *Haranari u* River, where the road's
path ends. They cleared a new garden here and started building a small
house. Yet before they even ate the bananas they had planted there,[7] they
left and built a new house, a little closer to Demini. After a few dry seasons passed, they advanced towards the outpost again, and this time they
built a much bigger house. Later, they built another and yet another,
closer and closer. Now they have been called the *Watoriki t^{h}ëri*, the people
of the Wind Mountain, for a long time.[8]

WHEN I was working on the Rio Mapulaú, the great man of the *Werihi
sihipi u* River people had once said he would give me one of his daughters
in marriage. But time had passed since that promise. Now I had already
been helping Amâncio at the Demini Outpost for many moons. Then,
one day, I decided to take rest time to go visit my people at Toototobi. I
was homesick for my sister, who still lived there, and though I was an
orphan, I also still had aunts whom I called "mother" and whom I also
missed.[9] So I happily set out on this long trip. I walked several days and
slept several nights in the forest. Then I stopped at the house on the Rio
Mapalaú, where my future father-in-law was still living at the time. Yet he
was not there. He had left for the highlands some time earlier for a *reahu*

feast held by his people in their old house at *Hapakara hi,* on the upper
Rio Lobo d'Almada. Only his wife and brothers-in-law had remained at
his *Werihi sihipi u* River's house. I only slept there one night and set off
again the next day. Finally, after another night in the forest, I reached
Toototobi. When I arrived, preparations for a *reahu* feast were also under
way in one of the people of *Sina tʰa*'s houses. Emissaries had just been
sent to invite the inhabitants of *Hewë nahipi,* on the Rio Jundiá, and those
of *Hapakara hi,* who had just finished their own *reahu* where my future
father-in-law has been invited. And so in the end I saw him arrive at Toototobi after just missing him on the *Werihi sihipi u* River! He made his
presentation dance with the other guests then settled in his *Sina tʰa* hosts'
house, where I had already been staying for a short time.

Now there were really many of us in this house and all were euphoric.
Plantain bananas and *rasa si* peach palm fruits had been gathered in
abundance. The hunters had smoked a large quantity of spider monkeys
and peccaries. The following night, the women started jubilantly making
their *heri* songs heard. Then one of the house's great men spoke out in
encouragement: "Young people, don't be cowards! Follow the ways of our
elders! Take the women by the wrist and sing with them! Begin to make
hakimuu!"[10] Put to the challenge by these words, several guests did as
they were told. I was one of them. I took hold of the arm of the daughter
of the great man of the *Werihi sihipi u* River and we danced for this entire
first night. Then we did it again for the following two! She was a beautiful
young girl who still had pointed breasts.[11] When the fires go out in the
middle of the night during the *hakimuu,* young people often take advantage of the dark to slip out of the house and copulate with their partner.
Yet this was not my case for I was frightened by the idea of becoming a
father too young. I only did this with my future wife much later, after her
father sent her to tie her hammock next to mine and we really became
closer.

After a few days, the *reahu* food ran out. The feast was going to end
and it was the time of departure. So the people of *Sina tʰa* began a *yãimuu*
sung dialogue with their guests before trading arrows, cotton, beads,
pots, and machetes. Then they gave everyone a bundle of cassava bread
and smoked game for the return trip. The people of *Hewë nahipi* and
Hapakara hi prepared to set off in the direction of the Rio Catrimani. My
rest time was finished and I decided to accompany them until the Demini Outpost, where they would stop over. All the guests had already taken

down their hammocks and packed up the goods they had just traded. We had started coming out of our hosts' house one after the other. But the moment we were about to go around it to take a path leading into the forest, my future father-in-law called me: "Davi, are you leaving?" I answered him: "*Awe!* I am going back to Demini, to work for FUNAI again!" He continued: "I intend to give you my daughter. Why don't you take her for a wife?" Surprised, I was unable to utter a single word. My own people at Toototobi had never offered me a wife. He was the first to truly make me this offering. He insisted: "Take her! Take her with you. I will join you later!" I felt very embarrassed and did not know what to say. In the end, I was only able to answer: "I don't know. I am willing to become her husband, but only if she likes me and she is willing to have me. But maybe there is already another man in her thought? Young people from your house probably already want to take her as a wife? If I marry her at their expense, they will be unhappy! They will angrily gather my footprints and have enemies rub them with sorcery plants!"[12] He reassured me, smiling: "*Ma!* No one will do that! She has no husband in view. She is truly alone!" I did not know what to say anymore. Without answering, I continued on my way to join the other guests. We made a last stop in the woods not far from the house we had just left, before beginning our long journey through the forest.

I had never thought of telling my future father-in-law: "I desire your daughter![13] I want to take her for a wife!" I did not know her at all. I had never approached her before these feast nights. Yet I already missed her. I had gotten attached to her when we danced together. She was really very beautiful. I also felt friendship for her father. And he had just proven very generous to me. He had given me his daughter on his own initiative. So I told myself: "*Hou!* If I don't respond to his offer, he will get mad at me and I will not see him again anytime soon. And if later I decide to ask him for his daughter, it will be his turn not to answer me!" I also feared his people would start speaking ill of me once they were faced with my refusal: "We gave Davi a wife, but he was scared to take her! What a coward! He is a sorry sight to see!" But on the other hand I was also worried that when I returned to the Demini Outpost Amâncio could send me to work far away, in the highlands or elsewhere. I did not want to take a wife, then reject her right afterward, like young men who are given wives too early often do. I also did not want to mistreat her by leaving her alone all the time. My thought was really confused.

Yet suddenly I decided to turn back. I came back to the Toototobi house alone and told my future father-in-law: "If you really want to give me your daughter, I accept her!" He remained quiet, but as soon as the night began to fall he sent her to tie her hammock next to mine. Then one of the elders from the people of *Sina tʰa*'s house encouraged me: "Do not fear to take this young girl for a wife!" I answered him: "*Awe!* I don't know what it is to be married, but I will try!" He retorted: "*Ma!* Her father really gave her to you, do not be afraid! You are not a weakling, you do not need to be fearful! You must really marry her, not only try!" I replied: "I am not scared! But I am concerned that if I stay here too long FUNAI will throw me out! The white people gave me rest time, like they do for themselves. This time is finished, that is what makes me anxious!" He continued to reassure me: "Do not be impatient! Do not fear these white people! You will come back towards them later! Let them wait for you, too bad for them!" Hearing these words, I reflected calmly and my thought eased down. It is true. I was young, and I had never thought of getting married before. It made me a little worried. In fact, since I had become a teenager, I was a little bit scared of women! My mother and my stepfather had often warned me against them: "*Ma!* Don't look at girls, it's dirty! If you eat their vulvas too young, you will be a bad hunter and you will never become a shaman! Wait to be an adult and then you can really get married!" Because of this, I had rarely approached young girls before. In fact, most of the time I tried to get away from them! Yet now the time had really come for me to have a wife![14]

WHILE WE were still in our *Sina tʰa* hosts' house, envoys from the *Xamatʰari* of the Rio Jutaí came to invite them to another *reahu* feast. So my new father-in-law and his people gave up the idea of going home and decided to accompany their hosts to this new feast. Freshly married, I had no choice but to join them, and so we set out all together. The *Sina tʰa* elders intended to appease their anger at the *Xamatʰari* by facing them in a chest-pounding duel. But I do not remember the grievances behind this quarrel. Had they been told bad words about themselves? Or maybe someone had wanted to take away one of their daughters? I don't remember anymore. Yet over the course of our journey to the Rio Jutaí, the people of *Sina tʰa* gave up on their idea of fighting. Maybe they changed their minds because I was there? In any case, once we were settled in forest

shelters close to their hosts' house, they only exchanged words of friendship with the emissaries who came to bring them baskets of smoked
meat and cassava bread. Yet during the *hiimuu* invitation dialogue they
held with them, the *Xamat^hari* men proved irascible and impatient to
confront them. The *Sina t^ha* elders did not react. In an effort to appease
their hosts, they declared: "*Ma!* We do not want to fight! We came to eat
your foods! We want to do our presentation dance and show you our
friendship! We did not come to pound your chests!" Finally, the *reahu*
feast took place without any fights!

Once this feast among the *Xamat^hari* ended, we returned to Toototobi.
Concerned about my long absence, I wanted to leave for the Demini Outpost very quickly. My father-in-law, seeing that I was in a hurry, told me:
"Take my daughter with you. I will join you later!" So I set off on the return journey with my young wife, accompanied by one of her young
brothers and a few inhabitants of Toototobi who were curious to see the
place where I worked. At that time, the FUNAI people employed me
from one place to another in the forest. I was constantly traveling! My
father-in-law had offered me his daughter so I could put an end to all this
moving around. He had told me: "Now that I have given you a wife, you
must live with us!" His intention really was to settle me with him. Soon
after, he left his old house on the Rio Mapulaú and began getting closer to
the Demini Outpost, building a new house on the *Haranari u* River. Yet
the fact that I had taken a wife did not instantly put an end to my comings and goings.

When I arrived in Demini, Amâncio did not hold my lateness against
me, for he knew I was with my people in the forest. But he needed me to
go back to work at the Ajarani Outpost at the beginning of the road for a
while. So I went there, and this time my young wife accompanied me. At
the time, she knew nothing about the white people's customs. She was
like our elders' daughters long ago. She was a very young girl, a *moko*,
and she was very fearful. She was always running away when a stranger
tried to speak to her! I had to teach her to wrap herself in a **dress** and eat
with a **fork** so that she could stay at the FUNAI outpost. We stayed at Ajarani together for a while. Then finally I had to bring her back to her people for a while because the FUNAI people sent me to yet another area.

At the time, I was still working as an interpreter. I was employed at the
Demini Outpost and I always came back to it. But often FUNAI called me
from Boa Vista on the radio and asked me to go help white people who

wanted to work in our forest but were afraid to be left alone there. These people knew nothing about us and they were even scared they would get arrowed![15] I constantly had to travel to accompany them. I was working hard, but I did not complain. By being an interpreter, I thought I could help my people more than the white people. I told myself: "All these inhabitants of the forest whom I visit on my trips are Yanomami like me. I must be at their side for they do not speak another language than ours and do not know what to do when the white people arrive in their homes. And the doctors won't be able to treat them without understanding them. So I must continue this work!" All these trips in our forest and to the city also made me better understand what was happening with our land. Little by little they allowed me to become an adult and made me wiser. It is because of them that I started to think: "You must protect your people! You must defend the forest!" Before this, I was still just a child, and I was a long way from thinking right!

During all my comings and goings, my young wife stayed alone with her father. As I said, he had given her to me thinking it would make me settle down, but I was still going away a lot! Yet he did not criticize me. On the contrary, he thought: "The white people chose him to work for them. That is why he travels a lot. Let it be so!" I had explained my work to him: "I made papers with the FUNAI people. And now if I don't answer their call, they won't give me any more money. Then when you ask me for merchandise, I won't have any and neither will you!" Instead of clearing gardens and hunting for him as compensation for my marriage, I gave him trade goods acquired in the **city**.[16] At the time, I bought many hammocks, pots, and metal tools for him and my brothers-in-law! I also offered him a brand new hunting shotgun. Yet neither he nor his sons were demanding. They were satisfied with little. Each time I left for a trip, I brought them back presents and they never complained about me. Yet if I had come back empty-handed, their words might have changed! My father-in-law would have been irritated and he would have told me: "Stop traveling if you can't come back with any trade goods! Where are the white people's axe blades, machetes, knives, fishhooks, and fishing lines? Do you really think it's a good thing to go wandering around their cities for nothing? You are pitiful, to leave your father-in-law in a ghost state, lacking tobacco!"

After a time, however, my trips finally became less frequent. The white people were starting to know us better. They were not as frightened of us.

They had realized that, on the contrary, it was the Yanomami who feared them! They must have started thinking: "We thought these Indians were fierce, but they are the ones who fear us! We don't need Davi anymore, we will travel alone!" And that is exactly what they did. Then I was able to live more quietly with my wife's father's people, the inhabitants of the Wind Mountain. My father-in-law had left the *Haranari u* River sometime earlier and he wanted to get still a little closer to the outpost so I could really settle with him. Now that the white people left me alone more often, I worked hard on building his new house.[17]

SEVERAL DRY seasons and several rainy seasons passed this way without my having to travel too much for FUNAI. Amâncio had long since left Demini and other white people from Boa Vista came one after another to replace him as head of the outpost.[18] They were all short-thinking people who hated the people of the forest. They spent their time getting angry, saying we were bad, and calling us lazy. In fact, all they thought of was running back to the city at the slightest opportunity. The first of them was a former gold prospector who had been chased out of the forest highlands in the time of the road.[19] He was a hard worker, but he did not like living in the forest. He was only interested in FUNAI's money.[20] And he missed his wife. He often thought of her with sadness and was always impatient to go see her. He did not stay long at the Demini Outpost. After him, another man arrived who used to drive Amâncio's pickup. He did not want to go live in the forest either. He constantly went back to Boa Vista to spend most of his time there. During the rare periods when he stayed with us, he worked very little and constantly gave me orders. He was also miserly with my father-in-law's people, the inhabitants of *Watoriki*. He even refused them the manioc from the outpost gardens they had planted themselves. That made them really furious! So he did not last very long either.

The next one was another one of Amâncio's **drivers**. This one truly hated us. He was also the laziest of all. He practically never came down from his hammock. He never did anything himself. He spent his time yelling and insulting us. Also, other white people later told me that he was sick with tuberculosis! The people of *Watoriki* soon developed an aversion to him. As for me, he made me work without respite and was never happy with what I did. No matter how well I obeyed him, he was constantly telling me off. He did not like me and constantly mistreated me whenever he could. He wanted me to be his cook, but he refused to eat what I prepared. Once I had boiled some turtle and I offered him some. He got angry, saying that was filthy and that he, a white man, would never eat that. These words were appalling to me and I yelled at him: "The game from the forest is not filthy! Cook for yourself or find yourself a wife who will agree to make your food! You only know how to sit around and sleep! You don't even know how to feed yourself! If you worked, I would keep cooking. You're just lazy. I won't do anything more for you! Stop deceiving the FUNAI people by claiming that you work here! You just lie to them to take their money! You think you can act like the chief here, but you are not an elder. You are just a know-nothing!" Hearing these words, he became enraged. He started insulting me and threatened me: "I'll kill you if you keep going!" Equally furious, I answered: "You don't scare me! I am a Yanomami! If you brag that you want to kill me, I will arrow you!" We nearly did kill each other that time! But my reaction surprised him and he calmed down, and finally I did not have to arrow him! Later, I complained about him to the FUNAI *delegado* in Boa Vista. This was a man who had friendship for us.[21] He listened to me and called this bad outpost chief to talk to him harshly: "You don't do anything! You just sleep with a full stomach and spend your time vociferating against the Yanomami! I don't want you at the Demini Outpost anymore!" This lazy man never came back to our forest.

But soon another man came to replace him, a fat one who was just as bad but also very violent. This one did not like us either and, on top of it, he was scared of us! From the moment he arrived, he always carried a **revolver** on his belt. He must have thought that the people of *Watoriki* were going to attack him and that he could defend himself with it! And he liked to show our daughters and our wives those pictures where white people copulate and show their penises and their vulvas. This made me very angry. I told him: "Stop showing our women that filth!" Then he also started to hate me. He even refused to let me use the outpost's supplies. This angered me and I protested: "Don't hide the food like that! We need to eat together. I am part of the FUNAI people, just like you!" He did not want to hear it. This angered me again, and also I hated that he wore a weapon among my people. So I told him: "If you want to live with us, hide that revolver! This is not a soldiers' house, you don't need it! The Yanomami don't like that. You're not going to go chase enemy sorcerers in the forest with that weapon, so put it away! Also, you're just a miser! If you continue like this, we don't want you here!" Then he retorted, mocking me: "I see! You, Davi, are the head of the outpost! So you're dismissing me from FUNAI?" I answered: "No, I am neither a chief nor an elder. I may still be young, but I won't accept what you're doing here! No one is aggressive with you. No one threatened you! You constantly display that revolver because you are hostile towards us! Your fear is a lie! You are the violent one and the one who wants to kill us, not the contrary!"[22]

Some time after this argument, this outpost chief abruptly went back to Boa Vista. The FUNAI *delegado* had changed and Amâncio had taken his place.[23] He ordered this fat man to stop behaving so badly towards us. This bad man tried to defend himself by claiming our argument had been caused by the lack of food. He added that he would not return to the forest without new supplies. Amâncio promised them to him and ordered him to return to Demini instantly. Despite this, the man refused. This time Amâncio got upset and threatened him: "If you refuse, you will no longer be head of the outpost!" And since he still had friendship for me, he added: "From now on, it is Davi who will replace you!" Then the fat man hated me even more. Like the one who had been fired before him, he was furious that the *delegado* sided with me! At the end, both of these bad men wound up staying in the city with their anger in their chest. They never came back to *Watoriki* and I still work there to this day! Once a FUNAI president did try to fire me, but he soon changed his mind.[24] I remained solid in my position, and there never was another silly white man to run the outpost in Demini.

I FIRST heard FUNAI people talk about closing our forest when I started working on the road. They called it *demarcação*.[25] Sometimes they told me: "We are going to close the Yanomami land and defend it. If gold prospectors, settlers, or cattle ranchers enter the forest, we will send them back home![26] If hunters come there to steal giant otter, ocelot, and jaguar skins, or arrow water turtles, we will expel them! This is a *terra indígena*. After the *demarcação*, they won't be able to enter it anymore." I liked these words. I told myself: "This is good! I also want our forest to be closed, like they say. There will be a big barrier where the white people's land starts.[27] It will prevent those we do not want from entering and will let in only those we invite to visit us. The road path in the forest will belong to us!" Yet I later understood that these words were twisted and that the FUNAI people of that time were not saying everything they thought. They claimed they would close our forest, that's true. But what they especially wanted, and they hid this from me, was to divide it into small parcels to encircle and trap us in there.[28]

Yet despite these lies, I kept this FUNAI talk about the demarcation of our land in me, and little by little it made its way in my thought. By traveling along the road, I was able to see the traces of destruction that the white people had left behind. I contemplated the wounded forest and, deep inside me, I thought: "Why did their machines put so much effort into tearing out all these trees and this soil? To leave us this path of pointy gravel abandoned in the sun? Why waste their money like this when in their cities many of their children sleep on the ground like dogs?" There were great patches of forest burned by the settlers and the cattle ranchers all along the road's path. The sun was blazing and the land stripped bare. I also told myself: "These white people are truly enemies of the forest. They do not know how to eat what comes from it. They can only clear it like *koyo* ants. And all this not to grow anything there! Just to sow weeds they abandon as soon as they become stunted and their cattle grows skinny!" Later I traveled by **bus** from Boa Vista to Manaus on another road, the one that crosses the Waimiri-Atroari's land. Once again I thought about these inhabitants of the forest who were very brave by refusing to give up their ancestors' land. But in the end this road still passed through their home, and the angry white people took revenge by making nearly all of them die.[29] Then their forest was cut up in every direction. These thoughts made me sad. I told myself: "The white people have no wisdom at all. They claim Brazil is vast. So why do they come from everywhere to occupy our forest and destroy it like that? Doesn't each of them already have a land, where his mother made him be born?" And then I also thought with sadness of the many elders I had seen devoured by epidemics since my childhood, and of all our people who had started constantly dying again since the road opened.

But at the time, other white people had also started talking about defending our forest. These were not people from the government. They were called CCPY.[30] Having come from far away, they worked alone, off on their own. In the beginning, we had not made friendship yet, and they did not speak to me. We only looked at each other from a distance, with suspicion. I worked in Demini with FUNAI, and they must have thought

I was hostile to them. I had nothing against them, but Amâncio, the head of the outpost, did not like them, and he always spoke badly of them. He did not want me to go visit them. He often told me: "Don't go listen to those people! They are outsiders. Beware of them! They want to take over your forest. This is why they are feigning to defend it."[31] In the beginning, the CCPY people only spoke about their project to the inhabitants of the Rio Catrimani, where they had begun working. They brought them maps on which they had drawn the image of our land. But the Yanomami in this region did not really understand the white people's things yet. They must have asked themselves what these big paper skins were that were being waved around in front of them by people talking about closing their forest!

In the end, I only spoke to the CCPY people much later, after Amâncio left the Demini Outpost. Sometimes I encountered one of them, Carlo, who lived in Boa Vista, because their house was not very far from the FU-NAI building. He always proved friendly. So, finally, I decided to go visit them and listen to their real words.[32] I began by telling them: "Do you want to protect our forest and never talk to me? What do you have to say to me about this? I do not want my thought to remain in oblivion!" They answered me: "Davi, you must defend your forest because if you do not do it yourself the white people will come to work there in ever-growing numbers and many more of your people will die!" This surprised me. FUNAI people had already spoken to me about closing our forest, it is true. But they had never told me that I had to fight for it my-self! Then I understood that these new words were right. I explained my own thought: "What you say is wise. It is true. But if you speak of protect-ing our forest alone, the other white people will refuse to hear you. They will call you liars. And if the Yanomami cannot listen to it in their lan-guage, they will remain deaf to it!" After this conversation, the CCPY people started helping me to travel and speak in the big cities to defend our land.[33]

At the time, the gold prospectors, whom the white people call *garim-peiros*, were starting to invade the Rio Uraricaá and Rio Apiaú in large numbers.[34] Then I had to leave my house of *Watoriki* again. But this time it was no longer to go help the FUNAI people. I started to travel to tell all the white people how the *garimpeiros* were turning our rivers into ponds of mud and were soiling the forest with epidemic fumes. During these trips, I heard other Indians defend their land with brave words for the first time. Listening to them, I understood that I could not stay quiet, waiting for others to fight to protect my people in my place. I decided to speak just like I heard them speak. It was by hearing them that I truly learned how to defend my forest. So my thought strengthened and my words increased. The white people who had become my friends encouraged me to speak, it is true. But they never taught me how to do it! Among my people, the elders teach us the way to utter right and wise words from childhood with their *hereamuu* speeches. Yet it was not my elders or the white people who taught me how to speak to protect the forest. I really figured it out alone, though at the beginning I had no idea how to go about it!

But before the *garimpeiros* arrived in our forest in great numbers and I started talking to the white people, I had become a shaman. My trips for FUNAI were less frequent. My thought had returned to calm. I had asked my father-in-law to make me drink the *yãkoana*. He had opened the *xapiri*'s paths to me and had made them build a spirit house in my chest. This was after the birth of my first son. I had become stronger and wiser.[35] I had already paid attention to the words about closing and defending our land that I had heard from the FUNAI people, then from the CCPY people. They had begun to make their way in my thought. To tell the truth, they did not leave me anymore. Having become ghost during the time of dream or under the power of the *yãkoana,* I often saw the white people splitting up our land, like they do with their own. This worried me a great deal and each time *Omama*'s image would instantly appear to me. Then I would tell myself: "What could the white people possibly want? Why do they mistreat the forest so much? This is not what *Omama* who created it wanted! If he sent them to live so far away after having made them come into existence, it was obviously so they would not devastate our land! We cannot accept for them to return to the forest to draw it into little plots like they did for theirs! It may be what their elders want. But if we give in to them, we will all die!"

We use our words this way to say that the ancient white people once drew their land to cut it up. First they covered it in crisscrossing lines, forming sections in the center of which round spots are painted.[36] This is how the shamans see it. These drawings of lines and dots, like jaguar skin paintings, appear to make it more beautiful. Yet afterward they are glued in a book and those who want to plant their food on these parcels must then give back their value. The white people claim these land drawings have a **price** and this is why they trade them for money. Yet *Omama* refused for the same thing to happen to our forest. When he created the white people's ancestors, he told them: "The land of the people of the forest will not be drawn. It will remain whole. If not, they will no longer be able to clear their gardens there or hunt how they want to and so they will perish. You can split up the land I gave you if you want to, but stay far away from theirs!" Despite these ancient words, the white people's thought remains full of forgetting. They do not know how to dream or how to make their ancestors' images dance. If they listened to these images' words, they would prevent them from invading our land. But on the contrary, their leaders constantly tell themselves: "We are powerful! We will possess the entire forest. Let its inhabitants die! They are settled there for no reason, on soil that belongs to us!" These white people only think about covering over the land with their drawings so they can carve it into sections and only give us a few pieces encircled by their mines and plantations. After that, they will be satisfied and they will tell us: "This is your land. We give it to you!"

Our forest is still beautiful and cool, even when the rain becomes scarce. Its *në rope* power of growth keeps the trees alive. It is rooted in the center of the ancient *hutukara* sky, where *Omama*'s metal is buried at the sources of the rivers. Beyond the forest, the white people's territory that surrounds us consists only of wounded lands from which the epidemic fumes come.[37] I have often traveled over the forest by airplane, and I have seen nothing but dead trees on its edges, even the buried seeds of which have been killed by fires. I saw the white people's land stretch into the distance, cut up in every direction and covered in low weeds. There is no more tree canopy there, and soon the soil will be nothing but sand.

Yet the white people do not hear our words. All they think about is making our land as bare and blazing as the savanna that surrounds their city of Boa Vista. This is their only thought about the forest. They probably believe that nothing can exhaust it. They are wrong. It is not as vast as it seems to them. In the eyes of the *xapiri*, who fly beyond the sky's back, it appears narrow and covered in scars. Its edges bear the wounds of the settlers and cattle ranchers' deforestation and fires. Its center is marked by those of the gold prospectors' mud ponds. Every one of them avidly ravages it as if he wanted to devour it. All this devastation worries us. The shamans clearly see that the forest is suffering and sick. They fear that it will finally return to chaos and that all the human beings will be crushed, as it happened once before.[38] Our *xapiri* spirits are very worried to see the land become ghost. They return from their distant flights singing songs bemoaning its wounds. I have often heard their voices lamenting when they took my image into the distances to show me this white people plunder.

My FATHER-in-law did not travel as far into the white people's land as I have. Yet he is a shaman elder and his spirits already know all these things. When I describe one of my trips to him, he simply says: "Your words are true! The white people's thought is full of ignorance. They constantly devastate the land they live on and transform the waters they drink into quagmires!" He was the one who truly made me wise by making me contemplate what the *xapiri* see. He often called me to say: "Come close! I am going to expand your thought. You must not get old without becoming a true shaman. Otherwise you will never be able to see the forest's image with the *xapiri*'s eyes!"[39] Then I squatted down and took the *yãkoana* in his company for a long time. Little by little, my eyes died under its power. Then once I became ghost, my father-in-law's spirits carried me into the sky's chest. They flew there at high speed, taking my image and my breath of life with them. My skin remained on the ground in our house while my inner part crossed the heights. On one side I looked upon the beauty of our forest and on the other the white people's land, ravaged and covered in drawings and gashes, like an old torn paper skin. In the darkness, the night spirit *Titiri* made these scars sparkle like scattered splinters of light. In the distance, I could even glimpse the mountains that the white people's ancestors cut up in the beginning of time to build their stone houses.[40] *Omama*'s[41] and the sky's spirits contemplated the earth like a vast image and told me: "The forest seems endless in humans' ghost eyes. Yet from where we see it, it is only a little spot on the land where we walk. Be careful, the white people might quickly be done with it! They will cut down all its trees, and once it is bare, it will belong to them!"

With my father-in-law, I made *napënapëri* outsider spirits dance, and they showed us the image of the paper skins with which today's white people claim to divide our land. These outsider spirits and *Omama*'s images go together. They are of the same kind, for it was *Omama* who created the white people's ancestors.[42] In their ghost talk, these spirits told us: "We are returning from distant lands that the white people have drawn and cleared. Be careful! Your forest is already covered in these same drawings. They want to take it over. They are very close and are already eating into its edges. If they advance any farther, the forest will wind up turning to chaos and you will perish with it. Defend your land by fixing images of our metal stakes around it. This way those outsiders will not be able to invade it!" The *napënapëri* spirits also mentioned the places where the white people build their machines and motors, on distant lands full of soiled waters, incessant noise, and epidemic fumes.

The image of *Remori*, the bee ancestor who gave these outsiders their ghost language, also danced for us. Returning from other devastated lands, he warned us: "Downstream of the rivers, the forest is sick because the white people are mistreating it so much. It has become other and many of them are dying of hunger there, or are devoured by the numerous evil beings who have settled on its barren soil!" Sometimes the images of **oxen** and **horses**, the white people ancestors' first pets, also came down and let us hear their worries about the burned land of the big **farms** alongside the roads. It is so. The journeying *xapiri* report everything they have seen to their shaman fathers, whether they come from arid and treeless lands, from vast lakes agitated by continuous storms, or from the great void beyond the sky. The inhabitants of our houses, knowing nothing about these faraway places, can then listen to their words through the shamans' songs. That is the way they come to know them in their turn. It is the same when the *xapiri* let us hear the voice of the *yarori* animal ancestors from the beginning of time. Today their ghosts live very far away from our forest. Yet the *xapiri* can come down to them. This is why they also bring us their images' words.[43]

This is how the spirits made me understand that the forest was not endless, as I once thought it was. I saw the charred tracks and cutting lines that encircle it in every direction. I know now that if the white people continue their advance, they will quickly make it disappear. They already claim that it is too big for us. Of course, this is just a lie. It is not so vast and it will soon be the only forest that remains alive. If we did not know anything of the *xapiri*, we would also know nothing of the forest, and we would be as oblivious as the white people. We would not think to defend it. The spirits worry that the white people will devastate all its trees and rivers. They are the ones who give the shamans their words. They always remain by our side and are the first to battle to protect our land. The *napënapëri* outsider's ancestor spirits fixed metal blades all around its edges so that the gold prospectors, settlers, and cattle ranchers do not approach our houses. *Omama*'s spirits planted a metal bar in its center, which is surrounded by storm winds that smash the gold prospectors' airplanes and helicopters in the forest. Thanks to these spirits' work the forest has not been completely invaded yet. But the images of the *napënapëri* and of *Omama* were not the only ones my father-in-law and I called down to keep the white people at a distance. When I came back from the city too worried, he would also call me to drink the *yãkoana* to help me and obscure the minds of the **politicians** who want to cut up and shrink our forest. Together, we would bring down the *mõeri* dizziness spirits to tangle up their eyes by scrambling their paper skin land drawings. It was so. My father-in-law is a great shaman who possesses innumerable *xapiri,* and it was him who taught me to make them dance to defend the forest.

I DO NOT have the wisdom of the elders. Yet since childhood I have always wanted to understand things. Finally, once I became an adult, the spirits' words made me wiser and supported my thought. I know now that our ancestors inhabited this forest from the beginning of time and that they left it for us to live in after them. They never mistreated it. Its trees are beautiful and its soil fertile. The wind and the rain keep it cool. We eat its game, its fish, the fruit of its trees, and its wild honeys. We drink the water of its rivers. Its humidity makes the banana plants, the manioc, the sugarcane, and everything we plant in our gardens grow. We travel through it to go to the *reahu* feasts we are invited to. We lead our hunting and gathering expeditions along its many trails.[44] The spirits live in it and play all around us. *Omama* created this land and brought us to existence here. He planted the mountains to hold the ground in place and turned them into the houses of the *xapiri,* whom he left to take care of us. It is our land and these are true words.

Seeing the white people tear the forest up with their machines and soil it with their epidemic fumes made me angry. Once they lived very far away from us, thinking that beyond them was only a great void. It is not true. In the beginning of time, *Omama* kept them far from our forest so that they could not get close to it. He kept their ancestors away from it, warning them: "This land is mine. You, people of *Teosi* who have no wisdom, you will live elsewhere, very far from it, in order not to destroy it. Only my children will stay here because they have friendship for it!" This is why the white people had so much trouble reaching us, even with their motorboats and then their airplanes. Our rivers are interrupted by countless rapids, and our forest is covered in hills and mountains, which are obstacles to them. We want to continue to live here alone with calmness of mind, like our ancestors in the past. We do not want to die before we get old anymore. We do not want our women and children to cry with hunger anymore. When we mix with white people, everything starts to go badly. They promise us goods when all they are thinking about is robbing our land. They open fire on us with their shotguns when they are angry. They begin to take our women. We constantly get sick, and we can no longer hunt or plant in our gardens. In the end, nearly all of us die from their *xawara* epidemics.

The *xapiri* of our long-ago shaman elders, who have friendship for the forest, refuse to let us allow its enemies, the gold prospectors, the cattle ranchers, and the **loggers**, to settle in it. These people only know how to clear its trees and soil its floor. They want to wipe us out to build cities in place of the traces of our abandoned houses. Yet this does not make us sad because the *xapiri* are always at our side to give us courage: "Many of you are dead but by defending your forest you will be many again! Your wives will give you many children! Your elders have disappeared but *Omama*'s words are still within you, always as new. You have wisdom and, alive, you will never give up your land!" Since the time of the road, I have often thought of all these things about our land. They constantly make words grow in me, words that refuse to open our forest to the white people. I want my children, their children, and the children of their children to be able to live in it quietly, like our ancestors could before us. This is my entire thought and work. I am a shaman and I see all these things with the *yãkoana* and by dreaming. My *xapiri* spirits never remain still. They travel relentlessly towards distant lands, beyond the sky or in the underworld. They return from them to give me their words and warn me of what they have seen. Through their words, I can understand all the things of the forest.

As I said, shamans do not sleep like other men. By day, they drink the *yãkoana* and make their spirits dance for all to see. But during the night the *xapiri* continue to make their songs heard in the time of dream. Sated with *yãkoana,* they never stop traveling, and their fathers travel through them in a ghost state. This is how the shamans can dream of the ravaged lands that surround the forest and the bubbling of the epidemic fumes that seep out of them. Only the *xapiri* truly make us wise because their images expand our thought by dancing for us. If I had not become a shaman, I never would have known how to go about protecting the forest. Ordinary people do not think about these things. When they see gold prospectors or other white people come to their home, their mind remains empty. Then they settle for smiling and asking for food and merchandise. They do not ask themselves: "What should I think of these white people? What are they doing in the forest? Are they dangerous? Should I defend my land and chase them away?" No, their thought remains planted at their feet, unable to advance. They simply tell themselves: "Why worry? The forest is so vast and cannot be destroyed. Instead I will try to obtain clothing and cartridges!" When our people's thought is mixed up like this, it becomes like a bad path in the forest. You have trouble following it through the tangled and dark vegetation, you stumble, you finally fall in a hole or a stream, you poke your eyes out on a thorn or get bitten by a snake. But I wanted to follow a well-cleared path whose light opens up far ahead of me. It is the path of our words to defend the forest.

CHAPTER 15

Earth Eaters

The *garimpeiros* and the father of gold

S INCE I HAD started working for FUNAI again, I had seen the white
people tear up the ground in the forest to open a road through it. I
had seen them cut down the forest's trees and set fire to it to plant grass
for their cattle. I knew the empty land and diseases they left along their
path. Yet despite all this, I still knew little about them. It was only once
the *garimpeiros* arrived where we live that I really understood what these
outsiders were capable of doing! These fierce men appeared in the for-

est suddenly, coming from all over the place, and quickly encircled our houses in large numbers. They were frenetically searching for an evil thing that we had never heard about and whose name they repeated unceasingly: *oru,* gold. They started digging into the ground in every direction like herds of peccaries. They soiled the rivers with yellowish mire and filled them with *xawara* epidemic fumes from their machines.[1] Then my chest filled up with anger and worry again when I saw them ravage the river's sources with the avidity of scrawny dogs. All this to find gold, so the white people can use it to make themselves teeth and ornaments or keep it locked in their houses! At the time, I had just learned to defend our forest's limits. I was not yet used to the idea that I also needed to defend its trees, game, watercourses, and fish. But I soon understood that the gold prospectors were land eaters who would destroy everything. These new words about protecting the forest came to me gradually, during my trips in the forest and among the white people. They settled inside me and increased little by little, linking up to each other, until they formed a long path in my mind. I used them to start speaking in the cities, even if in Portuguese my tongue still seemed as tangled as a ghost's!

I F W E L E T the *garimpeiros* dig everywhere like wild pigs, the forest's rivers will soon be no more than miry backwaters, full of mud, motor oil, and trash. They also wash their gold powder in the streams, mixing it with what they call *azougue.*[2] The other white people call it *mercúrio.* All these dirty and dangerous things make the waters sick and the fish's flesh soft and rotten. By eating them, we run the risk of perishing of dysentery, emaciated, pierced with pain, and seized with dizziness. The masters of the rivers are the stingray, electric eel, anaconda, caiman, and pink river dolphin beings. They live underwater in the house of their father-in-law, *Tëpërësiki,* with the rainbow being *Hokotori.* If the gold prospectors soil the rivers' sources, these beings will all die and the rivers will disappear with them. The waters will escape to return to the depths of the earth. Then how will we be able to quench our thirst? We will all perish with our lips dried out!

The *garimpeiros'* motors and shotguns will scare away the game and, in the end, they will starve us too. The peccaries used to be very numerous in our forest. After the gold prospectors arrived, their herds disappeared completely.[3] Soon hunters no longer encountered them anywhere, even by walking very far from their houses. The forest had become bad and was filled with *xawara* epidemic fumes. The shaman elders who knew how to make the peccary spirits' image dance had died of these deadly fumes. Then these spirits' mirrors shattered and their paths were cut off. Peccaries are human ancestors. They became game in the beginning of time by falling into the underworld when the sky collapsed. This is why they are so wise. The idea of being forced to live skinny and sick in a devastated forest made them angry. They returned under the ground, from where the sun's course passes, and the *xapiri* closed up the big hole through which they disappeared.[4]

Our elders of long ago did not die constantly and for no reason. It is different since the *garimpeiros* arrived among us. Most of our fathers and grandfathers were devoured by their diseases. In the highlands, many of our people now live in half-collapsed houses, covered in plastic tarps. The young people, orphans, no longer clear gardens or go hunting. They stay in their hammocks all day, burning with fever. This is why we do not want gold prospectors in the forest where *Omama* created our ancestors. These white people's thought is obscured by their avidity for gold. They are evil beings. They do not fear rain, heat or diseases. In our language, we call them *napë worëri pë*, the "outsider peccary spirits," because they relentlessly dig into the ground and burrow in the mud like wild pigs looking for earthworms. Because of this, we also call them the *urihi wapo pë*, "the earth eaters."[5]

The *garimpeiros* first appeared in our forest on the upper Rio Apiaú, near an old garden of the *Moxi hatëtëma*, the people Chico and I had looked for in vain in the time of the road.[6] The inhabitants of the Rio Lobo d'Almada warned us of their presence. One of their elders had just died. They had attributed his death to *Moxi hatëtëma* sorcerers. Seized with mourning anger, they decided to launch a raid to avenge him. A company of warriors set off for the upper Rio Apiaú. But, finally, they did not find any enemy houses there. Instead they came upon a vast camp of *garimpeiros!* One man of our group who was married to a woman of the Rio Lobo d'Almada people visited *Watoriki* soon after and gave us the news: "We found white people on the Rio Apiaú, they are digging in the ground and soiling the river! They are already very many!" Our house's elders then made several *hereamuu* speeches at night and decided to chase these earth eaters out of the forest.

For my part, I instantly called FUNAI people on the radio to ask for help. Amâncio, the former head of the Demini Outpost, was now the *del-egado* in Boa Vista. I thought I could count on him. Yet he answered me that I was an agent of FUNAI and that I should go chase away the gold prospectors on the Rio Apiaú myself![7] I was disappointed but it did not worry me because I knew that I was not alone and that my people would be by my side on this expedition. The following day, a large group of men gathered in *Watoriki* under the elders' guidance. Then we set out in the direction of the Rio Catrimani, which we reached after walking two days on the road's old path. There are several friendly houses in this area around the Fathers' mission, which welcomed us for the night. We asked them for reinforcement from a group of warriors. Then we traveled up the Rio Catrimani, and the Lobo d'Almada men who had spotted the white people's camp also joined us. In the end, we were really numerous! But the upper Rio Apiaú was still far away, and we had to sleep three nights and walk long days in the forest to reach it. Finally, we stopped in a clearing near the gold prospectors' camp. Once we had rested, we all covered ourselves in black body paint, as warriors do before they attack.

Then we started to encircle the white people's camp, bows taut, ready to shoot our arrows.[8] Everything was very quiet; most of the gold prospectors were working far away from there, floundering in their gold holes.[9] We only heard two men's voices coming from beneath the camp's blue plastic tarps. They must have started drinking *cachaça* quite a while earlier because they were already drunk and were talking like ghosts. We advanced in their direction, slowly, without making a sound. When they finally noticed we were there, they looked terrified. They stayed frozen and dumb. Then suddenly one of them tried to stagger over to his shotgun. He aimed at us and nearly fired but his companion, who seemed to be the great man of these *garimpeiros,* stopped him: "Don't fire! Don't fire!" He could see that there were many of us and that if one of us were wounded, they would instantly have been riddled with arrows.

Then I approached the two men and spoke to them. Like the other warriors, I was painted black from head to toe. I spoke to them in Portuguese: *"Boa Tarde!"* Surprised that I knew their language and afraid, they asked me: "Who are you? What do you want from us?" I simply answered: "We are inhabitants of the forest, we are Yanomami!" Still very worried, they tried to cajole me: "Are you hungry? Do you want something to eat? Take our food!" I refused instantly: "No, we do not want to eat, we want to talk!" They started to calm down a bit. Then little by little other gold prospectors began to arrive and I addressed them too: "We did not come to make war but to firmly ask you to leave this place. We want to convince you with our words, not our arrows. You are destroying the forest, thinking it is empty, but it is not. Our houses are many here and we drink the waters of these rivers you soil!"

Sobering up, the head of the *garimpeiros* started to lie to me: "We did not know that the Yanomami lived in this forest, otherwise we would not have come to work here!" I answered: "It is our land. You must have seen the abandoned old gardens in the forest near your camp. They were planted by Yanomami long ago. The forest *Omama* gave us extends all the way here. It saw our ancestors be born and our children are born under its cover. You cannot come and ravage it as you please. Call your men and all of you go back home! *Omama* never asked your ancestors to dig our forest's floor like peccaries and make its inhabitants die with your epidemics and guns! And what is the point of coming to work here? In the forest, you spend your days trudging through the mud and you are constantly sick. Why suffer like this? Brazil is so vast. There is no shortage of other land for you. Do not covet the people of the forest's land. Go work where you come from, far away from here!" I spoke to them harshly for a long time. Yet at the end, they merely answered me: "Yes, all right, we will leave, but later, first we have to finish our work here."

After this, we went to the gold holes where the other *garimpeiros* were working. This time it was our turn to be surprised: there were really very many of them there, far more than us! They had dug vast ditches bordered with huge gravel heaps all over the place to find the shiny dust they were relentlessly searching the streams for. All the watercourses were flooded with yellowish mud, soiled by motor oils, and covered in dead fish. Machines rumbled in a deafening roar on their cleared banks and their smoke stank up the entire surrounding forest. It was the first time I saw gold prospectors at work. I told myself: "*Hou!* This is all very bad. These white people seem to want to devour the earth like giant armadillos and peccaries! If we let them become more numerous, they will destroy the entire forest like they have started to here. We must absolutely chase them away!" Then I spoke to them more harshly: "Stop your machines and gather your belongings! You must leave immediately! You are turning over the riverbeds and soiling their waters, the marks you are leaving in the forest are dangerous!" But once again they merely answered: "Yes, later, once our work is done!" Then they added: "Be aware that as soon as news of this gold is heard in the city, other *garimpeiros* will arrive here too, then many others more! No matter how much you want to chase us away, it will be in vain!" This answer angered me greatly. Yet they were already too numerous and well-armed for our arrows to frighten them.

So, keeping my anger inside me, I was only able to retort that we would return in greater numbers, accompanied by FUNAI people and the **federal police** this time. Finally, we decided to leave without having been able to do anything. It was late afternoon. The gold prospectors had invited us to their camp to spend the night. But we feared they might try to attack us during our sleep. So we resolved to settle in the forest, far away from them. Worried, we remained on the alert until dawn, without really being able to sleep. The next day we set out for home. This was how my people and I tried to defend the forest from the gold prospectors for the very first time.

As soon as we returned to *Watoriki,* I called the FUNAI *delegado* in Boa Vista on the radio to tell him about our journey. I explained to him that there were already too many *garimpeiros* in the forest for us to expel them alone. Once again, I asked him for help and to hurry before their numbers on the Rio Apiaú increased to the point that it would become impossible to chase them away. Moons passed without any words coming back to me from the city. I was very worried and I thought that the *delegado* was not concerned with us anymore. Yet one day I received news that a group of federal police agents had landed on the airstrip at the Catrimani Mission. They were accompanied by several **military policemen** and a FUNAI agent. This message made me happy. From my side, I gathered all the men of *Watoriki* and we set out to join these white people who decided to help us. Once in Catrimani, we again called on the warriors of the houses near the Father's mission as reinforcements and we again started to head up the river together towards the *garimpeiros'* Apiaú camp. But this time the journey lasted much longer because white people do not know how to walk in the forest. They are very slow and always moaning that they are exhausted or thirsty! They are constantly asking if we will arrive soon or collapsing on the edge of the path! The federals' chief was complaining of a pain in his knees and we had to wait for him all the time. And also, we had to carry a large quantity of food for these white people to eat: dried meat, red beans, rice, manioc flour. They told me there was **400 kilos** [880 pounds] of it. We were the ones carrying all that, but they were the ones who weren't moving forward! The *garimpeiros*, they are certainly not weak like that! They are evil beings, peccary spirits, this is why they know how to walk in the forest all day with huge packs so well, feeling no pain or fatigue.

Finally, we arrived at the gold prospectors' camp after a **week**. They were settled in the same place and were even more numerous. But this time we were accompanied by the federals! I instantly headed for the plastic tarp-covered shelters with a young police officer from the South who had proved resilient in the forest. Seeing me, the *garimpeiros* thought that they were just dealing with a group of Yanomami visitors again. They approached without suspicion. But suddenly they distinguished my companion's uniform and the yellow letters on his vest: "Federal police." Then they got scared and instantly stopped moving. The police officer firmly declared: "Do not resist! We have come to expel you from Yanomami land!" The *garimpeiros'* eyes stayed on him, full of surprise and anger, but they did not dare protest. Before night came, we helped the federals to set up their camp near the gold prospectors' shelters. The very next day they began gathering the *garimpeiros* little by little and sending them walking back into the forest one by one to return to the city. But emptying this gold prospectors' big settlement took a long time! They had really become very numerous! So we slept on the Rio Apiaú for nights and nights before the noise of the machines and the murmur of men came to an end. But one day, the forest finally returned to complete silence and we were nearly out of supplies. The moment had come to leave. Frightened by the path we had followed on the way there, the federals did not want to walk in the forest anymore. They stayed on site and we quickly traveled to the Catrimani Mission alone with a message to call a helicopter that could transport them back to the city. This time we were satisfied. For the first time, we had succeeded in expelling the *garimpeiros* from our land!

Yet the head of the Apiaú gold prospectors had spoken the truth. Despite all our efforts, the *garimpeiros* soon came back to the forest and their numbers kept growing.[10] This made me more and more worried. I asked myself: "If they increase to the point where no one can expel them anymore, will we, inhabitants of the forest, all disappear?" On several occasions, I attempted to ask FUNAI to put together new expeditions to drive out the gold prospectors. I even traveled to Brasília to speak to its president.[11] It was in vain. He simply answered me: "I do not have any money to send all these gold prospectors back home, there are far too many of them! The places where they work have become like wasps' nests. If we try to chase them away, they will throw themselves on the policemen to take revenge! They are armed and very bellicose!" So I went back home sad, with my chest full of anger.

Two days after this meeting, evil words reached me. The *garimpeiros* had just murdered several Yanomami great men in the area of the *Hero u* River, near FUNAI's Paapiú Outpost.[12] They had begun by ravaging the upper Rio Apiaú's tributaries where we had tried in vain to chase them away. From there, they gradually spread upstream towards the highlands. They had finally come out on the *Hero u* River, where they suddenly found a lot of gold. From that moment on, they were seized with the frenzy of starving vultures.[13] You cannot eat gold, but you would have thought they wanted to devour all the earth in the forest to get it! This is how they came to kill the *Hero u* River leaders who stood in the way of their progress. The *Hero u* people had learned to search for a little gold with the Yanomami of the Rio Uraricaá, whom we call *Xiriana*.[14] The *garimpeiros* from the Rio Apiaú discovered the place where they were working. They were surprised to find so much gold there. Then they flooded in from every direction, in very large numbers, to dig in the neighboring streambeds. The people of the *Hero u* River worried and got angry seeing all these white people entering into their forest. So one day they decided to chase them out, like we did.

They set up a forest camp far away from the gold prospectors and left all their women and children there. The following day, a group of men set off in the direction of the holes where the *garimpeiros* worked. An elder whom I called father-in-law was at their head. He was probably telling himself: "I am brave! I will have no trouble making these outsiders flee!" All knew that most white people are not very courageous and that it would be easy to scare them, then steal their weapons. A vanguard of **four** seasoned warriors entered the gold prospectors' camp, finding only a small group of idling men. A few young people remained in ambush, on the edge of the forest. Once there, one of the four elders harshly asked one of the *garimpeiros* for his shotgun. But the white man did not let himself be intimidated and refused. The elder insisted and punched him in the face. Furious, the *garimpeiro* did not give in and violently pushed his adversary away, making him trip. Still not letting go of his shotgun, he used his other hand to take a .38 caliber revolver out of his belt and fired on the elder at close range as he was trying to get back up. Then the other *garimpeiros* approached the body sprawled in the mud to make sure he was truly dead. Seeing this, the murdered elder's son-in-law who was lying in ambush in the forest aimed his shotgun and killed one of the white men to avenge his father-in-law.

The gold prospectors instantly returned fire, killing two of the other elders who were still nearby. The last of them avoided their fire by throwing himself to the ground and rolling down the slope of a gravel embankment. But the moment he stood to run into the nearby forest, a bullet struck him in the back and he collapsed, paralyzed. A group of *garimpeiros* reached him and finished him off in the undergrowth with their knives. Then the young warrior who had avenged his father-in-law by firing on the man with the .38 caliber found himself face to face in the forest with the gold prospectors' great man, a big bearded man, and killed him too. Finally, another warrior also used his shotgun to shoot one of the *garimpeiros* who was running in the forest. At that point, most of the white men who were working in the surrounding gold holes had been frightened by the shots and left their machines to run off into the neighboring forest in every direction. This is the story the people of the Hero *u* River later told me.

As soon as the FUNAI *delegado* in Boa Vista heard the news of what had just happened, he sent me to the Paapiú Outpost. I arrived there alone. All the women's cheekbones were still black with tears and dust. The people of the Hero *u* River came out to meet me straight away, heavy with grief. Several times, they angrily used their fingers to show me the number of people who had died: "One elder, another, another, and yet another! Their bodies are still in the forest, where they were killed! We must go and get them!" They had only found the body of the man whom I called father-in-law and whom the gold prospectors had finished off with their knives after he was wounded. They had been able to hang it in the forest and begin mourning the deceased as is our custom.

The next day, a FUNAI agent and six federals arrived at Paapiú to help me. The place where the fight happened was several hours away from the outpost by foot. We set out instantly. When we arrived at the gold prospectors' camp, it was entirely empty. The policemen wanted to start out in search of the white people's bodies. The first *garimpeiro* had died on the placer and was buried there. We only found one other corpse nearby. It was the body of the gold prospector's bearded leader. We were not able to retrieve the last white man body lost in the forest. Then we found the place where the *garimpeiros* had buried the three Yanomami elders whom they had murdered. We found it quickly because it was a shallow grave. They had not had time to carefully conceal it. So we started to take the corpses out of the ground.

The men of the *Hero u* River people who had come with us helped. I asked the federals to stand guard so that these men could take care of their dead relatives without worrying. There were a lot of women too. All were crying. Once the three bodies were exhumed, their kin placed them in palm bags held tight by wood slats. They hung them in the forest to frames made of stakes ripped out of the gold prospectors' camp, then they all left for their houses downstream. Seeing the corpses being extracted from the ground, I was crying too. I thought with sadness and anger: "Gold is nothing more than shiny dust in the mud. Yet the white people can kill for that! Will they kill many more of us in the same way? And after that will their epidemic fumes eat those who remain down to the very last? Do they want us all to disappear?" From that moment on, my thought became really strong. I truly understood how much the white people who want to take our land are evil beings. Otherwise I might have remained like many of our people who ignorantly make friendship with the *garimpeiros* solely to ask them for rice, biscuits, and cartridges!

Soon after, the gold prospectors started to return to their abandoned camp. They were heavily loaded with equipment and supplies they had hidden in the forest. The federals approached them. I was at their side while the other Yanomami warily hid in the neighboring forest. The policemen ordered the *garimpeiros* to gather together and put down their loads and their weapons. They were all in the habit of working with a shotgun or a **rifle** slung over their shoulder and a revolver in their belt. They began to do as they were told, barely able to contain their anger. They did not say a word, but their eyes were very hostile. They gave me a threatening look and asked me: "Are you the chief of the Yanomami? Did you call the federals?" I did not feel safe, surrounded by these armed white people who had just killed several of our elders. Yet my outrage was stronger than my fear. I answered straight away that the Yanomami did not have a chief and that they had to leave our forest immediately. Speaking to the policemen, I added: "You must chase these earth eaters away because if they continue to invade our land in such large numbers we will all die, like our elders who we just pulled out of the forest's soil!" This is how it happened. Once the policemen had finished gathering the gold prospectors' weapons, they set up camp and waited for the helicopter and reinforcements they had called for. But I did not want to stay there. I knew that as soon as they could, the *garimpeiros* would try to kill me to get revenge. So I went back to the Paapiú Outpost with the other Yanomami and the FUNAI man. We had not eaten anything all day and we were exhausted. The following day a small airplane brought me home to *Watoriki*, with all the sadness and torment of what I had just seen.

Some time later, the *garimpeiros* returned to Paapiú and crammed into the highlands of our forest. They cleared landing paths all over the place for their airplanes and helicopters, which constantly crisscrossed the sky. They marched past the forest people's houses in dense columns, as numerous as *koyo* ants. The women became scared to come out, even to draw water! The forest emptied of any game, and the men stopped hunting. All remained sprawled in their hammocks, felled by relentless fevers. Unable to cultivate their gardens, they thought they would starve to death. The elders who used to speak with wisdom had been killed because they had bravely stood in the way of the gold prospectors or had died of malaria and pneumonia. Only a few orphans had survived, and now their thought was lost. They could only beg for food and clothes from the *garimpeiros*. All the trails in the forest had become peccary wallows, and the rivers were reduced to muddy ponds. Countless white people feverishly dug into the ground of the forest, which all stank of the motors' *xawara* epidemic fumes. Even the head of the FUNAI outpost at Paapiú had finally run off, as frightened by the gold prospectors' threats as by their diseases.[15]

THE GOLD prospectors reached our house at *Watoriki* the following dry season.[16] I was at the Demini Outpost with my wife and children when they appeared. Only one other Yanomami man was working there with me. Suddenly, a small group of *garimpeiros* came out of the forest, loaded down with packs of food and merchandise. They headed towards us immediately. They were all wearing torn shorts and shirts. Their feet were hidden in soccer shoes. They all carried shotguns and .38 revolvers. As soon as I saw them, I thought that our bodies might soon also be found in the ground, like those of the elders of the *Hero u* River. Yet despite my fears, I did not move, and I waited for them to speak first. They pretended to show us friendship: "We are here to help you! We are going to teach you to look for gold. This way you will be able to get a lot of trade goods. We will be generous with you!" I interrupted them to tell them I did not want to hear such lies: "I just saw how you killed our people at Paapiú! I saw how you eat the forest's earth and soil its rivers like wild pigs! The truth is that you are going to starve us and make us perish with your epidemic fumes! Go home!" Though they were surprised and angry to hear these words, they still continued to lie: "There will be few of us working! We will have food and medicine brought for you!" I retorted: "Nothing but more lies! There will be more and more of you and you will quickly forget these words of friendship! You will start drinking *cachaça* without respite. Then you will never stop trying to grab our wives and killing us! Then our youngsters will become lazy and ignorant because of your food! Leave, we do not want you here!" Hearing this, they got more and more angry, but since there were few of them and they were worried by my hostility, they finally left, continuing on their way to the end of the road, in the direction of the *Haranari u* River.

Yet some time later, another group of *garimpeiros* arrived at *Watoriki*. This time there were more of them, but their lies were the same: "We want to look for gold here with you! We are friends! Davi, we will make a great chief of you!" Hearing these kinds of words again made me furious. I answered them: "I do not know how to be a chief, as you say, and I do not eat gold! I have no use for this shiny dust in the sand. I am not a caiman to try to swallow it! I do not want anything from you and we will not let you work here!" This time we had not been surprised by the gold prospectors' arrival. All the men of *Watoriki* had gathered around them, bows and arrows in hand, their bodies painted black like warriors'. As for our women, they were yelling in anger and throwing their *hore kiki* sorcery plants at the outsiders in order to make them cowards.[17] We were ready to defend ourselves alone, without the federal police. Once again, I spoke to the *garimpeiros* harshly and informed them that we would arrow those who attempted to remain in our forest. They were nervous and hostile but did not dare to insist, and finally they left in the footsteps of those who had preceded them.

Time passed again. Then hunters from our house spotted a new group of gold prospectors who had just settled nearby. I instantly set off with a group of warriors. We slept one night in the forest to surprise them the next morning. They had just set up a camp and had begun digging the ground around it. They had also opened a clearing in the forest so that an airplane could drop their supplies. Once again, I spoke to them firmly to chase them away: "We do not want you to look for gold here! It is our land and we will defend it! Leave before you make us angry and before your mothers have to mourn your death!" They listened to me without knowing what to reply. The airplane they were waiting for had not come. Starving and ragged, they were a sorry sight to see. Finally, they said nothing and they left like the others, but this time in the opposite direction, towards the Rio Catrimani Mission.

A few days later, an airplane finally landed at the Demini FUNAI outpost. It was carrying the leader of these *garimpeiros* who were trying to work in our land. He had come to bring them rice, manioc flour, red beans, powdered milk, biscuits, coffee, and sugar. The other gold prospectors had told us about him. They had told us he was a fierce and fearless man. They called him Zeca Diabo, Zé the Devil.[18] But to us he was just another *garimpeiro*. As soon as he got out of his airplane, I told him: "Don't bother unloading your supplies! We have already chased all your people out of our forest!" Upset, he answered: "That isn't true! I want to hear them myself!" He stored all his sacks of supplies at the end of the Demini Outpost's landing path and instantly set off on foot looking for his men. He was accompanied by a Yanomami from the Rio Ajarani whom he had brought with him to Boa Vista. I also spoke very harshly to this ignorant man before they left. I was so angry that one of our people was helping a gold prospector!

Then a FUNAI airplane also landed at the Demini Outpost. The doctor who came out of it asked me about the sacks at the end of the strip. I described Zeca Diabo's arrival. He listened to me and told me I could confiscate his supplies and distribute them to all the people of *Watoriki*. I answered him: "Very well, we will hide them in the forest and eat them later! That way the *garimpeiros* will not be able to stay here!" The next day Zeca Diabo and his Yanomami guide returned to the Demini Outpost. They had walked a lot in the forest. The white man was a sorry sight. He was only wearing flip-flops and his feet were swollen and covered in blisters. His shorts had rubbed against his inner thighs and his skin was raw.

His guide was worried to see that we were so furious at him. As soon as they arrived, I told them: "All you can do now is walk back to where you came from! No airplane will come to get you!"

Our house's youth were very excited and menacing. They wanted to kill the *garimpeiros'* leader. But the elders did not intend to let them. They only wanted to frighten Zeca Diabo. Suddenly, the *garimpeiro* chief realized that his entire cargo had disappeared, and his eyes were pitiful to see. We had only left him his hammock. Everything had been hidden, even his clothes and his documents! He started yelling: "Where are my supplies? Where did you hide my things?" Then it was my turn to lie to him as I told him that FUNAI had taken them and that he would get them back in Boa Vista. He was not fooled, but he was increasingly nervous because he was beginning to worry about what we would do with him. He was alone among us. There was no more airplane to take him back and no radio to call Boa Vista.

When night came, he slept in the Demini Outpost. The next morning, a *garimpeiro* airplane flew over the landing path several times, flying lower and lower to land. But we posted ourselves in the middle of its way with our bows and arrows aimed at it. The pilot got scared and instantly turned back to Boa Vista. So Zeca Diabo slept three more nights under our guard. We really wanted to put him to the test! Finally, another airplane came to fly over the Demini Outpost. We were tired of this affair by now and we let him land. To be finished with it, we painted Zeca Diabo from head to toe with a black dye made of annatto pulp and soot.[19] We only left his shorts on. This is how we sent him back to the city, completely painted black! As soon as he saw the airplane on the landing strip, he started frenetically running in its direction, no matter how fierce he pretended to be. The man who was accompanying the pilot opened the plane's door without even cutting the motor. Zeca Diabo barely had time to climb aboard before the pilot, as afraid as he was, turned on the airstrip and hastily took off again! Zeca Diabo never tried to come back to our home, nor did any other *garimpeiro* for that matter!

I WAS UNABLE to sleep during this whole period of the *garimpeiros* invasion. Alone at the Demini Outpost, I could not stop thinking about those of our people whom they had murdered or who were dying of malaria.[20] I constantly thought of the forest, which had become as sick as the humans. My thoughts chased each other all night long without respite, until dawn. When I finally saw day break, my wife and children were sleeping peacefully by my side. I felt very agitated, and most of the time sleep escaped far away from me. And even when I sometimes managed to doze off, I never slept peacefully. As a young shaman, I had recently become other. I constantly saw *garimpeiros* attacking me when I dreamed. They were furious because I wanted to chase them off our land. I used to see them tell my name to city **sorcerers**, the *rezadores*, who possess evil spirits like our shamans do.[21] To get revenge, they asked these *rezadores* to weaken me and make me keep quiet. They told them: "We have to get rid of this Davi who claims to prevent us from working in the forest! He knows our language and he is our enemy. We have had enough of him, he bothers us! These Yanomami are dirty and lazy. They have to disappear so we can search for gold in peace! We have to smoke them out with epidemic fumes!" Then I saw the *rezadores'* hostile spirits come towards me in a helicopter. They were threatening me and trying to kill me. The **army** was also hostile at the time. They wanted to carve our land into pieces to let the *garimpeiros* in.[22] Then I would see the images of soldier spirits, with their steel helmets and their war planes, trying to take hold of me to lock me up and mistreat me. Yet my *purusianari* spirits were valiantly repelling these attackers. These are the *xapiri* of very aggressive warriors who possess white people's weapons.[23] They came down into my dream to battle the white soldier spirits. They tore out their paths and carried them off into the sky's chest. Then they abruptly cut them, and all of them were hurled into the void.

The spirits whom the *rezadores* possess are like our spirits, but they are other. Unlike our *xapiri*, they are not images of real evil beings starved for human flesh, like those of the *Koimari* bird of prey, the *Iramari* jaguar, and the *Mot^hokari* sun. Nor are they images of the *yarori* animal ancestors. They are much weaker and have the appearance of white people, with clothing and eyeglasses. They are armed with knives, shotguns, and revolvers. These are the weapons they use to do battle with us. They are also able to make their victims lose their mind and make them sick. Yet the *rezadores* only say their prayers to send their evil spirits in exchange for money. Us, we do not make our *xapiri* dance for money! We call them to defend our children and our forest without any gifts. In their own Portuguese language, the *rezadores* say that they can damage people— *estragar outras pessoas*. This is what they wanted to do with me. They sent me their spirits to wound me and plunge me into dizziness. If they had succeeded, my weakened and disoriented image would have fallen from the sky's heights and crashed to the ground. But it was in vain, for my *xapiri* were on watch and valiantly waged war on their evil spirits. No matter how the *rezadores* tried to harm me and make me lose courage, they never achieved their goal. They were far too weak to resist the power of my own spirits, who were always on alert and instantly repelled them. My body was in sleep here below, yet my *xapiri,* lying in ambush on innumerable paths, took care of my image in the sky's heights. As soon as the *rezadores'* spirits tried to approach it, they threw themselves at them and routed them. The *mõeri* dizziness spirits led them astray, then the moon spirits burned them with their fires while the spirits of the cotton master *Xinarumari* finally skinned them alive.

At the very beginning, when I did not yet know the *rezadores'* misdeeds, their evil spirits did succeed in shutting my image in a kind of prison, as if they were policemen. But my elders' *xapiri* were able to deliver it, then to avenge me. Their *koimari* bird of prey spirits strangled those of the *rezadores* with cords of burning cotton, while their anaconda spirits penetrated them to make their innards burst. Finally, they imprisoned them in a metal cloth that *Omama*'s image had given them to protect me.[24] It was a kind of heavy envelope that the *xapiri* slip on their victims like a garment. No key can open it and it has no holes. It sticks to the skin and you can never remove it. After that, these terrorized *rezadores'* spirits finally fled far away from me. Yet they often returned to attack me in my sleep. At that time, they left me no respite! I was often very sick because of them. They did all they could to frighten me and turn me into a coward. They really wanted to lead my thought astray and make me gutless to keep me quiet.

The white people say that there are a lot of *rezadores* in their cities and that they are powerful. I first saw one in Boa Vista. He was shown to me from a distance, and I was told that he was a shaman like me. Then, later, I met another man who prided himself on being a *rezador.* He was a former *garimpeiro* who had been hired by FUNAI to do some repairs on the Demini Outpost. One day while he was on break he came to visit our house at *Watoriki.* Suddenly, while he was watching my father-in-law make his *xapiri* dance, he told me: "I also know how to heal like he does. I am a *rezador!*" He bragged about it the same way several other times. One day I finally told him: "I would like to see how you call your spirits to

cure white people! Us, we do not make prayers with our eyes lowered to
paper skins. We become *xapiri* ourselves by drinking the *yãkoana*. Why
don't you try it too?" Suspicious, he hesitated, then finally he put his nose
forward. I blew a little powder into his nostrils. He got frightened right
away, because the power of the *yãkoana* is very strong. He quickly started
moaning and staggering around our house. He nearly collapsed to the
ground, and then suddenly he ran off to the outpost, whimpering. He
was already in a ghost state with so little powder! This *rezador* was not a
shaman, but just a lying little white man. We shamans do not see the *xapiri* with empty noses[25] and without our eyes dying from the *yãkoana!* We
do not see anything at all by merely humming with our eyes closed like
these *rezadores!*

AT THAT TIME, the white people had just killed Chico Mendes, who was
also defending the forest against the cattle ranchers.[26] So I became wary. I
knew that the *garimpeiros* hated me—those I had chased out of the forest
and all the others they had told about me. They had pronounced my
name all over Boa Vista. I often encountered their hostile gazes in the
street there. I could see that these eyes were those of enemies and I understood how they would have loved to make me die! I avoided going to
the city because I thought they would eventually kill me. When I had to
go there, I never traveled alone and I did not stay for long. I only slept in
the house of my friends from CCPY. Zeca Diabo, the one I had chased
out of Demini painted black, that one looked for me for a long time to get
revenge! But he never managed to catch me by surprise, and he probably
wound up getting tired, despite his anger. If I had not been so wary,
I would have stopped living a long time ago! Some gold prospectors
even threatened me in the middle of the street: "You can hide, but we
will find you when you are alone and kill you! We are tired of being prevented from working by you!" So I answered them with anger: "Do you
really think the Yanomami are cowards? To us, you are nothing but land
thieves. If you are truly valorous, don't just threaten me when I am alone
in the city. Come and kill me in the middle of my house. Let all my people
and the *xapiri* see you and hear you! Do not think you are courageous
simply because you display shotguns and revolvers! If you hate us so
much and you really want to eliminate us, don't just talk to me like this!
Massacre every last Yanomami with their women and children! Burn all

our houses with your bombs! Otherwise, stop speaking in vain like cowards do and go away!"

So, at the time, I was waiting for the day the *garimpeiros* would finally come and kill me at home, in the forest. At that time I lived alone in the FUNAI outpost at Demini with my wife and children. Our people lived nearby, in our big house of *Watoriki*. Often at night I would wonder if the white people would arrive to break down the outpost's door and be done with me. I never slept in peace; I was always on alert. I was worried for my people, but I was not really scared. Otherwise I would have hidden very far in the forest so that no one could find me! I only told myself that if I came to die, I would live again as a ghost on the sky's back, without any more torment. Also, I knew that my *xapiri* would never stop dancing by my bones, which would always have more value to them than gold, even once they were charred and crushed. I knew that they would take revenge on the white people who had murdered me. But finally the *garimpeiros* never dared to come kill me during my sleep! The spirit of the ancient ghost *Porepatari* must have dissuaded them from venturing into the forest at night![27]

Standing in the *garimpeiros'* way is dangerous. They are very many and they all carry knives, shotguns, and revolvers. They also possess **dynamite**, airplanes, helicopters, and radios. We only have our bows and our arrows. But despite this, I will never change my mind, I will always continue fighting them! And I will even continue when I am dead, through my *xapiri!* This is why I answered the gold prospectors who threatened me so vehemently: "Once you have killed me and you are in an *õnokae* homicide state, the *Kamakari* spirit will devour your eyes and make you blind.[28] Your intestines will decompose and I will not die alone! Do not think that your breath of life is longer than ours! Only that of stones has no end. I am not scared of dying. I am only tormented by the death of our children, whom your epidemics devour. But you should also be worried! If you make all the shamans perish, their evil beings will also attack your own people, and you won't be able to do anything about it. Your doctors do not know the *yãkoana*'s power and the *xapiri*'s anger. They will work in vain, and you will cry as much as we cry today!"

WHEN THE first *garimpeiros* arrived to look for gold in the forest's streams, there were not many of them, and the people of the highlands knew nothing about them. These people of the river headwaters lived very far away from the white people and only owned a few broken knife and machete blades. They felt impoverished. This is what led them to welcome the gold prospectors without suspicion. Coveting these outsiders' goods and food narrowed their minds. They did not think of anything else. This is how the gold prospectors were able to deceive them and make nearly all of them die before they reacted. If they had known what the white people would do, they would never have let them approach this way! That is what I think. It happened like it did in the past, at Toototobi, during my childhood. Most of the highlands' elders were happy about these newcomers' arrival. In their ignorance, they told themselves: "These outsiders might not be bad! They know how to be generous!" Then they started trading bunches of bananas with them to obtain rice and manioc flour. Then the young men also developed a taste for the white people's food and they became more and more lazy. They stopped carrying arrows, then they did not even make them anymore. In no time, the *garimpeiros* were feeding them everyday with their leftovers. But these young people did not realize that the white people were mocking them by calling them vultures and dogs. All stopped hunting and, little by little, working in their gardens. They became euphoric at the idea that they would never lack white people's food again: "These outsiders are not greedy! Their rice and their meat wrapped in metal are delicious. See all the food constantly coming down from their airplanes! Wouldn't it be good for them to settle by our side so they can continue to feed us?" Their thought was obscured before the beauty of the white people's big hammocks, metal pots, and shotguns. They did not even pay attention to their children anymore, and they let the *garimpeiros* take their wives. All day long, their thought was only occupied with words of merchandise. They never stopped begging, in a ghost language: "I want a knife, a machete, shorts, flip-flops, cartridges, biscuits, **sardines!**" Their ancient words about the forest and the gardens shrank in their minds until silence. You no longer ever heard them say: "Tomorrow at dawn, let's go arrow howler monkeys! Let's go in our gardens to plant banana plant shoots!" Little by little, they were becoming other and were pitiful to hear. Seeing them dried out thought.[29] Then suddenly all of them were seized by the gold fumes illness. Starving and burning with fever, they began to die one after the other. The young women also became fewer and fewer. Soon there were no old people and no children in their empty and cold houses. The survivors were emaciated and covered in skin diseases. Only then did they began to worry about the *garimpeiros'* presence! This was the state I found the people of the *Hero u* River in when I came to find the bodies of their elders murdered by the *garimpeiros.*

I am a son of the ancestors *Omama* created in the forest at the beginning of time. Seeing everything that I saw in the highlands made me very sad and angry. I was so distressed to see my people perish this way again because of their ignorance of the white people, without being able to defend themselves! This is why I started traveling towards distant lands to talk against the *garimpeiros* and their epidemic fumes. If the forest and the Yanomami had not been dying, I never would have made all these long trips! I would have quietly stayed in my house at *Watoriki*, with my people.

MAYBE THE white people think that we would stop defending our land if they gave us a large quantity of merchandise. They are wrong. Desiring their goods as much as they do themselves would only tangle up our thought. We would lose our own words and that would only bring us death. This is what has always happened since our elders first coveted them, long ago. This is the truth. We refuse to let our land be destroyed because it was *Omama* who brought us into existence. We simply want to continue living there as we wish, as our ancestors did before us. We do not want our forest to die, covered in wounds and the white people's waste. We are angry when they burn its trees, tear up its floor, and soil its rivers. We are angry when our women, our children, and our elders constantly die from epidemic fumes. We are not the white people's enemies. But we do not want them to come work in our forest because they cannot return the value of what they destroy. This is what I think.

I do not know how to make **accounts** like they do. I only know that our land is more solid than our life and that it does not die. I also know that it is the forest that makes us eat and live. It is not gold or merchandise that makes the plants grow or feeds and fattens the game we hunt! This is why I say that its value is very high and very heavy.[30] All the white people's merchandise will never be enough to exchange for its trees, fruits, animals, and fish. The paper skins of their money will never be numerous enough to compensate for the value of its burned trees, its desiccated ground, and its dirty waters. Nothing of this could return the value of the dead caimans and the vanished peccaries. The rivers are too **expensive,** and nothing can pay the value of game. Everything that grows and moves in the forest or under the waters, as well as all the *xapiri* and human beings, has a value far too **important** for the white people's merchandise and money. Nothing is solid enough to restore the sick forest's value. No merchandise can **buy** all the human beings devoured by the epidemic fumes. No money will be able to return to the spirits their dead fathers' value!

This is why we must refuse to give up our forest. We do not want it to become an arid land broken by muddy backwaters. Its value is far too high for it to be bought by anyone. *Omama* told our ancestors to live there by eating his fruits and his game, drinking the water of its rivers. He never told them to barter the forest for merchandise and money! He never taught them to beg for rice, fish in metal boxes, and cartridges! The breath of our life is worth more than that! To know that, I do not need to set my eyes on a paper skin for a long time the way white people do! I only need to drink the *yãkoana* and dream while listening to the voice of the forest and the song of the *xapiri*.

CHAPTER 16

Cannibal Gold

Xawara, the metal smoke

T HE THINGS that white people work so hard to extract from the depths of the earth, minerals and **oil**,[1] are not foods. These are evil and dangerous things, saturated with coughs and fevers,[2] which *Omama* was the only one to know. But long ago he decided to hide them very deep under the forest's floor so they could not make us sick. To protect us, he did not want anyone to be able to touch them. This is why they must be left where he has always kept them buried. The forest is the flesh and skin of our earth, which is the back of the old sky *Hutukara* that fell in the beginning of time.[3] The metal *Omama* hid in its soil is its skeleton, which the forest surrounds in humid coolness. These are our *xapiri*'s words, which the white people do not know. That is why these outsiders continue relentlessly digging the earth like giant armadillos, even though they already possess more than enough merchandise. Despite this, they do not think they will be contaminated like we are. They are wrong.

At night, I have often thought about those things of the underground that the white people so avidly covet. I asked myself: "How did they come into existence? What are they made of?" Finally, the *xapiri* allowed me to see their origin in the time of dream. What the white people call "minerals" are the fragments of the sky, moon, sun, and stars, which fell down in the beginning of time.[4] This is why our long-ago elders have always called the shiny metal *mareaxi* and *xitikarixi*, which are also our names for what the white people call the **stars**.[5] This metal beneath the earth comes from the old sky *Hutukara*, which collapsed upon the first people long ago.[6] Having become ghost during my sleep, I also saw the white people working with these minerals. They tore out and scraped off big blocks of them with their machines to make metal pots and tools. Yet they did not seem to realize that these fragments of the old sky were dangerous! They did not know that the thick yellowish metal fumes emanated from them are a powerful epidemic smoke that thrusts like a weapon to kill those who come near it and breathe it in.

I THINK that *Omama* was not really the one who created this metal.[7] He found it in the ground and used it to support the new land he had just formed before covering it in trees and disseminating the game throughout the forest. When he discovered it, he told himself that humans could use it to clear their gardens more easily. Yet as a precaution, he only left our ancestors a few harmless fragments of it, which they were able to use to make hatchets. He hid the hardest and most dangerous part of this metal in the coolness of the earth's depths, beneath the rivers. He feared that his muddle-headed brother *Yoasi* would put it to bad use. So he gave our ancestors the least harmful part of it, but also the least resistant. He told them: "Take these few pieces of iron to work in your garden and do not lust for any more! I will keep and hide the rest of it, which is dangerous and will only belong to the *xapiri!*" This other metal, *Omama's* concealed metal, is very heavy and burning hot. It is the real one. It is the most solid, but also the most fearsome. If he had not hidden it this way, *Yoasi* would soon have revealed its existence everywhere and the forest would have been completely destroyed long ago!

Yet, despite *Omama's* caution, *Yoasi* still managed to make the rumor of its existence reach the white people's ancestors. This is why they finally crossed the waters to come looking for it in the land of Brazil. The white people today do not dig in our forest without a reason. Though they do not know it, the words of *Yoasi,* creator of death, are within them. It is so. The *garimpeiros* are the sons-in-law and sons of *Yoasi!* These white people have become evil beings and are simply following in his footsteps. They are earth eaters covered with epidemic fumes. They think they are **all-powerful**, but their thought is full of darkness. They do not know that *Yoasi* put disease and death in these minerals. *Omama* hid them so that we would not be relentlessly tormented by tears of mourning. On the contrary, he gave us the *xapiri* so we could cure ourselves. We are his sons-in-law and his sons. This is why we are afraid of tearing these evil things out of the ground. We prefer to hunt and clear our gardens in the forest, like he taught us, rather than digging into its floor like armadillos and peccaries!

So far the *garimpeiros* have only been able to use their machines to suck up gold dusts from the bottom of the rivers. Yet these are but the children of metal. The white people do not yet know the father of gold,[8] who is buried much deeper, at the center of the highlands where *Omama* came into existence. Though the gold prospectors do not know it, it is this metal of *Omama's* that they really want to reach. In dream, I often see them destroying the entire forest as they search for it. They follow the trail of its debris in every direction. But it is always in vain because *Omama* buried it in the deepest part of the ground and the *xapiri* constantly lead their attention astray. As soon as they get near it, the *mõeri* dizziness spirits instantly disorient them and the giant armadillo spirits envelop them in impenetrable smoke. *Omama* buried this dangerous metal near the chaos being *Xiwãripo.* It is also surrounded by the spirits of the storm wind *Yariporari* and under the guard of the *napĕnapĕri* white ancestor warrior spirits. If today's white people were to try and tear it out with the big machines and bombs they first used to open their road in our forest, the earth will split open, swallowing up all its inhabitants.

This metal of the depths is so hard that it can even cut stones without being damaged. Other minerals such as gold, cassiterite, and even iron are closer to the earth's surface and are therefore more fragile. Knife blades chip and machete blades break. Pots easily get punctured and dented. This is why the white people relentlessly search for the real metal that does not deteriorate. They are not satisfied with the merchandise and machines built with the minerals they have torn out of the ground so far. Now they want to own objects that do not age and never get damaged. But all this will end badly because, as I have said, these things from the depths of the earth are dangerous. If the white people were able to reach *Omama*'s metal one day, the powerful yellowish fumes of its breath would spread everywhere like a poison as deadly as the one they call an **atomic bomb.**

THE MINERALS are kept in the coolness of the ground, under the earth, forest, and waters. They are covered by big hard rocks, little hollow pebbles, shiny stones, gravel, and sand. All this contains their heat, like a vaccine **refrigerator.** I have said that these things fallen from the first sky are very hot. If exposed, they would set the earth ablaze. Yet by cooling in the ground, they exhale an invisible breath that spreads in its depths like a humid breeze. But when the forest warms up under the sun, this breath can become dangerous. This is why it must remain captive in the cold of the ground where stones and sand hold its evil vapor in and prevent it from a escaping like a pot's lid does.

So *Omama* did not bury iron, gold, cassiterite, and **uranium**[9] without reason, only leaving our food above ground. Guarded by the cold of the *maxitari* earth beings and the *ruëri* rainy weather beings, the minerals are harmless. But if the white people tear them out of the ground, they will make the cool forest wind run away and burn its inhabitants with their epidemic fumes. Neither the trees nor the rivers nor even the *xapiri* will be able to contain their heat. Then, having no more fish to eat or flower sap to drink, the sun being *Mot^hokari*, who is also a jaguar being, will angrily come down towards the earth and devour the humans like they were smoked monkeys. And once the white people have finished extracting all the minerals, their heat will dissipate and the empty land will gradually get colder, for the minerals warm its depths. Doesn't a machete's blade, which burns hot in the sun, stay warm a long time before it gets cold? It is so. By day, the sun is very strong, and it does not let the sky's cold come down to the earth. At night, the heat of the earth's metal goes on pushing the cold back to the heights. Where the earth is very hot, there are stones and metals in its depths. However, where the ground is empty, it is very cold, the clouds are low, and you can barely see the sun anymore. This must be the case in those distant lands where the white people's ancestors have already extracted all the minerals.

OMAMA BURIED his metal in the center of the highlands, where he made the rivers spring forth. This is where the wind and the forest's coolness appear. This is where its value of growth comes from. When we shamans make this father of minerals' image dance, he appears to us as an underground mountain of iron, bristling with huge stems on every side. *Omama* set this metal mountain in the depths of the ground in order to hold the earth in place and prevent the thunder and lightning's anger from tearing it apart. He buried it like we bury the posts of our houses so they do not waver in the storm. Its metal rods penetrate the earth like the roots of a tree. It strengthens it like fish bones reinforce a fish's flesh and our skeleton reinforces our body. It makes it stable and holds it solid like our neck holds our head upright. Without it, the earth would start to waver and would finally collapse under our feet. This is not the case where we live because we were created in the center of the forest where *Omama*'s metal is buried. Yet where the white people live at our forest's edges, the ground is more crumbly and it sometimes shakes and tears apart, destroying cities.

IF THE WHITE people start tearing the father of metal out of the depths of the ground with their big tractors like giant armadillo spirits,[10] there will soon be nothing left but stones, gravel, and sand. The ground will become more and more fragile and we will all wind up sinking into it. This is what will happen if they reach the place where the chaos being *Xiwãripo* lives, who turned our ancestors into outsiders in the beginning of time.[11] The forest floor, which is not very thick, will start to break apart everywhere. The rain will never stop falling and the waters will begin to rise out of big cracks in the soil. Then many of us will be hurled into the darkness of the underworld, where we will drown in the waters of its big river *Moto uri u*. By digging so far underground, the white people will even tear out the sky's roots, which are also held in place by *Omama*'s metal. The sky will fall apart again, and every last one of us will be annihilated. These thoughts often torment me. This is why I carry *Omama*'s words in me to defend our forest. The white people do not think about such things. If they did, they would not unceasingly tear everything they can out of the earth. I want to make them hear the words the *xapiri* gave me in the time of dream so these thoughtless outsiders can understand what is really happening.

My father-in-law is a great shaman; I have said so. He is very wise. There are no others as wise as he in our *Watoriki* house. It was he who first saw the image of *Omama*'s metal and made it come down. Since then, our house's other shamans also began to make it dance. As for me, I first saw the father of gold and the other minerals in my dream. It happened when I was struck by malaria and I became ghost, burning so hot with fever that my image was carried to the deepest point of the underworld by the *Maxitari* earth spirit. This is why I can talk about *Omama*'s metal! His gigantic image is covered with epidemic smoke. He is a terrifying evil being who can fiercely slit our throat, lacerate our lungs, and dry up our blood. The white people must know this and give up on frenziedly looking for it. Though this metal may be the most beautiful and most solid they can find to build their machines and merchandise, it is dangerous for human beings.

To OUR ELDERS, gold was just shiny flakes on the sand of the forest's stream beds, like those we call *mõhere*.[12] They collected it to make a sorcery substance intended to blind people with whom they were angry. In the past, this metal dust[13] was highly feared. Today the young people no longer know how to use it. There is a lot of it in the stream near our house, but no one dares touch it for fear of losing their sight. My father-in-law, who is an elder, calls it *hipëre*,[14] the blinding powder. He also knows how to bring down the image of the father of this underground metal, *Hipëreri*. This is why we call the shards of shiny metal that the *garimpeiros* extract from the riverbeds *oru hipëre a*, the blinding sorcery of gold. When the white people tear minerals out of the ground, they grind them up with their machines, then heat them in their factories. During their work a fine dust emanates from these minerals, spreading like an invisible breeze in their cities. It is a dangerous sorcery substance that gets into their eyes and makes their eyesight worse and worse. Now even their young children have to shut their eyes behind pieces of glass to read their drawn paper skins!

The words of *Omama*'s image teach us to beware of gold and the other minerals. They are unknown and dangerous evil things that only bring disease and death. When it is still in the form of a stone, gold is a living being. It only dies by melting in the fire, when its blood evaporates in the big pots of white people's factories. As it dies, it lets off the dangerous heat of its breath,[15] which we call *oru a wakëxi,* the gold smoke. All the minerals do this when they are burned. This is why the fumes from metal, motor oil, tools, pots, and all the objects made by white people mix together and spread through the cities. These vapors are hot, thick, and yellowish like gas. They stick to hair and clothing. They get into eyes and invade chests. This poison fouls white people's bodies without their knowing. Then this evil smoke drifts far into the distance, and when it reaches the forest it tears up our throat and devours our lungs. It burns us with its fever and makes us cough endlessly, then weakens our flesh before killing us. In the past, we thought it appeared among us for no reason. But since then our *xapiri* spirits traveled to the white people's distant lands. They have seen all their factories and brought us back their words.

Now we know where these evil fumes comes from. It is the metal smoke, which we also call the minerals smoke. They are one and the same *xawara* epidemic smoke,[16] and it is truly our enemy. *Omama* buried the minerals so they would stay underground and never be able to contaminate us. It was a wise decision, and none of us ever dreamed of digging in the ground to pull them out of the darkness! These evil things remained buried, and our ancestors were not constantly sick the way we are today. Yet in their ignorance the white people started frenetically tearing the minerals out of the ground to cook them in their factories. They do not know that by killing them this way, they liberate the evil vapor of their breath. It rises in all the directions of the sky until it hits its chest. Then it falls back down on the humans, and this is how it ends up making us sick. Its poison is fearsome. We do not know how to fight it. This is why we are so worried. Though this epidemic smoke is still not very high over our forest, it is constantly expanding and accumulating. It has already spread all over the cities where the white people's factories are.[17] Now the *garimpeiros* stink up the forest with the fumes from their motors and the vapors from the gold and mercury that they burn together.[18] And before they sell their gold in the city, they keep it in iron boxes that exhale a bad smoke as they heat up in the sun. They store sacks full of cassiterite on the ground, and these also spread disease emanations all over. All these dangerous fumes are carried by the wind and come back down on the forest and on its inhabitants. All this mixes together to become a single *xawara* epidemic that spreads fever, coughing, and other unknown and fierce diseases that devour our flesh. The *xawara* that invades the forest turns us into armadillos getting smoked out of their dens! And if the white people's thought does not change paths, we fear it will kill us all, then poison them too.

WHEN THIS thick and muddy smoke reaches us for the first time, it is very dangerous for our children, women, and elders. Their flesh is still ignorant of its evil power, and it kills nearly all of them. This is what happened to our elders in the past with the Toototobi measles smoke, which nearly killed me too![19] Now we fear the *garimpeiros'* malaria, which is also very fierce. It is so. The people of the forest's breath of life proves fragile in the face of these epidemic fumes. It takes a long time before our flesh learns to harden and resist them. But this did not happen without reason. Our ancestors had never breathed their odor. Their bodies had remained cold. When these fumes appeared, our long-ago elders did not have any strength to defend themselves. All burned with fever and entered a ghost state at once. Then they perished rapidly, in great numbers, like poisoned fish in a dry pond. This is how the first white people made nearly all of them vanish.

In the distant past, our grandfathers also used sorcery plants like the one we call *oko xi, hayakoari hana,* and *parapara hi,* to send epidemic fumes hungry for human flesh to their enemies.[20] They were feared and would have decimated the white people too if they had burned in the middle of their cities! Our elders sometimes set off on secret *oka* sorcerer expeditions with these plants and burned them near the houses they wanted to contaminate. Once the smoke came down on their victims, it was not long before they died one after another. These long-ago *xawara*

fumes were really very dangerous! Those from old women's *oko xi* sorcery plants struck the most vigorous men first, then put a quick end to
the most beautiful young girls. Only the old people and a few emaciated
adults survived them. This is what I heard told in my childhood. Yet our
elders did not leave us the knowledge of these sorcery plants. It has been
lost. We do not know how to use them anymore. If we still did, I would
say so: "It is true, we possess evil plants that will one day serve to take revenge on those who decimated us!" But we no longer have any of all that.
We only heard our elders talk about them. Since our childhood we have
only known the white people's epidemic fumes, which devoured our kin!

Before the road and the gold prospectors arrived, the people of the rivers[21] were the first outsiders to make *xawara* epidemics burn against our
long-ago elders. Out of anger, they made unknown sorcery things explode in the air or heat up in metal boxes, which instantly spread deadly
fumes around them. But now it is no longer like that. The white people
spread their epidemic fumes throughout the entire forest for no reason
and without even knowing it, only by tearing the gold and other minerals
out of the earth. Yet these evil things' emanations are so strong that even
the smoke from cremating their victims' bones is poisonous. The survivors have barely breathed it before it also makes them die. Yet we are not
the only ones to suffer as a result of this **minerals' disease**.[22] The white
people contaminate themselves with it too, and it eats them as much as it
devours us, for the *xawara* epidemic has no preference in its hostility!
They may think that they are dying of common disease, but this is not the
case. Just like us, they fall ill from the smoke of the minerals and oil that
Omama hid under the earth and waters. They unwittingly make it spring
up everywhere by extracting and manipulating these evil things. They call
this **pollution**, but to us it is still *xawara* epidemic smoke.[23] Though they
also suffer from it, they do not want to give up their digging frenzy, and
their thought remains closed. All that matters to them is cooking the
metal and oil to make merchandise. This is why the *xawara* can relentlessly wage war on human beings. This is what our elders who are great
shamans say. These are the *xapiri*'s words, which they pass on to us.
These are the ones I want the white people to hear.

Today this disease is growing and spreading everywhere, and we are
constantly dying from it. The smoke from the minerals is increasing
wherever the white people are established. In the past, they lived very far
away, in their cities. Now they have gotten closer to us and we are becom-

ing fewer and fewer because their epidemic fumes are encircling us. Our *xapiri* relentlessly try to attack them and drive them far from the forest, but they always return. For we shamans, it is a great torment not to be able to push this *xawara* smoke away. If we all die, no one will be able to compensate for the value of our dead. They are far more numerous on the sky's back than we the living are in the forest. The white people's money and merchandise will not bring them back down among us! And the devastated forest will never be able to be restored either, it will be lost for all time.

What we call *xawara* are measles, flu, malaria, tuberculosis, and all those other white people diseases that kill us to devour our flesh. The only thing that ordinary people know of them are the fumes that propagate them. But we shamans, we also see in them the image of the epidemic beings, the *xawarari*. These evil beings look like white people, with their clothes, their glasses, and their hats, but are wrapped in a thick smoke and have long, sharp canines. They are the *t*ʰ*okori* beings of the cough, which slit our throats and chests,[24] and the *xuukari* diarrhea beings, which devour our guts,[25] but also the *tuhrenari* nausea beings, the *waitarori* scraggliness beings, and the *hayakorari* weakness beings. These evil beings do not eat game or fish. They only starve for our fat and thirst for our blood, which they drink until it has dried up. They know how to listen from far away to the voices rising from our villages to guide themselves to us. They approach our houses during the night and set up their hammocks inside but we are unable to see them. Before they attempt to kill us, they make us drink a fatty liquid that makes all of us sick and weak.

Then they look for the most beautiful and chubbiest of our children. They capture them to tie them up in big sacks and bring them to their home. At times they cut the throats of a few of them and put their bodies on iron spits to roast and eat them on the spot. If our *xapiri* do not act to rescue these children very quickly, they die instantly. After this, the *xawarari* epidemic beings tie up the elders and the women who have the weakest breath of life. First they cut one entire group's throats with their machetes, then they rest for a while before coming to get new prey. Little by little, they gather great quantities of corpses to roast them like game. They only stop killing once they think they have gathered enough human flesh to satisfy their appetite.

These evil beings are really fierce! As soon as they seize their victims' images, they decapitate them and tear them to pieces. They devour their heart and greedily swallow their breath of life. They throw down their guts to feed the hunting dogs that follow them.[26] After having sucked on their marrow, they also leave the bones of their victims for these starving animals to loudly chew on. This is why the *xawara* epidemic makes us feel such deep pain in our stomach and limbs! Then the *xawarari* beings cook their dismembered prey's bodies in big metal **basins**[27] like a pile of spider monkeys, sprinkling them with boiling oil. This is what makes us burn with fever! Finally, they store this cooked human flesh in big metal cases to eat them later. In this way they prepare a great number of human meat cans, like the white people do with their fish and their beef. Later, when they start to lack victuals, they send their **employees**[28] to hunt new victims among us: "Go get me nice and fat human children! I am so hungry! I would happily eat a leg!" Once they have eaten their fill, they leave us alone for a while. They are in no hurry! But once they are starving again, they come back to devour our children, women, and elders again and again, because they see us as their game. This is how the *xawara* epidemic decimates us little by little.

THE *XAWARARI* epidemic beings live in houses overflowing with merchandise and food, like gold prospector camps. This is where they cook the people of the forest's flesh. Remains of their cannibal meals hang everywhere in their homes, for they always keep the skulls and some bones of the humans they devour, as we keep those of the game we eat.[29] Their hammocks are made of human skin. To make them, they flay their victims' entire body before roasting them. The ropes that hold these hammocks in place are made with the skin from their victims' toes and fingers and are sewed with machines. They are truly fierce evil beings! When I was a young shaman, I was struck with horror when my image accompanied the elders' spirits to these man-eaters' houses. I remember my fears to this very moment!

The *xawara* epidemic thrives in the places where white people make their merchandise and accumulate it. Its smoke emanates from these goods and from the factories where the minerals from which they are made are cooked. This is why disease and death occur as soon as the people of the forest start desiring these objects. Exultantly accumulating clothing, pots, machetes, mirrors, and hammocks attracts the *xawarari* epidemic beings' attention, and they tell themselves: "These people like our merchandise? They have taken us for friends? Let's go visit them!" Then they arrive in the white people's footsteps, unseen by us, in their motorboats, their airplanes, and their cars. The big rivers, the roads, and the landing strips are their paths and doors into the forest.[30] They come to settle in our houses like invisible guests by escorting the white people's objects. Merchandise has the value of *xawara* epidemic.[31] This is why the diseases always follow it. They tear up our throats and chests and pierce our eyes and skulls with rods of metal.[32] It always happens in the same way. The *xawarari* epidemic beings constantly keep an eye on merchandise, no matter where it travels, even very far from the white people cities. When a loaded airplane flies in our forest's direction, they carefully follow its path. No sooner has it landed than they try to find the nearby human beings whom they can devour. Yet their future victims cannot see them. Only the *xapiri* can.

When we die by the power of the *yãkoana*, our spirits travel very high into the sky's chest. Then their gaze sweeps over the forest as if from an airplane. That is why they spot the epidemic smoke as soon as it grows and heads towards us. The *napënapëri* white people's ancestor spirits then instantly alert us: "The *xawara* epidemic is approaching and its smoke is glowing red! It is making the sky become ghost and is devouring all the human beings on its path! It must be driven away!" They are also the first to attack it, striking it with their huge metal bars. Only they truly know the metal fumes and are able to rescue their victims away from them. They are just like the white people friends who are at our side to defend the forest against the *garimpeiros*.

Many other *xapiri* follow them to fight the *xawarari* epidemic beings. The wasp spirits sting these man-eaters with their poisoned spears while the caiman spirits strike them with their heavy machetes. The aggressive *xaki* and *pari* bee spirits slash them while the *waroma* snake spirits pierce them. A great number of *õeõeri* and *aiamori* warrior spirits rush to riddle them with their arrows. The anteater and giant armadillo spirits wound them with their sharp knives. The vulture spirits tear them apart. The images of the evil beings of the anaconda and of the master of cotton, *Xinarumari*, seize them to choke them and skin them alive. The big *aro kohi* and *masihanari kohi* tree spirits crush them with the help of the stone spirit *Maamari.* The spirits of the *remoremo moxi* bee,[33] the *hõra* beetle, and the storm wind *Yariporari* tie the hair of their fumes to *Omama*'s airplane to drag them into the far-off places from which they came.

All the most valiant *xapiri* come down to fight the *xawara* epidemic and gather in an innumerable troop to confront it. They go to face it with courage and counterattack relentlessly, like the white people soldiers, never backing down. If they were scared of the *xawara* smoke, it would never stop eating every last one of us! When it really becomes too dangerous and they must save their people from death, our shaman elders even make dance the epidemic's own image, which they also call *Xawarari a.* This image is the epidemic's, but once it becomes a *xapiri* spirit, it bravely fights the white people's dangerous metal smoke and joins the *napĕnapĕri* spirits of their ancestors in their struggle against it.[34] This is how our great shamans sometimes succeeded in driving this fierce disease away and curing its victims in the past. Then their *repoma* bee spirits could finally open the earth and throw the moribund *xawarari* epidemic beings into the underworld, where their bodies would come crashing down to feed the *aõpatari* meat-hungry ancestors.

Yet most often the *xawara* epidemic proves tougher to fight than the *nĕ wãri* evil beings of the forest, and the *xapiri*'s efforts to destroy it remain in vain. Though they may strike it with all their might, it does not seem affected by their blows. Very high in the sky, its smoke then becomes far too aggressive and powerful. It no longer fears anything. The spirits' hands can no longer take hold of it, and their weapons are powerless to wound it. Though their assaults may make it back down, it constantly returns to the attack, always stronger. Even the *maari* rain beings can no longer chase it out of the sky! The *xawara* epidemic is very hard to fight because it is the trace of other people. It does not come from our forest. Its evil *xawarari* beings are more numerous than the *garimpeiros* and even than all the white people. For a shaman it is difficult to gather as many *xapiri* to fight them! This is why they fear these *xawarari* epidemic beings and sometimes lose the courage to face them. The *xawara* epidemic does not even hesitate to turn against the *xapiri* and capture them. It destroys their houses and locks them up in burning metal boxes. If other intrepid spirits do not come to rescue them, their breath of life dries up and they die. The epidemic beings roast their bodies before devouring them, just like they do with human beings. If this happens, the *xapiri*'s fathers, the shamans, die very quickly. Yet, in the end, their spirits always come back to life, for they are truly immortal.[35] Wherever the epidemic beings vomit their bones, they hatch again with their mirrors and instantly take body again.

TODAY THE *xawarari* epidemic beings are constantly increasing in numbers; this is why the epidemic smoke is so high in the sky's chest. But the white people's ears are deaf to the *xapiri*'s words. They only pay attention to their own speeches, and it never crosses their mind that the same epidemic smoke poison devours their own children. Their great men continue to send their sons-in-law and sons to tear out of the earth's darkness the evil things that spread these diseases from which we all suffer. Now the breath of the burned minerals' smoke has spread everywhere. What the white people call the **whole world**[36] is being tainted because of the factories that make all their merchandise, their machines, and their motors. Though the sky and the earth are vast, their fumes eventually spread in every direction, and all are affected: humans, game, and the forest. It is true. Even the trees are sick from it. Having become ghost, they lose their leaves, they dry up and break all by themselves. The fish also die from it in the rivers' soiled waters. The white people will make the earth and the sky sick with the smoke from their minerals, oil, bombs, and atomic things. Then the winds and the storms will enter into a ghost state. In the end, even the *xapiri* and *Omama*'s image will be affected!

This is why we shamans are so tormented. When the *xawara* epidemic goes after us and cooks our image in gas and oil in its iron pots, it makes us become others and dream without interruption. Then we see all these white people searching for the metal they covet. We see the fumes from the innumerable troops of *xawarari* epidemic beings who follow them, and we vigorously fight them. We are inhabitants of the forest, and we do not want our people to die. The white people probably think that *Teosi* will manage to make the *xawara* smoke from their factories disappear from the sky? They are wrong. Carried very high into its chest by the wind, it is already starting to soil and burn it. It is true, the sky is not as low as it appears to our ghost eyes, and it is getting as sick as we do! If all this continues, its image will become riddled with holes from the heat of the mineral fumes. Then it will slowly melt, like a plastic bag thrown in the fire, and the thunders will no longer stop shouting in anger. This has not happened yet because the sky's *hutukarari* spirits are constantly pouring water on it to cool it down. But we shamans fear this disease of the sky more than anything. The *xapiri* and all the other inhabitants of the forest are also very worried about it because if the sky catches fire, it will fall again. Then we will all be burned, and we will be hurled into the underworld like the first people in the beginning of time.

These are the words of our elders, who became great shamans long before us. This is what they saw and what their spirits' songs relate. As I have said, we shamans dream of everything we want to know. When we drink the *yākoana,* first we see the father of gold and minerals deep in the earth, surrounded by the fatty bubbling of his epidemic fumes. At night, once we have become ghost in our sleep, we dream about it at length when our *xapiri* take away our image. It was by becoming spirit with my father-in-law and the other elders of our house that I learned to know the gold epidemic, which we call *oru a xawara.* These great shamans taught me to think far away, and under their guidance *Omama*'s image allowed me to see all the things I am talking about. If I had stayed working for the white people by myself, if my father-in-law had not called me to live beside him, my thought would have remained far too short. This is why now I want the white people to hear these words too. These are things that we shamans speak about very often when we work together. We refuse to let the minerals that *Omama* hid underground be touched because we do not want any *xawara* epidemic fumes in our forest. My father-in-law often tells me: "You must tell the white people that! They must know that we are dying one after another because of this evil smoke from the things they tear out of the ground!" This is what I am now trying to explain to those who will listen to my words. Maybe it will make them wiser? But it is true that if they continue to follow this path we will all perish. This already happened to many other people of the forest on this land of Brazil, but this time I think that even the white people will not survive.

III

The Falling Sky

CHAPTER 17

Talking to White People

A bearded politician

O UR GREAT men—those we call *pata t^{h}ë pë*[1]—usually address the people of their house a little before dawn or shortly after dark. They exhort them to hunt and work in their gardens. They tell about the beginning of time when our ancestors turned into game and they speak with
wisdom. We call this *hereamuu*.[2] Only the oldest men speak like this. But

me, I had to learn to discourse in front of outsiders when I was very young. It is true! I was already addressing the white people harshly before I even dared to speak like my elders in my own house. My mouth was ashamed because if I had tried to exhort my people at the time, they would have shut me up with their mockery. They would have irritably declared that a young man cannot give advice to the elders, and no one would have acted upon my words. I truly would have been a pitiful sight! So I said nothing, fearing that they would mock me. I only tried to become as wise as my fathers and fathers-in-law and I thought I was still far from that. I quietly told myself that if I wanted to get there, my thought would need to remain set on the *xapiri* the great shamans had just given me.

When you're young, you don't know anything yet. You have a thought full of oblivion. It is only much later, once you've truly become an adult, that you can take the elders' words inside yourself. It happens slowly. White people's children must learn to draw their words by awkwardly twisting their fingers and constantly fixing their eyes on image skins. Among us, the young people who want to know the *xapiri* must overcome fear and let their elders blow the *yãkoana* powder into their nostrils. It is painful, and it also takes a very long time. Afterwards, they need to continue working by themselves, trying to link their thoughts together, as far as they can. This is why I was so zealous in studying the things the *yãkoana*'s power allowed me to see when I was a young shaman. But when I wanted to give my words to the people of our house, I did not dare to make a *hereamuu* speech. I merely passed on what I wanted to say during the sung dialogues on the first night of our *reahu* feasts. This is what we must do when we are not elders.

This is how it goes. One young person from the hosts and one from the guests stand face to face and answer each other in song on the house's central plaza. One by one, the young people are replaced by others, forming new pairs. Then, once they have all finished, the older men gradually take their place and follow the same way, without stopping until the middle of the night. This is what we call *wayamuu*. The words of this talk are very long. They are like the news on the white people's radio. We use them to tell what we heard when visiting other people. This way we sometimes inform our guests that angry people from distant houses want to challenge them with their clubs or even arrow them. We also describe the ills that have struck our own people. We talk of those bitten by a snake, those whose *rixi* animal has been wounded by an enemy hunter, those whose bones have been broken by *oka* sorcerers, as well as those devoured by the *xawara* epidemic. When the *wayamuu* words come to an end, the elders, hosts, and visitors squat face to face, very close to each other. Then they begin another sung talk, which we call *yãimuu*.[3] The *yãimuu*'s words are closer to us and wiser. They go deep inside us. During the *wayamuu*, we do not really reveal what we want to say yet. We still use a ghost tongue. When the elders really want to talk to each other and put an end to the hostility between them, they do it through the *yãimuu*. An angry visitor mentioning unpleasant rumors about his people that have been spoken by his hosts will be warned: "Forget these twisted words! Let's stay friends! My true words are beautiful! Don't listen to the words that these distant people have made become other! They are liars!" Then the visitor will calm down and answer: "*Haixopë!* This is good! This talk is truly right! I don't want to listen to that bad talk which could lead us to beat each other's heads or even to use our arrows. Let's be friends!" During the *yãimuu*, the elders also tell their hosts that they will invite them to bury their dead's bone ashes in a coming feast: "We want to finish this *pora axi* funerary gourd before your eyes! We will put it in oblivion together! This is what we want!" If we do not speak openly like this, people can get angry and claim to have been kept in the dark. It is also through *yãimuu* that we ask for trade goods during a feast: pots, hammocks, axe heads, knives, fishhooks, and machetes. A young man will ask for a wife and offer to work for his future father-in-law in the same way. The latter will answer: "Come settle in my house and take my daughter! But do not desire her without returning her value. When you live with me, you will have to sate my game hunger and work in my garden. Then I will truly give you a wife!"

Titiri, the night being, taught our ancestors to use the *wayamuu* and the *yãimuu* in the beginning of time.[4] He did it so that we could warn each other about our thoughts and use that to avoid fighting too much. But before that, the furious *Titiri* devoured *Xõemari*, the dawn spirit, so he would not come back from the downstream of the sky on his path of light at any time he wanted.[5] Since then the ghost of *Xõemari* can only interrupt the darkness once, at sunrise. Then *Titiri* told our ancestors: "Let this night talk remain in the depths of your thought! Thanks to it, you will truly be heard by those who come to visit you." This is why we have continued to speak out during our *reahu* feasts, first with the *way-*

amuu, then the *yãimuu,* from nightfall to dawn. To this day, this dialogue's words have never stopped growing inside us. *Titiri* made them multiply so that we could converse from one house to the next and think right. This is the heart of our talk. When we say things solely with our mouths, during the day, we do not truly understand each other.[6] We do listen to the sound of the words addressed to us, but we easily forget them. But during the night, the *wayamuu* and the *yãimuu's* words accumulate and enter deep into our thought. They reveal themselves in all their clarity and can truly be heard. This is why in the beginning I preferred to dialogue in darkness when I told the elders about distant things they did not know. This way they were not upset. They answered me without animosity even when I told them: "Do not desire what the white people eat! It is not good for us! It is old foodstuff that they leave to rot hidden in their houses! The *në rope* richness of the forest is here to feed us! We only need to hunt and clear vast gardens! This is where the real food comes from!" They answered: *"Awe!* You who defends our forest, when you give us your words like this, you warn us against the white people's bad things. It is good to make us vigilant!"

AT THAT TIME, it was so. My people already knew that I spoke my words about our land among the white people, very far from our forest. Yet in our house at *Watoriki,* they would tell me: "Later, when you are older, you can exhort us with your *hereamuu* speeches if you want to. But for the time being, settle for letting us hear them in the *wayamuu* and *yãimuu* talks. It is good this way!" This is how it goes among us. Once a man gets a bit older and has acquired some wisdom, he can try to exhort the people of his house. First he will try to give advice on hunting or garden work from time to time. If he puts his words together well and the young people follow his remarks, he will continue his attempt. But if no one reacts or some people rail at him, he will feel ashamed and stop instantly. He will tell himself: "The inhabitants of my house are hostile to me. They prefer hearing the real elders' speech! I must be patient and imitate these wise men's manner!" As time passes, if his people finally begin paying attention to his exhortations, his mouth will gradually lose its fear. Then he will speak with wisdom, following in the footsteps of those who preceded him in this task. He will begin his *hereamuu* speeches as a young adult and continue into old age.

This is the path I have tried to follow. Now I sometimes try to speak through *hereamuu*.[7] If the people of my house start to listen to my remarks, I will continue. Otherwise I will return to silence and keep quiet in my hammock. So far I have not often spoken in this manner. I always fear that exasperated elders will make me be quiet: "You speak these words because you probably think you have become a great man! It's not true! Your forgetful speeches irritate us. You are too young, why don't you work in silence to feed your wife and children instead of exhorting the others!" Yet my father-in-law did not prove hostile to my words, on the contrary. This gives me strength. Sometimes he tells me: "It is good that you speak this way because I am getting old. When I am no longer here, you will go on speaking in my place!" Then I answer: "Father-in-law, if ever you disappear and the *xawara* epidemic leaves me alive, then yes, I will remain to speak after you! Make your *Kãomari* falcon spirit[8] come down into me so I can become as eloquent as you are. Then it will be my turn to die because today the white people do not let us live for long."

Later I will probably have many sons-in-law. I will send them to work in my garden and hunt for me in my turn. Then I will truly be able to speak in *hereamuu* like the elders did before me. I will tell my daughters' husbands: "I am going to stay in our house. Go cut down the tall trees in my garden! Go arrow game and collect *hoko si* palm fruit for me!" Yet I would not like to have to give orders all the time. When they are wise, sons-in-law work alone, without their father-in-law having to ask them. So I will only give them instructions if they do not know what to do. I will tell them: "Clear a new garden and plant it with banana plants, manioc, and sugarcane so we do not have to suffer from hunger. I do not want to know the shame of having to beg for my food from other people!" And once I have really become older, I will also tell the young people what I experienced since my childhood. I will tell them about all the white people I met and all that I saw on my faraway journeys. This way they will become more aware.

If I settle for talking to them only with my mouth, without making true speeches like the elders, it will not work. They will certainly hear the sound of my voice, but they will continue to search in their thoughts and ask themselves: "What will happen to us? Will other white people come into the forest to take our land?" If I do not make the youngsters hear my words in *hereamuu*, these words will not enter deeply into their thought. They will not truly be able to understand things. If I want them to start thinking right, I will have to speak to them this way many times over. This is why I begin doing it by saying: "The bare lands that extend around our forest are those of other people. Do not try to go live there! You will be mistreated and will only come back with diseases that will devour your people. And do not go constantly wandering here and there in our allies' houses. Your visits will end up tiring and irritating them. Your hosts will become jealous over their wives and suspicious of their daughters. They will get angry at you. You will argue and they will want to hit you. Instead, stay quietly working among your own people!"[9] I also often explain to them that we have to put an end to hostility among our houses and stop mistreating each other in head-beating duels with clubs or arrowing each other. I know very well that some of them must think that my words are due to cowardice, but it is not true. What I want is for us to truly show our courage by defending our land against those who want to devastate it. Those are our real enemies! We inhabitants of the forest, we are the same people, we must be friends. This is only the beginning of my words to young people. Later, when I become an elder, my words will be longer and wiser.

To BE ABLE to make *hereamuu* speeches firmly, one must acquire the image of the *kãokãoma* loud-voiced falcon we call *Kãomari*. It gives the words of our exhortations their strength. It comes down into us by itself, even if we aren't shamans. Then we let it settle in our chest, where it remains out of sight.[10] It teaches our throat how to speak firmly. It makes the words rise out of it one after another, without getting tangled up or losing their strength. It allows us to deploy the words of a wise and alert thought in every direction. With it, our language remains steady, without faltering or drying out. Those who do not possess this image can only make clumsy speeches with words that are too short. Their talk is hesitant and their voice reedy. They express themselves like ghosts and say nothing. They are pitiful to hear. On the contrary, the great men whose chest is inhabited by the *kãokãoma* falcon's image truly know how to make long and powerful exhortations. They are skillful at convincing young people to follow their words. They never call them lazy, in order to avoid making them angry and reluctant. Instead they tell them: "We are all hungry for meat. Go arrow some game in the forest tomorrow! Follow a tapir's trail and we will all be satisfied with a full belly!" or else: "Clear big gardens, food will be abundant in our house! Your children will not whimper that they are hungry all the time! You will not be ashamed to send your wives into others' gardens![11] We will be able to call many guests to our feasts!" It is with the same skill that they send the women fish-poisoning when the rivers are drying up, enjoin the men to store their banana bunches in the house when they are becoming ripe, or advise them to prepare the smoked meats and cassava bread needed for a *reahu* feast. It is so. The great men exhort the inhabitants of their house during the night, and though those who dwell in the house may remain silent and appear to be sleeping, they listen. When the sun rises, their minds awaken and then they tell themselves: "*Haixopë!* These were good words! We are going to respond to them by following their advice!"

Yet often the elders only discourse with wisdom, without giving any instructions, merely so that those who hear them can become wiser. This way when an old man wakes before dawn, at the hour of the dew, he will name in *hereamuu* the old forests where his fathers and grandfathers lived as they gradually came down from the highlands.[12] He will speak about the place where he was born and those where he grew up. He will remember the house where he started hunting lizards and little birds with darts, the one where he became pubescent and his throat imitated the curassow bird,[13] as well as the one where he took a wife. He will tell what he observed of the lives of the long-ago elders during his youth: their trips from one house to another, the *reahu* feasts to which they invited each other, the raids they launched to seek revenge. He will explain how it was in the distant times when the white people's merchandise was scarce. And if he is at all unhappy with his sons-in-law's laziness, he will take the opportunity to add: "When I was young, things were not like they are today. I constantly hunted to get my elders their fill of game! I arrowed countless tapirs, peccaries, and spider monkeys. Everyone enjoyed rubbing their teeth with the meat I used to bring back.[14] Yet the highlands forest where we lived was steep and its vegetation tangled! At the time of my youth, I was a very good hunter! Today youngsters come back empty-handed too often. They themselves probably eat what little prey they are able to arrow!"

For their part, the shaman elders' *hereamuu* mostly recall the time of the *yarori* animal ancestors. They often open their speeches by saying: "In the beginning of time, our ancestors became others, they transformed into game; they turned into deer, tapirs, monkeys, and parrots."

Then they continue by relating the misadventures of one of these first
ancestors. They also tell how a woman sitting on the ground during her
period while she was traveling in the forest became a rock and how the
spider monkey ancestors tore her arms off by trying to stand her back
up.[15] They evoke the strident cries of Õeoeri, the newborn whom enemy
sorcerers abandoned on a *kaxi* anthill after they killed his mother.[16] They
report how the *hoari* marten ancestor chased away the honeybees when
their nests were still easily accessible at the foot of trees.[17] They describe
the way in which the *koyo* ant ancestor was secretly clearing a vast maize
garden in the forest in which to lose his mother-in-law.[18] The old shamans also make words heard about the places where their *xapiri* came
down, beyond the sky, in the underworld or on the white people's land.
This is how they instruct ordinary people, those who have only heard
about the first animal ancestors and have never seen all those distant
lands whose images they cannot bring down. Unsure what to think about
all this, they have to pay attention to the shamans' songs and learn what
these wise men were able to see after taking the *yãkoana*.

MY FATHERS and grandfathers grew up in the highlands, very far from
the white people and their roads and cities. When these outsiders began
to come up the rivers, long before I was born, our elders had already been
adults a long time. Their tongue had hardened in their own talk, and it
was very difficult for them to imitate white people's ghost talk. When
they met them, they only asked them for merchandise, using gestures
and a few tangled words. They never thought they had to defend their
land. They never suspected that one day the white people could invade
the forest to cut its trees, clear a road, and dig into its riverbeds to find
gold! They sometimes asked themselves why these people had come to
them. But they talked about it among themselves; their words never left
the forest.

Much later, I grew up and became an adult in my turn. I often lived
and worked with the white people, and little by little their words entered
into me. When I came back to the forest and saw that my people still did
not understand them, I told myself: "I am still young, yet I know a little
Portuguese. In the beginning of time, *Omama* gave us this land. I live
here now with my wife and children and I care for it, I carry it in my

thought. Isn't it up to me to defend our forest?" Then my mind continued on this path: "We are the sons and grandsons of brave men who were not afraid to arrow their enemies. *Õeõeri*'s and *Aiamori*'s images taught our ancestors courage, and they are still present among us! I do not want to behave like a coward in the face of the white people treating us so badly!" That is how, little by little, I decided to make them hear the thoughts of the forest's inhabitants and speak to them firmly all the way into their own cities. I was angry. I did not want my people to continue dying, devoured by their *xawara* epidemics. I intended to tell them how much their great men's minds were full of oblivion, despite their skill at making merchandise. If it were not so, why else would they want to destroy the forest and mistreat us like this?

The elders of my house encouraged me: "*Awe!* You will speak to the white people in *hereamuu*. Us, we cannot go so far to where they live, and anyhow they would not understand us. You, you know how to imitate their language. You will give them our words. Do not be afraid of them! Answer them in the same tone as they talk to us! Meanwhile, we will help you defend the forest and its inhabitants from a distance by making our *xapiri* dance and back you up!" Hearing these good words made me happy. For his part, my father-in-law added: "Despite the distance, my spirits will never take their eyes off you! If the white people attack you, they will bravely defend you!" He is a wise and good man. He has always taken care of me during my trips. So as I left to visit big cities I reassured my wife and children: "Do not worry! The white people will not kill me! If they try to hurt me, our elder will avenge me!" My thought remained calmer. I told myself: "This is good! I am going to defend our forest! I am going to speak to the white people forcefully, without fearing to make them hear my words!"

At that time, as I said, the white people ancestors' *napënapëri* spirits often came to visit me. Our house's shaman elders used to call them to me by making me drink the *yãkoana*. So they came down in large numbers, dancing with the image of *Omama*, who is their creator. They were joined by the image of *Remori*, the bee ancestor who gave these outsiders their humming talk, and the image of *Porepatari*, the ancient ghost who learned their way of speaking. *Porepatari* barters cat skins with the *napënapëri* for guns and cartridges. He is a great hunter who treks through the forest at night, invisible as the breeze. All we hear of him is his call: *"yãri! yãri! yãri!"*[19] He hunts jaguars, which, like him, are very fierce, but sometimes also arrows humans with his deadly arrow points dipped in curare. He is a great elder, a true inhabitant of the forest. He takes care of it and protects the *xapiri* who play in it. So *Remori* and *Pore-patari*'s images placed their spirit larynx in my throat so I could imitate the white people's talk. They taught me to repeat their words one after the other, more firmly and clearly. They put their *napënapëri* ancestors' language into me. If I had been alone, I would never have been able to make speeches in this outsiders' language!

THE VERY first time that I spoke about the forest far from my home was during an **assembly** in the city of Manaus. But it was not in front of white people: it was in front of other Indians. It was at the time when the gold prospectors were starting to invade the Rio Apiaú and Rio Uraricaá regions. Ailton Krenak and Alvaro Tukano, then **leaders** of the **Union of the Indigenous Nations**, invited me to talk about that.[20] They told me: "You must defend your people's forest with us! We need to speak together against those who want to take over our land. Otherwise we will all disappear, like our ancestors before us." I did not know how to go about it, and the breath of my words was still far too short! Yet I did not back down. With great anxiety, I tried for the first time to speak words about the *garimpeiros* who were soiling our rivers and killing us with their epidemic fumes. Some time later, the Makuxi also invited me to one of their big assemblies. It was in Surumu, on their land, in the savanna of Roraima state.[21] They encouraged me to speak again: "Come to our home to defend your forest, as we do for our land!" This time the meeting was much bigger. Many other people like ours had gathered there and also many white outsiders. I had no idea how I would be able to discourse in front of all these people sitting with their eyes set on me! At the beginning, I just listened to learn how the others would speak before me. So I paid attention to the Makuxi and the Wapixana speeches as they took turns speaking against the cattle ranchers: "These *fazendeiros* want to chase us off the land where our ancestors lived by claiming that it belongs to them now! We are surrounded by their **barbed wire** and their cattle. They burn our houses, insult us, and hit us. Later they will want to do the same to the Yanomami. But if we all join our words together against them, they will back down because they are liars!"

At the time, I was afraid to have to speak in front of a group of strangers, far from my forest, and to top it off, in the white people's language! My words were still scarce and twisted. I had never even made a *here-amuu* speech in my own house! I was worried and my heart was beating very fast in my chest. I did not know how to make the words come out of it one after another yet. I told myself: "How can I possibly do this? How do white people speak on these occasions? In what way should I start?" I anxiously searched for the beginning of the words I could make heard. My mouth was dry with fear. And finally it was my turn to speak! I felt all ashamed and I must really have been a pitiful sight to see. So I began by saying what was on my mind at that very moment: "I do not know how to speak like the white people! As soon as I try to imitate them, my words escape or get confused, even if my thought stays right! My tongue would not be so tangled if I was speaking to my people! But never mind, since you are lending me an ear, I will try. This way my words will become stronger, and maybe one day they will be able to worry the white people's great men." Then I continued by trying to follow the way of those who had preceded me. But I mostly said what I truly thought of the gold prospectors: "They are other people, earth-eaters, evil beings! Their thought is empty and they are full of epidemic smoke! We must prevent them from soiling our rivers and chase them out of the forest. Why don't they work on their own land? When I was a child, nearly all my elders died of the white people's diseases. I don't want this to continue!" I believe that these are the first words I said. Then little by little I tried to extend them and make them clearer. I probably only succeeded because the anger was inside me! In truth, the anger had had me in its grip for a long time, since my people had died in Toototobi and I had narrowly escaped the missionaries' measles epidemic myself.[22]

Some time later, my father-in-law and I invited the inhabitants of many Yanomami houses to our home at *Watoriki*. We wanted to hold the first Yanomami assembly there to talk about our land. Other Indians such as Ailton Krenak and Anine Suruí of the Union of the Indigenous Nations also came from very far away to join us. There were also Makuxi leaders and a few white people who were our friends.[23] Everyone took turns speaking to defend our forest. In the end, we did a *reahu* feast presentation dance and offered our guests a great deal of smoked peccary meat.[24] After this meeting, I was also a **candidate** to be a **deputy** to what the white people call the **Constituent Assembly** in Brasília.[25] At that time,

I also often addressed the other Indians of Roraima in assemblies as well
as on the radio. I did it to try the white people's **politics**, to learn something. But it did not last long and I did not win![26] Soon after, the gold
prospectors became very many in our forest highlands, destroying the
rivers' headwaters and killing their inhabitants with their diseases.[27] I
started frequently traveling to big cities, very far away from my home.
There, I joined other inhabitants of the forest who had come from all
over to speak against the gold prospectors, cattle ranchers, and loggers
invading our lands. From that moment on, I no longer needed to look for
my words. Each time I got angrier, and I wanted all the white people to
know what was going on in the forest. And so I learned to make long
speeches under their gaze, and the words to speak firmly to them increased inside me.

After Manaus and Brasília, I came to know São Paulo. It was the first
time I flew so far above the big land of Brazil. Then I understood how
vast the white people's territory is beyond our forest and I told myself:
"They have only gathered in a few cities scattered here and there! Everything between them, in the middle, is empty! So why are they so eager to
take our forest?" This thought never stopped coming back into my mind.
Finally, it made what remained of my fear of speaking to disappear. It
made my words more solid and allowed them to grow more and more. I
often told the white people listening to me: "Your land is not truly inhabited. Your elders jealously guard it to keep it empty. They do not want to
give any of it to anyone. They prefer to send their starving people to eat
our forest!" Then I added: "In the past, many of our people perished from
your epidemics. Today I refuse to let their children and grandchildren die
from the gold smoke! Chase the *garimpeiros* out of our home! They are
harmful beings whose thought is dark. They are metal eaters covered in
deadly *xawara* smoke. In the end, we will arrow them and if that happens
many more people will die in the forest!" It was difficult. I had to say all
that in another language than mine! Yet my wrath made my tongue more
nimble and my words less confused. Many white people started to know
my name and wanted to listen to me too. They encouraged me by telling
me that they thought it was good that I was defending the forest. That
made me more confident. I was happy that they understood me and that
they became my friends. At the time, I spoke in the big cities a lot. I
thought that if the white people could hear me, they would convince the

government to not let the forest be destroyed. It was with this single thought that I started traveling so far from my home.

SINCE THEN, I have never stopped talking to the white people. My heart has stopped beating too fast when they look at me and my mouth has lost its shame. My chest has become stronger and my tongue has lost its stiffness. If the words had gotten tangled up in my throat and I spoke with a reedy, hesitant voice, those who came to hear me would have told themselves: "Why does this Indian want to make speeches to us? We were expecting words of wisdom from him, but he isn't saying anything! He is far too scared!" This is why I always strive to talk with courage. I do not want people to think: "The Yanomami are idiots who have nothing to say. They are incapable of speaking to us. They only know how to remain frozen, their eyes lost, silent and afraid." This idea alone is enough to infuriate me and make me speak forcefully. The white people can go ahead and keep staring at me to try and intimidate me. It has no effect! My words thunder on as if nothing happened. First I tell them: "You who are listening to me, to us you are outsiders we call *napë pë*. I am the son of the people *Omama* created in the forest in the beginning of time. I was born of his sperm and his blood. Like my ancestors, I have the value of his image inside me. This is why I defend the land that he gave them."[28] Then I continue: "You are other people. You do not hold *reahu* feasts. You do not know how to make the *xapiri* dance. We are the few inhabitants of the forest who survived your fathers' and grandfathers' epidemic fumes. This is why I want to speak to you. Do not be deaf to my words! Stop your people from ravaging our land and making us die too!"

Talking about the death of my elders makes me sad, and the anger from mourning them comes back to me at once. My words harden and worry comes over those who are listening to me. They remain silent, their eyes are sad to see. I insist: "You don't understand why we want to keep our forest? Ask me, I will answer you! Our ancestors were created with it in the beginning of time. Since then, our people have eaten its game and its fruit. We want our children to grow up there laughing. We want to be numerous again and to live there like our elders did. We do not want to become white people! Look at me! I am imitating your talk like a ghost and I am wrapped in clothes to come speak to you. Yet at home I speak my language, I hunt in the forest, and I work in my garden. I drink the *yãkoana* and I make my spirits dance. I speak to our guests in *wayamuu* talk! I am an inhabitant of the forest and I will remain one! It is so!"

In the past, the white people used to talk about us without our knowledge, and our true words remained hidden in the forest. No one other than us could listen to them. So I started traveling to make the city people hear them. Everywhere I could, I scattered them in their ears, on their paper skins, and in the images of their television. They spread very far from us, and even if we eventually disappear, they will continue to exist far from the forest. No one will be able to erase them. Many white people know them now. Hearing them, they started telling themselves: "A son of the forest people spoke to us. He saw his people burn with fever and their rivers turn into bogs with his own eyes! It is true!"

Omama gave us our words to defend the forest. Their power comes from the image of our first ancestors from the beginning of time. We are their ghosts.[29] It was they who taught us the courage with which I speak firmly to the white people's great men. And so those who hear me worry when my words burst into their thought: "*Hou!* These people of the forest are not afraid! Their talk is hard and does not yield!" The first time that I spoke to the president of Brazil, I asked him to chase the gold prospectors out of our forest.[30] He answered me hesitantly: "There are too many of them! I do not have any airplanes or helicopters for that! I do not have any money!" He gave me his lies as if I were devoid of thought! I was carrying the wrath of my destroyed forest and dead relatives deep inside me. I answered him that by speaking with such twisted words he only wanted to deceive us and let our land be invaded. Then I added that he must be a weak man whose mind was full of oblivion and that he could therefore not claim to be a true great man.

WHEN I WAS younger, I often asked myself: "Do the white people possess words of truth? Can they become our friends?" Since then, I have often traveled to them to defend the forest, and I have learned to know a little of what they call politics. It really made me become more suspicious! Their politics is nothing but mixed-up talk. These are the words of those who want our death and to seize our land.[31] Often, the people who utter them have tried to deceive me by saying: "Let's be friends! Follow

our path and we will give you money! You will have a house and you will
be able to live in the city, like us!" I never lent an ear to them. My spirit is
only calm when I live in the beauty of the forest with my people. In the
city, I am constantly anxious and impatient.

White people call us ignorant because we are other people than they
are. But their thought is short and obscure. It does not succeed in spreading and rising because they prefer to ignore death. They are prey to dizziness because they constantly eat the meat of their domestic animals who
are sons-in-law of *Hayakoari,* the tapirlike being who makes people turn
other.[32] They constantly drink *cachaça* and **beer** that overheat their chests
and fill them with fumes. This is why their words become so bad and
muddled. We do not want to hear them anymore. For us, politics is something else. It is the words of *Omama* and those of the *xapiri* that he gave
us. These are the words that we listen to during the time of dream and
that we prefer because they are truly ours. The white people, they do not
dream as far as we do. They sleep a lot but only dream of themselves.
Their thought remains blocked, and they slumber like tapirs or turtles.
This is why they are unable to understand our words.

We do not have laws on paper skins, and we do not know *Teosi*'s words.
However, we have the image of *Omama* and that of his son, the first shaman. They are our laws and our government. Our elders did not have
books. *Omama*'s words and those of the spirits enter into our thought
with the *yãkoana* and dream. We have kept this law of ours deep inside us
since the beginning of time by continuing to follow what *Omama* taught
our ancestors. We are good hunters because he made the images of the
wakoa and *kãokãoma* falcons enter into our blood. We do not need to
teach our children to hunt. When they are very young, they start by arrowing lizards and small birds, then as soon as they grow they go hunt
real game. *Omama* also gave us the plants in our gardens, which he acquired from his underwater father-in-law. He taught us how to build our
houses and cut our hair. He taught us to host our *reahu* feasts and to put
our dead's ashes into oblivion. He passed down all the words of our
knowledge. The white people, they have their school for that. To us, what
they call **education** is the words of *Omama* and the *xapiri,* our elders'
hereamuu speeches, and our feasts' *wayamuu* and *yãimuu* talks. This is
why *Omama*'s law will always remain in us, in the depths of our thought
as long as we are living.

It is thanks to his law that we do not mistreat the forest the way the white people do. We know very well that without trees nothing will grow on its hardened and blazing ground. Then what would we eat? Who will feed us if we do not have gardens or game? Certainly not the white people, who are so greedy that they would let us die of hunger. We must defend our forest to be able to eat manioc and plantain bananas when our stomach is empty, to be able to smoke monkeys and tapirs when we are hungry for meat. We must also protect its watercourses to be able to drink and fish. Otherwise we will be left with streams of miry water covered in dead fish. In the past, we were not forced to talk about the forest with anger because we did not know all these white people land and tree eaters. Our thoughts were calm. We only listened to our own words and the *xapiri*'s songs. This is what we want to be able to do again. I do not speak of the forest without knowing. I contemplated the images of its trees' fertility and its game's fat. I listen to the voice of the bee spirits who live on its flowers and that of the wind beings who chase the epidemic smoke out of it. I make the animal and fish spirits dance. I call down the image of the rivers and the land. I defend the forest because I know it thanks to the power of the *yãkoana*. Only shamans' eyes can see its spirit, which we call *Urihinari,* and also the spirit of *Omama.* It is these spirits' words that I make heard. It is not just my own thought.

I ALWAYS think of all this when I go on a visit to the city. I have seen dangerous things with my *xapiri.* I want to warn the white people before they wind up tearing the sky's roots out of the ground. If their elders knew the talk of our *yãimuu* dialogues, I could truly tell them my thought. Squatting face to face, we would argue at great length, hitting each other's flanks. My tongue would be more skillful than theirs, and I would speak to them with such vigor that they would be exhausted. And so I would finally enmesh their words of hostility! Unfortunately, the white people know nothing of our ways of talking in the night. Sometimes when they listen to us do so during the *reahu* feasts, they ask themselves with confusion: "What are these songs? What are they saying to each other?" As if these were simply *heri* songs![33] Yet if they could understand me, I would tell them in *yãimuu* talk: "Stop pretending to be great men, you are pitiful to see! I will silence your bad talk! If your minds were not so closed, you would chase the earth eaters out of our forest! You claim that we want to cut out a piece of Brazil to live there alone.[34] These are lies to steal our land and confine us to it in little pens like chickens. You do not know how to do anything with the forest. You only know how to cut down and burn its trees, to dig holes in its floor and soil its watercourses. Yet it does not belong to you and none of you created it!"

All these words have accumulated in me since I have known the white people. Yet today I am no longer satisfied with keeping them deep inside my chest, as I did when I was younger. I want them to be heard in their cities, everywhere they can be. Then maybe they will finally tell themselves: "It is true! Our great men have no wisdom! Let's not allow them to destroy the forest!" I know that their elders will not easily listen to my talk because their thought has been set on minerals and merchandise for too long. Yet the people who were born after them and will replace them one day may understand me. They will hear my words or see their drawing while they are still young. These words will enter into their mind and they will have far more friendship for the forest. This is why I want to speak to the white people!

When I was a child, I did not think I would learn their language and even less that I would go make speeches where they live. Nor did I wonder about their thoughts or what they might say among themselves. I did not ask myself what their cities were like. I was simply afraid of them, and when they came close to me I ran away yelling. I liked it in the forest. I liked to listen to my kin's words and converse with my stepfather. Hearing him talk about hunting and *reahu* feasts delighted me. I was happy this way, and if the white people and their epidemics had not started devouring my people, maybe I would be still. Once I became an adult, the gold prospectors' sudden arrival made me think a lot. I told myself: "*Hou!* I did not know it, but the white people have always been the same, since long before my birth! They already wanted to tear rubber, para nuts, *masi* vines, and jaguar skins out of the forest, like they want to find gold in it today. Their avidity was what made most of our elders die long ago!" Today I do not speak of all that for no reason. I never forgot the sadness and anger I felt when my people died when I was a child.

CHAPTER 18

Stone Houses

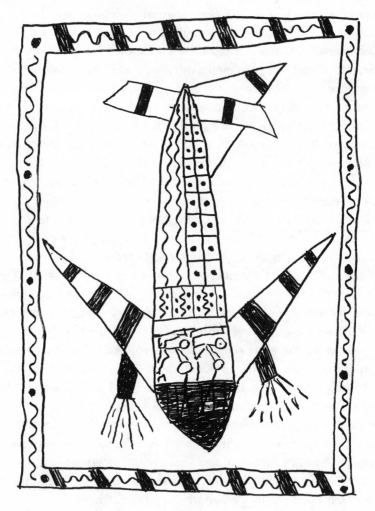

A big airplane

The journeys I made to defend our forest against the gold prospectors eventually led me far beyond Brazil. One day, white people who
had heard my name called me from a very distant land of which I knew
nothing, **England.** I accepted their invitation because I was curious to
meet these distant people who seemed to have friendship for us.[1] It was

the first time that I left our house at *Watoriki* to fly in an airplane for so long. It was so far away that I arrived in the land of the ancient white people, which they call **Europe.** Then I was able to see with my own eyes the trace of the houses of the first light-skinned outsiders, the *napë kraiwa pë*, whom *Omama* created very long ago with the first blood foam of the forest people of *Hayowari*.[2]

The friends who took care of me during this trip showed me a place where the white people's ancestors once lived and worked. I saw a great circle of stones planted in the ground there.[3] I instantly thought that they had been erected by *Omama* when he ran away towards the rising sun and that the wall they formed must have been remains of his former house.[4] These rocks are very tall and solid, like a house's great poles. They do look like huge stone posts. *Omama* chose to build his home this way because stone does not rot and thus never dies. But he did not do this work alone. The ancient white people all joined him, from the oldest to the youngest. They must have suffered a lot to lift and hoist these huge blocks! Since there were so many of them, *Omama* probably taught them to build stone houses so they would not destroy all the trees in their forest. This is what I thought. And after I saw all this, when the night came, the *xapiri* carried my image away during my sleep and sang for a long time about this line of rocks. They introduced me to the *Mot^hokari* sun being's passage from the downstream of the sky and the path by which *xapiri* from these distant lands come dancing to us. They showed me the place from which the *Yariporari* storm wind being repelled the epidemic fumes far from the ancient white people land. They also made me discover where these ancestors learned how to die and to bury their bones in graves closed by huge rocks.

Seeing the traces of these people who have been dead for so long upset me. It made me sad, as much as seeing the traces of our grandfathers' abandoned old gardens in the forest. Those who planted these big stones were the first outsiders whom *Omama* created with the blood foam of our *Hayowari* ancestors, who were carried away by the underworld waters. The first white people's land probably seems very distant from our own, but we should make no mistake. It is one and the same land. It only came loose in the time when the *Hayowari* inhabitants turned other. It was torn off by the power of the torrents that sprang out of the ground, and was then carried into the distance to be fixed where it is now. So it was very far away from us that our ancestors stuck these great stones in the ground after turning into white people. These rocks mark the limits where their drifting forest came to a standstill at the earth's edges, stopped by the sky's feet. Their alignment draws the outlines of the ancient white people's land. *Omama* wanted these stones to be arranged this way so that neither they nor their children could forget them. This land, which I named *eropa urihi a,* has been theirs since they were created there. No one other than them has lived there.

It has been a very long time now since *Omama* disappeared. Yet these rocks are still standing. This is exactly why long ago he chose to use such massive blocks. He wanted them to remain in place after his death so white people could continue to look at them and think: "These are the traces of *Omama* who created our ancestors!" He thought that without this their lost minds would vainly wonder how they came into existence. Yet today all the young people who come to see these stones without fearing the great wind that surrounds them seem perplexed. Their elders have lost the stones' words and cannot pass them on to them. So they observe them at great length without recognizing them. They only ask themselves how the ancients were able to lift up such a heavy load!

These rocks erected by *Omama* and the ancient white people must not be destroyed. These first outsiders' ghosts are still present around these big stones and also near the bones buried at their feet. If these rocks were brought down, their law would be abolished and forgotten. This law is the white people ancestors' knowledge. It is the memory and center of the thought of those who were born after them. It is to the white people what *Omama*'s words are to us. If this law, this **mark** of the beginning of time, no longer comes between them, they will lose all restraint. They will never stop mistreating the land and killing each other. The ancient white people who worked so hard to raise these stones did it so that they could be contemplated after their death and so that their memory would not be lost. They brought them from very far away, without machines. This is how they invented toil at work. We inhabitants of the forest were taught about the hard labor of gardens under the blazing sun by *Koyori,* the manioc ant ancestor.[5] If the present white people's machines knock over *Omama*'s big stones, their ancestors' ghosts will be furious. They will think that those who claim today to be great men on their land have no more wisdom.

ONCE I ACCEPTED to leave for England, I was concerned at the idea of going so far away from my people and hoped for the support of the shamans of my house. To tell the truth, I was very worried to fly to the land where *Omama* created the white people's ancestors. So before I left I asked my stepfather to be vigilant and to help me during this trip. He began protecting me by giving me his elder's words. He recommended that I only bring a few of my *xapiri* with me and that I lock all the others up in their house above our forest. Then he raised their paths very far up into the sky's chest so that they would not be torn by the airplane that would take me. It is true. Despite their power, the spirits are as light as down feathers. Without this precaution, they might have died suffocated or been carried to the earth's edges by the wind. And if they had gotten lost there or been captured by evil beings, it could have killed me. To keep me safe, the shaman elders of *Watoriki* worked to close all my spirits' mirrors so they would not move away until my return. Then they wove a solid covering over their house and surrounded it with a powerful wind to make it inaccessible.

Only the wisest and most resilient of my *xapiri* were able to follow me on this faraway journey. They were the image of *Omama,* who holds up the flight of airplanes with a metal path in the sky and created the first outsiders, and that of *Porepatari,* the ancient ghost who was the first to trade with them. But other *xapiri* also accompanied me to defend me, such as the caiman spirit with his big machete or the spirit of *Xinaru- mari,* the master of cotton, with his venomous tail. Without their protection, distant hostile shamans' evil spirits could have weakened me and made my head spin with dizziness or even made the airplane carrying me fall down. But knowing they were by my side, I did not fear anything, and I was able to keep my strength to make the white people hear my words.

But those of my *xapiri* who had to remain in the forest were worried to see me disappear far away, and they remained concerned throughout my trip. The moon spirit *Poriporiri* tried to keep the light of his eyes set on me, so I would not lose my way back. As for the spider monkey spirits, they constantly called out to be reassured about my fate. During the night, having become ghost under the effect of unknown white people food, I often heard their worried clamors in dream: "Where has our father gone? He will wind up getting lost! Let him come back to us very quickly! These outsiders will mistreat him! He will get sick!" Then the *xapiri* accompanying me reassured them: "*Ma!* He is here, with us! He is doing well! Do not be impatient! It is not so far, he will come back soon! Don't worry if you hear the voice of the thunders vociferate! It means they are angry about other shamans' death!"[6]

This is how I prepared to go to the ancient white people's land for the first time. Before this, I was not cautious enough! I traveled everywhere without thinking about what could happen to my *xapiri*. Once, I nearly even died of it, and it was on a trip not far away from my home. In fact, it happened in our forest highlands during a trip to the people at *Tëpë xina*, near FUNAI's Surucucus Outpost. I was accompanying white people who had come to take pictures of the people there.[7] But at some point I moved away from them because the shamans of the house invited me to take *yãkoana* with them. So I started to bring down my spirits and I no longer paid attention to the outsiders who had followed me. Then, suddenly, they pointed such a strong light in our direction that it blinded us all. I knew the white people, but I still knew nothing of their way of capturing images for their television, which we also call *amoa hi*, song tree.[8] It was frightening! My *xapiri* were still dancing close to the ground and were instantly attracted to the machine pointed at us. They were lured by the dazzling blaze that surrounded it. It reminded them of the gleaming lights of their paths and their house. They got lost following it and were instantly sucked into the image machine, where they remained prisoner. A few days later I returned to *Watorikɨ*. Having lost my *xapiri*, I got sick. I was seized with dizziness and very weak. I babbled like a ghost. My father-in-law worriedly asked me: "What did you do with your spirits? Did you give them away? Did they run away?" I was very anxious and I thought I would not survive. I knew that as soon as the *xapiri* leave their father, he remains empty and runs the risk of quickly dying. Yet one of my brothers-in-law—a great shaman elder who is no longer with us—understood that the television machine had swallowed them up like bright down feathers and that they remained stuck inside it. By working hard, he was able to rescue them and bring them back to my spirits' house in the sky's chest. He was really very clever, and after that I was able to recover quickly. But he warned me against my carelessness and since then I have always followed his advice: "Do not travel to the white people with your spirits anymore! They will capture them again and this time it will kill you!" This is why today only a few of my most powerful *xapiri* can accompany me and protect me on my journeys.

THESE DISTANT lands of the ancient white people are spirits' lands.[9] A great number of the *xapiri* who dance for us come from there. This is the case of the *napënapëri* outsiders' ancestor spirits, the *Remori* bee spirit who taught the white people their tangled language, and many others. This is why it is so dangerous to take an airplane to those areas where so many *xapiri* come from. For a shaman, flying towards the mirror-land[10] place of origin of the spirits he calls down and suddenly being faced with it would be risking an immediate death. But our *xapiri* are very wise and do not allow that to happen! We can never reach the place from which they come down. If a shaman headed for their mirror-land without knowing it, they would instantly make it disappear upon his approach. Then instead of reaching it, he would continue forward in space and would end up having his back to it without ever having seen it.

This is what happened to me during this first trip to the ancient white people's land. As I was about to touch down on it, I saw through the airplane's **window** that a vast mirror with blinding flashes of brightness was coming towards me at high speed. It was very frightening because at the time I knew nothing of these things. My eyes remained captivated by this intense shining light for a long time. I was prey to dizziness and overcome by a deep torpor. Then I felt myself losing consciousness. Yet at the very moment when I thought I would reach this land-mirror and just die, it turned on itself and settled behind me in the place of the land I had come from. It is true! Seeing me coming, the *xapiri* made their mirror-land tip over before me to allow me to pass over it without hitting it. And once the airplane was on the verge of touching down the ground, its black path was already resting on a new land that they had stretched before me. The vast dazzling mirror had suddenly vanished behind me and a different ground already replaced it before my eyes. Instead of losing consciousness and dying, I only felt a deep drowsiness. If these faraway spirits had not turned their mirror over when I arrived on their land, my dead body would soon have been brought back to *Watoriki* for my tearful kin to hang it in the forest and cremate my bones!

This is why it makes no sense to think that the *xapiri* do not exist on the white people's land. In this distant place as in our forest, the wind does not blow without a reason and the rain does not fall all by itself. But the beings of darkness and chaos are closer there. It is very cold. The night falls very fast and lasts a long time. Today's white people know nothing about the *xapiri* who inhabit their land and they never even think about them. Yet they always existed, long before the white people themselves were created. The spirits are truly numerous there! This is why I was seized by such dizziness as I approached their home. We shamans, we know these faraway *xapiri* because we bring them down in the forest by drinking the *yãkoana*. They live in the coolness of the highlands, away from the white people and their cities full of fumes. I saw the mountains where their homes are with my own eyes. Their peaks are covered in a dazzling whiteness like a huge heap of down feathers. I traveled there in dream when I was burning with malaria heat and discovered the pristine water spring where the spirits drink. They joyously bathe and frolic there despite the glacial winds that surround them, and their hands are as frozen as the water. This is why they are so good at curing fevers. This is also where the envelopes full of clear water come from, which white people sell to quench their thirst.[11] It is the same water as the water that comes down from our forest's rocky peaks. We shamans call it *mãu krouma u,* the water of the *krouma* frog, or *mãu pora u,* the water of the waterfall.

Even though today's white people of Europe have forgotten it, the *xapiri* who live on their land are the images of their ancestors who died long ago. These are the images of the first outsiders with ghost talk whom the shamans call *napënapëri*. These ancestors were the ones who passed down their words to the white people. They were the ones who put the tall stones of *Omama*'s house in place and created the first merchandise, paper skins, and medicine. I often saw their images dance during my first trip to this distant land. They came down in my dream in the form of ghosts, like the *xapiri* do in the forest. They reached me easily because I slept right where *Omama* created these first outsiders with the foam of the *Hayowari* people's blood in the beginning of time.

The *napënapëri* spirits want to keep the beauty of their mirror-land and protect it from epidemic fumes. Yet today's white people no longer know how to take care of it, and they know nothing of these images, which are those of their ancestors. This worries me too. In the past, their long-ago elders knew them and made them dance. They knew how to imitate their songs and build their spirit houses for the young people who wanted to become shamans. But then those who were born after them began to create the cities. Little by little they stopped hearing these spirits' words. Then the books made them forgetful, and they finally rejected them. As I have said, *Teosi* was jealous of the beauty of the *xapiri*'s words. He constantly spoke ill of them: "Do not listen to those spirits that soil your chest! They are inhabitants of the forest, they are bad! They are just animals! Stop calling their images, just eat them! Instead of listening to them, contemplate my words glued on paper skins!"[12] *Teosi*'s words of anger spread everywhere and chased the *xapiri*'s song from the white elders' thoughts. Their minds became tangled and confused, always in search of new words. Yet the spirits of this distant land are not dead. They still live in the mountains that *Omama* gave them for houses and they only come down from them for the shamans who are able to see them.

DURING THIS trip to Europe, I often slept in a ghost state after eating white people food I did not know. One day while dreaming I saw the image of the bee women of the beginning of time. They were shouting out their own names in every direction to attract the attention of the marten ancestor *Hoari* who collected their honey from one place to another.[13] Finally they made his head spin with their constant calling, and he tripped over a root. So he furiously cursed at them and made them flee all over the forest. Their images concealed themselves everywhere where honey is hidden today. This is why it is so hard to find honeybees' nests today. Some even fled all the way to the white people's land, where they keep them in large wood boxes to this day. Our elders have always made these bee spirits dance. It was they who came to speak to me in my dream. They wanted to tell me of their concern: "You who know how to become spirit, speak firmly to the outsiders who will listen to you! The white people truly lack wisdom! They must stop mistreating the trees in the forest! Soon there will be no more sweet-scented flowers to feed us and make honey. If this continues, we too will perish!" Bees are also *xapiri*; this is why their images spoke to me this way during my sleep. The next day, I revealed their complaint to the city inhabitants who had come to listen to me. I was upset to hear of these spirits' suffering and think that the white people mistreat them so. These animal ancestors feel threatened and, like us, they want to defend the forest where they came into being. Bees are very wise and work relentlessly in the flowers they go in search of far away, from tree to tree, to make their honeys. This is why these honeys are so tasty and both children and adults like it so much. Cutting down trees is also destroying the bees' paths through the forest. Without flowering trees, they will no longer know where to work, and they will flee from our land forever. This is why I told the white people: "You often claim to love what you call **nature**. Then do not settle for making speeches, truly defend it! You must help us to protect what still remains of the forest. All its inhabitants already speak to us with the fear of disappearing. You do not see their images dance and you do not hear their songs in your dreams. Yet we shamans, we know how to listen to the bees' distress, and they are asking us to speak to you so your people will stop eating the forest."

WHEN I RETURNED from this long stay on the ancient white people's land, I was happy to come back to my hammock in our house at *Watoriki*. Yet I had barely settled into it when I was seized by a violent dizzy spell. After I had flown so long in an airplane, the ground in the forest relentlessly spun around under my feet. I could only keep my eyes set ahead of me, staring like a ghost. My thought was blocked, and I was constantly drowsy. I did not want to look like a layabout, so I tried to go hunting. But it was no good, I couldn't see anything around me in the forest and I was incapable of spotting any game. I was so weak that I was tripping everywhere, and I was forced to lie down on the ground several times. Flying to this distant spirits' land had made me become other, and if the shaman elders who had long ago been my teachers had not protected me again, I might have died of it!

Shortly before my return, the *xapiri* who had escorted me on this long trip had returned to the forest as scouts to announce my arrival. Then they had laid down in the hammocks of their spirit house to recover their strength. Those I had left behind were happy to learn I was close by: "*Haixopë!* Father is on the way! He is finally coming back towards us! Nothing unfortunate happened to him! He is unharmed! *Aë! Aë!*" They were impatiently waiting for me because they were starving. So the shaman elders of *Watoriki* began to work by my side. They helped me to put my spirits' paths back in the sky's chest in order to chase away the sleep that had taken hold of me. I drank the *yãkoana* day after day in order to start feeding these *xapiri* I had left home so long ago. And once they had eaten their fill, they were happy to endlessly sing and dance for me. This is how I was able to recover, and I always do the same thing when I come home after a long visit to the white people. Without this, the dizzy spells would never leave me.

If a shaman does not properly feed his *xapiri* spirits, they suffer from hunger just like humans do. As I have said, the *yãkoana* powder is their food. Under no circumstances can we leave them abandoned in their hammocks, especially the youngest ones! If their father does not make them dance and sing as often as they want, they feel neglected. They get annoyed by it and begin to complain about him: "He is a layabout who just speaks ghost talk! He called us in vain! In fact, he forgets us and does not want us! Let's not allow him to mistreat us this way, let's go home!" If the shaman continues not taking care of them, they eventually really do escape his spirit house. However, if he often drinks the *yãkoana*, they are truly happy. They launch into their songs with such enthusiasm that the joy of their voices constantly draws other *xapiri* to come settle by him.

THE WHITE people's ancestors did not take care of the forest in which they came into being the way ours did. They cut down nearly all of its trees to open boundless gardens. I saw with my own eyes the little that remains of the trees there, only very small patches here and there. Yet *Omama* had taught them to build stone houses to avoid clearing everything. He had told them: "Wood posts rot and must often be replaced. Cut out tall rocks and plant them in the ground to build your houses. This way you will only work once and you will spare the trees that give you their fruit and whose flowers feed the bees." These long-ago outsiders started cutting into the rock with their axes. Then they became more ingenious. They built tools to cut out smaller stones and prepared a mud that glues them together when it dries. So they were able to build more solid and numerous stone houses. They were satisfied and they got the idea to draw the ground around each house. Then they discovered the beauty of merchandise and started making it relentlessly. After a time, there was so much merchandise that they had to build new houses to shelter and distribute it everywhere.[14] They also built some houses to store the food from their gardens. Once these stone houses had proliferated, they connected them to each other with tangled paths and gave them the name "city." This is how little by little the forest disappeared from their land along with the game that inhabited it. They only kept a few living animals in pens made of stakes and some other dead ones in glass cases so their children could contemplate them as **souvenirs**.[15] These are all the things I thought about as I walked in the ancient white people's cities. I was very far from my house, and for the first time I was seeing their old land with my own eyes. I strolled everywhere, most often without saying a word, carefully observing the houses and the people. My thoughts constantly spread in every direction. I really wanted to understand what I was seeing.

CHAPTER 19

Merchandise Love

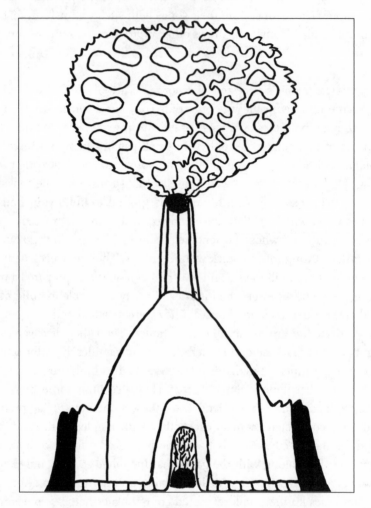

Factory smoke

I N T H E beginning the first white people's land looked like ours. It was a land where they were as few as we are now in our forest. Yet little by little their thought strayed onto a dark and tangled path. Their wisest ancestors, those whom *Omama* created and gave his words to, died. Their sons and grandsons had very many children in their turn. They started to reject the sayings of their elders as lies, and little by little they forgot them. They cleared their entire forest to open bigger and bigger gardens. *Omama* had taught their fathers the use of a few iron tools. They were no longer satisfied with them. They started desiring the hardest and most cutting metal, which *Omama* had hidden under the ground and the waters. They began greedily tearing minerals out of the ground. They built factories to melt them and make great quantities of merchandise. Then their thoughts set on these trade goods, and they became as enamored with them as if they were beautiful women.

They soon forgot the beauty of the forest. They told themselves: "*Haixopë!* Aren't our hands so skilled to craft these things? We are the only ones who are so clever! We truly are the people of merchandise![1] We will be able to become more and more numerous without ever lacking for anything! Let us also create paper skins so we can exchange them!" They made money proliferate everywhere, as well as metal pots and boxes, machetes and axes, knives and scissors, motors and radios, shotguns, clothes, and sheet metal.[2] They also captured the light from the lightning that fell to the earth. They became very satisfied with themselves. By visiting each other from one city to the next, all the white people eventually imitated each other. So the words of merchandise and money spread everywhere on their land. This is what I think. By wanting to possess all this merchandise, they were seized by a limitless desire.[3] Their thought was filled with smoke and invaded by night. It closed itself to other things. It was with these words of merchandise that the white people started cutting all the trees, mistreating the land, and soiling the watercourses. First they started all over their own forest. Now there are few trees left on their sick land, and they can no longer drink the water of their rivers. This is why they want to do the same thing again where we live.

IN OUR language, we have given the white people's objects the name *matihi*. We use this word to refer to merchandise, but it existed long before these outsiders arrived in the forest. This word is very ancient; it is a word from the beginning.[4] In the past, our elders used this name for other things. They called *matihi* the ornaments with which they adorned themselves for *reahu* feasts:[5] their bunches of scarlet macaw tail feathers, the toucan tails they hung from their armbands of curassow and agami crests, as well as the parrot and guan feathers they stuck in their earlobes.

They also hunted *sei si, hëima si,* and *wisawisama si* birds for the beauty of their feathered hides, which they referred to in the same way, *matihi.* Before a *reahu* feast, the great men never failed to exhort the youth of the house by exclaiming: "Go arrow some *matihi* so you won't look an ugly, poor hunter during your presentation dance!" The young girls would admiringly say of a young man adorned with feather ornaments: "How splendid he is! He is covered in *matihi!*" And the elders agreed: *"Awe!* He is a *matihi xio,* a very good adornment tracker!" It was so. For we shamans, this word has a high value, for it names goods that belong to *Omama* and to the *xapiri* he created. Seeing them makes our thought clear and strong. This is why the word that refers to this finery also has the value of the spirits: it evokes the beauty of the *xapiri* who possess it and it makes us think of them.[6]

When one of us dies, we also call the bones that we collect from his decomposed flesh *matihi.* We burn them and crush them in order to keep their ashes in a small *pora axi* gourd. We refer to this funerary gourd in the same way, *matihi.* The dead's bones and their ashes are not things we can accept to treat carelessly! This is why the power of this word, *matihi,* has always been attached to them. A guest who would get rid of the funerary bone ashes entrusted to him would instantly have to face the vengeance of the deceased's relatives.[7] If he declared: "I threw away the rest of his *pora axi* gourd in the forest, I really had no friendship for him!" and other people reported this talk to his kin, they would be seized with sorrow and anger and instantly want to fight. They would also be furious if the person in charge of burying these ashes next to their hearth during a *reahu* feast accidentally spilled them in the fire. You cannot mistreat a dead man's bone ashes with impunity! And when they are the ashes of a valiant and hardworking man or of a shaman elder who was good at repelling evil beings, we are even more careful with them. We do not call the ashes of our dead's bones *matihi* for no reason! Our ancestors gave us this strong word because the value that we give our bone ashes is higher even than the value the white people give the gold they covet so much.

As soon as they saw the abundance of strange objects in the white people's camps, our elders—who had never seen anything like it—were very excited.[8] For the first time, they admired new machetes and axes, shiny metal pots, large glass mirrors, rolls of scarlet cloth, enormous hammocks made of colorful cotton, and shotguns that make thundering noise. They told themselves: "All these things are truly magnificent!

These outsiders must be very clever if all that their hands touch becomes so beautiful! They must really be ingenious to possess such a vast quantity of precious objects!" Then they started to passionately desire the white people's **merchandise** and they called it *matihi*, as if it were feather ornaments or bone ashes. Later, once they knew it better, they attributed a name to each kind of merchandise so they could ask these outsiders for it.[9] They were euphoric and did not yet suspect that these trade goods carried *xawara* smoke epidemics and death with them.

MAN-MADE THINGS, especially those of the white people, can last far beyond our own existence. They do not decompose like our body's flesh. Humans get sick, age, and easily pass away. The metal of machetes, axes, and knives rusts and gets covered in termite filth, but does not disappear anytime soon. It is so. Merchandise does not die. This is why we do not accumulate it in our lifetime and we do not refuse it to those who ask for it. If we did not give it, it would continue to exist after our death and go moldy alone, abandoned on the ground inside our house. Then it would only serve to sadden those who survive us and mourn our death. We know that we will die, this is why we easily give our goods away. Since we are mortal, we think it is ugly to cling too firmly to the objects we happen to possess. We do not want to die greedily clutching them in our hands. So we never keep them for very long. We have barely acquired them before we give them to those who might in turn desire them. So merchandise quickly leaves us to get lost in the distance of the forest with guests at our *reahu* feasts or casual visitors. This way, all is well. We follow the words of our ancestors who never possessed all these goods brought by the white people.

When a human being dies, his ghost does not carry any of his goods onto the sky's back, even if he is very greedy. The things he made or acquired are left on earth and only torment the living by rekindling the longing for his presence. We say that these objects are orphans and that, marked by the dead man's touch, they make us sad.[10] If one of my children or my wife were to die, the things they used to handle would keep the trace of their hands. Then I would have to burn them while I mourned and make them disappear forever. As I have said, merchandise lasts a very long time, unlike humans. This is why it must be destroyed upon the death of those who possessed it, even if their kin come to lack it. It is so. We never keep objects bearing the mark of a dead person's fingers when he possessed them.

We are different from the white people and our thought is other. Among them, when a father dies, his children are happy to tell each other: "We are going to share his merchandise and his money and keep them for ourselves!" The white people do not destroy their deceased's goods, for their mind is full of oblivion. As for me, I would not say to my son: "When I die, you will keep the axes, pots, and machetes I happen to own!" I simply tell him: "When I am no longer, you will burn my possessions and you will live in your turn in this forest that I am leaving for you. You will hunt and clear gardens to feed your children and grandchildren on this land. Only the forest will never die!" It is true. We think it is bad to own a dead man's goods. It fills our thought with sorrow. Our real goods are the things of the forest: its waters, fish, game, trees, and fruit. Not merchandise! This is why as soon as someone dies we make all the objects he kept disappear. We grind up his bead necklaces; we burn his hammock, his arrows, his quiver, his gourds, and his feather ornaments. We crush his pots and throw them in the river. We break his machete against a stone, then hide the pieces in a termite nest. We try to not let any of his traces remain. We scrape the ground where he squatted and the place where he tied his hammock's ropes to our house's posts. This is what *Omama*'s words taught our ancestors, and we follow their path. This is not a recent thing. This is how the living can put an end to the sadness they feel when they see the objects and traces left by those who are no longer. Then little by little their pain is soothed and their thought can return to calm. If we did not do this, longing for the dead and the anger of mourning would have no end.

THE STONES, waters, earth, mountains, sky, and sun are immortal like the *xapiri*. These are things which cannot be destroyed and which we call *parimi*, eternal.[11] On the contrary, humans' breath of life is very short. We live a short time. *Xawara* epidemics, evil beings, and enemy sorcerers easily devour us. So we are concerned about our kin and those for whom we have friendship. We think that if they came to pass away, we would regret not having proved generous enough with them. We tell ourselves: "*Hou!* I lacked wisdom in being so stingy. I did not answer their demands and now this memory saddens me." And also, as I said, knowing that we ourselves will soon die, we do not want to leave behind goods which will only distress our kin when they see them.

This is why when a visitor from a friendly house asks us for trade goods, we do not refuse it to him. On the contrary, we tell him: "*Awe!* Grab that machete and own it in your turn! This way, if I come to pass away, will you mourn for me? Will you really lament yourself?" He answers us: "*Ma!* You are generous! Of us all, I will be the one who mourns for you with the most sorrow!" Finally, we add: "If a snakebite kills me, break the thing I have just given you and bury its pieces in the mud at the bottom of the river." In this case, we ask for nothing in return. We leave that for another occasion, later.[12] It is only if our guest covets our bow that we can instantly ask for his own bow in return. During such barters, the elders can also tell themselves: "My hair is already white and the white people are close. Their epidemic fumes will soon devour me and I will make my people sad. I am old and I am already a pitiful sight to see. Death will soon make me let go of my possessions, this is why I am giving you these things." Words like this are the ones we are accustomed to using about goods we happened to possess. The white people are other people. They accumulate merchandise in vast quantities and always keep it close to them, lined up on wood boards in the back of their houses. They let it age a very long time before they part with it with great parsimony. They spend their time dithering and making promises to refuse it to us. Or else they demand that first we work for them for a long time. In any case, in the end they give nothing or only part with a few worn objects, demanding even more work in exchange! They behave like a bad father-in-law who deceives his future son-in-law by making him work without ever giving him his daughter. He kindly promises her to him when she is still a child. But as soon as she has reached puberty he starts to find excuses to delay the moment when he will send her to tie up her hammock next to his or, worse yet, he finally gives her to another!

As I SAID, we Yanomami do not keep the objects that we make or receive, even if it leaves us impoverished. We soon offer them to those who ask for them. So they rapidly move far away from our hands to constantly pass from one person to another. This is why we do not truly possess any goods of our own. When we acquire a new machete from the white people, we always end up leaving it to a guest during a *reahu* feast soon af-

ter. We tell him: "I am an inhabitant of the forest, I do not want to own a
lot of merchandise. Take this old piece of metal which comes to us from
Omama. I have already used it enough. I will not refuse it to you. Take it
back to your home. You will clear a new garden with it. Then you too will
give it to someone else. Speak of me to the one who will receive it and to
his kin. I want to be regarded with friendship far from my home. Later it
will be my turn to ask you for something." Once this guest has returned
home, he will soon give away that machete to other visitors. Then little by
little it will reach unknown people in a distant forest. This is how our machetes with handles wrapped in wire come from Brazil to the *Xamat^hari*
of the Rio Siapa in Venezuela and that inversely many of their machetes
with large curved tips reach us.

The same thing happens when we obtain glass beads from the white
people. We keep them a truly short time before they vanish far away from
us. First we share them among ourselves. But as soon as we are invited to
a *reahu* feast by the people of the Rio Toototobi, we give them away to our
hosts to obtain other goods. Then the people of Toototobi go on a visit to
the *Weyuku t^hëri* of the upper Rio Demini, with whom they barter them in
their turn. Then the *Weyuku t^hëri* pass them on farther yet, upstream, to
other highland *Xamat^hari* who are their allies. They eventually also reach
the people of the Rio Siapa. In the end, like our machetes, these beads
will have traveled far away from us, accompanied all along by friendly
talk about us: "*Awe!* These people are generous, they are friends! They are
valiant, this is why they are able to show such generosity!"[13] When the inhabitants of these faraway houses hear these beautiful words, they tell
themselves it would be good to establish a new trail through the forest to
come visit us and barter directly for the goods they desire. Then they call
this trail the "generous people path."[14] They can then declare with satisfaction, pointing to the entrance to their house: "Here is a door of generosity! It opens on a path of merchandise!"[15]

These are our customs, both with our own objects and with the merchandise that comes to us from the white people. Yet the white people
often think that we are avid for their merchandise, just because we often
ask for it. But it is not true! No one among us wants their trade goods
only to pile them up in his house and see them get old under the dust!
On the contrary, we never stop bartering them with each other so that
they never stop on their course. It is the white people who are greedy and
make people suffer at work to extend their cities and accumulate their

merchandise there, not us! This merchandise is truly like a **fiancée** to them![16] Their thought is so attached to it that if they damage it while it is still shiny, they get so enraged that they cry! They are really in love with it! They go to sleep thinking about it like you doze off with the nostalgia of a beautiful woman. It occupies their thoughts even after they fall asleep. So they dream of their car, their house, their money, and all their other goods—of those they already possess and those they desire again and again. It is so. Merchandise makes them euphoric and obscures all the rest of their mind. We are not like them. Our thought is set on the *xapiri* more than on the objects that we want to barter, for only these spirits can protect our land and repel all that is dangerous far away from us.[17] If the white people could hear other words than the words of merchandise, the way we do, they would know how to be generous and would be less hostile towards us. They would not want to eat our forest with such voracity.

WE GENEROUSLY barter our goods in order to spread friendship among ourselves. If it were not so, we would be like the white people who constantly mistreat each other because of their merchandise. When visitors covet the goods we possess, we are sad to hear them complain about how they lack them so much and to see them desire them so strongly. So we quickly give away these things in order to gain their friendship. We tell them: "*Awe!* Take these trade goods and let's be friends! I got them from other people. These are not leftovers of my hand.[18] No matter, take them anyhow and later do not fail to pass them on to those who will visit your house in your turn!" Our mouth is afraid to turn down our guests' requests. Our hands are not as narrow as the white people's![19] If we possess two machetes, we give one of them away as soon as we are asked for it. If we have only one, we regretfully say: "*Ma!* I am as needy as you are! I cannot give it to you now because I won't have anything to work in my garden and my relatives will suffer from hunger." But we instantly promise that we will manage to get a new one soon in order to give it away on a future visit. If we bluntly reject our guests with stingy words, they go home unhappy and full of bad talk, which makes us unhappy.

When the path that leads to another house is not a merchandise path to us, we say that it has value of hostility.[20] In this case, we can wage war against the people it leads to as soon as we think that one of our people, a woman or an elder, could have been killed by one of their *oka* sorcerers.

But on the contrary, when we make first contact with the inhabitants of an unknown house in order to make them our friends, we exchange all that we own with them. We call this *rimimuu*.[21] If we behaved otherwise, they would think that we were hiding our hostility. They would instantly run away, fearing that the sole intention for our approach was to take their footprints to rub them with sorcery plants. When I was a child at *Marakana*, my fathers and fathers-in-law made contact this way with the *Weyuku tʰëri* of the upper Rio Demini, whom they had never seen before. They met them in the forest by surprise and made friends with them by quickly giving them most of the things they were carrying. This is our custom. We think that it is by acquiring a trace of another person that we become his friend.[22] In the past, our elders thought that the white people would behave the same way with them. But they were very wrong! On the contrary, these outsiders' great men just hurried their sons-in-law and sons into the forest without a word to tear latex, jaguar skins, and gold out of it! Us, we are different. We have never thought of sending our people onto the white people's land to destroy everything without saying a word!

When a man knows how to be generous, visitors and guests go home satisfied and joyous. However, if he proves greedy, they go back with their chests full of anger because refusing them goods amounts to declaring his hostility to them. Seized by resentment, they will want to take revenge against him with *hwëri* sorcery plants. Then they will angrily tell themselves: "If this man is so stingy, we will no longer go hang our hammock up in his house! We only want to visit generous men! What could he be thinking? Despite his avarice, he will not avoid dying! And when his ghost will have left his skin, we will not cry for him! We will not carry the anger and grief of mourning for him! Let him die alone with his merchandise!" Or if they are truly furious: "What a mean and bad man! He will not remain alive among his kin for long. He will die fast because some angry warrior will end up arrowing him soon!" Full of rage, they would even curse the stingy man to the evil beings, the night being *Titiri*, and the being of death: "Later, when you die, you will keep quiet, you will no longer move, and you won't be anything anymore!" But on the contrary, if an elder, the great man of a house, displays largesse with the goods he is able to acquire, it is said that he knows how to maintain the true path of a generous man. The people who received his goods praise his liberality and friendliness to those who ask for them later. Then these

people give these objects to other visitors in their turn, carrying his reputation for generosity even further.

So flattering words about this great man constantly spread through
the forest. They accompany the thoughts of many men and women, even
very far from his house. They keep him in mind as if they were in love
with him! They often say about him: "*Awe!* He is truly an unselfish man!
He knows how to distribute trade goods as soon as he has acquired them.
He does not settle for giving them away once they are worn out! He really
knows how to rid himself of what his hands touch!" Or else: "He is a
great man! He knows how to give without being miserly! Many are those
who ask him for his possessions, yet he never answers with bad talk of
greed! He only stops giving when he has nothing left and he is truly
empty-handed!" It is also said of great men who know so well how to content others with their generosity that their *nõreme* image of life is powerful and that it makes them wise and brave.[23] When one of them proves so
selfless that he really gives up all his goods, even the most beautiful and
the newest, people are nearly frightened by it. They exclaim: "This man
knows no greed! He is a true son of *Omama!* But his liberality cannot be
without reason. It is his bravery that makes him so generous. His image
of life is very strong. He is probably a truly fearless warrior!"[24] Finally
they ask themselves whether all this generosity might not be concealing
an aggressive will. It can even inspire joking: "This man is frightening! Is
he trying to deceive us? Is all this prodigality intended to foil our suspicion so he can more easily kill us with his arrows?"

WHEN A MISER dies, no one really mourns him. It is true. No one wants
to keep friendship or nostalgia for someone who has always ignored the
troubles of those who are in need. People simply comment on his death
by saying: "Good! He always provoked our anger with his refusals. Let's
not be sad! He had no generosity and was not concerned about us!" Then
we destroy and throw out the possessions he left without feeling the sorrow of his absence. But on the contrary, if it is a generous man who dies,
all are very affected by his death, and many are those who cry for him in
pain. If he was killed by enemy arrows or sorcerers' blowpipes, many will
also be ready to avenge him. At the memory of his generosity, his relatives and friends are left in torment and sorrow. They grieve for him at
great length, declaring how they mourn his loss. When their suffering is

too strong, they sob and clap their palms together or pat the deceased's hands and forehead. And if he was a shaman, his *xapiri* mourn him in the same way.

As I have said, as soon as someone dies, his kin begin to destroy everything he still owned. They cut down and tear out the plants in his garden, and they fell the trees he recently climbed on. They scrape the bark of the posts that supported his hammock and the soil he walked upon inside the house. They remove and burn the *paa hana* leaves that covered his home. They also cut his wife's and children's hair. They only keep a few of his possessions: his bamboo quiver, some arrow points, and feather ornaments. These belongings will be destroyed later, during the lamentations of the *reahu* feasts during which the deceased's bone ashes will be put into oblivion. Thus every trace of what he has touched must be eliminated.[25] Yet if they want to, those who mourn him can keep the goods he gave them before he expired. We also say that these objects are orphans, *hamihi*.[26] The people who own them truly have to take care of them and not give them to anyone, especially not distant visitors. So they keep them a long time, until they deteriorate or, sometimes, until their own death. In this case, they will be burned by their own kin. If a friend gives me a shotgun during a *reahu* feast and dies soon after, I will keep his weapon, for I am still alive. But if I too pass away, my wife and brother-in-law will destroy it. In the same way, if I die before my father-in-law, he can keep what I gave him to obtain his daughter in marriage. However, my wife will destroy all the things I touched and that remained in our house. This is how it must happen.

When we burn a generous man's bones, no matter the cause of his death, we also take special care of the bones of his hands. We consider them precious objects, for it is with his hands that he generously gave away food and goods. Seeing the bones of his fingers after his death makes us sad and nostalgic. This is why we are very careful not to lose any piece of them during cremation. All the men and women assembled in the house lament for the dead around the pyre, mentioning his hands as they burn his belongings: "*Osema*,[27] your hands give us great sorrow! We miss your generosity so much!" We call these lamentations *pokoomuu*.[28] His kin cry while remembering his past actions and praising his largesse, valor, and joviality. Then the guests invited for the ceremony sometimes consume a part of the still-warm ashes of his charred bones, taken from the bottom of the mortar where they have just been ground up.[29] They mix them in a banana soup pot whose contents they cautiously drink to the last drop.[30] It is especially the *Xamat^hari* who do this. We in *Watoriki* think it is dangerous to ingest fresh funerary ashes. But the *Xamat^hari* do it to acquire the image of the deceased's breath and thus to keep the imitation of his *nõreme* image of life.[31] To achieve that, our elders preferred to rub the valiant men's bone ashes on young boys' foreheads and chests mixed with annatto dye. It was their way to call the image of these warriors' bravery so that it imbued these children and made them courageous in their turn. I remember that they often did this for me when I was a child. It is so. After the cremation, the close friends of the dead who came from other houses also ask his kin for gourds of his bone ashes so they can bury them by their hearth in future *reahu* feasts. With these gourds, they take a few of his belongings so they can burn them and cry again, for they had much affection for the deceased. These are our customs when a very well-loved man dies, for he was valiant, good, and generous.

Every one of us carries in his thought men who were always prompt to give away their possessions. We do not appreciate misers, whether they are white people or some of us! As for me, I do not have a taste for possessing much merchandise. My mind cannot set itself on it. In the beginning it is attractive, yes, but it quickly gets damaged and then we start to miss it. I do not want to keep such things in my mind. For me, only the forest is a precious good. Knives get blunt, machetes get chipped, pots get black, hammocks get holes, and the paper skins of money come apart in the rain. Meanwhile, tree leaves can stiffen and fall, but they will always grow back, as beautiful and bright as before. The small amount of merchandise I possess is sufficient and I do not desire more. And after I have acquired it in town, I always wind up giving it away to friendly visitors from the Toototobi, Demini, and Catrimani rivers—to the point at which even my wife and children find themselves lacking it! These people tell me: "We come to ask you for these trade goods because we know that you are generous. If you were greedy we would stay home without saying a word!" So I answer them: "*Awe!* I give you machetes and axes to clear your gardens, matches to smoke out armadillos, hooks to fish, and pots to cook your game because we lack clay since our elders left the highlands. The white people are close to us now, but they are stingy. So do not blame me, I give you the little I struggled to tear out from them!" It is so. I only think about merchandise to hand it out. If I possessed as much of it as the people of the cities, I would give it to everyone who asked, telling them: "It is yours, take it and be satisfied! I only make it in such great quantities to share it widely!"

BUT THE white people are other people than us. They probably find themselves very clever to be able to constantly produce a multitude of goods. They were tired of walking and wanted to go faster, so they invented the **bicycle.** Then eventually they found it still too slow. Next they built **motorcycles,** then cars. Then they found that all that was still not fast enough, and they created airplanes. Now they possess a great number of machines and factories. Yet that still isn't enough for them. Their thought remains constantly attached to their merchandise. They make it relentlessly and always desire new goods. But they are probably not as wise as they think they are. I fear that this euphoria of merchandise will have no end and that they will entangle themselves with it to the point of chaos. They are already constantly killing each other for money in their cities and fighting other people for minerals and oil they take from the ground. But they do not seem concerned that they are making us all perish with the epidemic fumes that escape from all these things.[32] They do not think that they are spoiling the earth and the sky and that they will never be able to recreate new ones.

Their cities are full of big houses filled with piles of innumerable goods, but their elders never give them to anyone. If they were really great men, shouldn't they tell themselves that it would be wise to distribute them all before they make so many more? But this never happens! When we visit the city, do we ever hear any of the white people say: "Take all the machetes and pots you see! I do not want to let them get old here any longer! Share them with your people without asking anything in return and tell them about me!" On the contrary, they are used to greedily hoarding their goods and keeping them locked up. In fact, they always carry many keys on them, which are for houses where they keep their merchandise hidden. They live in constant fear that it could be stolen. They only give it away sparingly, in exchange for paper skins they also accumulate, thinking they will become great men. Overjoyed, they probably tell themselves: "I am part of the people of merchandise and factories![33] I possess all these things alone! I am so clever! I am an important man, a **rich** man!"

When I was young and I first visited the cities of Manaus, then Boa Vista, all this dust-covered, piled-up merchandise truly confused me. I wondered why these large quantities of axe heads and hammocks, made so long ago, were getting old this way, stacked on boards until they got moldy, in closed houses, without ever being handed out to anyone. It was only much later that I understood that white people treat their merchandise like women with whom they are in love. They only want to lock them up and keep them jealously under their gaze. The same is true of their food, which they constantly pile up in their houses. If we ask for some of it, they never accept to give anything without making us work hard for it. As for us, we are not people who are in the habit of refusing food to our visitors! When our gardens abound with manioc and bananas, we smoke a great quantity of game and invite the people of neighboring houses to a *reahu* feast in order to satisfy their hunger. As soon as they are settled in their hammocks, just after their presentation dance, we offer them large amounts of banana soup prepared in a hollowed-out tree trunk placed on the central plaza of our house.[34] We make them drink so much of it that their stomachs swell up and finally they vomit![35] We certainly don't tell them: "*Ma!* Do not ask us for food! First, you must work in our gardens! Bring us game! Go get us water and firewood! The value of our bananas is very strong! They are **expensive!**"[36]

The white people's food does not have as high a value as they claim! Like ours, it only disappears as soon as it is eaten to be transformed into excrement! Their merchandise is also not as precious as they say. It is only their great fear of lacking it that makes them weigh down its value. Once they are old and blind, they will really be a pitiful sight as they continue to cling to it. But anyway, when they die they will have to let go of all these goods! They will abandon them even if they do not want to, and their kin will quarrel endlessly to take possession of them. All this is bad! The white people probably think they will acquire a great reputation by making and handling so much merchandise. Yet it is not true. For it to be so, they would need to know how to be less petty. Then maybe distant people like us would eventually speak of them with contentment and carry them in their thought.

We people of the forest only have pleasure in the evocation of generous men. This is why we possess few things, and we are satisfied with that. We do not want to store great quantities of trade goods in our homes. It would tangle up our minds. We would become too much like the white people. We would constantly be concerned about merchandise: "*Awe!* I want this object! I also desire this one, and that one, and also this other one!" It would have no end! So keeping so little of it at our side is enough for us. We do not want to tear the minerals out of the earth nor make their epidemic fumes fall back on us. We just want the forest to remain silent and the sky to be clear so that we can see the stars when night falls. The white people already have more than enough metal to make their merchandise and their machines; land to plant their food; cloth to cover themselves; cars and airplanes to move in. Yet now they covet the metal of our forest to make even more of these things though their factories' foul breath is already spreading everywhere. Today the *hutukarari* sky spirits are still keeping it at a distance, far from us. But later, after my death and that of the other shamans, its darkness may descend to our houses so that the children of our children will stop seeing the sun.

CHAPTER 20

In the City

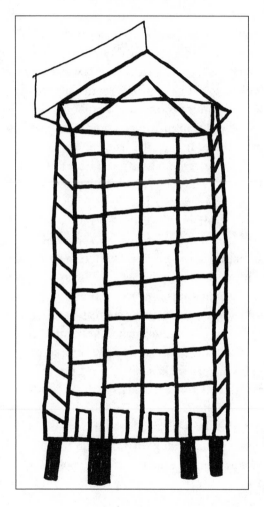

A skyscraper

B EFORE I knew the white people ancestors' land, I was used to traveling in dream, very far from the forest, and sometimes contemplated
the image of their cities in my sleep. In the night I often saw a multitude
of very tall houses sparkling with lights, the interiors of which seemed to
be entirely covered in game skins, as smooth and silky as a deer's coat.

When I woke, I was puzzled and asked the shamans of our house: "What are these strange things that appeared to me while I was sleeping? What will happen to me?" They answered: "*Ma!* Don't worry! One day soon, white people from distant lands will call you to them. They must be talking about you right now, that is why you saw their homes like that."

Much later, when I finally visited big cities, I remembered these old dreams and I told myself: "*Haixopë!* This is exactly how they appeared to me when the *xapiri* carried my image there!" At the time, I was still afraid to make such journeys for, as I have said, it is very dangerous to approach the lands where our spirits come from. Yet my father-in-law and the other shamans of *Watoriki* protected me, and despite my apprehensions I continued to go to these distant places to know the white people better and defend our forest there. In fact, if I had not come down from my hammock for that, no one else among us could have done it in my place.

So one day I had to leave for another city where I had been invited to speak on the white people ancestors' land. They called it "Paris." I only knew this place by the name my *xapiri* gave it: *Kawëhei urihi a*, the Shaking Land.[1] They named it this because as soon as I set foot outside the airplane, I felt that I was staggering. Though the ground appeared to be solid, I could only walk there with hesitant steps, as if I was walking on a mire that gave way with each of my steps. It was as if I was standing on a dugout canoe floating on the river. As soon as I arrived, I anxiously asked myself if this place was going to make me become other. It is true. Its ground is probably steady for those who live on it since childhood, but for the forest people who call their *xapiri* down from there, it seems to constantly waver. In fact, I think that its inhabitants imitated its image to create the shaking sliding paths on which they move in their **airport**.[2] Above this land, the sky is always low and covered by clouds. The rain and the cold never seem to stop. It is close to the edge of the terrestrial layer and *Titiri* and *Xiwãripo*, the underworld beings of night and chaos, are not far.[3] The white people may not know it, but the *xapiri*, they certainly know it.

In this city of Paris, a multitude of cars and **buses** ran squeezed between the houses the whole day long, making a deafening noise. The earth was full of long tunnels as endless as those of the big earthworms. Long metal **trains** constantly sped through them with a roar, on iron bars once torn out of the earth's depths. This is another reason why I felt that the ground always seemed to be shivering, even at night. For one who has always slept in the silence of the forest, this quivering is very worrisome. The white people, they do not seem to feel it, for they are used to never leaving the earth in peace. But I could not stop thinking that the land was becoming other there because of the noise and bustle relentlessly rocking it. This is why I so often became ghost at night during this trip. I barely slept at all, and during the day I had to meet unknown people all the time and talk to them at length. It was terribly cold and I was constantly dozing off. Yet I never complained. When I get anxious on these long trips far from my home, I never say a word of my worries to anyone because my *xapiri* have made me cautious. I merely think inside myself: "It is a distant land and these are different people, do not lament yourself about it!"

YET ONE NIGHT in Paris I felt even stranger. Shortly before my trip there, I had contracted malaria, and the fever was starting to burn me again. I was huddled up in bed in a hotel room at the top of a big building. I had just managed to go to sleep when suddenly I had the feeling of being carried away into a huge void. The ground beneath me abruptly gave way and split into big pieces. Then the place I was staying in came apart with a great crash. My fall lasted on and on. It was terrifying! But finally the *xapiri* who had come with me were able to stop my image at the last moment. They made a kind of **parachute** of light blow up above me to slow me down. Then suddenly *Omama*'s ghost caught me just as I was about to disappear into the underworld. All at once, I came to my senses in the middle of the night. I did not know where I was, and I nearly cried out in fear. Yet I managed to remain calm. I struggled to get up, without a word, then little by little I really woke up. I could see the things around me again. Then I told myself: "*Oae!* I'm still alive! The white people's ancient *napënapëri* spirits wanted to test my strength and wisdom! Long ago our ancestors opened their path from this distant land so they could come dance in our forest. That is why!" These outsiders' spirits were curious to examine my face, my eyes, and my hair, which are different than white people's. They also carefully inspected the feather ornaments of the *xapiri* escorting me. They told themselves: "*Hou!* Could he be one of the inhabitants of the forest, the children of *Omama*?" This is why they came to visit me and put me to the test.

Then over the following nights, I was able to visit in dream the place where these *xapiri* of the white people ancestors live, hidden in the cold of the high mountains. I came to know all sorts of foreign *xapiri* with magnificent dances, who had taken shelter in these heights since the white people stopped calling them. I was also able to contemplate the dazzling white *amoa hi* trees where they collect their songs. These are powerful *xapiri* that *Omama* only sends us sparingly. They know how to tear out and regurgitate diseases as well as our *ayokora* cacique bird spirits do. They are able to rout the evil beings of epidemic like no other. In my sleep, their images have often encouraged me to talk to the white people with ardor and courage. They told me: "Be vigilant! Give them your words in a voice that does not quaver and do not let yourself be deluded by vague lies. Your words must really defend the forest. If all its tallest trees are cut down and burned, they will never grow back. No matter how the white people try to replant other trees, they will never have the strength of the ones the *Në roperi* fertility being grew at the beginning of time. Only these first trees know how to make the wind and the rain move through their tops so that the plant and animal spirits can bathe and quench their thirst. Without them, the earth will die!"

One night in this city, the white people who accompanied me showed me a kind of house, very tall and pointy, made of metal, like a large antenna covered in vines of sparkling light.[4] I believe it was built to be admired by people who come from other lands, and that is exactly what they do. During the day, they contemplate it at great length and find it beautiful. They take one picture after another. Meanwhile, the people of the place must tell themselves: "*Ha!* How rich and clever we must be to have built such a beautiful thing!" That's all. No one thinks beyond that. Yet though no one knows it, this construction is in every way similar to the image of our *xapiri*'s houses, surrounded by a multitude of paths of light. It is true! This sparkling brightness is that of the spirits' metal. The white people of this land must have captured the light of the *Yãpirari* lightning beings to enclose it in this antenna. As I looked at it, I told myself: "*Hou!* These outsiders do not know the spirits' words, but they imitated their houses without even realizing it!" It baffled me. Yet despite the resemblance, the light from this house of iron light seemed lifeless. It was without resonance. If it were alive like a real spirit house, the vibrant songs of its occupants would unceasingly burst out of it. Its sparkling would spread their voices far and near. But this was not the case. It remained inert and silent. Only by making its image dance during the time of dream did I hear the voices of the white people ancestors' *xapiri* and of the *waikayoma* foreign women who live on their land and are covered in glass beads.

One day when we were in a car again, my white friends also showed me a tall stone sticking out of the ground in the middle of the city. They told me that the elders of this place had brought it back from another land where once they waged war.[5] Without answering, I thought to myself: "*Hou!* The white people do not have as much wisdom as they claim! They constantly tell us that it is wrong for us to arrow each other for revenge. Yet their ancestors were so bellicose they traveled great distances to plunder the land of people who had done them no harm! Despite what they say, the blood and the ghost of the war being *Aiamori* have divided and spread over their land as much as on ours."

Another time, I was taken to visit a vast house to which the white people gave the name of **museum**.[6] This is a place where they lock up traces of ancestors of the people of the forest who disappeared long ago. Here, I saw vast quantities of pottery, calabashes, and baskets; many bows, arrows, blowpipes, clubs, and spears; but also stone axes, bone needles, seed necklaces, bamboo flutes, and a profusion of feather ornaments and beads. These goods, which imitate those of the *xapiri*,[7] are really very old, and the ghosts of those who owned them are secluded with them. They once belonged to great shamans who died long ago. Their images were captured at the same time as these objects, when the white people seized them at war. This is why I claim that they are spirit goods. Yet these images of forest people ancestors so long confined in these distant houses can no longer dance for us. We cannot make their words be heard in the forest anymore, for their paths have been cut off for too long. In the din of their cities, the white people do not know how to dream with the *xapiri*.[8] This is why they know nothing of all this. But I instantly recognized these precious goods and was very worried. I told myself: "*Hou!* By locking them up and exposing them to everyone's eyes, the white people lack **respect** towards these objects that belonged to dead ancestors. Goods linked to the spirits and to *Omama*'s image cannot be mistreated this way!"

In glass cases arranged side by side, I could see an abundance of toucan tail ornaments next to the colorful feathered hides of *hëima si* and *wisawisama si* birds, which the *yawarioma* water beings, those great hunters, are accustomed to shoot with their *rihu u* white bamboo blowpipes.[9]

There was also a great deal of colorful glass bead finery belonging to the images of the *waikayoma* foreign women who first wove the glass bead armbands, belts, and aprons that our elders got from far away and also regarded as precious objects they called *matihi*.[10] To assemble their beads with red, white, blue, and yellow eyes, these spirit women had to hunt them and pierce their image with *ruhu masi* palm darts.[11] Today the white people make them in large quantities with machines. Those that the *waikayoma* darted were very different, for they were spirit goods. They were alive and looked like little children. As soon as these spirit women's darts touched them, they moaned in pain and cried like newborns: *"Õe, õe, õe!"* Then the *waikayoma* strung them one by one onto a thread that they pulled through their wounds. They made long strings of these beads, which they wore around their necks and crossed over their chests to display during their presentation dances. They owned enormous quantities of these bead-children with which they made all sorts of magnificent smooth and shiny finery. It was these *waikayoma* spirit women who taught our ancestors the name of the glass beads. When our shaman elders asked them where they came from, they simply answered: "We call them *õha kiki, topë kiki!*[12] They are spirits' goods! We arrow them on a distant land from which we descend to come to you."

Also in that city's museum, I saw the stone axes with which the long-ago inhabitants of the forest cleared their gardens, the hooks made of game bone with which they fished, the bows with which they hunted, the pottery in which they cooked their game, and the cotton armbands they wove. It really made me sad to see all these objects left behind by elders who disappeared in the distant past. But then I saw other glass cases containing the bodies of dead children, their skin hardened and dry. Finally all this made me very angry. I told myself: "Where do these dead come from? Aren't they ancestors from the beginning of time? Their dried out skin and bones are a sorry sight! The white people were hostile to them. They killed them with their epidemic fumes and their shotguns to take their land. Then they kept their bodies and now they exhibit them for all to see![13] What ignorant thought that is!"

Then suddenly, I began speaking harshly to the white people accompanying me: "These bodies must be burned! Their traces must disappear! It is bad to ask money to show such things! If the white people want to exhibit their dead, let them smoke-dry their fathers, mothers, wives, and children to show here instead of our forest ancestors! What would they

think to see their deceased exhibited like that to strangers?" Surprised
by my tone of voice, my guides asked me if I was truly furious. So I explained my thought to them: "*Awe!* Seeing all this really upsets me. The
white people should not treat these dead ancestors so poorly by arranging
them for all to see, surrounded by the objects they left behind when they
died. The same is true of these game hides and bones. These are animal
ancestors who danced for the long-ago shamans. They too should not be
mistreated in this manner. If the white people want, let them replace
them with chicken, horse, sheep, and cow carcasses!" In the end, those
who listened to me became embarrassed and tried to calm me by answering: "Don't be so angry! We only exhibit all this here so everyone can
know about it."

But I did not agree, and I continued: "It is bad to keep the goods of the
people of the forest who were eaten by the white people's diseases and
weapons long ago locked up in this faraway house! These people were
created in the beginning of time. They have always been the real holders
of the forest. Their objects belong to the *xapiri* and to *Omama*. It really
makes me very sad to see them displayed like this. I only want to look
at beautiful things and not things of death. I prefer to contemplate images of the sky, the sun, mountains, rain, day and night—everything that
never dies. We humans disappear very quickly, and as soon as our breath
of life is interrupted, we only inspire sorrow and longing. White people
could show all that in their museum if they wanted to, but only if these
objects did not come from ghosts. As long as we are alive, they can show
our pictures and belongings to explain our way of life to their children
and help us to protect the forest. But to exhibit the dried-up bodies and
abandoned objects of the forest's first inhabitants can only make me unhappy and torment me. It is truly a bad thing!"

The entire land of Brazil used to be occupied by people like us. Today it
is nearly empty, and it is the same thing everywhere. The first people of
the forest have nearly all disappeared. Those who still exist here and there
are only the remainder of the great number that the white people killed
long ago to seize their land. Then, their forehead still greasy with these
dead,[14] these same white people didn't feel afraid to become enamored
with the objects whose owners they devoured as enemies. Since then,
they have kept these goods locked up in the glass of their museums to
show their children what is left of those whom their elders killed. But
when they grow up, won't these children eventually ask: "*Hou!* These

things are truly beautiful, but why did you destroy those who possessed them?" Then their fathers will answer: "*Ma!* If these people were still alive, we would still be poor. They were standing in our way. If we had not taken over their forest, we would have no gold." Yet after all that the white people do not even worry about exhibiting the bodies of those they killed! We would never do something like that.

Finally, having seen all these things in the museum, I began to wonder if the white people had also started acquiring Yanomami baskets, bows and arrows, as well as feather ornaments because we are already in the process of disappearing ourselves. Why do they so often ask us for these objects, when the gold prospectors and cattle ranchers are invading our land? Do they want to get them in anticipation of our death? Will they also want to take our bones to their cities? Once dead, will we also be exhibited in the glass cases of a museum? This is what all this made me think. I told myself that if we give our curassow armbands, scarlet macaw tail feathers and annatto dye, our quivers and arrows to let them be locked up in the white people's houses and museums, we will end up losing our beauty and become poor hunters. Our toucan tail bunches, our parrot- and guan-wing feathers, our cock-of-the-rock and *sei si* bird feathered hides are precious goods that belong to the *yawarioma* water beings.[15] By taking them away, the white people also capture their images and keep them shut up very far from our forest. That is why, as I said, this will make us both ugly and clumsy at hunting.

LATER, WHEN I finally came home from this trip to Paris and was just back in our house of *Watoriki*, I really thought I was going to die. I was very weak and constantly dizzy. I couldn't come out of my drowsiness. Then I felt my legs get stiff and become numb. I pinched myself, but I couldn't feel anything. I stayed sprawled in my hammock, gradually becoming unconscious. I could no longer see my wife and children, though they were right beside me, or even my own hammock. I was in a ghost state, and suddenly my legs became completely paralyzed. I had returned to the forest, yes, but my image was still sleeping in the sky's chest. I knew that all this was happening to me because I had set foot once again on the land where the ancient white people's *xapiri* come from. I knew them, and I had already made them dance with my father-in-law. Yet having approached their place of origin had made me become other again, like the very first time I went to the land of Europe. Though *Omama's* image had protected me throughout this new journey, upon my return I was overcome with torpor. I had to stay in my house for days and days, lying by the fire to dry my flesh soaked with the humid cold of those distant places. Then, little by little, I started to drink the *yãkoana* again. The spirits who had accompanied me there woke up and warmed themselves. Having rested, they got their energy back, and I also started to feel better.

Now I had truly learned how dangerous this kind of trip could be for a shaman. Yet after a few moons, at the end of the dry season, white friends called me far away from the forest again. All my people were dying of malaria, and most of the white people who live around us remained deaf to our complaints. This is why I accepted once again to leave my home to go speak to the great men of another city, even bigger than all those I had previously known. Its inhabitants call it "New York." I wanted to gain their support to convince the government of our land of Brazil to prevent the gold prospectors from ravaging our forest and wiping out its inhabitants.[16] When I arrived in New York, I was surprised that the city looked like a dense group of rocky peaks in which white people live piled on top of each other. At these mountains' feet, multitudes of people moved very fast and in every direction, like ants. They started one way and turned around, then went the other way. They looked at the ground all the time and never saw the sky. I told myself that these white people must have built such tall stone houses after clearing all their forests and having started making merchandise in very large quantities for the first time. They probably thought: "There are many of us, we are valiant in war, and we have many machines. Let us build giant houses to fill them with goods that all the other peoples will covet!"

Yet while the houses in the center of this city are tall and beautiful, those on its edges are in ruins. The people who live in those places have no food, and their clothes are dirty and torn. When I took a walk among them, they looked at me with sad eyes.[17] It made me feel upset. These white people who created merchandise think they are clever and brave. Yet they are greedy and do not take care of those among them who have nothing. How can they think they are great men and find themselves so smart? They do not want to know anything about these needy people, though they too are their fellows. They reject them and let them suffer alone. They do not even look at them and are satisfied to keep their distance and call them "the **poor.**" They even take their crumbling houses from them. They force them to camp outside, in the rain, with their children. They must tell themselves: "They live on our land, but they are other people. Let them stay far away from us, picking their food off the ground like dogs! As for us, we will pile up more goods and more weapons, all by ourselves!" It scared me to see such a thing.

During this trip to New York, I had malaria attacks again.[18] There was also a lot of noise near where I was housed. Often people from across the street sang and yelled all night. It made me worried and agitated. I slept in a ghost state and was often overcome by dizzy spells. Then, as in the other big cities I visited, I saw the *xapiri* of these lands of the ancient white people come down in my sleep. They came one after another in my dream time, increasingly numerous. First I saw the images of the thunder beings dance, then those of the lightning beings and the jaguar ancestors. Often I also saw a noisy multitude of *ayokora* cacique bird spirits come towards me from their distant mountain homes. These *xapiri* know how to tear diseases out of the inner body and how to work beside white doctors. This is why they often appear in the dreams of shamans burning with fever.

One night the image of a young woman of the rivers, a sister of $T^h u\ddot{e}y$-*oma*, the wife that *Omama* fished out of the water in the beginning of time, appeared to me. Her eyes and black hair were very beautiful. I could clearly see her young breasts sticking out, but the bottom of her body looked like that of a fish. She was slowly pouring water on my feverish forehead and was bringing me back to life. Long ago this woman of the rivers went far away from our forests and got lost at the edge of the waters. This is why her image now lives under a big bridge in New York.[19] I saw that the white people know how to draw her and call her **Siren.** She stayed where the torrents that carried away our ancestors from *Hayowari* finally stopped far downstream to form what white people call the ocean.[20] This is the place we call *u monapë*, where all the rivers are clenched together now. If the watercourses were not held in place like that, they would all sink into the depths of the earth, and it would dry out for all time.

The height of the buildings was not what frightened me most in the city of New York. It was other things that appeared to me during my sleep. It is true. One night, I saw the sky catch fire with the heat from the factories' smoke. The thunders, the lightning beings, and the human ghosts who live on its back were caught in huge flames. Then the burn-

ing sky began crashing down to earth. Yes, that was truly terrifying!
Where the white people live, the sky is lower than in the forest, and they
constantly burn large quantities of minerals and oil. The fumes from
their plants never stop rising to the sky's chest. This makes it dry and
powdery, as inflammable as **gas**. Hardened and dried by the heat, it becomes fragile and tears into shreds like an old piece of clothing. All this
worries the *xapiri*. In my dream time, they tried to cure this dying sky by
turning the latch of the rains to drive back the furious blaze devouring it.
As they poured torrents of water on the flames, they cried out exaltedly:
"If you destroy the sky, you will all disappear with it!" But the white people remained deaf to their call. If the spirits do not continue to flood the
sky, it will end up completely charred. My father-in-law told me about
their endless work as soon as he started to make me drink the *yãkoana*,
before I truly became a shaman. It is so. But once again, I did not speak
to anyone about this dream because I was far from my house and my
people.

Another time in New York, I was surprised during the night by the
crashing and rumbling of the sky, which started to move heavily over the
city. I woke up with a start and got up. I stood there without moving, trying to keep myself from crying out in terror. But again I told myself:
"*Hou!* This is another land, I can't let myself get carried away by fear, otherwise the white people will think I lost my mind." Little by little, I tried
to calm down. Then the noise in the sky stopped, but I started to hear the
voice of its image, which the shamans call *Hutukarari*. She told me: "*Ma!*
It's nothing! I did that to test your vigilance. I do the same thing from
time to time so that the white people hear me, but it is in vain. Only the
inhabitants of the forest keep their ears open because they know how to
become spirits with the *yãkoana*. The white people's ears remain closed.
No matter how I try to frighten them to warn them, they remain as deaf
as tree trunks. But you, you really heard me. That is good."

AT THE TIME, I thought this city of New York must be the place where
the white people started tearing the metal out of the ground long ago, filling their houses with merchandise and inventing the paper skins of
money. I heard it said that this was where they built those shiny metal
things which pass in the sky like comets and which they call **satellites**. I
also saw that here more than elsewhere people's eyes are damaged by the

metal smoke and its powder, which makes them blind.[21] In the forest, we do not have factories or cars and our eyes are clear. In New York, so many people seem to have bad eyesight. Even the children and young people have their eyes shut in behind glass to see better. While I was staying in this city, I also told myself that the white people who had built it had mistreated these regions' first inhabitants the same way those in Brazil do with us today. Their land was beautiful, fertile, and full of game. The white people arrived and wanted to seize it right away. These forest people were in their way, so they quickly became hostile to them and they began to destroy them. These long-ago white people of the **United States** were truly bad and very bellicose; I saw it in a book.[22] It tormented me to think of all those people similar to us who had died in that country. I thought that many of them must have lived on the land of New York before the forest was replaced by stone houses. The white people in this place must have hated them as much as the gold prospectors and cattle ranchers in Brazil hate us. They probably told themselves: "We're going to do away with these dirty, lazy Indians. We're going to take their place on this land. We will be the real Americans because we are white. We are truly clever, hardworking, and powerful." Their elation for merchandise, roads, trains, and later airplanes constantly grew. It was with these lying thoughts that they began to make the people of the forest perish, before stealing their land and giving it one of their names, *America*. The gold prospectors and cattle ranchers want to get rid of us in Brazil with the same words: "The Yanomami are only inhabitants of the forest, unknown people! It doesn't matter if they die; they are useless, and we will work in their place."

Outside the city of New York, I was taken to visit what remains of the first human beings that the long-ago white people once killed in this country. They are called *Onondaga*.[23] I do not only call them "human beings"—*yanomae t^hë pë*—because they look like us, but also because they were the people who were created in the beginning of time on this land of the United States, like us in the forest of Brazil. I saw many feather ornaments in their houses. These are people who still possess *xapiri* and know how to make them dance. When I visited them, the men called me and I sat with them to listen to their words. They asked the women and children to step away. They burned tobacco and made the spirits come down. Their elders were hunters of great eagles that fly very high in the sky, like the fierce *mohuma* harpy eagle where we come from. They prepared magnificent headdresses with their feathers. They also hunted other game whose existence was unknown to me, **bears** and **buffalo.** Their shamans make the images of these animal ancestors dance to this day. The *Onondaga* also drink the sweet juice of the trees of their forest, like we drink wild bees' honey.[24] The land where their ancestors lived long ago was vast, but the territory the white people have left them is a narrow strip right next to a little town. They took me on a tour of the place. It made me so sad! They are encircled on a little patch of land. The settlers, the cattle ranchers, and the miners killed their ancestors. The *Onondaga* did try to fight back, but they only had arrows and were unable to defend themselves against so many guns. Once they were decimated and defeated, they only received this little enclosed plot. Then I thought: "*Hou!* This is what the white people also want to do with us and all the other inhabitants of Brazil's forest! This is what they have always done. They will kill all the game, the fish, and the trees. They will soil all the rivers and lakes, and they will finally take over what is left of our lands. They won't leave anything alive! They think that we are not human beings, and they all hate us! Yet even though we are other people than they are, we have a mouth and eyes, blood and bones, just like white people. We all see the same single light. We are all hungry and thirsty. We all have the same fold behind our knees so we can walk! But where do the white people get this fierce desire to destroy the forest and its inhabitants?"

These were all the things that woke me during the night in New York and my thoughts would move from one to the next without stopping until dawn. I also told myself: "The long-ago white people's elders drew what they call their laws on paper skins, but to them they are only lies! They only pay attention to the words of merchandise!" I was tormented and couldn't fall back asleep. The paths of my thoughts turned and turned in every direction, in ever more distant journeys. It is like this every time I have to sleep in a big city to speak to the white people. I am always searching for other words; words they do not know yet. I want them to be surprised and to open their ears. I think of our ancestors and the way they lived, I think of *Omama*'s words and those of the *xapiri*. I look for words from long ago. They are not always those I heard from the mouths of my elders. They are words that come from the beginning of time but that I go looking for deep inside myself. In the beginning, we did not know the white people, let alone their cities. Since my childhood, they have constantly increased in numbers and gotten closer to us while destroying the forest. The gold prospectors turned over our riverbeds and soon the **mining companies** will dig in the depths of the ground. The ranchers and the settlers relentlessly burn the forest's edges. This is why today I search for powerful words to say how much all this angers me. I do not want anything other than the forest and its game, the rivers and the fish, the trees, their fruit, and the wild honeys. I want all this so my children and their children will be able to live well after my death.

Getting to know the ancient white people's cities during my trips left me thoughtful. Their cities may be beautiful to see, but on the other hand the bustle of their inhabitants is frightening. Trains constantly speed by under the earth, cars on the **cement**-covered ground, and airplanes in the hazy sky. People there live piled up one on top of each other and squeezed side by side, as frenzied as wasps in their nest. It all makes you dizzy and obscures your thought. The endless noise and the smoke covering everything prevent you from thinking right. This is probably why the white people can't hear us! As soon as we speak to them, most of them answer us: "The people of the forest are just liars! We will continue to move our machines forward! We will take as many minerals out of the ground as we want!" Yet what we say about the earth and the sky are not lies. They are true words that the shamans heard from *Omama*'s image and the *xapiri*. The white people, with their mind set on merchandise, do not want to hear us. They continue to mistreat the earth everywhere they go, even under the cities they live in! It never occurs to them that if they mistreat it too much, it will finally turn to chaos. Their thought is full of oblivion and dizziness. This is why they fear nothing and think they are protected from everything. By visiting their ancestors' land, I understood that it was the place where all these bad things started. It is from these distant places that they gradually came to our forest to continue to mistreat the land there and to build new factories.

For me, it is not at all pleasant to live in the city. My thought is always worried and my chest short of breath. I don't sleep well there, I only eat strange things, and I am always afraid I will be hit by a car. I can never think calmly in the city. It is a worrisome place. People constantly ask you for money for everything, even to drink or urinate. Everywhere you go you find a multitude of people rushing in every direction though you don't know why. You walk quickly among strangers, without stopping or talking, from one place to another. The lives of white people who hurry around all day like *xiri na* ants seem sad to me. They are always impatient and anxious not to get to their **job** late or be thrown away. They barely sleep and run all day in a daze. They only talk about working and the money they lack. They live without joy and age rapidly, constantly busying themselves with acquiring new merchandise, their minds empty. Once their hair is white, they disappear, and the work—which never dies —survives them without end. Then their children and grandchildren continue doing the same thing!

Omama certainly did not want to mistreat us in this way. For the inhabitants of the forest, the white people's cities stink of burning and *xawara* epidemic.[25] The people work in a ghost state and constantly swallow the wind of factory and machine fumes. They go into their noses, their mouth, and their eyes, and stick to their hair. Their chest is dirtied by them. This is why white people are so often sick, despite all their medicine. Their doctors may open their chest, stomach, or eyes, but it doesn't help. The sperm of fathers whose flesh is tainted by this epidemic smoke becomes sick, and their children are born in bad shape. The metal smoke causes all this. In the city, you also never clearly hear the words that are addressed to you. You have to press together to understand each other when you speak. The roar of the machines and motors stands in the way of all the other sounds; the babble of televisions and radios cover all the other voices. White people are always worried because of all this racket through which they hurry all day. Their hearts beat too fast, their thought is seized with dizziness, and their eyes are always on the alert. I think that this endless noise prevents their thoughts from joining together. It is so. In the end, their thoughts become scattered and fixed at their feet, and this is how you become stupid. But maybe the white people like this din they have known since their childhood? However, for those who grew up in the silence of the forest, the city noise is painful. This is why my mind becomes blocked and fills with darkness whenever I stay there too long. I become restless and am no longer able to dream because my mind no longer stays untroubled.

I was born in the forest, which is why I prefer to live there. I can only hear the songs of the *xapiri* and dream with them in the calm it gives me. I like its silence, which is only interrupted by the loud call of the $h^w\tilde{a}ih^w\tilde{a}i$-*yama* birds, the rasping cry of the macaws, the weeping of the toucans, the howl of the *iro* monkey troops, and the peeping of the parrots. I like hearing these voices. When I come back from a trip among the white people, the dizziness leaves my eyes after a while and my thought becomes clear again. I no longer hear cars, machines, or airplanes. I only lend an ear to the *tooro* toads and *krouma* frogs that call the rain in the forest. I only hear the rustling of the leaves in the wind and the rumbling of the thunders in the sky. The ignorant words of the city politicians gradually vanish in the quiet of my sleep. I become calm again by going to hunt and making my spirits dance. The forest is very beautiful to see. It is cool and aromatic. When you move through it to hunt or travel, you feel joyful and your mind is slow-paced. You listen to the chirping of the cicadas in the distance, or the cries of the curassows and the agami herons, and the clamor of the spider monkeys in the trees. Your worries are eased. Your thoughts can then follow one another without getting obscured.

These are all the reasons that I want to continue living in the forest, like our elders who lived there before me. I am one of their grandchildren and I want to follow in their footsteps. Sometimes I imitate the white people's language and I own some of their merchandise. That is true. Yet I have no desire to become one of them. In their cities, one cannot learn things of the dream. People there do not know how to bring down the spirits of the forest and the animal ancestors' images. They only set their gaze on what surrounds them: merchandise, television, and money. This is why they ignore us and are so unconcerned that we die of their epidemic fumes. Yet we feel sorry for them. Their cities are vast and full of a multitude of beautiful objects they desire, but as soon as they are old or weakened by sickness, they suddenly have to abandon all that, which is quickly erased from their minds. All that remains is for them to die alone and empty. But they never want to think about that, as if they weren't going to disappear as well! If they thought about all that, perhaps they would not be as avid for the things of our land and so hostile towards us. These are the thoughts that occupy my nights in those big cities where I can never fall asleep.

CHAPTER 21

From One War to Another

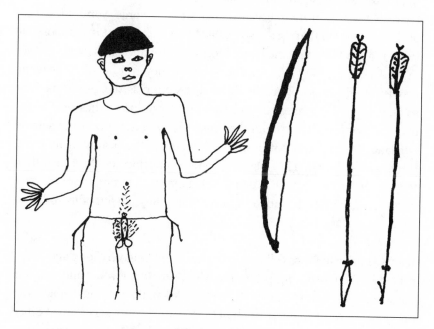

Warrior and bow

DURING MY distant trips to the white people's lands, I sometimes heard them claim that we are warlike and that we spend our time shooting arrows at each other.[1] Obviously, those who say such things do not know us, and their words are false or are just lies. It is true that our long-ago elders engaged in raids, just like the white people had their own wars. But theirs proved far more dangerous and fierce than ours. We never killed each other without restraint, the way they did. We do not have bombs that burn houses and all their inhabitants. When in old times our warriors wanted to arrow their enemies, it was a truly different thing. Above all, they tried to strike men who had already killed and whom we therefore call õnokaerima tʰë pë.[2] Seized with the anger of mourning for their dead kin, they carried out raids until they were able to avenge them.

This is our way of doing things. We only seek revenge when one of our relatives dies arrowed by enemies or in the hands of their *oka* sorcerers.[3] If warriors of another house kill one of us, the sons, brothers, brothers-in-law, and sons-in-law of the deceased will follow their trail to arrow them in return. The same is true if enemy sorcerers eat one of our elders. But we do not arrow one another all the time, for no reason! If this were the case, I would admit it, because I like straight words. Some white people have even said that we are so hostile to each other that it is impossible to let us all live on the same land![4] More lies! Our ancestors lived together in the forest for a very long time, long before they ever heard of the white people. Do these people really think that we are as dangerous as their soldiers when they wage war? No. They only try to spread these evil words because they fear that without their help they will not succeed in seizing our land. But they do not desire our forest for the beauty of its trees, game, and fish. No. They have no more friendship for the forest than they do for the beings who live in it. What they really want it for is to clear it all to feed their cattle and to tear everything they can out of its soil.

WAR VALOR, which we call *wait^hiri*, came into being very long ago. It appeared in the forest long before the white people knew us, and this was not without a reason.[5] The child warrior *Ōeðeri*, the valorous *Arowë*, and the bellicose spirit *Aiamori* taught it to us in the beginning of time.[6] Since then, these ancestors' images come down to us from the place where they lived long ago, on the land of the *Xamat^hari*.[7] *Ōeðeri* was a newborn at the time.[8] Marauding *oka* enemy sorcerers killed his mother just after she gave birth to him on the forest floor. They abandoned the orphaned baby on a nest of *kaxi* ants. Then, little by little, with the pain of their bites and in the rage of his tears, the child began to become other. He grew up very quickly and became an extremely valiant warrior. He attacked the house of the *Xamat^hari* who had murdered his mother so many times that he exterminated them all and got sick from the *õnokae* homicide state of having eaten so many enemies. Finally, under the guise of healing him, the ghosts of their shamans killed him in his turn. Since then, the raids continued among our ancestors of the highlands and the most aggressive warriors have been seized with the exaltation of arrowing each other like game. It is true that the *Xamat^hari* often raided each other's homes in these remote times. First they killed one or two men in a nearby house. Next the inhabitants of this house lamented their dead, then attacked for revenge. Finally the incursions against each other became endless. The sperm and blood of bellicose warriors passed on to their sons. Then they followed in their footsteps and grew up with this aggressiveness inside them. Because of this they were called the *Niyayopa tʰëri*, the War People. This was the name of the *Xamatʰari* who lived in the savannas beyond the headwaters of the *Hʷara u* River,[9] where *Õeõeri*'s ghost remained. They were not given this name without a reason! They were truly warlike! Their warriors felt exalted to kill and think about their enemies' mourning tears, like hunters happy to have arrowed their prey. It was they who taught our ancestors how to arrow each other, and they have continued since then. These War People's images still exist in the highlands, where their children continue to fight each other in their footsteps. The habit of attacking each other from house to house developed from them. Their image spread everywhere and the ghost of the *waitʰiri* war aggressiveness spread into our forest and even beyond, even to the *xapiri* whom we call *purusianari*[10] and to the white people's land. This is why we have all known anger and war since then.

WE DO NOT like what the white people call "war" in their language. They reproach the Yanomami for arrowing each other, but they are the ones who really wage war. We certainly do not fight with the same hardness as they do. If one of our people is killed by arrows or sorcery blowpipes, we only respond by trying to kill the enemy who ate him and is in an *õnokae* homicide state. This is different from the wars with which the white people constantly mistreat each other. They fight in great numbers, with bullets and bombs that burn all their houses. They even kill their women and children.[11] And it is not to avenge their dead, because they do not know how to mourn them the way we do. They simply make their wars for bad talk, to grab new land to tear minerals and oil out of its ground. Aren't the *garimpeiros* constantly fighting over their gold? They drink *cachaça*, and having become ghost, face off like chickens or starving dogs until they have killed each other. They do it out of jealousy for gold and do not lament their dead: they abandon them buried under the forest's floor. Yet it was not because of land, gold, or oil that *Õeõeri* brought the *waitʰiri* war valor into existence in the beginning of time. It was not because they coveted these things that the *Niyayopa tʰëri* taught our ancestors how to arrow each other. We inhabitants of the forest only go to war to avenge ourselves, out of anger for the mourning we feel because one of our people has been killed. We do not arrow each other about just anything, without good reasons! We lament our dead at great length, for several moons, for we carry their grief deep inside us and we are truly dedicated to avenging them.

White people may find themselves very clever, but their thought remains set on these bad things they always covet, and it is because of them that they insult and steal from each other, that they fight and eventually kill each other. It is also because of these things that they mistreat all those who stand in their way. This is why they are the ones who are truly fierce! When they make war on each other, they throw bombs everywhere and do not hesitate to set fire to the earth and sky. I watched them on television fighting each other for oil with their airplanes in a land with no trees.[12] Seeing all these blazes and their vast turmoil of black fumes, I worried that one day they could reach our forest and that our *xapiri* would not be able to disperse them. Later, I often saw this war again in the time of dream. It worried me a lot, and I told myself: "*Hou!* These people are truly warlike and dangerous! If they attacked us this way, they would rapidly kill most of us, and the epidemic smoke from their bombs[13] would instantly wipe out the few survivors."

WHITE PEOPLE bury their dead's bodies underground in places they call **cemeteries.** I have seen it with my own eyes. But since the beginning of time, our ancestors buried or drank the ashes of our dead's bones. White people do not wage war for their cemeteries. But on the contrary, we only go to war for the value of the funerary gourds of our deceased eaten by enemies. This is the only war talk that is clear to us. We are other people. We only arrow each other when we want to take back the value of one of our relatives' blood; only when we want to reciprocate the *õnokae* homicide state[14] of those who killed one of us. This does not happen all the time, and we do not attack other people for anything else. But as soon as the inhabitants of a house know where the warriors who killed one of them live, they quickly launch a raid to avenge him.[15] And the same is true if *oka* enemy sorcerers shattered the bones of one of their elders. As soon as visitors bring news indicating which house they may have come from, a group of warriors instantly prepares to set off in search of revenge.[16]

People mourn the deceased with much anger. His relatives burn his arrow points while painfully lamenting themselves. His bones are incinerated and several *pora axi* gourds are filled with their ashes. But the warriors who want to avenge him can also rub some of his bone ashes on the ground while imitating the jaguar's image. They do this to foil his killers' suspicion, so they can surprise them and kill them in their turn.[17] Then, covered in black dye,[18] they gather in the center of the house with their bows and arrows. Next they start to smash the ground with bunches of game bones, which they hold in their mouth by a vine, in imitation of the vulture's image.[19] To chase away the fear that could weaken them, shamans then call down by their side the images of the ancestors who first created war bravery in the beginning of time, then those of the *yarima* monkeys, who will make them alert in combat.[20] They bring down the *wainama* and *ōkaranama* war images, which will precede them on their incursions to weaken their enemies.[21] Then they make dance the images of man-eaters and scavenger animals who will help them to devour their enemies, like that of the jaguar, the vulture, and the *herama* falcon,[22] but also those of the flies and worms, as well as those of the *xaki, ōi,* and *wakopo* bees that feed on blood and carrion flesh.[23] They also make dance the images of beings of death who will go to the enemy's territory ahead of them, like those of the *yorohiyoma* and *hixākari* funerary spirits, the *ōrihiari* spirits of ill omen, and the *naikiari* spirits of hunger for human flesh.[24] Finally, before they set off, the warriors will practice their aiming skills by arrowing termite nests or packets of *hoko si* palm leaves representing enemies.[25] This is what our elders did. They sent all these images of death towards the people they were going to attack in order to kill them more easily. First, their *xapiri* destroyed the spirit houses of enemy shamans who could have stood in their way, then they debilitated their valorous warriors by sending all these dangerous images so they could not fight bravely anymore.

Later, after the first raid launched when the deceased's bones have been cremated, his children, wife, and brothers-in-law lament him again at a *reahu* feast, and the ashes from the top of his funerary gourd are buried by his widow's hearth fire.[26] The mourners then ask their guests who came from various houses to join the group of warriors to launch another raid to avenge their dead kin. They do that through an invitation dialogue we call *hiimuu*. If the warriors are unable to arrow a single enemy during these first incursions, everything is repeated in the same way over the course of several *reahu* with the ashes from the middle, then the bottom of the deceased's funerary gourd.[27] Finally, when all these ashes are finished and the anger of mourning ends, the incursions also come to an end.[28] This is how it happens. When a death is due to arrow wounds, we never bury the deceased's bone ashes until he has truly been avenged. But this can take a long time. Often, the raiders do not find the enemies they are looking for because they changed houses or have taken shelter in distant forest encampments. And even when they find them, they do not always succeed in reaching the reputed warriors they want to kill to exact their revenge. Sometimes the men of the attacked house are on the alert, and they instantly repel their attackers with a volley of arrows. It is so. As long as their hand hasn't reached the one they are looking for, as long as they haven't arrowed a man in *õnokae* homicide state, a man who has killed their relatives, the mourners tirelessly go back on a raid after every *reahu*.[29]

People truly carry the grievance of the funerary gourds[30] that hold their dead's bone ashes. This is why they so ardently wish to strike back at their enemies. Valiant warriors are incited to vengeance by orphans' sobs, the women's wails, and by the sorrow of all the deceased's kin. The pain and tears of mourning last for moons, as long as the funerary ashes are not all put in oblivion. For us, these words about *pora axi* ash gourds are truly powerful and have a very high value. Our ancestors have always carried them. They are still the words of the War People who live in the highlands of our forest. When an elder, a great man, has been arrowed by enemies or had his bones shattered by *oka* sorcerers, his people instantly go to war because of the grievance of his bone ashes. His sons, brothers, sons-in-law, and father-in-law mourn for him with great sorrow and want to take back the value of his blood. In doing so, they imitate what we were taught by *Õeõeri,* the child warrior who avenged his mother killed by *Xamat^hari* sorcerers in the beginning of time. Our ancestors followed in his footsteps and our grandfathers and our fathers after them. All this did not start yesterday!

Yet our elders of long ago did not launch such raids every day! Sometimes during my childhood I saw them go to war, it is true. But they only went out of anger and to avenge their dead relatives. They tried to arrow enemies after burying the ashes of their deceased to make the *õnokae* homicide state reciprocal. They sought to arrow the warriors who had eaten their kin and that was all. They did not arrow just anybody! The white people cannot say that we are bad and violent simply because we want to avenge our dead. At least we do not kill our people for merchandise, land, or oil, the way they do. We fight about human beings. We go to war for the sorrow we have for our brothers and fathers who have just died.

It is true that our elders of the past could be sometimes bellicose, but after a time, once the most aggressive warriors on both sides had been killed, they tried to send words of appeasement to their enemies through other houses. They told them they would no longer attack them and encouraged them to make friendship. So, tired by relentless raids, these enemies finally risked visiting our elders' house to attempt to be reconciled. We also call this *rimimuu*.[31] Despite suspicion, minds regained their calm and people were able to get along. Yet after several moons, bad talk could sometimes come back and someone else could be arrowed again.[32] New incursions were then launched for a while and finally stopped in the same fashion as before. Once all the great warriors in *õnokae* homicide state[33] were dead, the other men, less aggressive ones, always wanted to make peace in the end. But it was the older women who took the initiative to protect the people of their house, for women do not carry arrows. They traveled to the enemies' house and shouted as they got close: "Do not be afraid, don't run away! *Aë!* We are women, do not arrow us! *Aë!* We have come as friends! *Aë!*" This way they reestablished contact, and the men of their house could come sometime later to engage in a *hiimuu* invitation dialogue with their former enemies.[34] Then they spoke words of friendship and reaffirmed the end of hostilities: "*Awe!* Let us stop arrowing each other! Let us stop mistreating each other like that! Become our friends! We are tired of mourning our kin! We don't want to constantly make war anymore! Enough! It is pitiful that we can no longer clear a garden, hunt, or even draw water without fear of being arrowed. We want out children to stop crying with hunger and thirst."

Then fear came to an end on both sides and people started to think: "*Awe!* This is a good thing! I will be able to acquire their goods and we will become friends." They started bartering hammocks, pots, machetes and axes, knives, glass beads, cotton, tobacco, and dogs. After this first contact, they continued to visit each other and generously exchange their goods. This lasted for some time, then they eventually arranged marriages between young people of their houses and never stopped being friends. This is what our elders did when they got tired of arrowing each other, for if they had never wanted to put an end to their vengeances, they would have continued their raids endlessly and would all have died. In the past, the elders of my father-in-law's house made friends with my own elders in this way. At the time, these people lived on the upper Rio Catrimani, from where they often raided our houses on the upper Rio Toototobi.[35] Their incursions and ours in revenge lasted a long time, but finally the elders of both groups resumed peaceful contact, and we who are their children grew up and remained friends until now. This is why, in the end, I was able to marry one of their daughters!

This is how our elders used to arrow each other in the past. It is true that they often raided each other's houses, but it was in a period when the white people's elders also made many big wars, a time when I was not yet born. Yet our people only mistreated each other with *hwëri* sorcery plants and only fought each other using arrow canes from their gardens and curare arrow points from the forest. They did not burn multitudes of people with **rockets** and bombs. We are not a fierce people! Our elders—and today the people of the highlands—did not launch revenge incursions to kill a great many people.[36] If warriors conduct several raids one after another to avenge a dead man, it is just because they often take a long time before they succeed in arrowing their enemies, who are always on their guard. It is only over time that they are finally able to kill one or two reputed warriors of a house, then another who has joined them from an allied village. That's all. Once these men in *õnokae* homicide state are dead and their victim's bone ashes are buried, it is over. It is enough. Anger ends, thoughts regain their quietness. As I have said, it is these aggressive and valiant men who were primarily targeted in our elders' incursions. Yet in the heat of anger, warriors encircling a house could sometimes arrow other men, innocent of the death of the one they want to avenge.[37] But unlike white people, they would never kill women and children like the gold prospectors did with the inhabitants of *H^waxima u*.[38]

ONCE, A VERY long time ago, my grandfathers lived in the highlands, not far from the Orinoco headwaters. They had not yet known the *xawara* epidemics, so they were many and their houses were close together. At the time, they were at war with the *Hayowa t^hëri*, a group of *Xamat^hari* whose houses were located downstream from theirs, in the direction of the setting sun.[39] Tired of being constantly attacked by these enemies, my elders finally settled on the upper Rio Toototobi, at the edge of the land they had previously inhabited.[40] Then the *Xamat^hari* incursions came to an end. But later, shortly before my birth, the raids started again, this time towards the rising sun, against the people of *Amikoapë*, who lived at the headwaters of the Rio Mucajaí, and then, above all, against the elders of the people of the upper Rio Catrimani, whom our people called *Mai koxi*.[41] This is what I heard from my stepfather, my mother's second husband, during his *hereamuu* speeches in Toototobi when I was a child. Yet if the people of these old houses arrowed each other like this, it was certainly not to grab parts of the forest from one another! Far from that, these incursions were always due to the anger and sorrow caused by the bone ashes of their dead. The white people may call this "to make war," but we only say *niyayuu*, to arrow each other.

Finally, my elders arrived at *Marakana*, in the Rio Toototobi lowlands, where I lived when I was a very small child. At the time, they still often engaged in raids, especially against the *Mai koxi*. But they also sometimes launched incursions against the people of *H^w^axi*, near the sources of the Rio Parima, or the people of *Ariwaa*, a *Xamat^h^ari* group who lived on the upper Demini.[42] And they were also attacked by all these people. It was so in that time! Our shaman elders often called down the images of the *Niyayopa t^h^ëri* War People and made them dance. Then the men of our house became valiant and did not fear to go arrow the warriors or the enemy sorcerers who had eaten their relatives. Yet later, when I grew up, my elders finally stopped going to war.[43] They were now very far from the highlands inhabited by the ghost of *Õeõeri*, who had taught them the will for revenge. Only the people who live at the rivers' headwaters, far away in the hills, continue to arrow each other because that is where *wait^h^iri* war valiance was born. It is so. My stepfather often told me that when he was a young man he went to war to avenge his father, killed by *Xamat^h^ari* warriors. This is how he became a valorous man, by following in his turn the reclusion of the warriors who ate enemies, which we call *õnokaemuu*. Later, he also avenged my father, who was his friend, and the mother of my mother, whose bones were broken by *oka* sorcerers from the *Amikoapë* people once she went to the forest alone. Then he also avenged several of his brothers-in-law and fathers-in-law killed during raids launched by the *Mai koxi*. I remember very well that during my childhood in *Marakana* I saw him set out to go arrow enemies many times. He was a fearless warrior. But all this happened long ago. Now he is an old man, he stopped shooting arrows to his enemies a long time ago. He avenged the killing of many of our people in the past, yes. But he told me that he had taken back the value of all their deaths and that they have been **paid** for; that was enough. This is good like that.

OUR ELDERS certainly did not arrow each other because of women, unlike what the white people sometimes claim. As I said, they only went on a raid when all were seized with the anger of mourning and they wanted to take revenge on those who had eaten the dead person they lamented for. Sometimes they called on the warriors of neighboring houses to back them up on their incursions. They killed some enemies and the latter, now also in mourning, were not long in trying to get revenge too. So the raids between houses lasted for a while, until the bone ashes of both sides' dead had been put into oblivion. This is what would happen. It was a question of avenging the dead, not of fighting over women. I often saw this in my childhood. My stepfather, who was a much-feared warrior, never went to war over quarrels about women! He often led raids against distant enemies, but it was always from the anger of mourning, to avenge our house's dead. He would never have left alive enemy warriors in õno-kae homicide state after they had killed his people. He would not have rested until he had taken back the blood of our deceased by eating enemies in the same way they had eaten the people of our house. Since the beginning of time, the War People taught our long-ago elders this way of thinking.

Fights because of women are a different thing. When a guest tries to take away one of his hosts' wives during a visit or a *reahu* feast, people get truly angry, yes, but if the husband wants to fight, they will only take turns hitting each other on the head with long sticks, which we call *ano-mai*.[44] You don't go to war for that! It makes your scalp bleed a lot, but the skull is solid and you stay alive.[45] It is our way of putting an end to the angers caused by jealousy over women, for pain quickly calms the mind. This is truly what takes place between us when someone covets another man's wife. Men feign contempt about women by saying their genitals do not smell good, but this does not prevent them from furiously confronting each other to keep their wives! That is the truth, they carry them in their thought, and they do not hesitate to ardently struggle for them. White people also tell us that we are bad and aggressive for fighting to keep our women like this. Yet they do the very same thing among themselves! I often saw them! When a husband in the city realizes that his

wife is making friendship with another man, he goes into a rage. Furious, he insults his rival and instantly wants to fight. He also mistreats his
wife and sometimes he kills her. So why do they speak so badly about us?

When I was a child, our elders did not often fight over women. It happened, of course, but then a lot time passed before it occurred again. I
remember that once it happened over one of my younger sisters at *Marakana*. Her father had given her to a son-in-law who had worked a lot to
obtain her, but she preferred a young man from a neighboring house,
whom she found handsome and was in love with. She really desired him
and eventually she ran away with him. This obviously enraged the man
to whom she had been promised. Then the young woman's kin followed
the future husband's anger and fought her lover's relatives. They hit each
other hard on the head with long sticks, but they certainly did not kill
each other! They just took turns inflicting this sharp pain on each other
to appease their rage. They did not want to come to arrows because they
were part of the same people, and they only conducted raids against distant houses inhabited by other people.[46] This is how it often happened,
but our elders also vehemently quarreled and beat each other's heads
with clubs for many other reasons: for stolen bunches of bananas or merchandise; *rasa si* peach palm trees cut down in their gardens; forest shelters knocked down in the forest; or bundles of cassava bread discarded by
guests after a *reahu* feast. But in the same way, opponents only had to
make each other's scalps bleed and feel deep pain for their anger to calm
down. Dizzy and exhausted, they would then tell themselves: "All right,
that's enough!"

But when men fight out of jealousy for a woman, a very aggressive
husband can sometimes shoot an arrow at his rival or even his wife. That
is also true. In this case, his victim's relatives are seized with rage and
want to arrow him in the same way. Then finally if someone dies during
this fight, the dead person's kin will instantly engage in war retaliation
because of the fatal arrow's value of anger.[47] Furious at the man who has
become an *õnokae* homicide, they will instantly decide to get revenge and
put him to death. But this kind of thing rarely happens. The jealous husband truly has to be very bellicose and become enraged. This is what I
have always been told. It has never happened among my people, or at
least I have never heard my elders tell about it. For us, the true words
when a young bride runs away with her lover are those of a fight with
clubs between houses. Since no one usually dies in this case, the adver-

saries' anger comes to an end after a time. They stop mistreating each other and their thoughts finally calm down. No one considers going to war for all that.

Sometimes we fight in yet another manner, in this case solely to put an end to our annoyance with people from friendly houses. This mostly happens when they have spoken bad talk about us and their words have been reported to us either by a passing visitor or one of our people who married into their house or is staying there to find a wife. In such cases we invite these slanderers to a *reahu* feast, and as soon as we engage in a *yãimuu* dialogue with them, we ask them to confirm the rumor that reached us: "Is it true that you called us cowards?" If one of them has the courage to answer: "*Awe!* That is what I said! These are words of truth! You are cowards and you are afraid to confront us!," we are overcome with anger. We instantly start manhandling them, twisting their necks while continuing our sung talk.[48] Finally, stirred up by these exchanges of angry words, one of us will tell them: "*Asi!* I want to put your chests to the test." Then hosts and guests will pair up to confront each other, taking turns pounding each other's chest or slapping each other's sides.[49] In this case, if they are truly very angry, the opponents will put stones in their fists or will propose to hit each other with the dull edge of their machetes in order to really make each other suffer. This is how things go, it is true. But here too, nobody dies! We only fight this way to stop people from other houses from spreading false talk about us. This has always been our way. When anger plants itself in our chest, the images of the jaguar and the coati dance in us and make us aggressive and fearless.[50] Exalted, we are then quick to take our clubs to beat our heads or punch our chests. We mistreat ourselves this way because only pain can appease our rage. This was our ancestors' way of being, and it still is because if we only kept bitterly repeating bad talk, our anger would never truly disappear.

IF OUR elders had really killed each other the way the white people say, their war raids would never have ended, and they would nearly all have disappeared. This is not the case. Our fathers and grandfathers were once very many. What decimated them was not their own arrows, but the white people's epidemic fumes! Since these outsiders arrived in our forest, nearly everywhere people have stopped raiding each other's houses.[51] The great warriors of the past are all dead, devoured one after the other by the *xawara* epidemics. Of course, there are still valiant men among us, but they no longer have the will to go to war. This is the case here in our house of *Watoriki*. The words of warfare have not disappeared from our mind, but today we no longer want to harm ourselves this way.[52] We prefer to talk to try and contain our anger at each other. No one tries to arrow us anymore, and we do not try either. And when we sometimes think that one of our elders might have been killed by *oka* sorcerers, we merely ask ourselves: "What distant enemy could possibly come all the way here to blow *hwĕri* sorcery plants on us?" and it does not go much further than that. Men like me who grew up after our elders disappeared do not want any more deaths by arrow among us. Since the white people surrounded us, they have constantly destroyed us with their diseases and weapons. This is why I think we should no longer make ourselves suffer the way our elders did when they were alone in the forest.

I have never participated in a war raid. That is the truth. I do not know what it is to arrow a human being nor to remain stretched out in my hammock without moving like the *ōnokae* warriors do after they have eaten their enemies. I prefer for us to no longer live like this. Our elders had always possessed the words about war and the way of the *ōnokaemuu* reclusion of those who have killed in a raid. It was their habit to take revenge on their enemies long before the white people got close to us. But today our real enemies are the gold prospectors, the cattle ranchers, and all those who want to seize our land. Our anger must be directed at them.[53] This is what I think. Today it has become wiser to think about our mud-soiled rivers and our burned forests than to arrow each other. We must tell ourselves: "*Awe!* The *xawara* epidemic smoke is our real enemy! It is what transforms us into ghosts and does not hesitate to devour us by the houseful! Let our thoughts of vengeance be set on it! Let us forget the things of war!" Some of us in the highlands still have a taste for arrowing each other, it is true. But I am the one who travels to speak harshly to the white people and defend our land and our life, and I don't want that anymore. I say so to the people of all the houses I visit in our forest: "If you're angry, fight with words! Pound your chest with your fist! Make your heads bleed with your clubs! But do not think of arrowing and killing each other anymore! Only the *xawara* epidemic hates us so much that it persists in devouring us the way it does. Let us stop going to war and instead let us fix our thought on the white people's hostility towards us!" These are my words.

CHAPTER 22

The Flowers of Dream

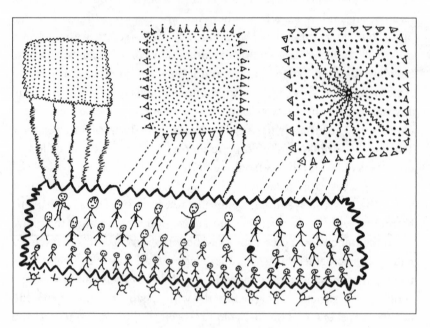

Mirrors and paths of the *xapiri*

THE POWER of the *yãkoana* powder comes from the trees of our land. So when the shamans' eyes die under its effect, they can call down the *urihinari* forest spirits,[1] the *mãu unari* water spirits, and the *yarori* animal ancestor spirits. This is why they are the only ones who truly know the forest. Our elders knew nothing of the way the white people draw their words.[2] But they have made these images dance since the beginning of time. For their part, the white people know nothing of the things of the forest, for they cannot really see them.[3] All they know of the forest are the lines of talk that come from their own minds. This is why they only have misleading ideas about it. The shamans do not draw any words about the forest or the sky, nor any outlines of the land.[4] They wisely avoid treating them as badly as the white people do. They drink the *yãkoana* to contemplate their images instead of merely reducing them to broken lengths of wavy lines. Then their thought keeps the words of what

they saw without having to **write** them. The white people never stop setting their eyes on the drawings of their speech, which they circulate
among themselves pasted on paper skins. This way they just stare at their
own thought and only end up knowing what is already inside their minds.
Their paper skins do not speak and do not think. They are simply there,
inert, with their black drawings and their lies. I much prefer our spoken
words! These are the ones I want to hear and that I want to continue
to follow. By constantly keeping their minds set on their own drawn
marks, the white people ignore the distant words of other people and
other places. If they sometimes tried to hear those of the *xapiri*, their
thought might be less narrow and dark. They would not work so hard at
destroying the forest while claiming to defend it with the laws they paint
on the skin of cut trees!

What the white people call paper, we call *papeo siki*, paper skin, or
utupa siki, images skin, because all this is made from tree skin. The same
is true for what they call money.[5] It is just other paper skins that they only
hide under this lying name to delude each other. I was told that the white
people make their paper by crushing wood with their machines. *Omama*
certainly didn't teach them such a thing! Maybe one day their ancestors
got tired of drawing on game skins and decided on their own to kill trees
to make paper? Since then, they have ground up huge quantities of wood
to make that paper. They don't care that these trees carry the food of the
bee spirits and all the winged game spirits. This is why I also call their
paper *urihi siki*, "forest skin."

I think that the liquid which white people call **ink** and which they use
to draw their words is also something that their elders started collecting
from the forest long ago. It is red or black and comes from the spirits' annatto dyes, which are other and far more beautiful than the ones we use
to cover our skin. It was *Omama* who put them inside trees in the beginning of time.[6] First, he showed them to the *xapiri* he had just created
so that they could paint and perfume themselves for their presentation
dances. These dyes have been among their precious goods ever since.
Next, *Omama* also taught our forefathers to decorate their bodies with
them during *reahu* feasts in order to imitate the *yarori* animal ancestors'
beauty. He wanted human beings to no longer exhibit the ugliness of
their grayish skin.[7] Since then, we favor these annatto skin paints.

Later, *Omama* also spread the beauty of his dyes inside the trees of the
white ancestors' land. Yet the white people soon spoiled them by twisting

their use in another direction. They began to cook them in factories to paint image skins and draw their words on paper skins. We are other people. We only draw on our bodies, as we were taught by *Omama* and our ancestors.[8] Out of jealousy for his brother, *Yoasi* made a mess of these ancient words, then placed them in the white people's minds. Then the white people stopped painting their skin and only used these dyes on their paper skin. I think that *Yoasi* is the one that the white people's ancestors called *Teosi*.[9] Yes, it is true. The white people are the people of *Yoasi!* On the contrary, we are the children of *Omama*, and we follow the uprightness of his words. This is why when we hold *reahu* feasts and make our spirits dance, we paint our bodies with black and red annatto drawings, cover our hair in white down feathers, adorn our armbands with scarlet macaw tail feathers, and stick parrot feathers in our earlobes.

OUR ANCESTORS' words have never been drawn. They are very old but they remain present in our thought to this day. We continue to teach them to our children who, after our death, will do the same with theirs. Children do not know the *xapiri*. However, they pay attention to the songs of the shamans who make them dance in our houses. And so, little by little, the adults' words come into them. Then, once they are adults, they can make them heard in their turn. This is how we pass on our history[10] without drawing our words. They live deep inside us. We do not let them disappear. When a young man wants to become spirit, he asks the shaman elders of his house to give him their *xapiri*. So they pass him words from very long ago which settle inside him to be renewed and increased again over time.

As for the white people, they constantly need to draw their words. This is also something that *Omama* did not teach them! I think this must be because their minds are really forgetful. Their ancestors probably created these drawings in order to be able to follow their thoughts. Long ago they must have told themselves: "Let's draw what we say and that way maybe our words won't escape further from us?" It is true. Their words do not seem to be able to stay in their minds for long. If they listen to a great number of them without making any drawing, they instantly vanish from their thought. But on the contrary, if they keep their painted marks, the next day, after they have forgotten them, they can think: "*Oae!* That's it! Things really are the way I drew them on this paper skin." This is their way. They do this all the time, otherwise they would quickly forget every-thing they said. They really love image skins, like their ancestors be-fore them, for they are other people. It must be a good thing for their way of thinking. This way they keep their old words by drawing them and give them the name "history." Then they contemplate them at length to fix them in their minds. Finally, they tell themselves: "*Haixopë!* This is the drawing of the words of our elders and what they taught us. We must fol-low their marks and imitate them in our turn." This is how young white people learn to think with their fathers' words. This is the way they will know how to build machines and motors like their fathers or that they will be professors, nurses, or airplane pilots. This is how they study.

We are inhabitants of the forest. We study by drinking *yãkoana* with our shaman elders. When they become spirit, they carry our images far away to battle the evil beings or repair the sky's chest. They teach us to know the *xapiri* and they open their paths to us. They send them to build our spirit houses. They teach us the words of their songs, then make them grow in our thought.[11] Without their support, we would get lost in the void or sink into the *mõruxi wakë* blaze.[12] We learn to think right with the *xapiri*. This is our way of studying, and so we have no need for paper skins. The *yãkoana*'s power is enough for us! It is the *yãkoana* that makes our eyes die and our thought bloom. It is true. With a living person's eyes, it is not possible to truly see things. The words that relate how the forest and the human beings came into being belong to *Omama*. They are very numerous. The great men reveal them to us during their *hereamuu* ex-hortations while they tell of the places where their fathers and grandfa-thers lived long ago. The shamans also set them out at great length in their songs when they become spirits. In fact, we never stop listening to them. This way they are strongly fixed inside us and never get lost. The young people who have so often heard them keep them in their minds. Once they are adults, they make them multiply within themselves again, then pass them on to others and it continues like this without end.

YET THE white people believe we are ignorant, simply because we do not have drawings to put down our words.[13] More lies! We would only be-come truly ignorant if we did not have shamans. Just because our elders did not have schools does not mean that they did not study. We are other people. We learn with the *yãkoana* and the spirits of the forest. We die by drinking the *yãkoana hi* tree powder so that the *xapiri* will carry our images far away. This way we can see very distant lands, go into the sky's chest, or descend into the underworld. Then we bring back these places' words to make the inhabitants of our houses hear them. This has always been our way of becoming clever. It would not be possible to draw the spirits' words to teach them, for their multitude is neverending. Nothing would come of it. When the white people study, they stare at old speech drawings. Then they report their contents to each other. But if they never see or hear the images of the beings of the beginning of time by themselves, they cannot really know them. As for us, without **pencil** or paper skins, we become ghosts with the *yãkoana* and go very far to contemplate the image of these beings in the time of dream. Then the *xapiri* teach us their words and this is how our thought can expand in every direction. We could not learn anything if we did not join our shaman elders to drink the *yãkoana* and call the spirits down.

These wise men give us their *xapiri*'s breath of life with the powder they blow into our noses, and it takes hold of us.[14] This is how we can join them when they become *xapiri* and discover so many unknown places. Then the other spirits we meet there joyously approach to join us and build their houses by our side. Their songs enter into us and become increasingly numerous, so much so that the *yõrixiama* thrush spirits and *ayokora* cacique bird spirits must sometimes conceal them in the heights of the sky to hide them from enemy shamans' jealousy. Without the spirits' words, we would be deprived of any knowledge and unable to say a word. We could always pretend to imitate the *xapiri* without having ever seen them, but it would lead to naught. A young shaman cannot talk about the distant lands they come from without first having had his image brought there by his elders' spirits. If this has truly happened, ordinary people who hear his songs will say: "These are words of truth! He has seen what he speaks of. The words of these songs come from very far away. They are truly other. How we wish we could know the places where the *xapiri* live the way he does!" The elders who first gave him the *yãkoana* will also listen to him, lying in their hammocks, and will also let him hear their satisfaction: "*Awe!* These are clear and beautiful words. Now you truly know things!" When he listens to these words, the young shaman will be happy. But if on the contrary he has drunk the *yãkoana* for no reason, only to try, he will only utter ugly and confused words and deceive his audience. In that case, the elders will be displeased and will complain about him:

"*Ma!* His tongue has remained a ghost tongue and his thought is only lies! He does not know any true word, and he is unable to tell about the distant lands from which the spirits come down. He did not see anything!"

The *xapiri* come from very far away, and their numbers constantly grow as they travel towards us. Their songs teach us the words of the unknown places they come from. If you want to acquire these words of wisdom, you have to answer the spirits as soon as you hear their clamor approaching. This is how they make us clever. By studying under our elders' guidance, we do not have any need to look at paper skins! It is inside our head, in our thought,[15] that these spirit words move from one to the next and expand endlessly. Ordinary people are not like this. They are content to live, sleep, and eat, nothing more. They feather their arrows and go hunting. They plant banana shoots in their gardens and that's all. They never think about the *xapiri*'s words. They fear the *yãkoana* and think that if they were to inhale it, they would die. Their thought is closed and short. It is the same thing with white people when they do not study.

WHITE PEOPLE do not become shamans. Their *nõreme* image of life is full of dizziness. The perfumes they rub themselves with and the alcohol they drink make their chest too odorous and too hot. This is why it remains empty.[16] They do not possess spirit houses or songs. Their elders have not fixed any feather ornaments or glass beads belonging to the *xapiri* on their images. They sleep and in their dreams they only see what surrounds them during the day. They do not really know how to dream because the spirits do not carry their images away during their sleep. On the other hand, we shamans are able to dream very far. The ropes of our hammocks are like antennas through which the *xapiri*'s dream constantly comes down to us. This is why our dream is fast, like television images from distant lands. We have always dreamed in this way because we are hunters who grew up in the forest. *Omama* put the dream in us when he created us. We are his children, this is why our dreams are so distant and inexhaustible.

AS FOR THE white people, they sleep near the ground, stretched out on **beds** where they restlessly move about. They have bad sleep and their dream is a long time coming. Yes, they have many antennas and radios in their cities, but they only serve for them to listen to themselves. Their knowledge does not go beyond these words that they address to each other everywhere they live. The shamans' words are different. They come from very far away and speak about things unknown to ordinary people. The white people, who do not drink the *yãkoana* and make the spirits dance, ignore their words. They do not know how to see *Hutukarari*, the sky spirit, nor *Xiwãripo*, the chaos spirit. They also do not see the images of the *yarori* animal ancestors nor those of the *urihinari* forest spirits. *Omama* did not teach them any of all that. Their thought remains filled with smoke because they sleep stacked on top of each other in buildings among the motors and machines.

We are other people. When our eyes are dead with the *yãkoana* powder during the day, we sleep in a ghost state at night. Then, as soon as we doze off, the *xapiri* start to come down to us. No need to drink the *yãkoana* again. Their clamor suddenly rings out in the night like the strident calls of a company of parrots in the trees. And at once the tangled multitude of their luminous paths appears in the dark, each time drawing closer to you and sparkling like moon slivers. Then you start to answer the *xapiri*'s songs and their value of dream quickly reaches you.[17] Your body remains stretched out in its hammock, but your image and breath of life fly off with the spirits. The forest then recedes at great speed. Soon you can no longer see its trees and you feel yourself floating over a great void, like in an airplane. You fly in dream, very far from your house and your land, on the *xapiri*'s paths of light. From there, you can see all the things of the sky, the forest, and the waters that the elders could contemplate before you. The spirits' day is our night, which is why they seize us by surprise, during our sleep. As I have said, this is our way of learning. We shamans possess the spirits' value of dream inside ourselves. It is they who allow us to dream so far away.[18] This is why their images constantly dance by our side when we sleep. By drinking the *yãkoana*, we do not slumber in vain. We are always prompt to dream. Having become ghosts, we relentlessly roam distant lands while making friendship with their inhabitants' *xapiri*. This is how shamans dream.

Ordinary men are different. Their thought is too often set on women, and by constantly breathing the fragrance of girls' *puu hana* leaf ornaments, their chests smell like penis. The *xapiri* are sickened and never look at this kind of man. They also dream, yes, but a little, only of things very close to them, and they forget them as soon as they wake up. They only see their hunting and fishing in the forest and their work in the gardens. They dream of jaguars, snakes, and *në wãri* evil beings. They see again their own presentation dances or their fights during *reahu* feasts. They recall their war incursions or their magic love charms.[19] They dream of the women they desire, the people from other villages with whom they are friendly, or also the dead kin they miss. They sleep in a ghost state and their images leave them, like that of shamans. Yet it never goes very far away. Only the very good hunters among them can dream a little farther.

As for the white people, they must only see their wives, their children, and their merchandise when they sleep. They must anxiously think about their work and their trips. They can certainly not see the forest the way we do! We shamans, we are different. We do not merely sleep. During our sleep, the *xapiri* never stop looking at us and wanting to talk to us. This is why we also see them and we can dream with them. They call us: "Father, do you hear us? Or are your ears really blocked? Answer us!" At that moment, we begin to dream and they appear to us in their intense brightness. Without them, we could never dream like this. Often, they wake us to warn us: "Father! An unknown being is approaching! Isn't this an evil being?" We answer: "What's going on? *Haixopë!* It's true! *Omoari*, the evil dry season being, is approaching our house!" Then they instantly go in his direction to fight him. Often, they also simply call us because they want to make their songs heard. We, their shaman fathers, are asleep, but they are awake and want to work. They think: "*Hou!* It is bad to sleep like this! We do not want this kind of laziness! We must do our presentation dance." It is true. If the *xapiri* did not have their eyes on us, we could not dream as far. We would be satisfied to sleep like axe blades abandoned on the floor of our house.

We people of the forest come from the sperm and blood of *Omama*, who was a true dreamer.[20] It was him who in the beginning of time planted the dream tree we call *Mari hi* on the land he had just created. Since then, it has sent us the dream as soon as this tree's flowers blossom.[21] This is how he put the dream in us, allowing our images to travel during our sleep. We possess it through the blood of our elders. As children, we often drink too much wild honey or hot banana soup. Once we are full, we fall asleep in a ghost state, and we start to dream while seeing unknown things. When we become young men, we spend our time walking in the forest, tirelessly tracking game. This is when our thought begins to truly focus on the *xapiri*. Little by little, we fall in love with them, as if they were young women! We start to see the images of the animal ancestors trekking through the forest with us in dream. First we see the images of the *wakoa* hawk and the *kāokāoma* falcon,[22] as well as images of the *yawarioma* water beings who are also very skilled hunters. Then we see the jaguar, peccary, spider monkey, and tapir spirits appear, as well as many other images of game we do not know yet. When the *xapiri* take such an interest in us, we see them dance as soon as we are asleep and we hear them sing very high in the sky's chest, all around us. These are our first dreams in their company. Then, once we are adults, we will be able to drink the *yākoana* with elders who really know them and will open their paths to us. As fine and translucent as spider webs or **fishing lines**, the *xapiri*'s shining paths will become fixed along our arms and legs. Then the spirits will come down along them to tear our chest and open a large clearing in it where they will do their presentation dance.[23] Our images will then be able to follow them in the time of dream, even farther than the white ancestors' land.

WHEN I WAS a child, I always flew when I dreamed, very high in the sky's chest or very far in the depths of the waters. This is why I later asked my stepfather to make me drink the *yākoana*. I did not become a shaman without a reason! My thought had never set itself on women or merchandise. I was always curious to know the *xapiri* better, for the images and the songs of dream they send us are very beautiful. These were my studies, from the beginning. The shamans who wear the spirits' ornaments and possess their songs dream with much cleverness. Seized by the power of the trees of the forest, they join them in their most distant journeys, flying towards bare flat lands only inhabited by magnificent *xapiri*. They can see the images of our ancestors who turned game in the beginning of time as well as those of *Omama* and his family. They see the epidemic fumes and the evil beings in the distance as soon as they approach to devour us. They can also travel to the white people ancestors' land and make their *napĕnapĕri* spirits dance.

While the *xapiri* seized my image, I was able to contemplate in the night all that my elders had known before me. I saw *Omama* piercing the ground with his heavy metal bar to make the rivers burst out of it with all the fish, caimans, and anacondas.[24] I saw him catch his wife *Tʰuĕyoma* and receive the cultivated plants from his underwater father-in-law. I saw the image of his son dance as he became the first shaman. I saw the people from the beginning of time light great fires of green leaves when the night did not exist yet so they could copulate by hiding in their smoke. I saw the animal ancestors make Caiman laugh to steal the fire out of his mouth. I saw Ant lose his mother-in-law in his huge maize plantation. I saw the forest burn in the beginning of time until nothing was left but savannas where trees no longer grow back. Often I also warily entered houses crowded with evil beings of the forest. I flew with terror in the great *wawëwawë a* void above the earth and the sky. I saw the spider monkey spirit, whom we call the son-in-law of the sun being, eat his father-in-law's fruits of heat without burning his mouth. I also saw him when he stopped the sky from falling and made rocky peaks collide to test their solidity. I glimpsed the bat spirits in the darkness, trembling with cold while they gnawed at the sky's edges and blew in their sorcery blowpipes. I saw the spirit of the giant *Simotori* beetle cut off mountain rocks to open his gardens. I heard the bee spirits endlessly chattering in the trees to defend the forest. Finally, long before I went to the white people's land, I saw the machines that run without feet in their cities, which the elders told me about.[25]

By flying with the *xapiri* like this, my dreaming has no end. It constantly roams the forest, the mountains, the waters, and all the directions of the sky and earth. The spirits' breath of life is in me; this is what makes me see all these things. They call me during the night and then I travel with them, constantly imitating their song. But when I sleep far from my house, I only contemplate their beauty in silence, for my voice could attract the malevolence of enemy sorcerers or hostile spirits. This is how I am used to journeying in the time of dream. But now the *xawarari* epidemic beings also often carry my image far away. Burning with fever and having become ghost, I battle throughout my sleep, fighting the white people and their soldiers who never stop provoking my anger. My *xapiri* valiantly attack them with their machetes and arrows to get revenge for how they mistreat the inhabitants of the forest.

WHEN WE truly want to know things we people of the forest try to see them in dream. This is our way of studying; I have said so. I also learned to see by following these ways. My elders did not merely make me repeat their words! They made me drink the *yãkoana* and allowed me to admire the dance of the spirits during the time of dream with my own eyes. They gave me their own *xapiri* and said: "Look! Contemplate the spirits' beauty! When we are dead, you will continue to make them come down after us. Without them, your thought would go on trying to understand things in vain. It will stay in darkness and oblivion." This is how they opened their paths to me and made my thought grow. Now I will grow old and try to pass on these words to young people in my turn so they do not get lost and are never forgotten. If I had not known the *xapiri*, I would have remained ignorant and would speak without knowledge. But thanks to them, my words can follow one another and spread everywhere. They can tell about all the unknown places from where the spirits come down. This is our way of becoming wise. We people of the forest never forget the distant places we have visited in dream. When we wake up in the morning, their images remain vivid in our minds. When we speak about them, we happily tell ourselves: "This is the beauty of the *xapiri*, which the elders knew before us! This is how they have made their songs heard and danced to present themselves since the beginning of time!" These images constantly come back in our thought and always remain as sharp. The spirit words that go along with them also stay inside us. They never get lost. This is our history. It is based on these words that we can think right. This is why I say that our thought is like the image skins on which white people keep the drawings of their elders' speeches.

These words come from our spirits' value of dream, and we make them heard by the people of our house. We do not deceive them the way the people of *Teosi* did long ago by repeating to us: "*Sesusi* will come down in the forest! If he wants to, he will arrive among us today or tomorrow!" Yet time passed and nothing happened. We shamans never speak like this. We never fool our people by looking at word drawings to be able to speak. No need to set our eyes on paper skins to remember the *xapiri*'s words! These long-ago words are stuck in our thought and rush to our lips to be sounded out, innumerable, as soon as we become spirits ourselves. This is why it is possible for us to so easily make them heard to those who listen to us. Now I try to explain these words on what I have seen in dream to the white people in order to defend our forest. If I did not have a spirits' house of my own and if I was unable to truly see anything, I would have nothing to tell them. My eyes would be a pitiful sight, my voice would be wavering, and they would soon realize that ignorance and fear numbed my mouth.

CHAPTER 23

The Spirit of the Forest

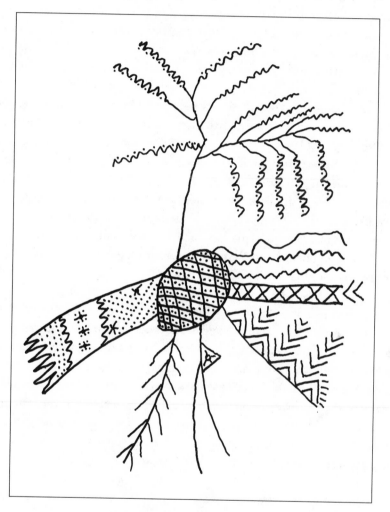

Urihi a, the land-forest

As I HAVE said, the shamans' thought spreads everywhere under the ground and under the water, beyond the sky and in the most distant regions. They know the innumerable words of these places and those of the beings from the beginning of time. This is the reason they love the forest and want to defend it. But on the contrary, the minds of the white people's great men only contain the drawing of the tangled words they stare at on their paper skins. Their thoughts cannot travel very far. They remain stopped at their feet, and it is impossible for them to really know the forest. This is why they do not worry about destroying it! They tell themselves that it grew by itself and that it covers the ground without reason. They probably think it is dead. But that is not true. It only appears silent and unchanging because the *xapiri* bravely protect it by pushing back the *Yariporari* storm wind that angrily arrows its trees and the *Xiwãripo* chaos being who always tries to make it become other. The forest is alive, this is where its beauty comes from. Doesn't it always seem new and damp? If it wasn't so, its trees wouldn't be covered in leaves. They could no longer grow or give humans and game the fruit they feed on. Nothing would grow in our gardens. There would be no humidity in the ground, everything would be dry and shriveled, for water is alive too. It is true. If the forest were dead, we would be as dead as it is! But it is truly alive. The white people may not hear it complain, yet it feels pain just like humans do. Its tall trees moan as they fall and it cries in pain when it is burned down. Finally, it only dies when all its trees have been cut down and burned and all that remains are their charred trunks lying on the dried-out ground. Then nothing grows back there, except a little grass.

The white people do not ask themselves where the forest's value of growth we call *në rope* comes from.[1] They probably think that plants grow alone, without a reason. Or else they take themselves for great workers, able to make plants grow solely through their own efforts. They even call us lazy because we do not destroy as many trees as they do. This bad talk makes me angry. We are not layabouts! The images of the *koyo* ant and the *waima aka* lizard live inside us,[2] and we tirelessly labor in our gardens under the sun. But we do not work the same way white people do. We care about the forest, and we think that clearing it without measure will only kill it. *Omama*'s image tells us: "Open your gardens without making them go too far. Cut up the wood of fallen trunks for the fires to warm you and cook your food. Do not cut the trees just to eat their fruit. Do not damage the forest for no reason. Once it is destroyed, no other will replace it! Its richness will escape forever and you will not be able to live on this land anymore."

On the other hand, the white people's great men think: "The forest lies here without reason, we can therefore mistreat it as much as we want. It belongs to the government."[3] Yet it was not they who planted it and if we leave it to them, they will only damage it. They will chop down its tall trees and sell them in the cities. They will burn those that remain and soil all the rivers. The land will soon be naked and blazing hot. Its value of growth will leave it forever. Nothing will grow there anymore, and the game that comes to feed on its fruit will also flee. This is what happened when the road crossed the Rio Ajarani[4] people's forest. And it happened again when the gold prospectors burst into the highland people's land. By digging in the rivers' beds, clearing their banks, and filling the forest undergrowth with the smoke from their motors, they chased its richness away and made it so sick that *Ohiri*, the being of hunger, settled right in the middle of it. The game died or took refuge very far away in the hills. Fishes cannot be found anymore in the waters, nor shrimp, crabs, stingrays, electric eels, and caimans.[5] Their angry images fled far away, called back by the other *xapiri*. It is so. After I grew up, I often saw the bad marks left by the white people in the forest. They truly don't care that its trees are replaced by weeds and its rivers by trickles of miry water! They probably think that it does not matter, as they will later cover its floor with the concrete from their cities.

We were born in this forest: we grew up there and we became shamans there. Unlike the white people, we take care of it, like our elders before us, because without it we could not live. This is why the spirit of hunger always remained far away from it. We want our children and our grandchildren to keep on feeding themselves from the forest when they grow up. We do not cut many of its trees, only enough to open our gardens. We plant manioc, banana plants, yams, taros, sweet potatoes, sugarcane, papaya trees, and *rasa si* peach palms. Then after some time we abandon our old gardens and let the tangled vegetation overrun them and the trees grow back little by little. If you always replant in the same plot, the plants no longer yield. They become too hot, like bare land that has lost its forest smell. They become stunted and dry up. Then nothing comes anymore. This is why our elders moved around in the forest, from one garden to another, when their plantations declined and game became scarce around their houses.

BUT WE DO not clear our gardens just anywhere in the forest. We always choose a place where the image of *në rope* value of growth lives, where the earth is beautiful, dry and a little raised, protected from flooding. We avoid the ground that is too low and damp, overrun with vines and palm trees, where the plants that feed us have trouble growing. We choose places where we see that a garden is laid down on the forest ground.[6] So we prefer patches of forest where *poroa unahi* and *himara amohi* cacao trees, *wari mahi* kapok trees, *mahekoma hi* shrubs, and *krepu uhi* and *mani hi* trees grow, but also big *ruru asi* and *irokoma si* leaves. If we clear a new plot there, it will yield a lot of food. The gardens' *në rope* value of growth is always present in the forest's soil, like it was for our elders. It will only escape if the white people destroy all the trees with their machines and make all the forest inhabitants perish.

Deep soil is red and bad. Plants cannot grow stronger there. The forest's value of growth lives in the part of the soil at the surface. A damp breath of life comes out of it, which we call *wahari a*.[7] This cold exhalation comes from the darkness of the underworld, from its great river *Motu uri u* and from *Xiwãripo*, the chaos being. It belongs to the spirit of the forest, *Urihinari*. Its coolness mostly spreads during the night, for during the day it returns into the soil as soon as the sun becomes hot. This breath lives on because the earth's back is covered in leaves and protected by trees. We say that the forest is the earth's skin. If the white people tear it away with their tractors, there will soon only be gravel and sand left in the depths of the earth, and its dampness will disappear. This cool moisture from the ground is a **liquid** like sperm. It fertilizes trees by penetrating into their roots and seeds. This is what makes them grow and flower. If it dries up, the earth loses its smell of growth and gets barren. It no longer yields any food. But when this liquid impregnates the soil, it becomes black and beautiful. It releases a strong smell of forest. This liquid is also a food; this is why it makes the plants we eat grow. Garden patches were put down on the forest ground by the *Koyori* ant ancestor. So they take his image's fertility, and the plants that we cultivate there grow vigorously. It is so. The food we plant only yields well where the *Në roperi* image of growth dances, where the *koyo* ant spirits, the bat spirits, and the giant armadillo spirits come to play. When the forest is bad, no garden lies on its ground, and we say that it is a forest that became other.[8]

As I have said, the forest did not grow alone, without reason. Its *në rope* value of growth makes it alive and gives it its richness. The shaman elders often told me about it, and since my eyes learned to die under the power of the *yãkoana* I have also been able to see its image. It is the true owner of the forest and it knows how to prove generous. Yet if it decides to flee, nothing grows anymore, the ground becomes too hot, and the forest quickly takes value of hunger. The earth's skin is beautiful and sweet-smelling, but if you burn its trees, it dries out. Then the soil breaks up in friable clumps and the earthworms disappear. Do the white people know this? The spirits of the big earthworms own the forest earth. If you destroy them, it instantly becomes arid. Red soil appears below the black soil, out of which only shoots of bad plants and grass can grow. We never tear away the earth's skin. We only cultivate its surface, because that is where the richness is found. In doing so, we follow our ancestors' ways.

The trees' leaves and flowers never stop falling and accumulating on the ground in the forest. This is what gives it its smell and its value of growth. But this scent disappears quickly once the ground dries up and makes the streams disappear into its depths. It is so. As soon as you cut down tall trees such as the *wari mahi* kapok trees and the *hawari hi* Brazil nut trees, the forest's soil becomes hard and hot. It is these big trees that make the rainwater come and keep it in the ground.[9] The trees that the white people plant, the **mango trees**, the **coconut trees**, the **orange trees**, and the **cashew trees**, they do not know how to call the rain. They grow poorly, scattered around the city in a ghost state. This is why there is only water in the forest when it is healthy. As soon as its soil lies bare, the *Mot^{h}okari* sun being burns all its watercourses. He dries them out with his burning tongue before swallowing up all their fish and caimans. Then when his feet come close to the ground, the earth starts to bake and increasingly hardens. The mountain rocks become so hot they split and shatter. No tree can sprout out of the soil anymore, for there is not enough dampness left to keep seeds and roots cool. The waters return to the underworld and the dry earth crumbles. The wind being, who follows us in the forest to cool us like a fan, also flees. We stop seeing his daughters and nieces playing in the treetops. A stifling heat settles everywhere. The fallen leaves and flowers stiffen on the ground. The cool smell of the soil is consumed and vanishes. No plant will grow any longer, no matter what you do. The forest's *Në roperi* image of growth is angry and escapes far away. It goes toward other lands, among other people, or even on the sky's back with the ghosts. So where the white people have eaten the forest, they also end up suffering from heat, hunger, and thirst. Their ancestors did not give them any words of wisdom about the forest. This is why they finally can only escape it and return to the city.

THE FOREST belongs to *Omama*; this is why it has a very long breath of life, which we call *urihi a wixia*. This is what keeps up its breathing. On the contrary, humans' breath of life is very short. We do not live long; we usually die very fast. If it is not destroyed, the forest never dies. It is not like humans' bodies. It does not rot and disappear. It always becomes new again. The plants that feed us can grow because of the forest's breathing. So when we are sick to the point of being in a ghost state, we sometimes borrow the forest's breath of life so it supports and cures us. This is what the shamans do. The forest breathes, but the white people do not notice. They do not think that it is alive. Yet one only has to look at its trees with their leaves always shiny. If it did not breathe, they would be dry. This breath of life comes from the center of this earth we walk on, which is the back of the old sky *Hutukara*.[10] It spreads all over its surface and even in its rivers and streams. We call the image of the earth *Maxitari*. Wherever its damp coolness flows, the forest is beautiful, the rains abundant, and the wind vigorous.[11] The *xapiri* are happy to live in it because they were created together. And everything remained like this until now because our ancestors always knew the words I am giving here. But white people who clear the forest probably think that its beauty came to be for no reason? It isn't true! They only ravage it without worrying because they cannot see it with shamans' eyes. In the places that they occupy, all that remains are savannas and a soil that has lost its breath of life. But so long as we live here, this will not happen!

IN THE BEGINNING of time, there was no game in the forest. Only the *yarori* ancestors, who were human beings with animal names, existed. But the forest, which was still very young, soon turned to chaos, and they became other. These animal ancestors began to paint themselves with annatto dye and gradually changed into game.[12] Since then we, humans who came into being after they did, eat them. Yet at first we were all part of the same people. The tapirs, the peccaries, and the macaws that we hunt in the forest were once also humans. This is why today we are still the same kind as those to whom we give the name of game.[13]

The spider monkeys that we call *paxo* are humans like us. They are spider monkey humans, *yanomae tʰë pë paxo pë,* but we arrow them and smoke them when we gather the game for our *reahu* feasts. Despite this, in their eyes, we are still their fellow creatures. Though we are humans, they give us the same name they give themselves. This is why I think our inner part is identical to that of game and that we only attribute to ourselves the name of human beings by pretending to be so. Animals consider us their fellow creatures who live in houses while they are people of the forest. This is why they say "humans are the game that live in houses."[14]

They are really very clever! This is why they can understand us and hide when we approach. They find us frightening and think: "*Hou!* They are our fellow creatures but despite that these humans are so greedy for our flesh! They seem like evil beings! Yet they are people like us!" Armadillos, turtles, and deer are other humans, yet we still devour them! It is true. We who did not become game, we eat our people, our brothers, the tapirs, the peccaries, and all the others. In the beginning of time, there was no game. Our ancestors lived starved for meat and devoured each other. This is why they became other. They transformed into game so we could eat them properly.[15] It was so.

When animals see us hunting in the forest, they also call us *kõaa pë.* They call us that because we eat our own prey too often, though it is bad to do so.[16] As soon as we try to approach them, they see us from a distance and shout: "*Hou!* The *kõaa pë* are approaching to arrow us! How disgusting! They devour the game they have just shot themselves! Their mouths are so dirty!" Then they escape before we can even see them. It is true! By always eating their own leftovers, bad hunters move through the forest in a slumber. Though their eyes are open, they see nothing. Nor do they hear the game's voices. They are constantly seized with dizziness and lose all will to make arrows and hunt. They give off a sickly, nauseating smell.[17] The game fears it will get dirty by being in contact with them. This is why it never shows itself to these men. It remains apart and observes them from a distance while they wander in vain through the forest. These poor hunters only have one desire: to stay in their hammock and sleep. They venture into the forest from time to time, but they do not kill any game there. Their arrows get lost in the treetops and they leave them there out of laziness. They don't know the forest anymore, and the game doesn't like them.

But on the contrary, as soon as the animals see in the forest a hunter who generously gives away all the prey he arrows, they fall in love with him. So they go to meet him joyously, exclaiming: "*Pei! pei! pei! pei!* Here is the *Kãomari* falcon spirit! Here is a true *yawarioma* water being! This is our friend *Urihinamari*, the forest spirit! Look! A great hunter approaches!"[18] This is why game so easily shows itself to good hunters. These men do not need to see the game from a distance. It comes toward them to show itself to them. It feels nostalgia for the hunters the way a man misses a woman he is in love with. This is why it lets itself be arrowed without effort and is happy about it. It does not tell itself: "*Hou!* I am going to be killed, I will be in sorrow!" But on the contrary, if it is wounded by a clumsy hunter and has to escape in pain, it will be angry. It is so. A great hunter is always accompanied by the images of the *kãokãoma* falcon and the *yawarioma* water beings. They never leave him. He goes to sleep dreaming of them and wakes up happy to be thinking about them. They do not live in his chest. They accompany him from a distance, in the heights, without him knowing where they are. It is they who guide his arrows without his knowing. So he always comes back from his hunts bearing many catches.

Game can also get disgusted with humans if, after eating it, we carelessly throw its bones in the forest and its cooking juices into the streams. Its nostalgia for us instantly disappears, and after that we always come back from the hunt empty-handed. Our elders were far wiser than we are. Their wives carefully kept many game bones hanging in their hearths: monkey arm bones, peccary jaws, and curassow and great tinamou sternums. But this is no longer the case. Today we have become forgetful of the long-ago people's ways, and this makes us pitiful hunters compared to our fathers. It is so. These words are the little I know about game. They are the ones I heard from my elders when drinking the *yãkoana* and that I now give to my sons: "If you do not eat your prey, the game will be your friends. If you do not have any consideration for it, it won't like you either and you will always be clumsy at hunting." These words have always been inside us. This is why we do not kill game without restraint. But the white people do not know them and their elders wiped out the animals that lived on their land.

IN OUR VERY old language, what the white people call "nature" is *urihi a*, the forest-land, but also its image, which can only be seen by the sha-

mans and which we call *Urihinari*, the spirit of the forest. It is thanks
to this image that the trees are alive. So what we call the spirit of the forest consists of the innumerable images of the trees, of the leaves that
are their hair, and of the vines. It is also those of the game and the fish,
the bees, the turtles, the lizards, the worms, and even the *warama aka*
snails.[19] The image of the value of growth of the forest we know as *Në
roperi* is also what the white people call "nature." It was created with it
and gives it its richness. For us, the *xapiri* are the true owners of "nature," not human beings. The toad, caiman, and fish spirits—these children of the underwater being *Tëpërësiki*—are the masters of the rivers.
The macaw, parrot, tapir, and deer spirits, and all the other game spirits,
are the masters of the forest. It is so. The *xapiri* are constantly moving
around the entire forest without our knowing. Coming down from the
mountains to the lowlands, they make the winds rise up as they run and
play among the trees,[20] whether it is the dry season breeze, *iproko,* or the
rainy season wind, *yari.* The *maari* rain spirits also come down from the
sky to cool the earth under their downpours and chase away the *xawara*
epidemics.[21] If the *xapiri* stayed far away from us and the shamans never
made them dance, the forest would become far too hot for us to stay alive
for long. The *në wãri* evil beings and the *xawarari* epidemic beings would
settle next to our houses and would never stop devouring us.

The *xapiri* are friendly to the forest because it belongs to them and
makes them happy. The white people, they find "nature" beautiful without knowing why. On the contrary, we know that what they call "nature"
is the forest as well as all the *xapiri* who live in it. *Omama* created their
houses and paths there. He wanted us to protect them. The bee spirits
open their paths in the forest trees to reach the flowers for their honey.
The game spirits happily frolic in the coolness of its cover. The tapirs, the
spider monkeys, the peccaries, and the deer are fond of the shade of its
leaves and the breeze of its undergrowth. They like to quench their thirst
in its streams. When the heat is too strong, the game's images also suffer.
If the white people ravage the forest and destroy its hills and mountains,
the *xapiri* will lose their homes. Furious, they will flee far away from our
land, and human beings will remain there at the mercy of all ills. Even
with all their medicine and machines, the white people won't be able to
do anything about it. The spirits fear places that are too hot, such as the
savannas that surround our forest in the distance where *Mot^hokari,* the
sun evil being, truly lives. They are also afraid of the city and its stink of
car, airplane, and helicopter fumes.[22] They like to have fun as they run

through the forest and they revel in the cool dampness of its scent. They really enjoy its beauty and its fertility. They live under its cover and they feed on it; this is why they want to defend it, just like its human inhabitants. But the white people ignore that. They cut down and burn all the trees to feed their cattle. They dig in the beds of the watercourses and destroy the hills to look for gold. They blow up the big rocks that stand in the way of opening their roads. Yet hills and mountains are not simply put down on the ground, as I have said. These are spirit dwellings created by *Omama*![23] But these are words that the white people do not understand. They think that the forest is dead and empty, that "nature" is there for no reason and that it is mute. So they think that they can take it over to destroy the houses, paths, and food of the *xapiri* as they wish. They do not want to hear our words nor the spirits'. They want to remain deaf.

Yet even the *në wãri* evil beings that also live in its depths want to defend the forest. Their houses are located in places where our paths never go, and at the bottom of the big rivers and lakes. They are as numerous as the *xapiri* and are as furious as they are at the white people who devastate their paths and the animals they feed on. It is true! When they do not attack human beings, the evil beings of the forest eat game. They open their prey's belly, devour their innards, and collect their fat in *horokoto* gourds. This is why we sometimes encounter thin, sick monkeys and tapirs when we are hunting. As for the dry season being *Omoari*, he is fond of the honeys that abound in the dry season. He also smokes large quantities of fish and caimans from the dried-out streams to delight in them.[24] This is why he will want to take revenge on the white people who cut the trees of the forest and soil its rivers. As I have said, the dry weather does not occur without a reason. It begins with the arrival of *Omoari*, who comes to set fire to *Toorori*, the toad being of the wet season.[25] But after several moons, once *Toorori* has been able to gradually moisten his burned and stiffened skin, he comes out again. He spreads his waters in the forest to take revenge on *Omoari*. Then, frightened by the cold and dampness, *Omoari* runs away with his daughters and sons-in-law, the butterfly, cicada, and lizard beings. So the rainy season begins with the revenge of *Toorori*, who comes back to life and turns the lock of the waters to scare away *Omoari* and make the forest cooler and more pleasant. He repels the heat of the *xawara* epidemic, and the plants in the gardens start to grow again. The trees and the game recover from the dry weather, and the humans feel revivified. This is what the shamans know. And this is why, if the white people destroy the forest, the starving and angry *Omoari* will no longer leave it. The arid and blazing land that remains will forever be his domain.

OUR ELDERS knew how to call the images of *Omama* and the metal he owned in the beginning of time. This is why we continue to make them dance to defend the forest.[26] The white people speak of protecting the forest. In our language, which is the language of the spirits, we talk about *Omama*'s metal, for without it the earth would die out and its inhabitants with it. When we make dance the image of this metal of the sky, which is also the metal of "nature,"[27] we see it as a gigantic iron mountain, smooth and shiny, as tall as a rocky peak. *Omama* uses it to chase away the evil beings and the epidemic beings from the forest. The sharp blades that the macaw, parrot, and caiman spirits possess are also made of it.[28] The image of this metal is truly a very strong weapon for the shamans, for it is the power of "nature."[29] It is the spirits of the forest, the sky, and the storm wind all at once. It is surrounded by swirling gales that repel the dangerous fumes and lead the earth-eating white people's thought astray. This is why when the shamans of a house do not know how to make this metal's image come down and dance, its inhabitants constantly fall ill and die.

When they think their land is getting spoiled, the white people speak of "pollution." In our language, when sickness[30] spreads relentlessly through the forest, we say that *xawara* epidemic fumes have seized it and that it becomes ghost. The shamans must then all work together with the help of *Omama*'s spirit to replace its soiled image. First they tear away the forest's rotten floor and throw its shreds to the ends of the earth. After that they bring down the image of another ground, which is clear and healthy, to put down in its place. Then little by little they spread a fresh forest over this new land, covered with the shiny, fragrant paintings of the *xapiri*.[31] This renewal must also take place when a great shaman dies. This happens because the spirit of his ghost—whom we call *Poreporeri*— wants to avenge his father's death by rotting the land around his former house. My father-in-law and I have often made the image of *Omama* dance to tear up and then renovate our land made sick by the gold prospectors. *Omama* created our ancestors in the forest and gave them the *xapiri* to protect themselves from the *në wãri* evil beings. This is why today his image also defends us against the white people's epidemics and speaks angrily at their lack of wisdom: "You must stop ravaging the forest where my spirits, children, and sons-in-law live! The land where you were created is big enough! Keep living on your ancestors' traces!"

These words come from what the inhabitants of the cities call "nature." Yet they don't want to know anything about them. Their ears remain closed and their thought filled with smoke. They tell themselves that we must be ignorant and liars. They prefer contemplating the word drawings of the endless merchandise they desire. The beauty of the forest leaves them indifferent. They only repeat to us: "Your forest is dark and tangled! It is bad and full of dangerous things. Do not regret it! When we have cleared it all, we will give you cattle to eat! It will be much better! You will be happy!" But we answer them: "The animals you raise are unknown to us. We are hunters, we do not want to eat domestic animals! We find it nauseating and it makes us dizzy. We do not want your cattle because we would not know what to do with them. The forest has always raised the game and fish that we need to eat. It feeds their young and makes them grow with the fruit of its trees. We are happy that it is like this. They do need gardens to live, the way humans do. The earth's value of growth is enough to make their food flourish and ripen. As for the white people, they wipe out the game with their shotguns or scare it away with their machines. Then they burn the trees to plant grass everywhere to feed their cattle. Finally, when the forest's richness has disappeared and the grass itself no longer grows back, they must go elsewhere to feed their starving oxen."

In the beginning of time, our ancestors were still few in number. *Omama* gave them the plants of the gardens he had just got from his underwater father-in-law *Tëpërësiki.*[32] Then they began to plant and cultivate them while taking care of the forest. They did not tell themselves: "Let's clear all the tress to plant grass and dig in the ground to tear metal out of its depths!" On the contrary, they started eating the food plants from its soil and the fruit from its trees. This is what we are still doing to this day. Far from our land, the white people's ancestors became very numerous and lived with *Yoasi,* who taught them to destroy everything. As for our ancestors, they remained in the forest with *Omama,* who did not tell them to burn all its trees, nor to turn over its earth or spoil its watercourses. On the contrary, he gave them clean and beautiful land and rivers. He taught them to grow the plants of their gardens to soothe their children's hunger. He pierced the ground so the waters of the underworld could flow out of it and quench their thirst. He told them: "Eat the game, fish, and fruit of the forest! Eat what your gardens yield: bananas, manioc, sweet potatoes, yams, taros, and sugarcane." It is so. *Omama* gave them good words and made them think right. He taught them to take care of the forest in order not to chase its value of growth away. And so its beauty has been able to persist to the present day.

SINCE THE beginning of time, *Omama* has been the center of what the white people call **ecology.** It's true! Long before these words existed among them and they started to speak about them so much, they were already in us, though we did not name them in the same way.[33] For the shamans, these have always been words that came from the spirits to defend the forest.[34] If we had books like they do, the white people would see how old these words are! In the forest, we human beings are the "ecology." But it is equally the *xapiri*, the game, the trees, the rivers, the fish, the sky, the rain, the wind, and the sun! It is everything that came into being in the forest, far from the white people: everything that isn't surrounded by **fences** yet. The words of "ecology" are our ancient words, those *Omama* gave our ancestors at the beginning of time. The *xapiri* have defended the forest since it first came into being.[35] Our ancestors have never devastated it because they kept the spirits by their side. Is it not still as alive as it has always been? The white people who once ignored all these things are now starting to hear them a little. This is why some of them have invented new words to defend the forest. Now they call themselves "people of the ecology"[36] because they are worried to see their land getting increasingly hot.

As I said, our ancestors never thought to clear the forest or dig into its ground without measure. They simply thought that the forest was beautiful and that it must continue like this forever. For them, the words of "ecology" were to think that *Omama* had created the forest for human beings to live in it without destroying it. That is all. We are inhabitants of the forest. We were born in the middle of the "ecology" and we grew up in it. We have always heard its voice because it is the voice of the *xapiri* who come down from the mountains and hills of the forest. This is why we understood these new white people words as soon as they reached us. I explained them to my people and they thought: "*Haixopë!* This is good!

The white people call these things 'ecology'! As for us, we say *urihi a*, the forest, and we also speak of the *xapiri*, for without them, without 'ecology,' the land gets warmer and the epidemics and the evil beings get closer."

IN THE PAST, our elders were not able to make their words about the forest heard by the white people because they did not know their language. And when the white people first arrived to their houses, they surely did not speak about "ecology." They were more eager to ask them for jaguar, peccary, and deer skins! At that time, these outsiders did not possess any of these words to protect the forest. Such words appeared in the white people's cities not long ago. They must have finally told themselves: "*Hou!* We have soiled our land and our rivers, our forest is dwindling! We have to protect the little that remains under the name of 'ecology.'" I think they got scared by so badly ravaging the places where they live. In the beginning, when I was young, I never heard the white people talk about protecting "nature." It was much later, when I got angry and started making speeches about gold prospectors and their *xawara* epidemics that these new words suddenly came to my ears. I think that in Brazil it was Chico Mendes[37] who made them spread in every direction, for I first heard them when the white people started talking about him a lot. At that time, I was often shown his image on paper skins. Then I told myself: "This must be that white man who thought with wisdom and revealed to his fellows these new words of 'ecology'!" Before him, the people of the city did not care much about the forest. They did not talk about it much and never worried that it might be destroyed.

Chico Mendes was a white man, but he grew up like us, in the middle of the forest. He refused to cut down and burn all its trees. To live, he only extracted a little bit of their sap. He had taken the forest for a friend and loved its beauty. He wanted it to remain the way it was created. He constantly dreamed of it, worried to see it eaten by big cattle ranchers. This is probably how, one day, new words came to him to defend it. Perhaps *Omama*'s image put them into his dreams? Lying in his hammock at night, he must have told himself: "*Haixopë!* The forest feeds us with its fruit, its fish, its game, and the food plants of its gardens. I must speak to the other white people firmly and stop them from devastating it! I am going to stop those who want to clear and burn by using the words of the 'ecology' against them!" When these words were reported to me for the first time, I instantly thought: "This man is truly wise! His breath of life and his blood are like ours. Maybe he is a son-in-law of *Omama,* like we are?" Then I wanted to talk to him, but right before I could meet him, forest-eating white people killed him in an ambush. I had barely heard his clever words and he was already dead because of them! I had never listened to a white man say such things. What he said about his forest was beautiful and true. My thought was ready to receive such words, and it answered them at once. Thanks to them, I knew how to speak to the people of the big cities to defend our land. I think that Chico Mendes's words of wisdom will never disappear because after his death they spread in many other people's thought like they did in mine.

Before then, I had only met a few white people who cared about the forest and wanted to protect its animals. The first time was when I had just started working at the Demini Outpost and the FUNAI people asked me to join them to travel up the Rio Catrimani.[38] I was very young at the time. These white people wanted to defend the peccaries, the caimans, the otters, and the jaguars from the hunters who were wiping them out to pile up their hides. These were new words for me because in my child-hood the SPI was always asking our people for game skins. In those days, our elders hunted a great deal in vain, just to trade with these outsiders. But it had become different with the FUNAI men. Since I had started working with them on the Rio Mapulaú,[39] I had often heard them say that they wanted to expel from our land the white people who destroy game to skin it and those that exterminate turtles and river dolphins with their **harpoons.**[40]

During this trip up the Rio Catrimani, I saw the places downstream where these white hunters and **fishermen** had settled. These are also peo-ple who are constantly invading our forest. With FUNAI and the federal police, we stopped several of their canoes on the river to seize their jag-uar and giant otter skins. We also forced them to throw all their turtles back in the water. Their eyes were furious, but they did not protest be-cause they were scared of the police. I did not yet know the white people very well at the time. But I understood that those I was traveling with truly wanted to protect the animals and trees of the forest. It was the first time I heard such words. They made me think. I started to tell myself: "*Haixopë!* I am also going to defend the game so it does not disappear! Like us, the animals are inhabitants of the forest and they are not so many. If we let the white people hunt on our land, our children will soon cry in hunger for meat.[41] They speak the truth! The trees of the forest are beautiful and we eat their many fruit. It hurts to cut them down in vain." After this trip, time passed and I became a grown man. My ideas on the forest continued to develop little by little until, much later, I listened to Chico Mendes's words. This is how I learned to know the white people's words about what they call "nature." My thought became clearer and higher. It spread a lot. I began to understand that it was not enough to protect only the small area where we live. So I decided to speak up to defend the entire forest, including the one human beings do not inhabit[42] and even the white people's land very far beyond us. In our language, all this is *urihi a pree*—the great forest-land. I think it is what the white people call the entire world.[43]

Once the talk about "ecology" appeared in the cities, our words about the forest also began to be able to be heard there. The white people started listening to me and telling themselves: "*Haixopë!* So it's true, the inhabitants of the forest's ancestors already possessed 'ecology'!" Our words were then able to spread very far from our houses, drawn on image skins or captured on television images. This is why today our thoughts are no longer as hidden as they once were. In the past, we used to be as invisible to the white people as the turtles on the forest floor. They had never even heard our name sounded out. It is no longer so. When I was still young, I decided to go far away from my home to bring these words of my elders out of the forest's silence. At the beginning, I did not know much. Yet by drinking the *yãkoana* and becoming a shaman, my image traveled with the spirits of the forest and I became wiser. With them, I understood that our land could truly be destroyed by the white people. Then I decided to defend it, and I thought: "This is good! Now that the white people have invented their 'ecology' talk, they must not only repeat it in vain to make new lies out of it. They must really protect the forest and all those who live there: the game, the fish, the spirits, and the humans." I am a son of the first inhabitants of the forest and these words have become mine. Now I want to make them be heard by the white people so they can take them up.

In the past, our elders did not say: "We are going to protect the forest!" They only thought one thing: "The spirits left us by *Omama* and his son, the first shaman, are taking care of us!" It was good that way. The *xapiri* already owned the "ecology" while the white people did not yet speak of it. The spirits have always fought the *në wãri* evil beings, calmed the thunders, and prevented the earth from turning into chaos and the sky from falling. It is also they who call the rain beings to clean the forest when it is so hot that humans become ghosts or who make them go back up into the sky's chest when the dry season is too slow to come and women and children are hungry for game. It is they who open the storm wind door underground to push the epidemic fumes far from the forest. These rain and wind beings are all fathers of the "ecology," just like the spirits of the forest and sky.

It was so. Our ancestors knew the *xapiri*'s words, but not those of the "ecology" that the white people created much later, on their own and far away from us. I had never heard them either when I was young. But since the spirits knew the "ecology" before the white people gave it this name and our shaman elders already knew these things, I quickly understood this new talk. As soon as I listened to it for the first time, I told myself: "Didn't the white people who utter these words also get them from *Omama*'s mouth? Aren't they the image of the first outsiders whom *Omama* created long ago from the blood of our ancestors? Aren't they *Omama*'s sons-in-law?"

WHEN THEY speak about the forest, these white people often use another word: they call it **"environment."**[44] This word is also not ours and until recently we did not know it either. For us, what the white people refer to in this way is what remains of the forest and land that were hurt by their machines.[45] It is what remains of everything they have destroyed so far. I don't like this word. The earth cannot be split apart as if the forest were just a leftover part.[46] We are inhabitants of this forest, and if it is cut apart this way, we know that we will die with it. I would prefer the white people to talk about "nature" or "ecology" as a whole thing. If we defend the entire forest, it will stay alive. If we cut it down only to protect small parcels that are leftovers of what was ruined, it will yield nothing. With leftover trees and leftover watercourses, leftover game, fish, and humans who live there, its breath of life will become too short. This is why we are so worried. If today the white people start talking about protecting "nature," they must not lie to us again like their elders did.

We shamans simply say that we are protecting "nature" as a whole thing. We defend the forest's trees, hills, mountains, and rivers; its fish, game, spirits, and human inhabitants. We even defend the land of the white people beyond it and all those who live there. These are words of our spirits, and they are ours. The *xapiri* are truly defenders of the forest, and they give us their wisdom. By making them come down and dance, our elders have always protected all of "nature." And we who are their sons and grandsons do not want to live in a leftover of the forest. In their land, the white people have already cleared nearly all of the trees. They only keep a few patches of forest, which they have enclosed in fences. I think that now they intend to do the same thing in our land. This makes us sad and worried. We do not want our forest to be ravaged and the white people to finally leave us a few scattered little patches of what remains of our own land. There will soon be no more game nor fish, no wind nor coolness in these scraps of sick forest with muddy rivers. All the forest's value of growth will be gone. The *xapiri* do not want to see us live in shreds of forest but in a vast and undamaged forest. I do not want my people to live in a leftover of the forest, or for us to become leftovers of human beings.

Unlike the white people, we do not eat the forest and its earth. We feed on its game, its fish, the fruit of its trees, the honey of its bees, and the plants of our gardens. This is how we sate our wives' and children's hunger. It's very good like this. You cannot fill your belly by clearing and burning all the forest! This only attracts *Ohiri,* the hunger spirit, and the man-eater beings of the epidemics, nothing else. If we mistreat the forest, it will become our enemy. The white people elders of the past already destroyed a big part of it by following the words of *Yoasi,* the one who put death into our breath. Today their sons and grandsons must finally listen to the clear words of *Omama,* who created the forest and the *xapiri* to defend it. What the white people call "nature's protection" is actually us, the forest people, those who have lived under the cover of its trees since the beginning of time. Yet the gold prospectors' and the cattle ranchers' thought is that of evil beings. They never stop calling us ignorant and useless, but contrary to what they believe, we probably lack a lot less knowledge than they do. We have friendship for the forest because we know that the *xapiri* spirits are its true owners. The white people only know how to abuse and spoil it. They destroy everything in the forest—

the earth, the trees, the hills, the rivers—until they have made its ground
bare and blazing hot, until they even make themselves starve. On the
contrary, we never die of hunger in the forest. We only die of the white
people's *xawara* epidemic fumes in our houses.

WHEN I REALLY started defending the forest, white friends invited me
to Brasília to give me what they call a **prize**.[47] There were many of them
looking at me and listening to my words. They wanted to hear what a son
of the forest's inhabitants had to say. They knew I was angry at the gold
prospectors who were eating our land. After my speech, they told me they
were happy to have heard me. This way they made my words more solid
and helped them to spread farther than they had been able to before.
They also probably thought that with this prize, the gold prospectors
might hesitate to kill me. So they protected me from death. At that time,
my path out of the forest was still narrow. They made it bigger and gave
me courage to fight. My words kept multiplying and became stronger.
They began to be heard by white people who live in more and more distant lands. So I told those who had given me this prize that I was glad to
receive it but that at the same time I was sad because my people were dying. I also told them that even if this prize had great value for the forest
people, it was more important for the white people to make this value live
in their chests.

Yet many are those who continue to ignore what we say. While our
words sometimes reach their ears, their thought remains closed. Maybe
their children and grandchildren will be able to hear them one day? Then
they will tell themselves that these are words of truth, clear and right.
They will see how beautiful the forest is and understand that its inhabitants want to go on living there like their elders before them. They will
become aware that it was not the white people who created it, not it nor
its inhabitants, and that once they are destroyed their governments will
not be able to bring them back into being. If the white people became
wiser at last, my mind could become calm and joyful again. I would tell
myself: "This is good! The white people have found some wisdom. They
have finally become friends with the forest, the human beings, and the
xapiri spirits." My travels would then come to an end. I would have done
more than enough speaking far away from my house and filled more

than enough paper skins with the drawing of my words. I would only go visit the white people's land once in a while. I would tell my friends there: "Don't call me so often anymore! I only want to become spirit and continue to study with the *xapiri*. I only want to get more knowledge." Then I would hide in the forest with my elders to drink the *yãkoana* until it made me very thin again and I forgot the city.

CHAPTER 24

The Shamans' Death

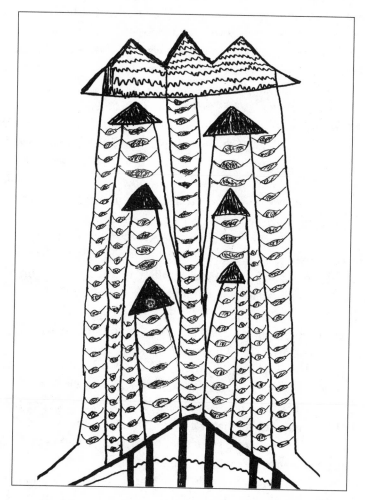

Spirits' house of a great shaman

W HEN A SHAMAN becomes very old and wants to stop living, or when he is very sick and dying, his spirits finally abandon him. Then he remains alone and empty before dying away like a firebrand in the hearth. Once left to decay, his spirit house will collapse by itself. This is how things go. The *xapiri* flee far away from their father as soon as he dies. They return to where they lived in the past, in all the hills and mountains of the forest and on the sky's back. They will only come back to visit human beings much later, in order to dance for a new shaman, often the son of the dead man they left.[1] Yet not all the spirits move away from their dying father as hastily. Some remain by his side until his last breath. They only leave in the grip of anger, at the moment he dies. These stronger *xapiri* are the jaguar spirit, who supports the dying man and gives him courage, the moon spirit, who always keeps his eyes wide open on him, and *Aiamori*, the spirit of war bravery, who only leaves his dying father at the very last moment.

But other *xapiri* even refuse to leave once the deceased shaman's ghost has left for the sky's back. These are very powerful and hostile spirits. *Omoari*, the dry season spirit, defends his dead father's tobacco with his arrows, protesting against his grieving kin who want to destroy his plantation: "No! Do not burn this tobacco, it belongs to me! Do not attack me, I did not cause his death!" There are also the spirits of the *Koimari* bird of prey evil being, *Poreporeri* ghost, *Mot^hokari* sun being, *Titiri* night being, and many others too.[2] These *xapiri* are very dangerous and want to stay by their dead father's traces at all cost. This is why when a shaman elder dies we abandon and burn the house where the ashes of his bones were buried. We rebuild a new one farther away, to continue to live there safe from his hostile spirits.

If we did not do this, we would be unable to avoid relentless attacks from these evil *xapiri*. Yet when we try to keep them at bay, they angrily protest: "*Ma!* Do not burn our father's house! We are innocent of his death! Move away! We want to continue living here in silence!" Then they try to rebuild their own houses in the area and, regaining strength, they attack the human beings that walk through the nearby gardens. This is what the *Omoari* dry-season spirit does by arrowing children in the chest, the anaconda spirits by eating women's vulvas to make them die while giving birth, the *remori* bee spirits by tearing out people's tongues or making their eyes swell, the *mõeri* dizziness spirits by hitting the back of one's neck with their clubs, and the *motu uri* underworld water spirits by drowning us in the rivers.

It also takes a great deal of effort to dissuade the ghost of a shaman who has just died from returning in his father's footsteps and harming humans. This is why shortly after such a death we pour the deceased's bone ashes in a hole in the ground dug at the foot of one of the house's posts, near its hearth. We also add tobacco, banana soup, and *yãkoana* powder over it in order to calm the ghost's anger. Then we close the hole with a stone that we cover in earth, carefully packing it down with our heels. After that the other shamans of the house try to get the ghost's gaze away from his relatives. They tell him: "Cast your eyes in the distance and go away! You will find what you need to eat and women with whom to copulate in that direction, in our enemies' houses! That is where you will find those responsible for this death!" If we did not do that, the living would not remain in good health for long. It is true. The ghosts of shamans who owned very tall spirit houses are very aggressive. Their hostile spirits do not hesitate to crush humans and devour them like game to avenge their father. We are very afraid of them. But this is not the case with ordinary people's ghosts. These simply return to the sky's back, nothing more. Their death cannot be avenged by the *xapiri*, only by their kin.

But one must not believe that our elders passed away simply of very old age. No, they were all devoured by the white people's sicknesses, one after the other! Since my childhood, most of my kin disappeared this way, and the anger of these losses has never ended. Today most of our great shamans have died. The *xawarari* epidemic beings destroyed their spirit houses one after the other, and they were unable to get revenge. This is what made them perish. In the past, our elders never thought such a thing could happen. As soon as the gold prospectors invaded the forest, the oldest shamans tried to repel their epidemic fumes, which were voraciously attacking our women and children. But despite all their efforts, they failed. Instead the *xawarari* evil beings, furious at their attack, killed them wherever they could. This is why older shamans, once so many in our forest, are now so few in our houses. Young people still make the spirits dance here and there, but our most reputed shamans no longer make their songs heard, especially in the highlands. This is why we feel so worried.

The dead shamans' hostile *xapiri* become more and more numerous, furious to see their fathers' spirit houses destroyed by the white people's ignorance. They constantly ask the spirits of the living shamans: "Who ate our father? Tell us! Do not be afraid!" To divert their vengeance, these spirits answer: "*Ma!* Do not mistreat the inhabitants of the forest who are our people! None of them killed your father! Your anger does not come from here! Go devour white people instead! They are the true killers in õnokae state!" This is the reason these hostile *xapiri* angrily knock trees down on the gold prospectors, swell the waters to drown them, and make the earth slide to bury them. They even crash their airplanes into the forest. It's true! The spirit of the ancient ghost *Porepatari* suddenly stands up in their path in the sky and throws them into the void. This happened on the upper Rio Mucajaí when a very great shaman of the *Hero u* River died of malaria.[3] Several gold prospectors' airplanes crashed in the treetops at that time. I saw their wrecked shells abandoned in the forest with my own eyes.

If they continue to be so hostile towards us, the white people will finally kill our few remaining shaman elders. For us these men who have known how to become spirits for so long have a very high value. They constantly drink the *yãkoana* to cure us and protect us. They push back the evil beings, prevent the forest from being torn apart, and consolidate the sky as soon as it threatens to fall. In the beginning of time, *Omama* taught them to become spirit through his son before he left to go downstream of all the rivers. Much later, the white people he created there with the foam of our ancestors' blood returned to the forest where we live. They became more and more numerous there and started eating our people with their weapons and their *xawara* epidemics. Then nearly all of our great shamans died. It is very frightening, because if they all disappear the earth and the sky will end up turning to chaos. This is why I would like the white people to hear our words and dream about all that they say: if the shamans' songs stop being heard in the forest, white people will not be spared any more than we will.

IT IS TRUE. The shamans do not only repel the dangerous things to protect the inhabitants of the forest. They also work to protect the white people who live under the same sky. This is why if all those who know how to call the *xapiri* die, the white people will remain alone and helpless on their ravaged land, assailed by a multitude of evil beings they don't know. Their doctors will not be able to do anything about it, no matter how many and clever they are. Then, little by little, they will be wiped out like we will have been before them. If they persist in devastating the forest, all the unknown and dangerous beings that inhabit and defend it will take revenge. They will devour them relentlessly with the same voracity with which their epidemic fumes devoured our people. They will burn down their land, smash their houses in storm winds, or drown them in floods of mud and water. All this could truly happen one day if all the remaining shamans die and their orphan *xapiri* run away, enraged by their fathers' death. Then there would soon be nothing left in the forest—in what white people call "nature"—save the *në wãri* evil beings. They already make their threats heard: "*Ma!* If the inhabitants of the forest all die, we will stay to avenge them! We will not let the white people who devoured them survive!" This is what the *xapiri* sometimes tell me during my sleep, after I have drunk the *yãkoana* all day long. Then I think: "*Haixopë!* If the white people's *xawara* epidemics kill us all, our death may well be avenged by the *në wãri* evil beings!" The white people can doubt my words: "Does he really hear these words of the spirits he claims to pass on?" Yet they are true words! I am a shaman, like my elders, and the *xapiri* always come to me in the time of dream. I contemplate their beauty and I listen to their songs in the silence of the night, lying in my hammock, and this is how my thought spreads and becomes stronger.

Without the shamans, the forest remains fragile and does not stay in place on its own. The waters from the underworld soften its soil and always threaten to rise up and tear it apart. Its center is only steady because the weight of the mountains holds it in place. But the edges of the terrestrial layer loudly sway in the void, pushed by great storm winds. If the *xawarari* epidemic beings continue to invade our land, the shamans will all die and no one will be able to stop the forest from turning to chaos anymore. *Maxitari*, the earth being, *Ruëri*, the cloudy weather being, and *Titiri*, the night being, will get angry. They will mourn the shamans' death and the forest will become other. The sky will soon be covered in dark clouds, and the sun will never rise again. It will never stop raining. A gale of wind will blow unceasingly. The forest will no longer know silence and calm. The furious voice of the thunders will endlessly rumble while the lightning beings' feet never stop landing on the earth. Then the ground will split open, and all the trees will collapse on top of each other. In the cities, the buildings and the airplanes will also fall down. This has happened before, but the white people never ask themselves why. They do not worry about it much. They only want to continue digging in the ground looking for minerals until they meet *Xiwãripo*, the chaos being. If they get that far, this time there will be no more shamans to push back *Titiri*, the night being. The forest will become dark and cold and will remain so forever. It will no longer have any friendship for us. Giant wasps will swoop down on humans, and their sting will turn them into peccaries.[4] The gold prospectors will die one after another, bitten by snakes fallen from the sky or devoured by jaguars appearing from everywhere in the forest. Their airplanes will be caught in the tall trees and break up. The soil will soak up water and start to rot. Then the waters will gradually cover the entire earth and the humans will become other, just as it happened in the beginning of time.[5]

When the white people tear dangerous minerals out of the depths of the earth, our breath becomes too short and we die very quickly. We do not simply get sick like long ago when we were alone in the forest. This time, all our flesh and even our ghosts are soiled by the *xawara* epidemic smoke that burns us. This is why our dead shaman elders are angry and want to protect us. If the breath of life of all of our people dies out, the forest will become empty and silent. Our ghosts will then go to join all those who live on the sky's back, already in very large numbers. The sky, which is as sick from the white people's fumes as we are, will start moaning and begin to break apart. All the orphan spirits of the last shamans will chop it up with their axes.[6] In a rage, they will throw its broken pieces on the earth to avenge their dead fathers. One by one they will cut all its points of support, and it will collapse from end to end. For this time there won't be a single shaman left to hold it up. It will truly be terrifying! The back of the sky bears a forest as vast as ours, and its enormous weight will brutally crush us all. The entire ground on which we walk will be carried away into the underworld where our ghosts will become *aõpatari* ancestors in their turn. We will perish before we even notice. No one will have the time to scream or cry. The angry orphan *xapiri* will also smash the sun, the moon, and the stars. Then the sky will remain dark for all time.

Our spirits are already talking about all this, even if the white people are convinced all these words are lies. The *xapiri* and *Omama*'s image try to warn them: "If you destroy the forest, the sky will break and it will fall on the earth again!" They do not pay any attention to them, for they do not drink the *yãkoana*. Yet their skill with machines will not allow them to hold up the falling sky and repair the spoiled forest. They do not seem to worry about disappearing either, probably because they are so very many. But if we peoples of the forest are no longer, the white people will never be able to replace us there, living on the old traces of our houses and abandoned gardens. They will perish in their turn, crushed by the falling sky. Nothing will remain. It is so. As long as there are shamans alive, their *xapiri* will be able to quiet the sky when it threatens to come apart and hold back its fall. If all the shamans die, the sky will break apart for good, and no one will be able to do anything. That is why, for us, what the white people call **"future"** is to protect the sky from the *xawara* epidemic fumes to keep it healthy and strongly fastened above us.

LATER, PERHAPS all of us who inhabit the forest will die. But the white people should not think that we will perish alone. If we disappear, they will not live very long after us. Even if they are very numerous, they are no more made of stone than we are. Their breath of life is as short as ours. They can get rid of all of us today, but later, when they want to settle in the places where we lived, they will be devoured by all sorts of fierce evil beings. And as soon as they break the ancient spirits' mirrors by ravaging the forest ground, these angry *xapiri* will also take revenge on them. They are already comforting us: "Do not be afraid! Do not fear death! Even if the white people think they can increase in numbers without any limit, we will put them to the test. We'll see if they are as powerful as they think! We will throw them into the darkness and storm! We will smash the sky and its fall will carry them away!" This is what the *xapiri* tell us now when their images speak to us in the time of dream. They do not lie. They are valiant warriors who never warn us in vain. The death of so many of their fathers puts them in a fury and fuels their desire for revenge.

THE GREAT shamans' hostile spirits always want to devour humans. It is so, even though the white people do not suspect it. This is why we try to warn them: "Finally get some wisdom! Give up on our forest because once we are all dead and you try to build cities on our forgotten traces, you will end up destroying yourselves. The orphan *xapiri* of the shamans that you have killed will take revenge on you. Then you will search for the cause of your torments in vain, and your thought will remain lost!" My father-in-law, who is a true elder, often speaks to me about all this, and his words made me think right. All these bad things could truly happen if the songs of the spirits are no longer heard rising up from our houses.

This is why, like all shamans, I am not scared of dying. I do not fear the weapons of the gold prospectors and the cattle ranchers.[7] I know that after my death my *xapiri* will soon be able to find the one who killed me. Then his kin will have to shut his body in a wood box and hide it in the ground. He will really be pitiful! On the contrary, my ghost will not be crushed by the weight of earth. It will return to the sky's back, where it will be able to live again. My bones will not remain abandoned in the forest's dampness. My people will burn them and put their ashes in oblivion in our house with their guests. Only my most dangerous *xapiri* will stay behind to avenge me. As I have said, this is what happens when shamans die.

I often listen to the words of my spirits, who angrily ask themselves: "Why are the white people so hostile to us? Why do they want us to die? What do they have against us who do not mistreat them? Is it simply because we are other people, inhabitants of the forest? Do not worry, they may kill you, but they themselves will not long remain safe!" It is so. We are sad at the idea of disappearing, yet our thought stays quiet because we think that the *xapiri* are innumerable and will never die. We suffer now, but we know that these spirits will never abandon our traces to the cold and that the fumes from our burning bones will become a *xawara* epidemic for those who have killed us. This is why after mourning our dead a great deal, our pain is soothed as soon as their ghosts have returned to the sky's back. Then our thought regains its calm and its strength. We can then laugh and joke again.

Today there are few great shamans still alive in the forest, but there are more and more angry orphan spirits of dead shamans. This is also why we want the white people to stop mistreating the forest. It makes us angry, yes, but we will not arrow them for that. They think that they are so clever and powerful with their image skins, their machines, and their merchandise. Yet later, if the hostile spirits of the countless dead shamans attack their cities, they will be pitiful for having ignored our warnings, and they will not understand what is happening. With these words, I simply want to make them aware that the evil things that they like to tear out of the ground will not make them rich for long. The value of all the dead shamans will be very high, and they will certainly not be able to compensate for it with their paper skins. As I have said, no price can buy the land, the forest, the hills, and the rivers. Their money will be worthless against the value of the shamans and of the *xapiri*. They must become aware of this! Since *Omama* gave the spirits' words to our ancestors, we have kept them to protect ourselves. If the white people do not make us all die, we will continue to call them to strengthen the forest and prevent the sky from falling again.

THE GHOSTS of the dead shaman elders and their hostile spirits have already started taking revenge on distant lands by causing constant droughts and floods there. The *Hutukarari* sky spirit, the *Yariporari* storm wind spirit, the *Mot^{h}okari* sun spirit, the *Maari* rain spirit, the *Yãpirari* lightning bolt spirit, the *Yãrimari* thunder spirit, and the *Xiwãripo* chaos spirit are already furious at the white people who mistreat the forest so much. It is so. The forest is wise, its thinking is the same as ours. This is why it knows how to defend itself with its *xapiri* and its *në wãri* evil beings. The only reason it does not yet turn to chaos is that a few great shamans are still making their powerful spirits dance to protect it. But today, as I said, more and more angry orphan *xapiri* are released as their fathers are devoured one after the other by the *xawara* epidemics. The spirits of the living shamans can still contain them for the time being. But without their work, the forest and the sky will not be able to stay in place very long and remain as quiet and calm as we have known them until now.

Some time ago, a great shaman whom I called father died in Ajuricaba on the Rio Demini. As soon as he passed, his orphan spirits started cutting up the sky in a fury. Huge pieces of its carved chest started to come loose and shake with a rumble, about to break off and collapse onto the earth. You could hear loud muffled cracks succeeding each other over the forest. All the inhabitants of the deceased's house were crying in fear. I started working with my father-in-law and two other shamans right away. We hastily drank *yãkoana* and brought down multitudes of *xapiri* from everywhere to help us. So we instantly sent them onto the sky's back in order to secure its broken pieces with iron vines. It was truly frightening. I really thought that this time the falling sky would drag us into the underworld! To successfully repair it, we finally had to call the images of the chaos being *Xiwãripo* and of *Omama*'s metal.

The first time my stepfather made me drink the *yãkoana*, I already saw the image of the sky breaking and falling. I also heard its complaints: "Later, if there are no shamans left in the forest to hold me up, I will collapse on earth again, like in the beginning of time! But this time the people whom I will allow to live on my back will not be those earth-eating outsiders who are so hostile!" Since then, I have often heard it in dream making frightening cracking sounds and threatening to come apart. The orphan spirits of the dead shaman elders have been hacking at it for so long! It is covered in wounds and jagged with loose broken patches. If all the remaining shamans are finally devoured by the *xawara* epidemic, it will end up falling like it did the first time, when it was still young and not very solid. It might take some time, but I think it will really happen one day. The white people are increasingly burning the sky's chest with their metal fumes, and the *xapiri* are constantly trying to cure it by pouring torrents of mountain water on it.[8] Yet if there are no more shamans left in the forest, it will soon burn up until it becomes blind. Finally, it will suffocate and, becoming ghost, will suddenly start falling onto the earth. Then we will all be carried away into the darkness of the underworld, both the white people and the rest of us.

After that, it may be that in a very long time other people will come into being where we once existed. But they will certainly be other inhabitants of the forest and other outsiders. These are our elders' words about the "future." The white people should also think of all this when they dream. Maybe they would finally understand all those things of which the shamans often speak among themselves. But they should not believe that we are only concerned with our houses and our forest or the gold prospectors and cattle ranchers who are destroying it. Beyond our own fate, we also worry about the entire world, which could well turn to chaos. Unlike us, the white people are not afraid to be crushed by the falling sky. But one day they may fear that as much as we do! The shamans know a great deal about the bad things that threaten human beings.

There is only one sky and we must take care of it, for if it becomes sick, everything will come to an end. This may not take place right now, but it could happen later. Then it will be our children, their children, or the children of their children who will die. This is why I tell the white people these words of warning that I have heard from very great shamans. Through them, I want to make them understand that they should dream further and pay attention to the voices of the forest's spirits. But I know very well that most of them will remain deaf to my words. They are other people. They do not hear us or do not want to listen to us. They think this warning is just a lie. But this is not the case. Our words are very old. If we were ignorant about these things, we would remain silent. But on the contrary, it is the thought of the white people, who know nothing of the *xapiri* and the forest, that seems full of oblivion. In any case, even if they do not listen to my words while I am alive, I am leaving the drawing of these words on this paper skin so that their children and those who are born after them can one day see and understand them. Then they will discover the thought of the Yanomami shamans and know how much we wanted to defend the forest.

Words of *Omama*

WHEN I WAS young and not yet a shaman, I did not know how to dream. I was ignorant and I slept like a stone on the ground. I was unable to see the things of the forest in my sleep. Later I understood that I should not forget the words of *Omama*, which have come down to us from the beginning of time. So I asked my elders to give me the *xapiri*'s songs so I could truly dream. Before then, what I was able to see when I dreamed was too close. I did not yet possess the dream of the spirits inside me, which allows the shamans' image to travel in the distance. I could see neither the things of our ancestors' time nor what the thunder, sky, moon, sun, rain, darkness, and light truly were. I knew nothing. It was only after I had been drinking the *yãkoana* for a long time that I was finally able to know the images of all these things. As I said, this is how the inhabitants of the forest study, by becoming spirits and dreaming. The white people are other people. The *yãkoana* is not a good thing for them. If they start to drink it on their own, the angry *xapiri* will only tangle their thought and make their stomachs fall in fear. Its image, which we call *Yãkoanari*, only has friendship for the people of the forest.

Once I became a shaman, I started to know the *xapiri* better and to be able to expand my thought. Since then, I have constantly called them and made their images come down. I rarely ever sleep without answering their songs during the night. I always see them dancing with joyous clamors in my dreams. When I was a youth and I still knew nothing of the spirits, I sometimes thought that the elders might be singing for no reason. I even asked myself if they might be lying. But later, after having felt the dangerous strength of the *yãkoana* for myself, I realized this was not so. I began to understand that they were really answering the *xapiri*'s songs. I told myself: "If they did not really see the spirits, they would fear the power of the *yãkoana* and stop drinking it in vain. They do not lie; they truly report the words of the distant lands where their spirits come from. This is the truth!"

Our elders open the *xapiri*'s paths towards us and have their houses built for us. Then, if your chest isn't dirty and you answer the spirits' songs well, they feel happy and settle by your side for a long time. It has been so since our ancestors were created and began imitating *Omama*'s son, the first shaman. And since then elder shamans went on blowing the *yãkoana* in younger men's noses with their *xapiri*'s breath. This is why we are able to see the spirits to this day. A great shaman who owns many *xapiri* cannot jealously keep them for himself. If a young man asks for them, he cannot refuse him. He must open their paths and pass their breath of life on to him. This is how I was able to receive my father-in-law's spirits when I wanted to become a shaman myself. He proved generous towards me because his *xapiri* are so very many and their spirit house is higher than the sky.

As adults, we live for a long time during which we are still young. Then little by little we age, and unless enemy sorcerers break our bones or the *xawara* epidemic devours us, we get very old and die rightly. We do not pass on so soon. During all this time, the *xapiri* protect us and cure us of all our sicknesses. This is why I often tell my elders: "I want your spirits so I can cure the people of our house when you are no longer here. You are great shamans, do not be greedy. The *xapiri* are not like us; they never die. You cannot refuse them to me!"

SINCE MY father-in-law started giving me his *xapiri,* I have acquired a lot of their words. Yet there are still many that I have to learn because the *xapiri* are innumerable and their songs have no end. This is why I will go on studying with my father-in-law again later.[1] Several other shamans will do it with me. He will call us to him to drink the *yãkoana* with him. He will then give us new spirits so that we are not lacking them after his death. Today there are not many great shamans like him left in our forest. The gold smoke nearly completely emptied it. Even in our house at *Watoriki,* my father-in-law is the only elder who survived. This is why I am eager to continue taking the *yãkoana* with him. As soon as I need to travel a little less to defend our forest, I will ask him for a great many more of his *xapiri*.

For the moment, he tells me: "You are young, wait a while. You are still too weak and hungry for meat. Thinking about the white people's misdeeds too much has tangled your thought. Do not be impatient. Later, the more ancient *xapiri* will come to you, and little by little your spirit house will grow. Soon you will be able to make dance the *hutukarari* sky spirits, the *maamari* rocky peak spirits, and the *yãrimari* thunder spirits, then the *warusinari* insect spirits of the new sky and the evil *koimari* bird of prey spirits." I answer him: "*Haixopë!* These are good words! I will ask for other *xapiri* when quietness has returned to my thought." For now, the white people are still trying to invade our land. I have to travel and speak to them harshly everywhere I can in their cities. But if I was like the other inhabitants of the forest and I did not know these outsiders' ghost talk, I would much rather stay in my house to keep drinking the *yãkoana* and making the *xapiri* dance!

My father-in-law and the other *Watoriki* shamans also tell me: "Do not ask us for new spirits at this time! If you receive dangerous *xapiri* from us, you will want them to attack the white people. That will not be a good thing. Your thought is worried, and once you are angry you will become aggressive. As soon as these outsiders make you hear their bad talk, you will want to send your evil spirits to devour them and their kin will seek revenge!" These are wise words, I admit it. If I call new *xapiri* without being able to answer their songs well, they will be unhappy and will want to mistreat me. And it is true, if I am angry too often, my evil *xapiri* could attack the white people despite me. This is why I hesitate to ask my elders for new spirits at this time. I will do it later, when my distant travels come to an end. For now, it is hard for me to stay quietly at home. The white people who are my friends often call me to defend the forest, very far away from my house at *Watoriki*. So sometimes I worry by thinking that I could grow old without learning anything new. My thought is still searching. To truly know the spirits, I would still need to study a long time with the *yãkoana*.[2] But I will do so later, when my mind has calmed down—as soon as it is no longer darkened by the twisted words of the gold prospectors and the cattle ranchers. When they finally stop soiling the rivers, cutting down trees, and spreading their epidemic fumes, I will quietly drink the *yãkoana* in the silent forest, and my attention will be focused on the spirits' words again.

DESPITE MY trips to distant lands, my *xapiri* are not angry. They continue to dance when I call them down and to live in their spirit house. Yet some of them who were once settled in its highest part, above the sky, finally left me. They probably ran away because of the strange food the white people gave me to eat. This fatty and salty food weakens the chest, and its frightening smells chase the spirits away. Even the water of the cities is strangely bland, though it is so clear. And also, because I keep traveling so far away, some of my *xapiri* were carried away as down feathers by the blast of the airplane motors.

Besides, I have not yet been able to obtain the *xapiri* that our elders appreciate above all others—those of the *ayokora* cacique bird.[3] As I have said, these spirits are truly the most skilled at curing. They see under sick people's skin and know how to extract the harmful things from their body without cutting it up, unlike white doctors. As soon as an evil being introduces his fever and pain things into someone, the *ayokora* cacique bird spirits can instantly tear them out and have their father, the shaman, spit them out. In the time when the white people's medicine did not exist among us, these *xapiri* were truly favored by our elders. This is why I want to own them in my turn. I do not want to be a shaman and deceive my people! I often saw my elders cure with those powerful spirits, and I want to truly follow in their footsteps.

Yet my father was an ordinary man and did not know how to make these *ayokora* cacique bird spirits dance. So they never came to visit me of their own will. I had to ask my Toototobi stepfather to make them come down to me.[4] He too proved very generous and gave them to me without delay. To do so, he made me drink the *yãkoana* for several days without stopping. Then I was finally able to see these *xapiri* set up their hammocks in my spirit house during the time of dream. Yet a few days later, before I could fully recover my strength, the white people called me to their land again. During this trip, I stayed sitting in an airplane for a long time again. Because of that, the paths of my new *ayokora* cacique bird spirits must have been severed without my knowledge. When I returned to the forest, my stepfather tried to bring them back to me. But it was in vain. I had not been careful enough. The *yãkoana*'s effect on me was recent and I was still weak. Their paths did not withstand flying so high and so far. Without this carelessness, I could have kept these powerful spirits until today.

Some time later, a Toototobi shaman whom I call brother-in-law came to visit me at *Watoriki*. He is a childhood friend and he himself owns a great many of these *ayokora* cacique bird spirits. We drank the *yãkoana* together to cure a man sick from a *hwëri* sorcery plant. This harmful thing was badly burning his image. So I wanted to tear it out of his body. To do so, I tried to call the *ayokora* cacique bird spirits I had acquired right before my last trip. My brother-in-law was at my side and was guiding me. I really tried very hard, but I was unable to spit this evil plant onto the floor of our house. It must have fallen in another place, out of our view. I thought I still possessed these healer cacique bird spirits. But they had already fled. My mouth had become sterile. All that remained with me was a clumsy spirit whom we call *Ayokorari xapokori a*.[5] This *xapiri* also spits up harmful things, but by making them land far away, where you can never see them appear. The tube of his throat[6] is twisted and he regurgitates these things behind him, on his own mirror, and not through the mouth of the shaman, who is then unable to make everyone see them. It is much different with the real *ayokora* cacique bird spirits. Those are the ones I want to keep in my spirit house. I often saw them come to settle in it in the time of dream. Yet each time I woke up, and they had not truly settled there. I think hostile shamans must have sent me bad ones in order to take their place. If this is the case, my elders will have to chase these sterile cacique bird spirits away so I can one day make real ones dance.

These powerful *xapiri* only come to live in a shaman's spirit house when his mouth is not salty or burned by game. And they do not stay there easily, for they are very wild. It is so! They came to me at first, but finally they changed their minds and went back to the distant places they came from. Their paths were torn like spider webs and were carried away by the wind. I am sad about it, and though I try to listen for them, I can no longer hear their songs. I have often visited the places where the white people ancestors lived at the edges of the earth. As I said, it is from there that the *ayokora* cacique bird spirits come down to us. To go there, I traveled in airplanes for a long time, very high in the sky's chest. I think this is what finally destroyed the paths by which these magnificent *xapiri* came to me. So they gave up visiting me. They must have thought: "*Hou!* The white people destroyed our tracks! We can no longer return to our father!" Then they turned back and I lost them. My Toototobi brother-in-law still owns his *ayokora* cacique bird spirits because he has stayed in the forest since our childhood. He never went to where the white people live.

Yet my father-in-law at *Watoriki* had warned me: "Don't go to those distant lands so often! And if you really have to go there, do not eat those outsiders' food! They stink of **onion, garlic**,[7] and burned grease! If you continue traveling like this, you will end up making your *xapiri* run away from you!" But despite this warning, I still often had to go to big cities and eat strange things there. The sickly sweet smell of the white people's soaps and fabrics penetrated my skin, which also soaked up the smoke from their **cigarettes** and cars. The *ayokora* cacique bird spirits fear all these odors, which they find disgusting. Aren't the nests of the *ayokora* cacique birds in the forest hung high up in the trees, far from our fires and our women's honey leaves? They only smell and eat the things of the forest. This is why their spirits are so sensitive to the city's fumes. It is also what drove them away from me.

These were the *xapiri* I most wanted, and now that they have left me I feel like I am lacking them and I am sad about it. Yet I will try to call them again later, with the help of great shamans who, like my stepfather, really know them. I will study a long time with the *yãkoana* and will focus my thought on them again. I will visit the shamans of the *Xamat^hari* on the upper Rio Demini, who are also willing to generously give their most beautiful spirits. I will stay among them for one or two moons, when the food of the gardens and the honey of the forest trees are abundant. They will blow the strong powder they take from the seeds of the *paara hi* tree into my nose. I will become thin and weak, just like the first time I drank the *yãkoana*. Then maybe the *Ayokorari* spirits will come to dance for me again and allow me to cure like our great shaman elders? This is what I really want. What I love above all is to contemplate the beauty of the spirits and learn through their words. The elders already blew the *yãkoana* into my nostrils when I was younger.[8] But I do not want to be lazy. I will present my nose to them again several times. Then the things of the forest will truly be clear to me. For the moment, I do not see them well enough yet. I will have to drink the *yãkoana* often to truly achieve this. If I do not continue on this path, I fear that my thought will become confused. I have to protect myself. By going to the white people too often, I will begin to become ignorant. But on the contrary, I want to become as wise as my elders and our ancestors before them.

THE *XAPIRI* have always danced for us, and I find their images and songs magnificent. This is why I wanted to bring them down in my turn. If we lied and were not able to see them with the *yãkoana*, we would not be shamans. But we really do bring down their images. We take care of their dwellings and we constantly study their words. This is how it was for our ancestors, and we are following in their footsteps. This has been the people of the forest's way of being since the beginning of time. We must not forget it. If we do not feed the spirits with the *yãkoana*, they sleep in silence and our thought remains closed, I have said so. We become incapable of seeing. This is why I always carry the *xapiri*'s teachings in my thoughts. They extend into the distance, one after the other, and never end. White people are surprised to look at us become spirits with the *yãkoana*. They think that we are losing our minds and that we are singing for no reason, like them when they become ghost with the alcohol they drink. Yet if they understood our language and cared enough to ask themselves: "What do these songs mean? Which forest do they speak of?" they might understand the words that the *xapiri* bring us from the edges of the earth, the sky's back, and the underworld they come from. But once again, the white people prefer to remain deaf because they find themselves too clever with their paper skins, their machines, and their merchandise. Yet for we who are shamans, all these things' value is far too short for us to set our thought on them! What the *xapiri* teach us has far more weight and strength than all the white people's money. The value of the spirits' songs is truly very high! Can you lift the earth and the sky? No? Well, that is the measure of their weight! Their words are the ancient words of *Omama*. For us, that is what you call the "future." It is to think that our sons and our sons-in-law, then their children and their grandchildren, will become shamans in our place and make the *xapiri*'s words be heard in the forest in their turn. By always making these ancient words new, they will prevent them from disappearing. So if the white people do not make us all die or forever tangle up our thought, they will continue to spread in the forest without end.

We know the *xapiri*'s valor and strength. Before the city's medicine reached us, it was always they who cured our long-ago elders. Shamans die one after another, but the spirits never disappear. This is why I defend their words against the white people's hostility. If our shaman elders had died without passing on their images to their sons and sons-in-law, our ignorance would be pitiful. And if today the *xapiri*'s voice was reduced to silence, the thought of those who will live after us would fill with oblivion and darkness. Unable to become spirits, they will live for no purpose. They will no longer be able to cure the sick, to prevent the forest from re-turning to chaos, or to hold back the falling sky. If we forgot the *xapiri* and their songs, we would also lose our language. Deep inside ourselves, we would become other. By imitating the white people, we will only suc-ceed in becoming as ignorant and fearful as their dogs. This is what I think! Without the *xapiri,* we will end up disappearing. This is why we will not stop making their images dance as long as we are alive.

When they were still alone in the forest, our forefathers had a great deal of wisdom. They preferred the words of the spirits' songs to any other thoughts. Today the talk that constantly pours in from the city stands in the way of our elders' voices. The *xapiri*'s words have weakened in young people's minds. I fear they are too concerned with white people things. They are sometimes even afraid of the power of *yãkoana* and are scared to become shamans. They are worried at the idea of seeing the *xapiri* and dread their fierceness. So I try to prevent their thought from closing before they become adults. I tell them: "Do not be cowards! Later, you will take wives and your children will be born. If you cannot make the *xapiri* dance, how will you cure them? Become shamans like your el-ders! If you behave with uprightness, the *xapiri* will come to you easily. They are beautiful and powerful! Do not be afraid of them!"

Before the white people arrived in the forest, there were a great num-ber of spirit houses in the sky's chest. Today many of them are empty and charred. The *xawarari* epidemic beings devoured so many of our elders! So sometimes, when my thought is overcome with sadness, I ask myself if there will still be shamans among us later. Maybe the white people will be able to confuse the minds of our children and grandchildren to the point that they will stop seeing the spirits and hearing their songs? Then, without shamans, they will live helpless, and their thought will get lost. They will spend their time wandering on the roads and in the cities. They will be contaminated there by sicknesses that they will pass on to their wives and children. They will not even think about defending their land anymore. Sometimes at night these thoughts torment me until dawn. Yet

I always wind up telling myself that as long as there are shamans such as me and other sons of elders alive, as long as the *xapiri* protect our forest, we will not disappear. We will strive to make our sons and sons-in-law drink the *yãkoana*. This way they will be able to make the *xapiri* dance like our fathers and grandfathers did before us and thus the spirits' words will never be lost.

The *xapiri* would not come to do their presentation dance without the *yãkoana*. This is why the shaman elders want to make young people inhale its powder. This way they give them their breath of life and their spirits' path so that they can see and call them in their turn. Then the *xapiri* continue to come down to them in the same way they did for our ancestors since the beginning of time. Nothing has changed. This is why their words have no end. They are very old but always new. They are strong words that never age; words of bravery that relentlessly avenge us. Before they expire, the oldest shamans give their spirits to their sons and sons-in-law. Then these younger men do the same before their own death. It has always been this way. The *xapiri*'s songs pass from one shaman to another endlessly throughout time. This is why we become spirits today, just like our forebears long before us.

THE *XAPIRI*'S words are as innumerable as they are themselves, and we have passed them on to each other since *Omama* created us, the forest's inhabitants. In the past, my fathers and grandfathers carried them in their minds. I listened to them throughout my childhood, and today, having become a shaman myself, it is my turn to make them grow within me. Later, I will give these same words to my sons and, if they want, they will continue in the same way after my death. This way the *xapiri*'s teachings are constantly renewed and cannot be forgotten. They only increase from shaman to shaman. Their history has no end. Today we still follow what *Omama* taught our ancestors in the beginning of time. His words and those of the spirits he left us remain within us. They come from a very distant time, but they never die. On the contrary, they grow and strongly fix themselves inside us one after the other, and this way we have no need to draw anything to remember them. Their paper is our thought, which since very ancient times has become as long as a big, endless book.[9]

There are no other words than that of the *xapiri* to defend against the ills afflicting us. We fear the epidemic fumes, the evil beings, and enemy sorcerers. We are anxious about the sky's fragility and the *Xiwãripo* chaos being's eagerness. We are worried that the forest could be torn apart by great floods of the rainy season or that it could be burned by blazes of the dry season. We are afraid of jaguars, snakes, and scorpions. If all these harmful things did not exist, we would not be so concerned. Yet they often threaten us, and only the *xapiri* can valiantly hold them back. This is why shamans work so much for the people of their house.[10] But one must not believe that they care only about their relatives and the forest they live in. That isn't true! The *xapiri* try hard to defend the white people the same way they defend us. If the sun is obscured and the land is flooded, the white people will no longer be able to perch in their buildings or travel the sky's chest sitting in their airplanes. If, on the contrary, *Omoari*, the dry season being, settles on their land for good, they will only have trickles of dirty water to drink and they will die of thirst. This could truly happen to them! Yet the *xapiri* bravely fight to defend all of us, as many of us as there are. They do so because human beings seem alone and at a loss to them. We are mortal and this weakness makes the spirits sad. This is why they see us as ghosts, even if we are still alive.[11]

OMAMA WAS not a shaman himself. Yet he was the one who created the *xapiri* and made his son into the first shaman. We do not want to forget his words. They are the words our elders left us, and we want to keep them. They are the only ones that are clear for us to hear. We do not really understand the white people's words. They always seem worrisome to us. As soon as we try to imitate them, our mouth becomes twisted to utter them while our thought gets lost on the way to searching for what they might mean. What is truly beautiful to know for us are the images and songs of the *xapiri*. These are magnificent and very old things that we see and hear by drinking the *yãkoana*. We got them from our ancestors in a time when they lived alone in the forest, far from the white people. These outsiders own other words inside themselves, about *Teosi* and merchandise. This is why they ignore our words and speak so many lies about us.

When we become spirits, *Omama*'s image comes to us first. Then all the other *xapiri* follow, doing their presentation dance. *Omama*'s image gives us the words with which he makes our thought grow. You only have to listen to the shaman elders' songs to hear his voice. This is how

Omama continues to take care of us and warn us against the white people's obscure thoughts: "When they arrive bearing all their merchandise, they seem clever and generous. But beware! They never take long to become greedy and begin mistreating us. If they truly wanted to make friends with the people of the forest, they would not behave this way."

When a young shaman does not yet know *Omama*'s image, his elders clear its path and make it come down for the first time. As soon as he sees it, its beauty dazzles him and his thought instantly opens. Then he thinks with admiration: "*Haixopë!* Here truly is *Omama*, whose name I only knew! How graceful he is with his thick brown hair adorned with a black saki tail headband and covered in shining white down feathers! How beautiful is his skin covered in annatto dye gleaming in the light! The scarlet macaw tail feathers in his armbands and the blue *hëima si* feathered hides in his earlobes are so splendid! We are so ugly compared to him, and how gray our skin looks!"

You only truly become a shaman when *Omama*'s image comes down to you. Without that, the other *xapiri* would not want to approach you. But if *Omama*'s image sends us the spirit women of whom he is the father and father-in-law, all their suitors crowd after them, pushing each other aside to dance for you and joyously build their new house. This is what happened when my stepfather made me drink the *yãkoana* for the first time. I had absorbed such a great quantity of powder that I was on the point of becoming other. The breath of the *xapiri* whom the elders had given me with the *yãkoana* had made me die. I was truly in a ghost state. It was exactly at that moment that *Omama*'s image appeared to me. Then I instantly became a spirit myself, in the same way as his son long ago. It is so. If you do not become other with the *yãkoana*, you can only live in ignorance. You limit yourself to eating, laughing, copulating, speaking in vain, and sleeping without dreaming much. You never think about the things of the beginning of time. You never ask yourself: "Who were our ancestors who turned into game? How did the sky first fall? How did *Omama* create the forest? What do the *xapiri*'s songs and words truly say?" But if you often drink the *yãkoana*, like *Omama* taught us to, your mind never remains empty. Your thoughts can expand and multiply in the distance, in every direction. This is how we really gain wisdom.

Despite all this, the white people have often threatened us so that we would reject the *xapiri*.[12] They told us: "Your spirits lie! They are weak and delude you! They belong to *Satanasi!*" When I was still very young, I was afraid of these outsiders' threats, and I sometimes doubted the *xapiri* because of them. For a time, I let myself be deceived by their bad talk, and I even tried hard to answer *Teosi*'s words. But all of that is long over! It has been a long time since I let myself be fooled by the white people's lies and since I thought: "Why wouldn't I try to become one of them?" I became a man, then my children grew up and they had children in their turn. Today I never want to hear any ugly talk about the *xapiri* again! *Omama* created them after drawing our forest, and since then they have taken care of us. They are valiant and magnificent. Their songs make our thoughts grow and keep them strong. This is why we will continue to make their images dance and defend their houses as long as we are alive. We are inhabitants of the forest, and this is our way of being. These are the words I truly want to make the white people understand.

How This Book Was Written

I WAS TRAINED in anthropology at a time and place (University of Paris X Nanterre in the early 1970s) when the norm was to view the ego as suspect, and all subjective or reflexive thoughts as unseemly intrusions on theory.[1] So it is with a little embarrassment, as well as relief, that I here transgress that positivist convention. At the same time, I have no intention of giving in, late in life, to a kind of postmodern "introspection-by-proxy" in which, under the pretext of deconstruction, the narcissistic verbosity of the ethnographer finally overwhelms the voice of his interlocutor.

So, while recognizing that "every ethnographic career finds its principle in 'confessions,' written or untold,"[2] it is only with the purpose of shedding light on the way in which Davi Kopenawa's exceptional account was collected and edited that I will describe my own ethnographic journey. I will do my best to clarify the circumstances and adventures that brought us together and the personal affinities that made his words so captivating, as well as the choices that governed the way I have put them in written form.[3] Explaining this history is the least I can do in return for the trust Davi Kopenawa has placed in me by disclosing his memories and most intimate reflections with such intensity.

An Anthropologist's Education

As a child, I was thrilled by the naive exoticism of narratives about "Extraordinary Voyages" or adventurous expeditions to the Amazon. When I was about ten years old, living with my parents in the Rif Mountains in Morocco, watched over by a Tarafit-speaking Berber governess with intriguing facial tattoos and heavy silver jewelry, I was captivated by the heavy old editions of *The Mighty Orinoco* or *Eight Hundred Leagues on the Amazon* by Jules Verne. I recall even more vividly how I devoured *Journey to the Far Amazon: An Expedition into Unknown Territory,* published in French the year I was born.[4] (Much later, I realized that this 1948–1950 expedition was the first to cross the Serra Parima, passing through the northern end of Yanomami territory.)

Adolescence and the move back to France turned my innocent passion for the old-fashioned mythology of Amazon explorations into a desire for knowledge. Claude Lévi-Strauss's *Tristes Tropiques* awoke my intellectual curiosity and led me to other books in the Terre Humaine series. At that point the series consisted of just sixteen volumes, starting with *The Last Kings of the Thule* by Jean Malaurie and ending with *Piegan: A Look from Within at the Life, Times, and Legacy of an American Indian* by Richard Lancaster. It also included *Sun Chief: The Autobiography of a Hopi Indian* (Don C. Talayesva), *Affable Savages: An Anthropologist among the Urubu Indians of Brazil* (Francis Huxley), *L'Exotique est quotidien* (George Condominas), *La Mort Sara* (Robert Jaulin), *The Four Suns: Recollections and Reflections of an Ethnologist in Mexico* (Jacques Soustelle), *Ishi* (Theodora Kroeber), and, of course, *Yanoama* (Ettore Biocca). I read them with feverish avidity, one after another.

Starting out with undergraduate studies in sociology (the events of May 1968 and radical high school days not far behind me), I was quickly led by my hunger to unite my new ethnographic passion with the reality of fieldwork to the Colombia *llanos*. I was just twenty years old. One of the daughters of the famous Colombian anthropologist Gerardo Reichel-Dolmatoff was a student in Paris. In the summer of 1972, she steered me from Bogotá towards one of her friends, a doctor who was working in a small village near the Sierra de la Macarena. From there, I traveled up a wide muddy river in a small motorboat, perched on piles of boxes of soft drinks, to San José, an Amazonian hamlet, some four hundred kilometers southwest of the Colombian capital. People told me about the Guayabero Indians who lived very nearby.[5] A few days later, itinerant salesmen dropped me off near one of their villages, on the bank of the Rio Guaviare, and brought me to the first confrontation between the romantic vestiges of my juvenile imagination and the complexities of the actual Amazon.

At first glance, nothing distinguished the Guayabero from the poorest peasants in the region, and my unexpected arrival apparently inspired nothing more from them than profound indifference. A man dressed in rags caked with soil only pointed a finger, silently, towards the center of an oval of palm huts to a small cement structure with low walls and a sheet metal roof. Then he disappeared, content with resolving the problem of my incongruous presence by depositing me in this small building used for hulling rice. Puzzled, I awkwardly hung my hammock there, and having neither food nor goods to barter, I contemplated the unforeseeable consequences of my visit throughout a sleepless night in a raging tropical storm.

At dawn, an elder in tattered clothes, his face covered by a masklike pattern of bright vermilion red annatto designs, suddenly appeared at my side. With a smile, he handed me a gourd bowl of boiled fish with manioc bread. This silent gesture of generosity and the proud elegance of his painted face above the rags touched me deeply. The enigma of this encounter was instantly etched in my mind as the challenge of an alien wisdom, which I could almost touch yet could not comprehend.[6] A few days later, I traveled back to San José in a small dugout canoe accompanied by a man seeking care for his young daughter whose fever was barely kept under control by my last aspirins. Stunned by the intensity of these brief experiences, I vowed to acquire the intellectual and material means to begin a genuine anthropological fieldwork and to get involved in a long-term commitment alongside the Amazonian Indians.

When I returned to France in March 1973 after a brief stay in Brazil, I quickly completed my degree in sociology and, the following year, defended my master's thesis in anthropology at the University of Paris X on historic documents about ancient Peru. Although the early 1970s were an extraordinarily fertile time to study social sciences in Paris, the snobbish atmosphere of academic hothouses bored me, and I could not forget the undecipherable smile of the old Guayabero Indian. I could have been seduced by the diatribes against anthropological scientism and Western ethnocide[7] that dominated French Amazonian ethnography at that time, but the rhetoric of their main authors was so full of eighteenth-century Indianist ethnocentric exoticism that it repelled me too. The generic denunciations of ethnocide, made with few true commitments to solving real local problems, frequently struck me as smug and blind to what was actually politically at stake. In short, I found myself living on a third bank of the anthropological river.[8] I was intrigued by Anglo-American anthropology and was avidly reading the works of Lévi-Strauss at a time when most French Amazon ethnographers ignored or rejected the theoretical contributions of his work to praise only its elegiac tones.[9] I looked forward to just one thing: finally beginning "my" own fieldwork and finding my own way in Amazonian ethnography.

First Fieldwork

Patrick Menget taught one of my first courses in anthropology at University of Paris X. He had studied in the United States with David Maybury-Lewis, founder of the famous Harvard Central Brazil Project, and he had worked in Brazil since the late 1960s, among the Ikpeng (Txicão) along the Rio Xingu.[10] Menget was a passionate teacher and always very concerned for his students' professional future. He introduced me to the pioneers of "Anglo-structuralist"[11] modern Amerindian ethnography (David Maybury-Lewis and Peter Rivière) as well as to Brazilian sociological studies of interethnic contact (Darcy Ribeiro and Roberto Cardoso de Oliveira).[12] Reading these authors allowed me—thankfully—to bypass the early 1970s Paris school of Amazonist studies.

In 1974, Patrick Menget recommended me to colleagues at the University of Brasília. They were looking for doctoral candidates willing to participate in an action research project among the Yanomami recently reached by the Perimetral Norte branch of the Trans-Amazonian highway in Northern Brazil.[13] It was an opportunity that went beyond anything I could have hoped for. Neither rumors of Yanomami bellicosity, which the Amazon specialist circle reveled in, nor news of endemic onchocerciasis (river blindness) in Yanomami territory could dampen my enthusiasm.[14] I dove into the American and French ethnographic literature about the Yanomami, which at that time focused mainly on communities located in Venezuela.[15] I was immediately startled by the degree to which these ethnographies depended on the old European "savage" stereotypes, either noble or ferocious.[16] In France, the Yanomami were viewed in the idyllic light of their love stories in works such as Jacques Lizot's "Histoires indienne d'amour" (1974); in the United States they had become (in)famous as the "Fierce People," predisposed to almost Hobbesian warfare, as in Napoleon Chagnon's book (1968). But these ethnographies were disconcerting in other ways as well. Their sociological concepts were mostly borrowed from Africanist studies and were particularly ill-suited to Amerindian societies. Besides, their presentation of Yanomami cosmology was limited to a few scattered monographic fragments, as if it was only a minor imaginary side-issue of a reified, genealogically based social organization.[17]

Lacking financial means of my own or access to grants through the French university system, I rather audaciously applied for a grant from the Brazilian embassy in Paris to allow me to go to the University of Brasília. It proved the right thing to do. My request was generously fulfilled, and I soon found myself ready to start work, first in the Brazilian capital, then in Boa Vista, the main city in the state of Roraima. My first fieldwork among the Yanomami thus began in March 1975 on the upper Rio Catrimani, a tributary of the right bank of the Rio Branco, in the northernmost region of Brazil, near the Venezuelan border. I began my work for the FUNAI Perimetral Yanoama project and was assigned by my University of Brasília professors in the Catrimani Mission area.[18] This Catholic mission, run by a young and enthusiastic Italian priest from the Consolata order,[19] did not seek converts at all, but concentrated solely on healthcare and social work among the Indians, who had been reached by the Perimetral Norte highway road works the year before.[20] Workers had spread both tattered clothing and diseases among the three Yanomami groups located nearest the missionary outpost and the zone mapped out for the roadway. Despite the mission's best efforts, nineteen Indians had already died from an outbreak of measles.[21]

My first glimpse of the Yanomami along the highway—just before arriving at the Catrimani Mission—drowned what was left of my exotic dreams in a vat of acid. Men and women waded through the red mud, half-dressed in dirty and torn tee shirts bearing commercial logos or election slogans. They wove their way through the deafening roar of bulldozers, smiling at the roadway workers and begging for food, clothing, old metal cans, and plastic bags. I later learned that they were the survivors of a small community of the *Yawari*—a Yanomami subgroup from the Rio Ajarani basin—the first to suffer the effects of the opening of the highway. Since 1973, so many of them had been wandering the route that the missionaries dubbed them the *Estrada t^hëri* (a mixture of Portuguese and Yanomami meaning "people that live along the highway").[22]

Shaken by this harrowing sight, I finally reached the mission, which harbored the bucolic atmosphere of a small tropical country house, with its birdcages and henhouses among the coconut palms and guava trees.

As soon as I arrived, I saw a group of Yanomami women sitting on the ground with their children. Their bodies were painted with a luminous vermilion red dye, and their arms were decorated with bouquets of fresh leaves. They chatted in the shadow of an enormous Saurer Berna military truck parked nearby. Suddenly, a small group of curious men, joyously waving bows and arrows much larger than they were, yanked me out of my bewildered state. I was instantly engulfed in shrill crescendos of mocking laughter. I learned later that the joke was on me, for my presence had immediately sparked the Yanomami's refined art of nicknaming.[23] Standing in front of the mission's health post, a gangling young apprentice ethnographer, I knew I was being scrutinized with lighthearted but caustic humor. I felt terribly alone and absolutely ridiculous.

My first visit to the big cone-shaped communal house just a few hundred meters away was hardly more glorious. I folded myself in half to cross the threshold of the low entryway and then took a few halting steps inside. Howling dogs immediately encircled me. I froze, blinded by the darkness and half-asphyxiated by smoke from the hearths. A vigorous bite on my right knee brought me back to reality. Shadows moved around me amid loud shouting. Brands in hand, my future hosts drove away the enraged animals. When calm was finally restored, they did not make the slightest attempt to hide their amusement at the entertainment I had just provided, but kindly led me to the area where the young unmarried men were hanging their hammocks.

A few days later, I accompanied the inhabitants of the community on a week-long hunting trip to prepare a *reahu* alliance feast and funeral ceremony. I thus began my "field baptism"[24] with day-long hikes through the forest during the height of the dry season, anxiously staggering after hunters who were completely indifferent to my presence, grasping not a word of their language, quenching my thirst with a little muddy water scooped from dry riverbeds, and with no provisions except for a few bananas.

I spent my first month working at the Catrimani Mission in the company of a young volunteer from England, Nicolas Cape, supervised by Alcida Ramos, a professor at the University of Brasília who was one of the two anthropologists leading the Perimetral Yanoama project. During this period, I served as a research assistant and guinea pig for putting together a practical guidebook to the Yanomami language.[25] While learning the language, I was involved in multipurpose social service activities and improvised paramedical help. Accompanied by my English teammate, or as an assistant to the Italian missionary, I tried to do as much as possible to minimize the misfortunes my Yanomami hosts could encounter along the nearby road works. In the absence of any official healthcare, I occasionally administered treatment for respiratory infections that ravaged the region. This was my job each time I stayed at the mission, throughout the period of my initial fieldwork.

Then, bureaucratic sabotage by the FUNAI military head all but paralyzed the Perimetral Yanoama project, and I soon found myself on my own.[26] I had had an opportunity during that first month to visit a communal house located about two days hike upstream, called *Makuta asihipi*.[27] I had accompanied a hunter who had come to the mission in search of help for several cases of pneumonia. This was how I came to administer intramuscular injections of antibiotics for the first time in my life, with my stomach knotted in fear, but, thankfully, with successful results. The house and its hosts, with whom I lived for a week, captivated me. I decided to make this more isolated community the base camp for my future ethnographic research along the upper Rio Catrimani. I stayed there eleven months, interspersed with a few short visits to the mission, where I participated in a rapid vaccination campaign in November 1975 and some inspection rounds with a team of the Perimetral Yanoama project along the highway. Then, suddenly, in February 1976, I learned that everyone involved in this project, which had meanwhile been renamed Plano Yanoama, was banned from staying in Yanomami territory by the military dictatorship's generals. I had to drop everything and return to Brasília.[28]

The Ethnographic Pact

This first fieldwork forced me again to confront the unsettling challenge of cultural otherness and its attendant political and ethical questions. These had been temporarily shoved aside after my brief encounter with the Guayabero Indian. How could I render what I knew of the Yanomami world without indulging in exoticism? How could I jointly analyze the catastrophic impacts of Amazonian "development" on their society and environment and reflect on the implications of my own presence as an engaged observer in the midst of this situation? With this unsettling first experience as a starting point, I began to grasp three inseparable imperatives that would guide my ethnographic work. The first, of course, was to be scrupulous in doing justice to my hosts' conceptual imagination; the second, to think rigorously through the sociopolitical, local, and global context in which their society was embedded; and the third, to maintain a critical overview of the framework of the very act of ethnographic observation itself.

But something more fundamental than vigilance about epistemology was at stake in the intense "ethnographic situation" in which I became involved. The nature of this stake would later become more obvious when I began working with Davi Kopenawa. Ethnographic literature is shot through with brief conventional remarks about anthropologists' so-called "adoption." These had always struck me as products of self-deception. What does it mean to be "adopted" by one's hosts, when they see themselves increasingly subjugated by a threatening outside world from which the ethnographer is essentially an emissary, no matter how laughable or inoffensive he may seem at first sight?[29] It quickly became clear to me that the Yanomami had half-heartedly accepted my incongruous presence only as a safety precaution, just as they would have with any other foreign visitor who was not overtly hostile.

The elders of the Rio Catrimani area had already suffered through several lethal epidemics during first contact. They usually attributed them to vengeful sorcery by white people, whose sexual or economic greed had been frustrated. Since then, they thought it wise to put up a good face and watch what happened whenever a napë (an "outsider," an "enemy") barged in. This is the context in which my interlocutors, converted into "informants" (according to the profession's evocative police jargon), reluctantly agreed to collaborate in my first clumsy attempts at ethnographic "investigation." This collaboration was cautious, sparing with words, and revealed nothing more than I could understand from their answers to my questions—which is to say, very little, for a very long time.

Of course, budding ethnographers are generally not typical outsiders. That they willingly endure physical hardship, not to mention their obvious dislocation from their own world, and show such humility and tenacity in their desire to learn makes them intriguing. All that induces some sympathy, along with bemused commiseration. Then, on this basis of demonstrated good intentions, and with the help of the medicine and trade goods they make available, they end up convincing even the most skeptical that they are exceptional individuals of their kind. After a period of observation, the relationship between the ethnographer and his or her supposed "informants" begins to take a different turn. The latter, gaining trust in their uninvited guest, begin to assess his or her ability to serve their needs as a mediator between two worlds. Having earned such credibility, without realizing it or acknowledging it, the apprentice ethnographer seals an implicit solidarity pact with his involuntary hosts. From now on, the data collected will become both the product and the stake of this ethnographic relationship.

By teaching their cultural knowledge, the ethnographer's hosts certainly agree to take charge of his new socialization according to rules they consider more appropriate to the human condition. But, beyond any sense of complicity or empathy the strange novice may have inspired, their fundamental aim is to transmit this knowledge beyond him, out to the world that he always represents, despite himself.[30] They care about much more than cultural pedagogy. They seek, as much as possible, to reverse the unequal exchange that has always been the basis of the ethnographic relationship.

In short, the instruction provided by our supposed "informants" is first and foremost provided for diplomatic purposes. Their patient education is intended to push us towards a transition, transforming us from the unexpected ambassadors of a threatening outside world into well-meaning translators who can make their otherness understood and broker a potential alliance. In the best case, while the ethnographer believes he is "gathering data," those who have accepted his presence are seeking to reeducate him as a "truchman" in their cause. (The term *truchman* meant, during the brief sixteenth-century France Antarctique period in Brazil, "young men, stationed, voluntarily or not, in Tupinamba and in other Tupi ethnic groups allied with the French, to learn the language and serve as intermediaries in negotiations—commercial, diplomatic, etc.—between colonizers and natives.")[31]

This portrait of the ethnographer as a "reversed truchman" is a far cry from the conventional fiction in which unfriendly natives wind up revealing their secrets to a dauntless anthropologist who manages to become their confidant and needs only to write a fleeting homage in the acknowledgments of his monograph to settle his ethnographic debt. The tacit pact I have just outlined is more demanding for both sides of the ethnographic relationship and implies that the anthropologist take his or her responsibility much more seriously. What's at stake for his interlocutors is accepting a process of self-objectification through ethnographic observation in a way that allows them to claim cultural recognition and empowerment in the opaque and malignant world trying to subjugate them. At the same time, the ethnographer must fully assume his political and symbolic role of "truchman" for the other side to pay back the debt of knowledge he has incurred, and this without giving up his own intellectual curiosity (on which the quality and effectiveness of his mediation will largely depend).

This has been my view since my very first fieldwork: that the ethnographic "data" given to us can only properly be considered as such in exchange for this mutual engagement and complex cross-collaboration. The prospect of a long-term commitment alongside the Yanomami took shape for me as I thought about how to offer something in return for the gift of their knowledge. I would not claim that this ethical road map is anything but my own way of practicing ethnography (and there are surely as many ethnographic styles as there are ethnographers). I raise this issue here mainly to clarify the genesis and nature of my collaboration with Davi Kopenawa, which is a further-elaborated variant of this "ethnographic pact." Our partnership has certainly been atypical, in more ways than one. Today, however, indigenous organizations in Brazil and elsewhere increasingly suggest to young anthropologists at the beginning of their fieldwork that they engage in such projects of collaborative ethnography.[32]

From False Start to Close Collaboration

From the time I returned to France in mid-1976, I itched to get back to the Yanomami, despite the Plano Yanoama's sorry end and the discouraging political situation in Brazil. Of course, my ethnographic work still seemed too rudimentary to me.[33] But above all I was worried about my former hosts along the upper Rio Catrimani. I was sure that it would not take long for them to visit the Perimetral Norte highway, and I knew they had not been vaccinated yet. At the end of 1977, I joined the French military service as an overseas aid volunteer, became affiliated with the anthropology department at the University of Brasília, and obtained research funding from the anthropology department at the University of Paris X Nanterre. I finally got back to fieldwork in March 1978.

Interrupted by the rainy season, the Perimetral Norte road works had been abandoned since the end of 1976. Quiet appeared to have finally returned to the forest. Under this apparent calm, however, were tragic losses and imminent new dangers. As I had feared, a measles epidemic had decimated the population along the upper Catrimani between December 1976 and February 1977.[34] Moreover, in July of 1977, FUNAI had carried out aerial reconnaissance flights over the entire region occupied by the Yanomami. Their reckless work became the pretext for the rapid promulgation of several decrees between December 1977 and July 1978 designed to carve up Indian lands and transform them into an archipelago of twenty-one forest islands encircled by five to thirty-five kilometer-wide corridors to facilitate colonization.[35]

Very upset by this news, I managed to spend six months in the forest between March and August 1978. For most of the time, I stayed again with the *Makuta asihipi* community, which had relocated on the right bank of the Rio Catrimani and taken a new name, *Hewë nahipi*.[36] It was not easy. The new bishop in Roraima, doubtless eager to stop outsiders from meddling in his mission, first tried to deny me access to the Catrimani region while the local FUNAI outpost viewed my return with extreme hostility. The *Hewë nahipi* communal house was under construction and mostly open to the wind on a site infested with black flies, and it was the rainy season. For months I lived in the mud and humidity, itching from head to foot. Most important, the results of my census gradually revealed that at least sixty-eight inhabitants, including a large number of my friends, had succumbed to a measles epidemic the year before.

In addition to documenting this tragedy, I began new ethnographic research into Yanomami rituals and cosmology, planning to remedy the symbolic poverty that then plagued most studies of Amerindian social organizations. The resulting data provided essential materials for my doctoral thesis,[37] but I had to leave my hosts earlier than expected because I had come down with labyrinthitis.[38]

Recovering in Brasília, I stayed with a friend, Mémélia Moreira, an intrepid antiestablishment journalist for *Jornal de Brasília* (and later *Folha de São Paolo*), and met Claudia Andujar, a very talented photographer who had begun working among the Yanomami in the Rio Catrimani area in 1974. Together, we edited one of the first papers contesting the military's plan to dismember Yanomami territory. Andujar then invited me to join her in São Paolo, where we worked with Carlo Zacquini, a progressive Catholic brother from the Catrimani Mission. The three of us developed the proposal for a large, continuous Yanomami territorial reserve and launched a campaign in Brazil and internationally to win public opinion against the Brazilian generals' ethnocidal initiative.[39]

In late 1978, São Paolo, the economic capital of Brazil, was the site of a huge political movement and media campaign against the government's plan to seize Amerindian lands, in the guise of a decree to "emancipate" supposedly "acculturated" Indians.[40] This unprecedented political protest marked the convergence of the nascent Indian movement with sectors of the intelligentsia (lawyers, journalists, academics) in resistance to the military dictatorship, which was beginning to wane. This was the political context in which Claudia Andujar, Carlo Zacquini, and I created the Comissão Pró-Yanomami (CCPY), an NGO that fought for almost thirty years to defend the rights of the Yanomami until they founded their own organization—the *Hutukara* Associação Yanomami. Davi Kopenawa has served as president of that organization since its founding in 2004.

I had not heard of Davi Kopenawa during my first field trip in 1975. Recently hired by FUNAI, he was then working along the upper Rio Negro, far from Yanomami territory in the state of Roraima, where I was located at that time (Chapter 13).[41] We finally met in early 1978, in the communal house of the "people of *Waka $t^h a$ u*" near the Catrimani Mission, during a big *reahu* alliance feast and funeral ceremony. In an unusual gesture, the faraway inhabitants of the Rio Toototobi had been invited. Davi Kopenawa was born in the Rio Toototobi region, but since December 1977 he had been working as an interpreter at the FUNAI Demini Outpost, only sixty kilometers west of the Rio Catrimani. So he did not pass up the opportunity to come visit his people at Catrimani.

Davi Kopenawa was twenty-two years old at the time. Dressed in a short-sleeved shirt and tight blue jeans, his hair carefully combed with a part on the side, taciturn and circumspect, he was obviously uncomfortable on the Catholic mission's territory. The head of the FUNAI outpost for whom he worked in Demini was in charge of a new "Plano Yanomami," which was created as a replacement for the Plano Yanoama, whose members had been expelled in early 1976 with the blessing (and undoubtedly the assistance) of this FUNAI agent.[42] This man was furthermore in open conflict with the Italian Catrimani missionaries.[43] He had set up the Demini Outpost at the farthest abandoned road works of the Perimetral Norte highway, at kilometer 211. There were no Indians in the area. He therefore tried by any means he could to lure there one of the Yanomami communities located near the Catrimani Mission, and he dreamed of expelling the missionaries from the region.

Having been warned by this xenophobic FUNAI agent, Davi Kopenawa saw me as he saw the missionaries, as a dangerous foreigner, hungry for the local riches.[44] He understood that he should keep his distance from me to protect himself.

> Yes, I did indeed see you for the first time at the Catrimani Mission. But I knew you already because I had heard people speak about you. Amâncio [the head of the Demini Outpost] told me before I left: "Don't go near those foreigners, they are very dangerous! They want to take advantage of you and steal the riches from your land! They want to take your picture and your words to exchange them for money. If you give those to them, they will take them far away and then they will come back and seize your forest! I'm warning you, be careful!" That's why I stayed in the background and avoided talking to you!

As an ex-member of the Plano Yanoama who had been expelled two years earlier by the military, I had not much love for the head of the FUNAI Demini Outpost. He zealously served the military,[45] and I knew he had made a bad name for himself among the Waimiri-Atroari Indians in early 1975.[46] My animosity against him somewhat extended to Davi Kopenawa. I was aware that he had been working as that FUNAI agent's interpreter for just over a year. My distrust was further reinforced by rumors circulating about him, which I had naively bought into after I arrived in Boa Vista in early 1978.[47] As he walked towards the communal house near the mission, anxious and deep in thought, I was unable to see anything in him, just a little younger than I was, except an assimilated Indian in the pay of FUNAI, the state indigenist administration run by the Brazilian military.

> They must have said things to you about me: "Davi has become a white man! Don't think that he is a real Yanomami!" But they lied to you and you let yourself be fooled. Just like me, you weren't well informed at that time. Those who lied to you by pretending that I had become a white man no doubt wanted to make fun of me or were very hostile to me! It's true; I wore clothes, shoes, a watch,

and glasses. I wanted to imitate white people. But inside, I was still
a Yanomami and I continued to dream with the *xapiri*. Those who
spoke to you about me in this way tricked you. They probably
thought that if I became friends with you, I would end up talking
to you and that wouldn't be good.

When I walked into the central living area of the *Waka t*^{h}*a u* communal house a few minutes later, I saw Davi Kopenawa sitting in a hammock in the dim light, silent and serious. He was earnestly observing a
collective cure, right next to a group of shamans from his native region.
The latter appeared to be doing everything possible to impress their hosts
with the exuberance of their performance. I was immediately captivated.
It was my first experience of the elaborate chants, sumptuous orna-
mentation, and impressive choreography of such shamanic opera.[48] After
placing my tape recorder in a strategic location, I impetuously grabbed
my camera and took one shot after another of this dazzling performance.
My flash went off a number of times and the blinding light seemed to
freeze the shamans in fits and starts in the darkness of the communal
house. Suddenly, Davi Kopenawa slipped over to me and made me realize, with a few unforgettable words, how disruptive and discourteous I
was being.

> The Toototobi people were invited to a feast by those of the *Waka*
> *t*^{h}*a u* River, and they had settled down in the house of their hosts.
> They were making their spirits dance together. At that moment, I
> saw you and thought: "*Hou!* Why doesn't he stop taking photos
> with all those lightnings? That's not good!" So, I approached you
> for the first time and I told you: "Stop taking pictures of these men
> who are becoming spirits! Their *xapiri* are here, even if neither you
> nor I can see them! You are going to tangle their paths and lose
> them!" That's how it was. I was still young then, yet I already car-
> ried the spirits in my thoughts. That's why I wanted to protect the
> shamans by speaking to you in that way.

That intervention immediately demolished my prejudice against him.
In his sensitivity to the work of the shamans, Davi Kopenawa, a FUNAI
interpreter and allegedly "acculturated" Indian, put me—the novice ethnographer and supposed defender of indigenous authenticity—in my
place as nothing more than a disrespectful tourist. The sudden collapse

of my stereotypes knocked my ethnographic pride down several pegs and allowed me to understand, for the first time, the remarkable capacity of my interlocutor to articulate the thinking of the two worlds between which he was constantly shuttling. Seeing my apologetic response and my interest in his ideas, he took the trouble to explain to me a few fundamentals about Yanomami shamanism. This impromptu lesson, dispensed cautiously and in Portuguese, aroused my curiosity about that rich and intellectually complex universe that my competence in Yanomami would not enable me to approach for several more years.

I then crossed paths with Davi Kopenawa several times during my fieldwork on the Perimetral Norte in 1978 and 1979, and later, in 1981, in his native region along the Rio Toototobi.[49] By then I had dropped all preconceptions about him and made efforts to win his trust. Although he spoke more freely with me, he still displayed a certain cautiousness. But this reserve gradually diminished in the early 1980s under the influence of the many Yanomami with whom I had maintained friendships since 1975. In addition, he became sufficiently intrigued by the CCPY's political struggle to defend Yanomami land rights that, in 1983, he came to visit our NGO headquarters in Boa Vista (see Chapter 14).[50] This initiative was preceded by a long, thoughtful and scrupulous observation of his future interlocutors. All this already prefigured the shrewdness, both audacious and considered, that would later be the foundation of Davi Kopenawa's ethnopolitical career. He was then twenty-seven years old. His first son had been born the year before, and he had just begun his initiation as a shaman, under the tutelage of his father-in-law (see Chapter 5). After this initial contact, CCPY, then coordinated by Claudia Andujar from São Paolo, sponsored his participation in the July 1983 assembly of the Union of Indigenous Nations, the first Brazilian Amerindian political organization, founded a few years earlier (see Chapter 17).

Davi Kopenawa's association with the CCPY's campaign for legalization of a Yanomami territory consolidated with a series of international trips, beginning in 1989 (see Chapters 18 and 20). The activities of CCPY intensified over the course of the 1980s, both in legal defense of Yanomami lands and in protest against the terrible health consequences of the gold rush in the region (see Chapters 15 and 16).

My friendship with Davi Kopenawa grew closer through our meetings and work together in the field. From 1985 to 1987, I stayed many times at his father-in-law's communal house, near the FUNAI Demini Outpost, of which Davi Kopenawa had been in charge since 1984 (see Chapter 14). This community, *Watoriki*, had pulled together survivors of the most isolated groups from the upper Rio Catrimani, whom I had met during my first stay in 1975–1976.[51] Its inhabitants knew me well, and I still count them among my closest Yanomami friends. Beyond these regular stays in *Watoriki* at the end of the 1980s alongside Davi Kopenawa, his family, and his in-laws, two intense common commitments strengthened our bond in the early 1990s.

First, we began working together in November 1990 to establish CCPY healthcare services along the Rio Toototobi, work that was coupled with a struggle against the negative cultural influence of the resident missionaries from the New Tribes Mission.[52] Later, in 1993, we became particularly close as members of the official inquiry of the Brazilian federal police and Federal Prosecution Service into the infamous Haximu Massacre. In this attempt to exterminate a Yanomami community, sixteen people—mainly women, children, and old people—were savagely assassinated by gold prospectors in the area between the upper Orinoco in Venezuela and the headwaters of the Rio Mucajaí in Brazil (see Appendix D).[53]

So, through the struggle to defend Yanomami rights during the 1980s and 1990s, Davi Kopenawa and I forged the mutual respect and close working relationship that was the vital basis for the project that ended in the writing of this book. In 1991, in an interview with the head of a special commission of inquiry established by the American Anthropological Association (AAA) into the situation of the Yanomami in Brazil, Davi Kopenawa described his conception of the role of anthropologists:[54]

There are only two anthropologists [who have helped us], one, who is not Brazilian, is called Bruce Albert, and there is another, called Alcida Ramos. These anthropologists have helped us. They have helped by writing reports and delivering them. Bruce speaks our language, he is helping very much. We like what he is doing. He is working for the Yanomami, in support of the Yanomami, and carrying out much news about what they are experiencing. Bruce is going to our area and Alcida is going to Auaris [the area of a northern group of Yanomami]. When they get to the communities where they are going, the Yanomami who live there tell them what is happening, so Bruce and Alcida know, they write it down and after-

wards send it to the news media so other people also hear about it.
. . . That's the way it is, it takes a courageous person, a courageous
anthropologist not to just do some research and then leave, we Indians need anthropologists with courage, anthropologists who can
speak our language. We need anthropologists to come to us with
news of what the whites are doing, of what the (Brazilian) government is saying, of what foreign governments are saying.

The Adventures of a Manuscript

In the late 1980s, the Yanomami territory in Brazil was devastated by an
unprecedented gold rush.[55] Davi Kopenawa was distraught at the epidemiological and ecological catastrophe—one he felt could foreshadow his
people's imminent demise. He had started to develop a shamanic prophecy about the smoke of gold, the death of the shamans, and the falling
sky with his father-in-law and mentor, the *Watoriki* "great man." At the
time, I was still unaware of all this, since from mid-1987 until December
1989, all of my Brazilian colleagues and I had been banned again from
Yanomami lands by the military, who, this time, supported the invasion
of gold prospectors.[56] It had become very difficult for me to communicate
with Davi Kopenawa during this troubled period.

But taking advantage of a quick trip to Brasília on December 24, 1989,
he left a recorded message for me on three audiotapes. He had been staying at Alcida Ramos's, and had just seen a television newscast on TV
Globo about the gold rush, which had then reached its zenith. Journalists
exposed the gouging of the land by the *garimpeiros* and their systematic
devastation of rivers in the Yanomami forest highlands. Shocked by these
pictures of wreckage in the historic heart of his people's territory, Davi
Kopenawa remained mute and pensive for a long time. Finally, he said
somberly, in Portuguese: "The white people do not know how to dream.
That is why they destroy the forest this way." Alcida Ramos was puzzled
by this enigmatic statement and suggested that he record his thoughts
about what he had just seen for me in the Yanomami language. I was in
France for a short holiday at the time and was soon to return to Brasília.

Davi Kopenawa decided to use that recording to describe his view of
the Yanomami's tragic situation and to launch an appeal to me. And so he
did, in his own language; an anguished tale of illness, death, violence,
and destruction caused by gold prospectors' feverish greed. His account

was interspersed with shamanic reflections that grew out of sessions with his father-in-law in *Watoriki*. Finally, he asked me to help him spread his message and to establish an emergency healthcare program among the Yanomami. This recording was the foundation of the political and literary pact that led to this book.

Starting in January 1990, I accompanied several teams of humanitarian doctors along the upper Rio Mucajaí; and then later, towards the end of that same year, through the Rio Toototobi area, this time with Davi Kopenawa.[57] We began in February to rework the content of his December 1989 message through four hours of interviews in Brasília. In the beginning of March we participated together in a filmed interview in the Yanomami language for a São Paolo NGO and condensed in it the issues of his previous talks in order to reach a wider audience. The text of this video interview, translated into Portuguese, was widely spread in Brazil by the civil rights movement Ação Pela Cidadania (APC), which had mobilized at that time in defense of the Yanomami.[58] It was later translated into French and published in a special issue of *Ethnies* dedicated to the 500th anniversary of the conquest of the Americas. Claude Lévi-Strauss wrote an eloquent comment about it in that issue's foreword.[59] Its main themes also appeared in English in an interview given to the head of the AAA Yanomami commission of inquiry, which was published in the United States in *Cultural Survival Quarterly*.[60]

The impact of this first attempt at joining shamanism with ethnopolitics made both of us realize the potential of our cultural collaboration in defense of the Yanomami cause. It also significantly changed the nature of our occasional ethnographic interviews, for it spurred Davi Kopenawa to ask that I go beyond the bounds of my usual research and turn his words into a book. In 1991, our collaboration was well established, as he made it clear during his interview with the AAA's representative in Boa Vista: "I am working with Bruce [Albert] for him to write down the shaman's knowledge about *Omama* and make a book out of it so that the whole world can know it."[61]

This declaration was followed by numerous recording sessions of conversations conducted in numerous locations while running a political obstacle course in the early 1990s. We recorded in NGO offices and forest camps, in communal houses and ministry corridors in Boa Vista, Toototobi, Brasília, and *Watoriki*. In early 1993, I finished a typescript in French taken from more than five hundred pages of Yanomami language transcriptions of some forty-three hours of recordings. I then published a study about it in a special issue of *L'Homme* dedicated to indigenous Amazonia.[62] But this first version of our work, written in haste and focused mainly on the shock of the gold rush years, did not seem to me to do justice to the complexity of Davi Kopenawa's thinking or his exceptional personality.

I thus decided to resume my work on the text, enriching the main themes that had emerged in the early interviews by filling in gaps and clarifying certain chronological, biographical, or ethnographic points through new recordings. I also embarked on related documentary research, which is addressed in numerous endnotes in this book. But most important, I decided to start a new set of interviews about Davi Kopenawa's initiation and vocation as a shaman, which was clearly the basis for all of his recent political-cosmological comments. The precision required to translate these materials led me to research Yanomami shamanism in depth. I had only touched on this subject in my earlier work, which was more focused on social and ritual organization. From then on, recordings were made every year, driven by my insatiable curiosity and predilection for (probably excessive) detail. Meanwhile, so as not to renege on the agreement to bring out his story, I continued to publish numerous translations of excerpts of Davi Kopenawa's thoughts in response to requests and political events in the latter half of the 1990s.

We recorded this supplementary material each time we were able to meet.[63] It was transcribed with the same strict standards as the earlier recordings, but in a more irregular way. The end result was more than six hundred new pages of transcription (excluding the documentary notes) from fifty hours of recordings. This additional material required a complete overhaul of the initial manuscript. The enormity of the task seemed insurmountable, especially since my increasing immersion in the intellectual and poetic subtleties of Yanomami cosmology made it difficult to pull myself back from the fascination of an endless succession of shamanic interviews with Davi Kopenawa. I risked losing myself in chronic unfinished work.

My interlocutor and accomplice put up with all my questions and meanderings in the maze of this seemingly endless manuscript with heroic patience, unwavering trust, and extraordinary generosity. At every meeting, he asked good-naturedly how my work was going, and often told me that squeezing the infinite multiplicity of the spirits and the force of their words into a printed book (modestly noting, by the way, that he himself knew only a little of the spirits' songs) could be an impossible task. He deeply understood how much I was struggling to face with him the challenge of this very unusual intellectual and writing experiment between our two worlds. He only occasionally suggested, with an ironic smile, that I should try not to die before finishing my work.

Towards the end of the year 2000, I met Hervé Chandès, director of the Cartier Foundation in Paris, who had discovered Claudia Andujar's remarkable photographs of the Yanomami in 1998 at the São Paolo Biennal. Shortly thereafter, Hervé Chandès and I traveled together to the Yanomami lands to attend a large political gathering in Davi Kopenawa's village. During our freewheeling conversations there, we decided to organize a series of meetings between shamans and contemporary artists from many countries. This experimental intersection of perspectives would apply Lévi-Strauss's famous description of art as the ultimate refuge of the savage mind in our society on behalf of the Yanomami cause.[64] The resulting exhibition, *Yanomami. The Spirit of the Forest*, was on view in Paris from May through October 2003, with an early version of Chapter 6 of my manuscript published in the exhibition catalogue.[65] Hervé Chandès was a continuous source of encouragement with the book, even sending two chapters to Jean Malaurie, the editor of the prestigious Terre Humaine series, and then kindly introducing me to him.

This unexpected meeting with Jean Malaurie at his home in Normandy, in Madame Bovary country, among his Inuit dog sleds and memories from the shamanic enigmas of the Whale Alley Siberian archeological complex, was decisive.[66] Mired in an ever-growing and inextricable text, I found myself before a man some thirty years my senior who had seemed almost mythic to me in my adolescence. His enthusiastic interest in my manuscript seemed providential, and it restored my energy. My first impression of this man of science, letters, and adventure was of his aristocratic impetuosity. But through our conversations I also discovered a man both warm and visionary, who had been profoundly moved by the Inuits' metaphysics, and for more than half a century had deeply committed himself to their cause.[67]

We both shared the experience of solitary fieldwork in far places, a friendship with shamans, and a certain rebelliousness. Thanks to his advice and perspicacity, I found the narrative center of gravity for my interviews with Davi Kopenawa. A subtle go-between tying together different worlds and spheres of knowledge, Jean Malaurie guided my shamanic writing adventure, finally, to safe harbor. My labyrinthine manuscript acquired a structure that satisfied me in the end, if not as a definitive text, then at least as an acceptable one. Indeed, as Jorge Luis Borges wrote: "The concept of the '*definitive* text' corresponds only to religion or exhaustion."[68]

Beyond the First-Person Narrative

As I said at the outset in "Setting the Scene," this book is by no means a direct translation of an autobiographical account belonging to a Yanomami narrative or ritual genre, which would require an anthropological discourse analysis. In such studies, the places of the narrator and of the transcriber/researcher are clearly distinguished, with the narrator's words usually encased in the heavy framework of the scholar's analytical metalanguage. This kind of approach has resulted in reference works in Amazonian ethnography since the 1990s. Some examples of this are the studies of Ellen Basso on the Kalapalo, Suzanne Oakdale on the Kayabi (two groups in the Rio Xingu basin in Brazil), or the book by Janet Hendricks about the Shuar in Ecuador.[69]

Nor did I intend to take the life story of a "privileged informant" as material for a book that would be organized according to an ethnographic plan utterly foreign to the narrator, and with all traces of editorial work largely obliterated. This classical model of ethnobiography was traditionally the basis for many North Amerindian published life stories, such as those analyzed by David Brumble, as well as autobiographies of Australian aborigines, such as those studied by Fanny Duthil.[70] The convention of an "absent editor" or "ghostwriter" that underpins these works requires, as we know, staging a fictitious absence of fiction.[71] The writing "I" of the editor is tucked away beneath the "I" that apparently belongs to the narrator to create a "hyper-real" effect and to provide the reader with the illusion of a literary face to face, an unmediated encounter with the narrator.

The writing strategy of this book differs in several significant ways from these two styles. First of all, the narrator, Davi Kopenawa, initiated the book project, and he is named as the first author. Second, the division of labor between narrator and editor was clearly defined and jointly conceived. The text has been edited through a lengthy collaboration based on an explicit editorial agreement reinforced by thirty years of friendship and ethnographic research. Third, Davi Kopenawa entrusted me with the mandate to bring his words to the largest possible audience in the written form used in my world. This automatically ruled out producing a heavily annotated literal translation intended for specialized scholars. Finally, this book is explicitly the result and intersection of our political and cultural meeting. It is as much a product of Davi Kopenawa's shamanic and ethnopolitical views as of my own experimentation with a new form of ethnographic writing based on what I called the "ethnographic pact."

Still, one could not unreasonably object that this book is constructed in such a way that it partakes, obliquely, of the same two writing strategies that it claims to reject. Echoes of the first approach are evident in my attention to details and my great respect for Davi Kopenawa's spoken words, as well as in the ethnolinguistic and documentary endnotes. And my decision to limit signs of my own presence and explanations of my editorial work to the beginning and end of the book is not unlike traditional ethnographic "ghostwriting."

However, my choice to be a "discreet editor" rather than an "absent editor" was not intended to simulate the nonexistence of an intermediary between the narrator and the reader, as in the classic ethnographic life stories. Rather, my purpose was to shift the balance of the writing process in favor of the narrator's words. Too many recent ethnobiographies tend to weight it the other way, submerging their narrators' accounts beneath the interpretive omnipresence of their editors. I think this is true of ethnographic discourse analysis as well as more orthodox ethnographies and even texts purporting to be postmodern critiques.[72] My own editorial strategy, as far as possible, was guided by the search for a compromise that tempers the hierarchical relationship embedded in the "ethnographic situation" and the textual production that flows from it.[73] I thus tried to create a breach in conventional ethnographic writing so that Davi Kopenawa's voice would be heard more directly. By playing a discreet editorial role, I hoped to avoid neutralizing the singular otherness of his words or dissolving their poetic quality and conceptual impact into patronizing academic theory.[74]

This book consists of autobiographical stories and shamanic reflections and is told in the first person, in Davi Kopenawa's inspiring and memorable voice. Yet this first person encompasses a double "I." The

words in this text are truly the narrator's own words, rendered as faith-
fully as possible from a huge body of audio recordings. This narrator had
a limited writing experience, however, and so the "I" in this narrative also
belongs somehow to me, his editorial alter ego.[75] This book is thus a
"written/spoken textual duet"[76] in which two people—the author of the
spoken words and the author of their written form—produced a text
working together as one.

THIS SHARED identity, in which two authors cohabit the same "I," could
cast some doubt on the faithfulness of the written word to the oral source
material. But distrust about the double "I" of such a "first person heterobiography"[77] is misleading, since complete identity between the narrator's "I" and the editor's "I" is always an illusion. Such convergence does
not even exist in conventional autobiographies, given the gap between
the "I" of memory and the writing "I" of the present. A life story, written
in collaboration or not, always implies a multiplicity of "I"s because, as
Philippe Lejeune correctly emphasizes, "one is always several when one
writes, even all alone, even about one's own life."[78]

Finally, the ethnobiographical process does not imply that only the
narrator has to "become other" through the editor's duplicating "I" of the
written text. The editor's "I" must also be capable—as Balzac put it—of
"becoming someone other than himself through the sheer exhilaration
of his moral faculties,"[79] just as the narrator has agreed to "become other"
by allowing the ethnographer to recreate his own words in writing. The
listening required for ethnobiographical writing demands an immersion
in and deep empathy for the ideas, history, and personality of the narrator. This sort of "lyrical depersonalization"[80]—in which the other is now
an "I"—is the only way an ethnographer can legitimately claim to render
the narrator's account in writing "in his stead." In this instance, my
thirty-year friendship with and admiration for Davi Kopenawa, rooted in
lengthy close acquaintance with the Yanomami and their language, is a
bond that sustained a unique experience of cross identification. I became
his alter ego by trying to bring out the richness of his thoughts just as he
committed himself to become mine through the trust he placed in my
way of putting his words in writing.

But, in this case, the narrator's "I" is split by more than the autobiographical process. As he often emphasizes himself, Davi Kopenawa's ac-

count usually embodies a multiplicity of other voices, and it works as a true narrative mosaic. Beyond his own reflections and personal memories, he continually refers to the values and history of his people, and he recounts them to us as such. The narrator's "I" is here inextricable from the "we" of the traditions and group memory to which he is trying to give voice. It is the collective "I" of an autoethnographer motivated by an intellectual, esthetic, and political drive to reveal his people's cosmology and tragic history to the white people capable of understanding.

In addition, through direct or quoted speech, Davi Kopenawa frequently talks about the teachings of the two elder shamans who were his primary mentors. The first is his mother's second husband, now deceased, who raised him in Toototobi from the time he was a baby and who first introduced him to shamanism and encouraged his interest in the vocation (see Chapter 3). The second mentor is his wife's father, who initiated him later as a shaman in Watoriki. It was under his tutelage that Davi Kopenawa elaborated his complex shamanic critique of the white people's world (see Chapter 5 and Appendix C).[81] This second mentor, whose presence is felt throughout this book, played indeed a decisive role in Davi Kopenawa's creative elaboration of his prophetic discourse on the gold rush in Roraima. Over the course of their many shamanic sessions together, these two men gradually created a shared interpretive space in which the mythological knowledge of one melded with the ethnopolitical skills of the other. Beginning in the late 1980s, this alliance set the basis for a poetic and powerful cosmological-ecological discourse that was decisive for expelling the gold prospectors and campaigning for legal recognition of Yanomami lands.[82] In this sense, Davi's father-in-law—to whom I am also bound by a friendship of more than thirty years—could really be considered another co-author of this book. He sat in, as a listener or commentator, on most of our recording sessions in Watoriki. From his perspicacious and ironic point of view, Davi Kopenawa and I have both walked off the beaten path at the halfway point between two worlds. He regarded us with amused perplexity, as eccentric cultural brokers between the shamanic knowledge of the elders and the improbable curiosity of white people.

Finally, the "I" of Davi Kopenawa's account also embodies the voices of many shamanic "images" of animal ancestors and cosmological beings. He refers to and quotes these entities throughout his stories and reflections to express or explain their point of view. This narrative multiplicity reflects the fact that Davi Kopenawa never thought that our conversations were merely ethnographic interviews. In his view, our exchanges were, if subtly, shamanic intercultural polylogues, alternating with traditional sessions about white people conducted with his father-in-law and other *Watoriki* shamans. As the Brazilian anthropologist Eduardo Viveiros de Castro observed:

> If shamanism is essentially a cosmic diplomacy devoted to the translation between ontologically disparate points of view, then Kopenawa's discourse is not just a narrative on particular shamanic content—namely, the spirits which the shamans make speak and act—; it is a shamanic *form* in itself, an example of shamanism in action, in which a shaman speaks about spirits to Whites, and equally about Whites on the basis of spirits, and both these things through a White intermediary.[83]

Constructing the Text

To edit an account as rich and complex as Davi Kopenawa's in French, from recordings made at different times entirely in his language, during turbulent events, is, as one can imagine, a delicate and challenging task. The journey from an oral "source" discourse to a "final" written account is long, even—and undoubtedly particularly—if the writer insists on being faithful to the words and style of his "model." Editing this type of "heterobiography" does not simply consist of transcribing and then translating, in installments, a pre-existing mental text that was progressively recited in a series of recorded interviews. Rather, the editor has to confront the outgrowth of a vast, fragmentary, and protean oral pretext, produced by means of a long-term dialogue through which he must find a unifying thread. It is this complex work of *alter ego* and *alter auctor* that, weaving an enduring narrative trail, allows the narrator's words to extend beyond the time and place in which they are spoken. I will try now to clarify the writing choices I made to manage this challenging narrative adventure.

Recordings

The primary sources for this book are two large collections of recordings made with Davi Kopenawa, the first from 1989–1992; the second from

1993–2001. The first series of interviews began with the issues that were addressed in that seminal recording by Alcida Ramos in Brasília in December 1989. Those core topics were developed into a series of monologues in which Davi Kopenawa spoke at length about his shamanic reflections on the smoke of gold, the death of shamans, and the destruction of the forest by gold prospectors. These first accounts became the thematic kernel of this book. They were a significant component of its first manuscript, edited in 1993, and later condensed in three long chapters in the present version (Chapters 15, 16, and 24).

After completing this work we began a series of interviews in which I asked Davi Kopenawa to comment on and explain numerous points from his previous narratives. He then developed his line of thought in two directions, which became the main axis for his later account. These new recordings initially dealt with the history of his group's contact with white people, as told to him by his elders, and then with his own direct experience of such contacts. He described the still-tentative advance by those outsiders in the forest from a period predating his own childhood to its later full-scale invasion by gold prospectors when he was in his thirties. These narratives provided the primary material that was edited for Chapters 9 and 14 in the second part of the book. After that Davi Kopenawa recounted his memories and shamanic impressions of his first trips to England, France, and the United States (where he had been invited to speak in defense of his people). This time, his personal experience of contact inverted the collective history of the white people's encroachment in the forest told by his elders. He described his own trajectory as a contemporary Yanomami shaman, moving out from the forest to the white world in Brazil and, then, to what is still called the "First World" there. These new recordings provided the essential material for the third part of the book.

The record of two final collections of reflections that highlight particularly original dimensions of Davi Kopenawa's account completed this work. The first theme of these reflections consisted of a historical counter-anthropology of the white world, built on cultural issues perceived by Davi Kopenawa in his journeys to distant places, as deeply conflicting issues opposing his world and our own (merchandise, war, writing, and "nature"). The breadth and singularity of these shamanic meditations on the "People of Merchandise" were key to our development of the third part of the book.

The second theme that emerged from these new conversations was much more difficult to deal with, undoubtedly because some aspects of it already infused everything we had already recorded. To clarify and organize these data required us to go all the way back to the sources of Davi Kopenawa's shamanic thought. This time, after all the ground we had covered, the challenge consisted in returning to the traditional foundations of his own thinking to explain his vocation, his initiation and work as a shaman, and, above all, the intricate ontology on which all that rests.

The complexity of the images and concepts that Davi Kopenawa tried to convey to me in our long discussions about the spirit world forced me to admit that the task we had set ourselves would require much more work on my part than I had imagined. Once the material that came out of these last interviews was arranged in a few short chapters, I had to acknowledge that I was still a long way from providing a truly satisfactory rendering of what Davi Kopenawa wanted our book to make clear. So I decided to rework everything we had already done. I could not just touch up the huge body of work we had pulled together, but had to develop all these new shamanic motifs to give them a much more prominent place.

The second wave of our conversations, which took place from 1993 to 2001, initially addressed most of the themes that were brought out in my first manuscript. I tried to give a thicker texture and brighten the tones of Davi Kopenawa's account by providing supplementary details and new wordings, so we rediscussed and rerecorded many sections of the same accounts and commentaries, this time arranged in chapters. The textual variants superimposed themselves in successive layers, in the same way that one creates, *per via di porre,* depth and density to design and color in a painting. I had used this strategy of "thick translation" before, with the narrators of a large collection of myths assembled in the 1970s and 1980s along the Rio Catrimani and the Toototobi.[84] The results were so rich that I was convinced that any single version of an account could give only a poor sense of the narrative talent of my interlocutors.

Finally, I put a considerable amount of work into new recordings of and notations about the initial parts of the 1993 manuscript concerning Davi Kopenawa's shamanic experience. This involved in-depth dialogic ethnographic research, which allowed me to develop this topic—the living source of Davi Kopenawa's account—and to give it the space it deserved in the final text of the book. This material, consisting of a large series of narrative sequences and autoethnographic explanations, pro vided the essential core for Part I, as well as material for several other chapters.

Transcription and Translation

These conversations with Davi Kopenawa were all conducted in the Yanomami language from 1989 to 1992, and from 1993 to 2001. They began as spontaneous talks on several themes and were gradually reworked through the format of nondirective interviews. I transcribed all the audio recordings using a writing system adapted for the Yanomami language. This same system has been used by the CCPY since 1996 in their Yanomami education project.[85]

Before being translated, the two collections of transcriptions were all read over again in Yanomami to prepare them for indexing. Rereading provided the first opportunity to prune the transcribed material. I began by eliminating interventions I had made to follow up on points or to refocus discussion, as well as questions I had raised to elicit more detail or to develop certain topics and, of course, the digressions that sometimes resulted from my poorly formulated questions. I also cut back all the least interesting repetitive sections, parenthetical comments on everyday incidents that were off the subject, parallel conversations with occasional listeners, or passages that were just too difficult to use because of the poor quality of the recording.

It would have been a formidable task to attempt a written translation of more than a thousand pages of transcripts in the order in which they were recorded. I therefore decided to translate the indexed thematic sections as I went along, in accordance with their relevance to my editorial work. I then melded different versions of the same stories and commentaries to make the French text more substantial and concise. I was committed to following Davi Kopenawa's words as closely as possible, but a literal word-for-word translation was, as I said, out of the question. Under the guise of accuracy, the result would have been a perfectly unreadable text in broken French. A literal translation also would have made Davi Kopenawa's voice sound like a stereotype of a colonial "native." My knowledge of Yanomami language and society were already acceptable (for a white person) by the time I had edited the first version of the manuscript and had certainly improved by the further work. By this time, having listened to him for hundreds of hours while meticulously transcribing our interviews, I had become very familiar with Davi Kopenawa's unique way of speaking. This experience allowed me to propose a translation that falls midway between a literal translation, which risked becoming a caricature, and a literary transposition that would have been much too far from Yanomami language constructions.[86]

The text in this book therefore closely follows Davi Kopenawa's own wording and speaking style, although I did trim sentences to make them flow better, and removed hesitations and repetitions, phatic expressions and questions. I also eliminated most interjections and onomatopoeias, retaining only those that helped keep the tone of the original narrative to the degree that was compatible with readability. Sometimes I elaborated on the elliptical form of certain ethnographic descriptions that Davi Kopenawa had not taken pains to explain because he credited me with some knowledge of the subject matter. Similarly, I sometimes completed his accounts or commentaries with minor additions, to clarify the tone of his remarks or to emphasize the forcefulness of his nonverbal expression. Within the limits of ordinary French speech, I also played with the classic tenses for verbal narrations, which do not correspond directly with those of Yanomami grammar, to accentuate the liveliness of his ideas.

I did not try to search for French equivalents of the conceptual and esthetic richness of the most unique words and expressions of Davi Kopenawa's speech. On the contrary, I always looked for formulations that approximated the metaphors woven into Yanomami language. I tried to avoid the intrusion of overly dissonant expressions from the French idiomatic repertoire. I used many endnotes to define and discuss key concepts of Yanomami cosmology and rituals so that their distinctive features would not be obliterated, despite everything, through the process of translation. I also thought it useful to indicate the rare occasions when Davi Kopenawa used certain Brazilian Portuguese words or expressions. I left a great number of ethnobiological, ethnogeographical, or Yanomami ontological terms as they were, for lack of plausible equivalents. These terms are made explicit in their context, and they are also explained in endnotes or glossaries.

Finally, within the constraints that I imposed on myself for accuracy, my work of literary elaboration largely consisted of trying to recreate a tone and a turn of phrase that would do justice to Davi Kopenawa's unique verbal expressions and to the emotions that suffuse his speech. Davi Kopenawa is a shaman, a Yanomami intellectual, and he expresses himself in his language as such. It was not possible to resort to an improvised form of exotic colloquial language. But recourse to a specialized and affected style that would imitate our standard academic discourse was not an option either. Steering clear of those two perils, I sought to keep together ethnographic accuracy and esthetic concerns, make the text readable, convey the poetic and contrastive conceptual effects of Yanomami speech, and bring out the voice of the narrator, at times indignant, jovial, or poignant. It is now up to the reader to decide how well I have succeeded in this challenging search for the right balance between the echo of a voice, documentary fidelity, and the "pleasure of the text."[87]

Organization and Composition

After rereading, trimming, and thematically indexing all the transcriptions of the interviews that took place between 1989 and 1992, I sketched out the broad outlines of a tentative synopsis for the first draft of the book. I began by using the key stages of Davi Kopenawa's life, describing his successive encounters with representatives from the encroaching frontier (military [the Boundary Commissions], Evangelist missionaries, FUNAI agents, roadway workers, gold prospectors, and NGO members) in the order he had gradually told me about them. However, two groups of thematic interviews did not fit into this chronological schematic framework at all. One involved the prophetic interpretation of the gold rush, which Davi Kopenawa had recently elaborated with his father-in-law; the other concerned the first accounts of his shamanic experience, which underlay these elaborations and required greater emphasis.

The interviews from the second period of recording sessions (1993–2001) were conducted according to my first tentative chronological framework. But the breadth of this new collection of shamanic comments and reflections yielded so much new material that I had to recast the book's structure almost completely. Apart from the work involved, this caused no significant difficulty, for neither Davi Kopenawa nor I had ever intended to limit ourselves to the ethnocentric conventions of classic autobiographies and their too-simple "plot of predestination."[88]

Through this reworking of my 1993 first manuscript, I was finally able to organize the book in three sections, each with eight chapters, to make the twenty-four chapters alternately chronological and thematic. Then this hybrid structure became even more complex. I organized the chronological material, mixing episodes from Davi Kopenawa's personal life story (such as Chapters 3, 5, 6, 12, 14, 17, and 20) and episodes of the collective historical experience of his people (Chapters 9, 10, 11, 13, and 15). In the same way, I arranged the autoethnographic thematic material to vary between explanations of traditional knowledge (Chapters 2, 4, 7, 8, and 21–23) and personal elaborations about it (Chapters 14, 16, 18, 19, and 24). I chose chapter and part titles to make good use of expressions borrowed from Davi Kopenawa himself, and to introduce their content as accurately and as evocatively as possible. The title of the book was also selected in this way. "The Falling Sky" refers to a myth explaining the cataclysmic end of a first humanity, which the Yanomami think prefigures the fate of our world, invaded by the deadly smoke of metals and fuels.

The strategy I used to edit the content of the chapters echoes the one I used to order them. I followed in each the same hybrid chronological and thematic chain that links one to another throughout the book. Although this arrangement remained imperfect, it was an attempt to blend pure narrative parts with more ethnographic ones. This alternating structure became increasingly elaborate, all the more so since Davi Kopenawa's accounts and reflections are expressed in different discursive registers, and in direct or quoted speech. The threads of his narratives are woven through memories of personal experiences, as well as through accounts of historic events, dreams or visions, and myths or shamanic prophecies. Similarly, his explanations of his own society and his astute comments on the worrisome practices of ours are also combined in several discursive domains: autoethnographic description, comparative cultural observation, conjectural history, social critique, and political exhortation.

Gradually piecing together the fragments of indexed transcriptions to assemble this book required scrupulous attention to detail in selection and arrangement. This proved to be even more complex than the translation and written rendering of the narrator's voice. I avoided arbitrary criteria external to the transcribed material and let the process be driven by inferring possibilities of arrangement and symbolic resonance from Davi Kopenawa's speech. Through an almost hypnotic immersion, listening and rereading countless times, I built an architecture for the manuscript as its coherence and harmony unconsciously took shape for me. I explained the successive drafts and formats to Davi Kopenawa and made sure, through lengthy sessions when we revised the text together in March 2008 in *Watoriki*, that we were in complete agreement on its final version.

Lastly, I framed the twenty-four chapters with two introductions (my

"Setting the Scene" and Davi Kopenawa's "Words Given") and two con-
clusions (Davi Kopenawa's "Words of *Omama*" and the present chapter). These frames are complementary in many ways. Arranged in chias-
mus, the double parts create an echo between the co-authors' voices on
both sides of the book and break out of the convention of the paired anthropologist's prologue and epilogue that are the readers' guides in traditional ethnobiographic texts.[89] In this modest textual rebellion my primary goal was to highlight the exceptional personality of the narrator,
Davi Kopenawa, an extraordinary witness to his people's recent history
and ambassador of Yanomami shamanic knowledge, and allow him to
give his own perspective on the pact and project that supported our common effort.

—*B. A.*

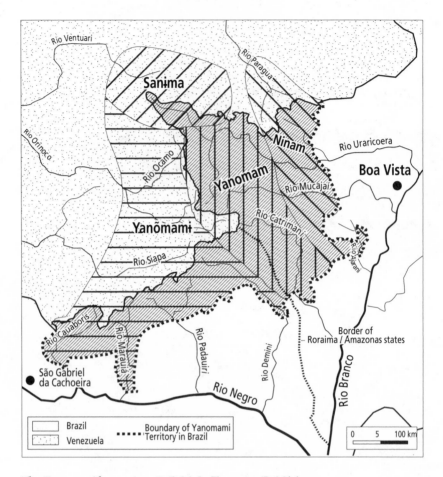

The Yanomami languages. © F.-M. Le Tourneau/P. Mérienne

Ethnonym, Language, and Orthography

"Yanomami" is a simplification of the ethnonym *Yanõmami tëpë* (which means "human beings") and is how members of the western branch of this Amerindian group refer to themselves. This term was first adopted in Venezuela as a name for the entire ethnic group. It also began to be used in the late 1970s in Brazil by anthropologists, nongovernmental organizations, and by the government's indigenous agencies. In the latter country, the Yanomami were previously referred to by different regional names, such as Waika (Guaica, Uaica), Xiriana (Xirianã, Shiriana), Xirixana (Shirishana), Yanonami, and Sanumá, among others, as well as by generic names such as Yanomama or Yanoama.

Today, the term "Yanomami" also refers to the family of languages spoken by the members of this ethnic group. It is an isolated linguistic family in the Amazon composed of four major languages, each subdivided into several dialects, whose mutual intelligibility varies greatly.[1] The American linguist and former missionary Ernesto Migliazza first described this language group ("Yanomama family") more than forty years ago.[2] The distribution of the languages in Brazil is as follows:

1. *Yanomam*: along the Rio Parima, Rio Uraricoera, upper Rio Mucajaí, upper Rio Catrimani, and Rio Toototobi (Roraima state);
2. *Yanomami*: along the Rio Demini, Rio Aracá, Rio Padauiri, and Rio Cauaboris (Amazonas state);
3. *Ninam* (or *Yanam*): along the middle Rio Mucajaí and the Rio Uraricaá (Roraima state);
4. *Sanima*: along the Rio Auaris (Roraima state).

In his 1994 study, Henri Ramirez proposed changing the classification of these four languages ("Yanomami family") as follows:[3]

1. *Yanomam* and *Yanomami* are considered two "super dialects," eastern and western respectively, of one and the same language ("Y division");
2. *Ninam (Yanam)* remains classified as a different language ("N division");
3. *Sanima* also remains classified as a different language ("S division");
4. A claim is made for the existence of possibly a fourth new language ("A division") in the Rio Ajarani and Rio Apiaú areas and along the lower Rio Mucajaí and Rio Catrimani.[4]

The language spoken by Davi Kopenawa and his in-laws' group from *Watoriki* belongs to the same Yanomami dialect (the eastern Yanomami-c dialect, according to Ramirez, 1994: 35) spoken in the Rio Toototobi, Rio Catrimani, and Rio Uraricoera regions, although there are some small linguistic differences among these areas. This dialect is designated by the people of *Watoriki* as *yanomae t^{h}ë ã*, "*yanomae* speech," and its speakers call themselves *yanomae t^{h}ë pë* ("human beings").

This language has seven vowels. Six of them are familiar to English speakers: *a, e* (pronounced like the vowel sound in *fate*), *i, o, u* (pronounced like the vowel sound in *food*) and *ë* (pronounced like the vowel sound in *but*). The seventh vowel, *i* (barred *i*), is a high central unrounded vowel whose sound is between *i* and *u*. All of these vowels can be nasalized, which is marked with a tilde (for example, *maũ u*, "water"). Long vowels are indicated by repetition of the letter (for example, *xaari*, "straight").

There are thirteen consonants:

the voiceless occlusives: *p* (bilabial), *t* (apico-alveolar), and *k* (velar). The first two are often pronounced interchangeably with their voiced counterparts *b* and *d*;

the aspirated apico-alveolar occlusive *t^h* (pronounced like a *t* followed by a light puff of air);[5]

the fricatives *s* (lamino-alveolar) and *ʃ* (lamino-postalveolar); the latter is written as *x* using Brazilian Portuguese orthography and corresponds to *sh* in English spelling;

the rounded glottal fricative h^w, a phoneme specific to the *yanomae*
   dialect (pronounced as an aspirated *h* with rounded lips);[6]
the apico-alveolar vibrant *r*, which is pronounced similar to the medial consonant *t* in the English word *ditto* or, when it occurs next
   to a nasalized vowel, like the medial nasal in *many*;
the nasal consonants *m* (bilabial) and *n* (apico-alveolar);
the approximants *w* (bilabial) pronounced as in *wax* or *web*, and *j*
   (dorso-palatal), written as *y* and pronounced as in *yak* or *yoga*.

The Yanomami languages are polysynthetic. Affixes are added to root
words to express all forms of grammatical relationships. For example,
they employ a system of some fifty suffixes that can be linked in numerous combinations of up to eight or nine, one after the other, starting from
the same root verb, to specify different dimensions of the expressed action.[7] Traditionally, these are not written languages. The first convention
for the transcription of a Yanomami language was introduced in Brazil in
the 1960s in the context of local missionary schools opened by evangelical organizations (New Tribes Mission, Unevangelized Fields Mission).

Starting in the 1990s, a new writing system based on the International
Phonetic Alphabet and the Portuguese alphabet was introduced throughout much of the Yanomami territory in Brazil.[8] This orthographic system
was intended to be as compatible as possible with the one already adopted
among the Venezuelan Yanomami. It has been spread among the Brazilian Yanomami mainly through the bilingual education program run
by the nongovernmental organization Comissão Pró-Yanomami (CPPY)
and through the work of linguists and anthropologists associated with
the CPPY (including this author). The writing conventions established by
this education program are now currently in use in the majority of Yanomami schools in Brazil and were adopted by the official education system. Moreover, it is widely used in village newsletters and documents
published by the principal Yanomami association in Brazil, *Hutukara*,
which was founded in 2004. This is the orthographic system that I also
used in this book to transcribe Davi Kopenawa's words.

—B. A.

The Yanomami in Brazil

Few Amerindian groups of the Amazon have become as famous, in professional anthropology and far beyond, as the Yanomami, and none have been as controversial. German ethnographers published the first modern studies about the Yanomami of Venezuela and Brazil in the mid-1950s. In Venezuela these studies were undertaken by the expedition organized by the Frobenius-Institut of Frankfurt under the direction of Otto Zerries.[1] In Brazil they emerged from the field trips of Hans Becher in Amazonas state.[2]

It was not until the late 1960s, however, that the Venezuelan Yanomami became widely known as a result of the American anthropologist Napoleon Chagnon's ethnographic writings on their social and political organization. Chagnon's *Yanomamö: The Fierce People,* published in 1968, sold probably several million copies.[3] Due to the dramatic reviews of Chagnon's claims about Yanomami violence, the group developed an enduring reputation among English-speaking readers for being "primitive" and warlike.[4]

In France, the Yanomami had the reverse image as "affable savages," thanks to the French anthropologist Jacques Lizot's *Le cercle des feux* (1976). Although this book was translated into English as an explicit response to Chagnon's theories,[5] it had little impact in the Anglophone world. The Yanomami's negative reputation began to wane in the late 1980s for a different reason. The spectacular gold rush in Roraima state in northern Brazil (1987–1989) was then at its peak. The world's attention was suddenly drawn to the local Yanomami who were dying by the hundreds from malaria, pneumonia, and acts of violence resulting from the invasion of their lands by hordes of illegal gold prospectors.[6] Once seen as exemplars of neo-Darwinian theories of "primitive" warfare, the

Yanomami were now transformed by the international press into para-
digmatic victims of the modern devastation of the Amazon.[7] This turn-
around reached its high point in 1993 with worldwide coverage of the
Haximu Massacre (see Appendix D) in which sixteen Yanomami, mostly
women, children, and old people, were massacred by *pistoleiros* (hired
killers) in the pay of the gold prospectors.[8]

The Yanomami made international headlines again in the early 2000s
in the wake of an intense polemic against what was described as unethi-
cal professional behavior by anthropologists, geneticists, and reporters
who had worked with the Venezuelan Yanomami from 1960 to 1970.
The very title of Patrick Tierney's book, which launched this controversy
in the United States—*Darkness in El Dorado: How Scientists and Journalists Devastated the Amazon*—shows the seriousness of the accusations
and the overall polemical tone. An essential complement (and correction)
to this book is *Yanomami: The Fierce Controversy and What We Can Learn
from It*, which takes up the issues of the polemic in greater depth and in a
more balanced and complex way.[9]

The Yanomami are considered to be very different genetically, anthro-
pometrically, and linguistically from their immediate neighbors, the
Ye'kuana Carib-speaking group.[10] Based on this data, researchers hypoth-
esized that the Yanomami could be descendants of an ancient Amerindian group ("proto-Yanomami") that dates back a thousand years and
who lived in relative isolation over a very long period in the highlands of
the upper Orinoco and the headwaters of the upper Rio Parima (Serra
Parima).[11] This theory holds that a process of internal differentiation began there about seven hundred years ago and that this is the origin of Yanomami languages and dialects in use today.

According to Yanomami oral tradition recorded by several anthropologists since the 1960s, as well as the earliest written references to this
ethnic group, which date from the late eighteenth century, the historic
center of their territory was indeed located in the Serra Parima range,
which rises 1,700 meters and defines the border between Brazil and Venezuela. To this day, this region is still the most densely populated area of
Yanomami territory.

Despite the first incursions into the upper Orinoco, Rio Branco, and
Rio Negro basins during the second half of the eighteenth century, map-

ping this area was left to the imagination for a very long time, and the present Yanomami territory was generally depicted as nothing but a vast desert expanse. Yet, at the beginning of the next century, on this imaginary blank map, Yanomami communities that were at war with several nearby ethnic groups and later with each other probably began to disperse through successive fissions from the Serra Parima highlands towards the surrounding plains. This territorial expansion and splintering of villages then increased from the mid-nineteenth to the mid-twentieth century. Experts generally attribute this population surge to the acquisition of new cultivated plants (such as bananas) and metal tools (or at least iron fragments) through trade or raids against neighboring Carib-speaking or Arawak-speaking peoples who were themselves in direct contact with the colonial frontier. The gradual decline of the latter groups—decimated by Portuguese slavery, violence, and epidemics since the mid-eighteenth century, notably during the colonization of the Rio Branco in Brazil— meant that huge forest areas were emptied of their original population and started to become available to waves of Yanomami migration.[12] The current configuration of Yanomami territory is the result of this long historical process of demographic growth and residential division among multiple warring communities migrating toward open land, as well as the interruption of this process by the encroaching white frontier. Thus the Yanomami expansion, radiating outward from what was probably a modest original territory located between the upper Rio Parima and the Orinoco, was gradually interrupted between 1940 and 1960 by the establishment of the first SPI outposts in Brazil and, above all, through the founding of missions by various religious organizations.

Some 15,500 Yanomami live in Brazil along the tributaries of the upper Rio Branco in the western part of Roraima state and along the tributaries on the left bank of the Rio Negro in the northern part of Amazonas state. Their vast territory of 96,650 square kilometers, legally sanctioned by presidential decree in May 1992, is home to diverse natural habitats, ranging from dense tropical forest lowlands to mountainous forest regions and high-altitude savannas. The scientific community views the region as a priority for protection of biodiversity in the Brazilian Amazon.[13] Its surface area represents nearly one percent of the remaining tropical forest on the planet.

At the end of the nineteenth century, the Yanomami in Brazil traded and were at war with approximately a dozen neighboring Amerindian groups: Carib-speaking groups (Ye'kuana, Purukoto, Sapara, Pauxiana) and isolated language groups (Makú, Awaké, Marakana) in the north and east, as well as Arawak-speaking groups in the south and west (Bahuana, Mandawaka, Yabahana, Kuriobana, Manao, Baré).[14] Their first contacts with white people occurred during the early twentieth century (1910–1940) on the fringes of their territory, most notably with collectors of forest products (such as balata latex, piassava palm fibers, and Brazil nuts), soldiers from the Brazilian Boundary Commission (CBDL), *sertanistas* from the Indian Protection Service (SPI), and foreign travelers like the famous German ethnographer Theodor Koch-Grünberg (1872–1924).

SPI outposts were opened in the 1940s and 1950s, generally to provide support to military expeditions demarcating the border with Venezuela.[15] They were followed in the late 1950s and early 1960s by several missions established by American evangelical organizations (New Tribes Mission, Unevangelized Fields Mission, and Baptist Mid-Mission) and Italian Catholics (Salesian and Consolata orders). These outposts and missions became the first permanent points of contact on what has since become the periphery of Yanomami territory. They formed focal points in the areas where they were placed, and the local Yanomami population gradually converged to settle around them. They began regularly providing the surrounding villages with manufactured objects, in particular, highly coveted metal tools. They also became the entry point for a series of deadly epidemics and infectious diseases (measles, flu, whooping cough, and so on), to which the hitherto isolated Yanomami were highly vulnerable. In the course of these epidemics, populations nearest the outposts and missions were able to get some medical care and thus limited their demographic losses. More isolated villages, on the other hand, were decimated. This process gave rise to the polynucleation of the region, with sites closest to outposts and missions becoming more densely populated and surrounded by vast areas that were sparsely occupied or uninhabited. Since then, this multipolarity has characterized the pattern of the Yanomami population's spatial distribution.[16]

In Brazil, regular and permanent contact between the Yanomami and forest product collectors, state indigenists, and missionaries continued across their territory until the early 1970s. Then, the Brazilian military's geopolitical plans to occupy the Amazon suddenly thrust the Yanomami into a new era of much more intense contact with the regional economic frontier, notably in the western part of Roraima state (which was only a federal territory at the time).[17] Thus, in 1973, a 235-kilometer stretch of the Perimetral Norte highway (BR-210) was opened across the southern edge of Yanomami territory as part of a "National Integration Plan." This plan was launched in 1970 under General Emílio Garrastazu Médici's administration (1969–1974) to control and populate the northern border area of the Amazon region. Then the first fifty kilometers on both sides of the highway were opened to agricultural colonization in the wake of a new public project to develop the Amazon, the Polamazônia Project, undertaken by General Ernesto Geisel's government (1974–1979). The presence of the road workers and, later, the influx of colonists along the highway caused an unprecedented epidemiological shock among the Yanomami population and led to significant demographic losses. In 1976, however, the Perimetral Norte construction in the Yanomami territory along the Brazil–Venezuela border was suddenly abandoned for lack of international loans. Nothing was left but a track of red gravel that was gradually taken over by the forest. All that remained for the Yanomami in the Rio Ajarani and Rio Apiaú regions, where the roadwork began and agricultural colonization had expanded, was a situation of social and sanitary degradation. Its effects are still evident there forty years later.

Coincident with the opening of the Perimetral Norte, a systematic inventory of natural resources in the region was begun as part of the RADAM project,[18] which revealed the mining potential of the Serra Parima. Shortly after this survey was completed in Roraima (1975), a first wave of illegal prospectors (garimpeiros) invaded the upper Rio Parima area via the airstrip of an evangelical missionary outpost located near the Surucucus plateau. The sensationalized news of mineral wealth in the highlands of Yanomami territory in Roraima state then unleashed a series of invasions by gold prospectors over the next decade. This quickly grew into one of the most dramatic gold rushes of the twentieth century. Between 1987 and 1989, no less than ninety clandestine landing strips were opened, notably in the headwaters region of the main tributaries of the Rio Branco (Rio Uraricoera, Parima, Macajaí, and Catrimani). The number of gold prospectors in the area at the time was estimated to be 40,000.[19]

During this period, trade and conflict with gold prospectors was the main form of contact between the Yanomami and the regional frontier. The number of *garimpeiros* in Yanomami territory in Roraima was approximately five times larger than the indigenous population. This massive invasion inevitably led to catastrophic ecological and epidemiological consequences on a much larger scale than the one suffered during the 1970s period of highway construction and agricultural colonization. In just three years, epidemics of malaria and respiratory infections claimed the lives of nearly 13 percent of the Yanomami population in Brazil (according to the Brazilian Ministry of Health's own statistics). The devastation of the upper streams of the main rivers in the region, and their systematic pollution with mercury and motor oil discharge and other refuse, also caused considerable damage to the local environment and deprived the Indians of most of their means of subsistence.[20]

In 1990, only after the decimation of the Yanomami had provoked an international scandal, FUNAI and the Brazilian Federal Police began a series of expulsions that managed to control the influx of *garimpeiros* in Yanomami lands for a period of time. Despite these initiatives, however, the gold prospectors maintained their activities in the more isolated areas of the Yanomami territory along the Brazil–Venezuela border. This insidious invasion continued to expose the Indians to chronic contamination and dramatic violence—the most extreme example being the 1993 Haximu Massacre.[21]

In addition to the *garimpeiros'* interest in the Yanomami highlands, which endures as long as the price of gold in world markets is high, other existing and potential economic activities (agricultural colonization, cattle ranching, forest exploitation, and industrial mining) could pose serious mid- and long-term threats to the integrity of the Yanomami population and the tropical forest they live in and seek to preserve. Thus, despite official recognition in 1992, more than half of the surface area of the Terra Indígena Yanomami is already claimed by some six hundred titles or requests for titles on file with the Brazilian Ministry of Mines and Energy.[22] Since 1978, agricultural colonization plans by federal and regional administrations—expanded by a parallel land-grab movement—escalated forest clearing and settlement at the southeastern edge of Yanomami territory. This process has already reached the border and even encroached on Yanomami lands.[23] In addition to their predatory exploitation of natural resources in the surrounding forest (through hunting, fishing, and harvesting lumber), the local settlers and cattle ranchers engage in large-scale clearing and burning in a region where each dry season is more extreme than the last. This has sparked giant forest fires, such as those that occurred in 1998 and 2003, and these fires have had a lasting negative impact on the region's biodiversity.[24]

—B. A.

Watoriki

Yanomami local groups are generally made up of real or classificatory cognatic kin living together in a cone-shaped communal house (larger houses are truncated cones) called *yano a* or *xapono a (kami t*ʰ*ëri ya-maki)*.[1] Each of these social units considers itself economically and politically autonomous. Ideally, members prefer to marry within their community, to the degree that demographics and their Dravidian kinship system permit.[2] Despite this autarkic ideal, each local entity is linked to several similar nearby groups through marriage and ceremonial interactions (mostly funeral and warfare rituals). These allied communities form webs that vary in composition and stability and are opposed in a state of structural hostility to other similar multipolar networks. This hostility is expressed in different ways: raids, sorcery, and aggressive shamanism.[3] The borders of these clusters of allied local groups are unstable and not sharply defined, but clearly separate friends *(nohimotima t*ʰ*ë pë)* from enemies *(napë t*ʰ*ë pë)*, visitors *(h*ʷ*ama pë)*, and warriors *(wai pë)*. These small sociopolitical galaxies partially overlap and gradually form a wider social, political, and ceremonial network that ultimately connects all the communities throughout the Yanomami territory, "the land-forest of human beings" *(yanomae t*ʰ*ë pë urihipë)*.[4]

The "inhabitants of the Wind Mountain" *(Watoriki t*ʰ*ëri pë)* are one of the nearly 250 local Yanomami groups that exist in Brazil today. Davi Kopenawa has lived in this community with his family and his in-laws since the late 1970s, and this is where most research and interviews were conducted for this book. *Watoriki* is located in the far northeast of Amazonas state, in a region bordered by the Rio Catrimani basin (a tributary of the Rio Branco) to the east and the Rio Demini basin (a tributary of the Rio Negro) to the west. It is situated in lowlands south of the Serra Parima

range, which defines the border between Venezuela and Brazil. Established less than 200 meters above sea level, the *Watoriki* communal
house is flanked by the Serra do Demini, a cluster of steep hills and rocky
peaks rising more than 700 meters. The area is covered by lowland tropical forest, typical of the Amazon and the Guiana Shield, which forms a
thick canopy of medium-sized trees and a sparse scattering of spectacu-
larly tall individual species, such as masaranduba *(Manilkara huberi)*,
cedrorana *(Cedrelinga catenaeformis)*, kapok tree *(Ceiba petandra)*, jatoba
(Hymenaea parvifolia), and Brazil nut tree *(Bertholletia excelsa)*. The understory is generally rather clear, except in areas susceptible to flooding,
which are colonized by thick groves of prized fruit-bearing palm trees.[5]
On the hillsides, vegetation is sparser and grows closer to the ground due
to the steepness of the slopes, and some cliffs and ridges are completely
barren. Generally, the region has ferralitic soil—red or yellow with a clay-
like texture that is common to tropical forests—but there are also sandy
zones around large outcroppings of rocks and hills.

The large communal dwelling that provides shelter for the *Watoriki*
community in its present location was built in 1993 when there were
eighty-nine inhabitants. Today there are nearly one hundred and seventy-
four (54 percent are less than twenty years old), including sixteen shamans. The structure and the roof of the house have been regularly renovated at this location since the 1990s. The dwelling is surrounded by
approximately thirty hectares of cultivated gardens, primarily banana and
manioc. It is located near the end of the section of the Perimetral Norte
highway that was opened in the southeast area of Yanomami territory
in 1973. Road construction was abandoned in 1976, and the forest has
largely taken over. However, one section of the highway has been converted into a landing strip (kilometer 211), which provides the only access
by airplane to and from Boa Vista, the capital of Roraima state, 280 kilometers away. The FUNAI outpost in Demini, which Davi Kopenawa was
entrusted to run in the 1980s, is located along this airstrip, 2.5 kilometers
east of the *Watoriki* communal house. Today it is basically a health clinic,
run by the Special Secretary for Indigenous Health (SESAI).

The *Watoriki* house is an imposing ring-shaped structure, approximately 70 meters in diameter.[6] Its circular sloping roof surrounds a large
open-air central plaza *(yano a miamo)*, and the exterior is walled off with
slats of wood that are approximately 1.25 meters high. This dwelling, cov-

ered with leaves from small understory palms *(Geonoma baculifera)*, was built in a large clearing *(yano a roxi)* as a precaution against trees falling from the adjacent forest. Its inhabitants include some thirty family groups, living side by side under the large circular roof. Each group has its own space where family members hang their hammocks around a fire. The fire almost never goes out. It is used for cooking during the day and for heating at night.

This large dwelling has four main entrances *(pata yoka)* that are separated on the inside from adjacent hearths by small corridors made of palm wood slats. Each of these openings is referred to generically as a "path door" *(periyo yoka)*. They are called, more specifically, "garden door" *(hutu yoka)*; "hunting door" *(rama yoka)*; "guest door" *(hʷama yoka)*, which is the entryway for visitors from allied villages; and "outsider's door" *(napë yoka)*, which leads to the health clinic. Other, much smaller doors *(wai yoka)* are for everyday use and allow members of the different family groups more privacy to go outside.

During tropical storms, the earth floor in the central plaza can flood. The water is channeled to the outside by two small drainage ditches. The ring-shaped roof is constructed in two parts: a wide sloped roof that tilts towards the outside clearing and covers the hearth areas, and a narrower sloped roof that leans in the opposite direction and encircles the central plaza. The main roof is five meters tall at its highest point and slightly overlaps the secondary roof to keep the rain from getting in while allowing smoke from the hearths to get out.

The area covered by the main roof forms a circular band some ten meters wide. Slightly less than half the area is used strictly as living space. This ring of households is subdivided into three concentric circles. The first *(yano a xikã)* is a more feminine space, located between the back of the house and the circle where hammocks are hung. Women store firewood, pots of water, and baskets in this area. Food, utensils, and tools are also stored there on palm slat shelves. The adjacent area *(yahi a or nahi a)* is the family space where hammocks for couples and their children are hung around a fire. The next area, located between the ring of hearths and the central plaza *(yano a hehã)*, is generally reserved for political and ritual activities (meetings of the men, presentation dances by guests, ceremonial food preparation, shamanism). This area is open to everyone as a passageway between family hearths in the course of everyday life. Finally, the central plaza, usually reserved for the ceremonial activities pertaining to the funerary and alliance feasts *(reahu)*, is also used on a daily basis by small children as a playground.

The exterior wall of the house is usually made of slats from the trunk of a palm *(Socratea exorrhiza)*. It protects inhabitants from the wind and cold night air. The large opening of the central plaza allows light to filter under the ring-shaped roof, which gently illuminates the living areas. The high palm-thatched roof shades and insulates to maintain a comfortable temperature in the house, even on the hottest days. The interior of the large *Watoriki* communal dwelling offers a harmonious feeling of spaciousness and intimacy to inhabitants and visitors alike. It is simultaneously welcoming and comfortable.

Davi Kopenawa's father-in-law, Lourival, almost eighty years old, the preeminent doyen and shaman of *Watoriki*, says that in his childhood (probably during the late 1930s), his elders stayed primarily in two areas along the upper waterways of the Rio Mucajaí: *Xioma* and *Mrakapi*. These areas are also remembered as the sites of the first epidemics introduced by white people through the intermediary of other ethnic groups, from whom the Yanomami acquired metal tools. During the following decades (1940s–1950s), the ancestors of the current inhabitants of *Watoriki* gradually migrated south across the small tributaries of the upper Rio Catrimani. By the 1960s, they arrived on the upper course of one of the large tributaries on the river's right bank, the Rio Lobo d'Almada. They lived there until the early 1970s.

At the time they occupied a site there called *Hapakara hi.*[7] But one part of the community decided to migrate much farther south. They settled along the *Werihi sihipi u* River,[8] a small tributary of the Rio Mapulaú, in response to an invitation from the Toototobi evangelical missionaries. This move was motivated mainly by the desire to get closer to the mission outpost in order to gain direct access to a new source of trade goods. During the early decades of the twentieth century (and probably earlier), fragments of metal tools could only be obtained through a complex chain of intervillage trade along the upper Rio Mucajaí or by traveling a long way to other Amerindian groups located farther north, notably along the upper Rio Parima. From the mid-1960s, with the opening of the Catrimani Catholic mission, south of the Rio Lobo d'Almada, the group's trading network had shifted towards the Yanomami communities located downstream. This considerably enhanced opportunities to acquire man-

ufactured objects. However, the community's location on the periphery
of the Catrimani mission's sphere of influence (four houses away from it)
made it politically difficult to take full advantage of the situation. In the
end, moving towards the Rio Mapulaú appeared to provide a more advan-
tageous trade position. From this location, there was only one intermediary village connected with the Toototobi Mission (the house where Davi
Kopenawa's relatives lived).

However clever this plan may have been politically, it proved to be disastrous in terms of health. In 1973, an epidemic took the lives of a great
many members of the group after they moved into their second house
along the *Werihi sihipi u* River. After this tragedy, the survivors temporarily returned to live in their former dwelling, *Hapakara hi,* on the upper
Rio Lobo d'Almada. In 1974, however, some of them turned back again to
settle near a temporary FUNAI outpost that had recently opened in the
Rio Mapulaú area. This did not bring an end to their troubles. In 1976,
a measles epidemic spread from the Catrimani mission upstream and
decimated the members of the group who had stayed on the Rio Lobo
d'Almada. Then, in 1977, FUNAI closed its outpost on the Rio Mapulaú
and abruptly abandoned the remaining inhabitants of the *Werihi sihipi u*
River to their fate. The group, which had dwindled to about twenty people, found itself completely deprived of access to trade goods and, in the
context of ever-greater epidemiological threats, quite simply threatened
with extinction.

At this time, Davi Kopenawa had begun working as an interpreter at
Demini, the new FUNAI outpost that had opened to the south of Yanomami territory at the far end of the Perimetral Norte highway. The head
of the outpost planned to establish a model indigenous agricultural colony. Of course, to carry out this project he needed to attract a Yanomami
population, which was lacking in the area. The nearest communities to
the east and west were, respectively, fifty and eighty kilometers away.
Lourival, Davi Kopenawa's future father-in-law, had become—after the
death of his older brother during the 1973 epidemic—the great man
*(pata t*ʰ*ë)* in his community. After that tragedy, he knew that assuring his
people's access to white people's medicines was as crucial as obtaining
metal tools.

When he learned that a new FUNAI outpost had opened at kilometer
211 on the Perimetral Norte highway and that the young Yanomami interpreter he had met in 1973 on the *Werihi sihipi u* River worked there, he

understood that this was a much more promising situation for his group than prior moves had afforded. Settling near white people was always dangerous and unpredictable. He decided to pick up and move again towards the new Demini Outpost, but this time through a cautious and gradual migration. In 1978, he began to build a new house along the Rio Ananaliú, a tributary off the right bank of Rio Demini. This first approach made visits to the Demini Outpost easier and enabled the group to protect itself from the dangers of ill-considered contact with outsiders, the consequences of which it had so cruelly endured. In the 1980s, no less than four sites were then successively occupied by the community on the distant periphery of the outpost. Finally, in 1993, the *Watoriki* house was established on the site where it is located today. A skilled strategist, Lourival had meanwhile granted one of his daughters' hand in marriage to Davi Kopenawa, a few years before initiating him in shamanism. In so doing, Lourival thus made Davi Kopenawa simultaneously his disciple and his dependent,[9] the relationship between father-in-law and son-in-law being the fulcrum of Yanomami political authority.

Thus, by moving closer to the Demini Outpost, Lourival managed to leverage the role that his new son-in-law, Davi Kopenawa, was expected to play for FUNAI to the advantage of his own community. This subtle ethnopolitical shift gradually undermined the domination of a series of white FUNAI agents at Demini. Things got to the point at which the agency had no choice but to put Davi Kopenawa in charge of the outpost. Lourival's strategy was, this time, a complete success. Having patiently mastered the terms of interethnic relations through the traditional political game of kinship relations, he succeeded in winning material advantages for his group through an association with a FUNAI outpost while neutralizing the paternalistic dependency that such a relationship traditionally implies.

Moreover, by initiating Davi Kopenawa in shamanism, he was able to work with him to elaborate a cosmological and political interpretation of the white world's encroachment, one that proved very effective for defending Yanomami rights in general and his community's rights in particular. Davi Kopenawa's statements against the gold prospectors, the mining companies, and the cattle ranchers were thus based as much on information taken from white people's legal and ecological discourses as on the shamanic knowledge transmitted by his father-in-law. By the end of the 1980s, these statements were carried by all the major media in

Brazil and internationally. Finally, just before the UN summit on the environment in Rio de Janeiro in 1992, when the Amazon was viewed as a major symbol of the world crisis of predatory development, Davi Kopenawa compelled the Brazilian government to pass a decree that officially guaranteed the Yanomami land rights.

—B. A.

The Haximu Massacre

The original version of this report was written in 1993 at a time when contact between Yanomami from the highlands and white people was still relatively recent. The text was edited after the conclusion of the official Brazilian inquiry into the massacre of sixteen Yanomami by gold prospectors in the Rio Orinoco headwaters (Venezuela). Davi Kopenawa was a Yanomami observer to this inquiry, in which I participated as the interpreter and anthropological adviser to the Brazilian federal police and federal prosecutors working on the case. This account is based on direct testimony by survivors, interviews with arrested gold prospectors, and the forensic investigators' report. It was published in the Brazilian daily newspaper Folha de São Paulo *on October 3, 1993, under the headline* "Antropólogo revela os detalhes da chacina dos Indios Ianomâmis" *["Anthropologist Reveals Details about Slaughter of Yanomami Indians"]. It was also published in Venezuela in* El Nacional *on October 10 and 11, 1993.*[1]

The *Garimpeiro* Trap

Illegal gold prospectors *(garimpeiros)* generally infiltrate Yanomami territory in small groups. Isolated in the forest and few in number, they fear the Indians and, as soon they run into any of them, try to buy their friendship by offering an abundance of food and manufactured items. In most cases, the Yanomami have had little or no experience with white people. They take this show of generosity as a promising start by an unknown community to establish an alliance *(rimimuu)*. This misunderstanding develops while they are still unaware of the ecological impact and health consequences of gold prospecting. The *garimpeiros* are still nothing to them but strange "earth eaters," and they disparagingly compare them to peccaries scavenging the muddy forest soil.

Feeling more confident after having established good relations with the Indians, the gold prospectors then increase their numbers daily in the forest. It seems less and less essential to them to be generous. The Yanomami, apparently "pacified" by the gifts they have come to expect, no longer seem to constitute a threat. Over time, they have become simply bothersome visitors whose incessant demands are exasperating. Increasingly irritated by their presence, the gold prospectors push them away with false promises, impatient gestures, and even threats.

The Indians begin to realize how the intensification of gold prospecting in the forest is affecting them. The rivers are polluted, game is becoming scarce, and infectious diseases are spreading. Endemic malaria and pneumonia associated with continual flu epidemics are decimating communities located near gold prospecting sites. The Indians now view their demands for clothing, tools, ammunition, and food as necessary compensation for the harm caused by the intruders. The gold prospectors' repeated refusals are thus seen as sudden and incomprehensible signs of hostility.

The situation is then completely deadlocked. The Yanomami have become dependent on the economy that gravitates around the gold prospectors just at the very moment that the miners themselves no longer feel the necessity of buying peace with the Indians. This double bind has been the source of most of the conflicts between the Yanomami and gold prospectors since the 1980s. Once such a trap is laid, the smallest incident can lead to overt violence. Given the disparity of force and weapons, the Indians always end up being the main victims.

The predatory behavior of the *garimpeiros* forecloses any possibility of enduring coexistence with Indian communities in the forest, particularly when these communities have almost no experience of contact with white people. Modern gold prospecting in the Brazilian Amazon is a heavily mechanized activity with an inexhaustible and eager supply of manpower coming from all over the country. Generally, there is no interest in recruiting a local Indian workforce. So, in the best of cases, the *garimpeiros* view the Yanomami as a nuisance; in the worst, a threat. If the Indians don't die of malaria or pneumonia, and if they can't be kept at a distance through gifts and promises, all that's left is to intimidate them, and if that doesn't work, exterminate them.

Murders on the Upper Orinoco

Around mid-1993, tensions escalated between the Brazilian *garimpeiros* on the upper Rio Orinoco and the Yanomami living along one of its small tributaries, the $H^w axima\ u$ River. More than a year went by, and visits by Indians to the gold prospectors' campsites were frequent and their demands for food and trade goods were growing more and more insistent. During one of these encounters, two mining bosses promised to give hammocks and clothing to one of the Indian group's leaders to get rid of him. This promise, like so many others, remained unfulfilled. Exasperated by the white people's lies, the young chief went with his brothers-in-law to claim what he considered his due from the camp where one of the mining bosses was staying. The man wasn't there. A violent discussion ensued with one of his employees, who tried to drive the Indians away. But the Yanomami fired a shotgun and the *garimpeiro* fled. Furious, the Indians used their knives to cut up the gold prospectors' hammocks, threw their blankets and radio into the woods, and made off with their cooking pots. Then they returned to their communal house.

Already fearing for their safety after an earlier conflict, the gold prospectors had forcibly taken back several shotguns they had previously given the Yanomami. After this last incident, they decide to resort to terror to dissuade the Indians from being so bold. They plan to kill any who may return to bother them again in order to set an example. This decision by the *garimpeiros'* bosses directly precipitates the events that lead to what will be later called the Haximu Massacre.

On June 15, six young Yanomami go again to one of the *garimpeiros'* camps to ask for food and clothing. On the suggestion of their elders, they are prepared, if the opportunity presents itself, to take back the shotguns that had been seized from them. Yet, they are begrudgingly given nothing but a little bit of manioc flour and a scrap of paper with a message written for the *garimpeiros* in another camp located upstream. They are told that they will be given what they want up there. The unsuspecting youth go to the indicated camp and find a group of gold prospectors playing dominoes. A cook receives them and reads aloud the message: "Have a good time with these idiots!" She quickly tosses the note into the fire. Then she offers the Indians some sugar, rice, and a few pairs of shorts before sending them back where they came from. Forewarned and encouraged by the cook, the *garimpeiros* prepare to execute their premeditated plan and assassinate the youth as soon as they've turned their backs. Fearing, however, that other Indians might be lying in ambush nearby, they reconsider and decide to kill the Indians a bit later, on the path leading back to their village.

Less than an hour away from the gold prospectors' camp, the Yanomami youth stop in the forest to eat some of the provisions they have just obtained. Six *garimpeiros* armed with shotguns suddenly appear. They invite the Indians to go hunting with them before visiting another nearby camp. Surprised by this untimely invitation, the youth refuse. But they finally allow themselves to be convinced by the apparently friendly insistence of these white people. They set off, in single file, interspersed between the gold prospectors.

At a certain point, the Yanomami hunter at the end of the line leaves the group and walks off the trail to defecate in the woods. The only Indian with a shotgun, he leaves his weapon with one of his companions and tells the troop not to wait for him as he heads into the undergrowth. Yet the gold prospectors stop where they are on the path and silently surround the other young men, who are squatting. Suddenly, one of the *garimpeiros* grabs the arm of the Yanomami holding the firearm and shoots him in the stomach at point-blank range with a sawed-off shotgun. Then three other Yanomami youth are assassinated. One of them covers his face with his hands and begs, "*Garimpeiro,* friend!" He is executed with a shot in his skull. The two others are killed as they try to escape.

Warned by the shots, the Yanomami who had gone into the woods throws himself in the nearby upper Rio Orinoco and manages to escape. Meanwhile, three gold prospectors encircle the youngster who had been walking at the head of the line of hunters. They take aim and fire their guns at him, one after the other, as if this were target practice. Thanks to the tangled undergrowth and his agility, he manages to evade the first two shots. The third one, however, hits him on his right flank. The shooters quickly reload their weapons to finish him off. He takes advantage of the moment to escape, and he too throws himself in the river. In a state of shock, he hides along the bank, barely keeping his head above water. From his hiding place, he sees the gold prospectors bury the bodies of his three companions assassinated on the trail. One of the *garimpeiros* goes looking for the fourth body, climbs down to the riverbank, and suddenly sees the Indian youth's face in the stream. He races back for his gun. The wounded young man quickly manages to give him the slip. Seriously injured, it takes him several days to make it back to his house on the $H^w axima\ u$ River.

In the meantime, the first escapee has already made it back to his family's home and spreads the news of the *garimpeiros'* attack. He quickly retraces his steps with a group of the victims' relatives to look for the bodies of his four assassinated companions. Along the way, they find the wounded adolescent who tells them about his escape and that the corpses were buried (an unbearable desecration for the Yanomami). The group finally arrives at the scene of the crime and exhumes three bodies. The fourth can't be found. Mortally wounded while trying to escape, he also must have sought refuge in the river, where he probably drowned and was carried downstream by the current. The three exhumed bodies are taken deeper into the woods and incinerated. Once the bodies have been cremated, the victims' relatives carefully collect the charred bones and put them in baskets to carry back to their village.

In the days that follow, they organize a collective hunt for game that they will offer to guests invited to the funeral ceremony to prepare the dead's funerary ashes. The charred bones will be ground to a powder and placed in gourds sealed with beeswax. After a week to ten days of hunting, envoys are sent to invite three neighboring allied groups to the ceremony: the people of *Hoomoxi, Makayu,* and *T^{h}oumahi.* Then, after the cinerary gourds are prepared, a group of warriors, including people from the $H^w axima\ u$ River and their allies, leave, as is customary, to avenge the deaths they just commemorated. Yanomami war expeditions are aimed at taking vengeance only against male enemies who are held responsible for homicides (whether they killed with arrows or sorcery). Women and children of enemy communities are never targeted.

On July 26, after a two-day hike, the Yanomami warriors camp overnight in the forest, hidden near the clandestine gold prospecting sites on the upper Rio Orinoco. The next morning, under a driving rain, they creep closer to the lean-to that serves as a kitchen for a gold prospector's shack. Two *garimpeiros* are chatting near the fire. One of the Indians lying in ambush behind a tree fires his shotgun at one of them. Hit in the head, the man collapses on the ground and dies instantly. A second shot wounds the other man in the back as he tries to flee. The warriors then gather around the body of the dead gold prospector. They complete their act of vengeance by riddling his body with arrows and opening his skull with an axe. Then they make off with everything they can find in the deserted campsite, notably a shotgun and cartridges.

Preparations for the Massacre

This act of vengeance by the Yanomami makes the gold prospectors absolutely furious. They bury the corpse of their companion in the kitchen lean-to and then immediately abandon camp. They find the wounded *garimpeiro* hiding in the forest and carry him to a landing strip, a two-day walk away. Then they begin plotting their revenge, which will reach a new threshold in the escalating violence and end in absolute horror. Two back-to-back meetings are held, drawing all the men from all the mining camps in the area. They decide to put an end to their troubles with the Indians once and for all through terror. For the first time, the idea of exterminating the $H^w axima\ u$ River community is raised. Following the meeting, they meticulously plan their punitive expedition. Volunteers are recruited on the spot. Weapons and ammunition are assembled (200 shotgun cartridges and a few boxes of bullets for revolvers).

The entire operation is backed, if not directly run, by the four principal gold mining entrepreneurs in the area, all of whom are well known in Boa Vista, the capital of Roraima state. These *garimpo* bosses give their workers time off and supply them with weapons and ammunition. All the preparations for the expedition and the assassins' meetings are held in their camps. Fifteen gold prospectors armed with twelve- and twenty-caliber shotguns, thirty-eight-caliber revolvers, machetes, and knives set out on their mission of extermination. Several of them will have a direct hand in the imminent massacre. Four professional hired killers *(pistoleiros)*, employed as bodyguards by the gold prospecting bosses, lead the sinister band.

Since the raid by their warriors, the inhabitants of the $H^w axima\ u$ River have been camping at a distance from their two communal houses for several days as a precaution against a counterattack by *garimpeiros*. Yet they decide to retrace their steps because they are expecting an invitation to a feast at their allies' house in *Makayu* and so head in that direction. They spend one night at their $H^w axima\ u$ River dwellings and leave the next morning to continue their journey, stopping at one of their old gardens between the $H^w axima\ u$ River and *Makayu*. They set up there a large camp of temporary forest shelters and wait, as is customary, for an emissary from their hosts.

Three young warriors, dissatisfied with the results of their recent raid, suddenly decide on their own to go back and attack another group of *garimpeiros*. The man at the head of this improvised expedition has good reason to seek vengeance. His older brother was murdered by the gold prospectors, and his body, washed away by the waters of the Rio Orinoco, was never found. The appropriate funeral ceremonies could not appease his grief and anger. Still enraged, the three young warriors set out. After a two-day walk, they arrive near the gold prospecting area. Under the cover of foliage and noise from the motor-pumps, they manage to surprise a gold prospector at work. The man remains unaware of their presence until the very instant one of the Indians aims at him. The gun jams. Another warrior rushes to his companion's side and fires a shot. The *garimpeiro* covers his face with his forearms and manages to escape, only slightly wounded. The three warriors immediately flee to rejoin their group.

While this impromptu raid occurs, the column of fifteen gold prospectors is already en route to implement their terrible plan to wipe out the Yanomami community. The three young Indians head back to the forest camp via long detours off the marked trails. They thus detect no sign of the advancing *garimpeiros*, even though they are numerous. The *garimpeiros* finally arrive at the $H^w axima\ u$ River on the morning of the third day of their trek. The two large Yanomami communal houses are empty. The gold prospectors eventually discover the trail leading to the old garden where the house's inhabitants are camped out. They begin to track down the group they intend to exterminate.

The night before, emissaries from *Makayu* had arrived at the forest camp of the $H^w axima\ u$ River people. The latter, alarmed by the return of the three young raiders, decide to scale down their plan to participate in the feast of their allies. They decide that only able-bodied adult men will accompany the *Makayu* messengers. Women, children, and the elderly will stay in the encampment so the group is not slowed down on its journey. The men think they will be back very soon. In accordance with the Yanomami's code of conduct, they believe that white people will only seek revenge on warriors.

The Massacre on the $H^w axima\ u$ River

The next morning, the women, accompanied by their small children and one elderly man, go to gather wild fruit a short distance from the old garden. Around noon, there are just nineteen people left in the camp. Most of them are stretched out in their hammocks, including the three young warriors, who are following a ritual of seclusion after their raid. Children play between the shelters. A few women chop wood. The atmosphere is peaceful. Meanwhile, the *garimpeiros* are already making their way into the garden and are lying in ambush along one side of the encampment. One of them suddenly opens fire on the Yanomami. The fifteen other men start shooting simultaneously and continuously with shotguns and revolvers as they advance towards their victims.

The three young warriors, one elderly man, one woman, and three little girls (six, seven, and ten years old) manage to escape this unrelenting barrage, thanks to the complicated layout of the forest shelters and the dense vegetation. Two of the little girls and one of the young warriors are wounded by shotgun blasts in the face, neck, arms, and flanks. The ten-year-old girl is critically wounded, shot in the head with a bullet from a revolver. She will later die from her injuries. From their hideouts, the fugitives hear their relatives screaming in terror over the sound of gunshots that continue to bombard the encampment. After several minutes, the shooting stops. The merciless *garimpeiros* then finish off their victims with machetes and knives. They massacre the wounded who are unable to flee, as well as several children that the gunshots missed.

Twelve Yanomami are savagely assassinated: two adult women, an old man, a young woman in her twenties who was visiting from *Hoomoxi*, three adolescents, two baby girls (a one-year-old and a three-year-old), and three boys ranging in age from six to eight years old. Several of these children are orphans; their parents died from malaria introduced by the gold prospectors. The young woman from *Hoomoxi* is first struck by a shotgun blast from approximately ten meters away, then shot again with a revolver at less than two meters' distance. One of the old women, who is blind, is kicked to death. A baby sleeping in a hammock is wrapped in a scrap of cotton fabric and knifed to death.

The *garimpeiros* understand that they have killed only some of the group. They decide to terrorize the survivors, and so proceed to mutilate or dismember the bodies of their victims. They grab the shotguns they find in the camp and make their way back in the forest, firing a signal flare to deter anyone in pursuit. A few hours later they arrive back at the communal houses along the $H^w axima\ u$ River. Feeling safer in the dwelling than in the forest, they decide to spend the night there. The following morning, they pile up everything the Indians have left behind, notably a dozen or so aluminum pots. They blast them with gunshots and then smash them with machetes. Next, they burn the two houses to the ground and then head back to their own camps. Several weeks pass with no news of their crime filtering out of the forest. Then, in mid-August, the assassins suddenly hear news on the radio about the massacre they committed. They immediately decide to flee the area. They make it to a clandestine airstrip located a few days' walk away. They threaten to kill anyone who could report them and force the pilots of single-engine aircraft parked there to fly them to Boa Vista. As soon as they get to the city, most of the killers disperse to other parts of the country.

The Cremation

As soon as the shooting stops, one of the surviving three young warriors runs to the group of women gathering wild fruit in the forest and urges them to hide. He then quickly goes back to the blood-drenched camp and looks in vain for his own shotgun. He gives up pursuing the killers and rejoins the women, sending a few of them on to *Makayu* to spread word of the tragedy. Three terrified young women go as fast as they can, making it there in a few hours. They arrive in tears and describe the horror they left behind, a scene strewn with the corpses of women and children mutilated with machetes and knife wounds.

The grief-stricken and enraged men of the community race back to the forest camp shelters, arriving at nightfall. They huddle together with the survivors of the massacre, but the stench of blood is so unbearable that they are forced to set up a new camp a half-hour away. Darkness compels them to put off cremating the bodies until the next day. The night is pierced with the sounds of wailing over the deaths, cries of terror, and livid harangues by the group's leaders. At dawn, they begin to gather the mutilated bodies. Suddenly, the little ten-year-old girl with the serious head wound emerges from the undergrowth where she had been hiding, screaming in fear. Her mother rushes towards her, crying desperately.

The funeral cremation begins. Each corpse is placed on a pyre improvised from nearby fallen branches. The bodies of the adults are burned on the ground of the encampment where they were killed. The children's bodies are taken to the clearing where the group had spent the night.

As soon as the funeral pyres die down, the burning hot, charred bones are hastily retrieved and put in baskets or aluminum pots. Everyone is convinced that the gold prospectors could return at any moment to finish off the group's survivors once and for all. Not even this danger, however, can dissuade the $H^w aximu$ River people from carrying away the bones of their dead relatives. There is nothing more precious to the Yanomami than the bone ashes of their deceased. The women keep watch over them constantly, near their hearths, or carry them along when they travel. Danger is so imminent, however, that countless bone fragments—riddled with buckshot and splintered by bullets—are left behind in the crematory fires. Only the dismembered body of the young female visitor from *Hoomoxi*, who had no relatives among the people of the $H^w axima\ u$ River, is left behind and not cremated. The cinerary gourd of one of the four youth assassinated earlier by the gold prospectors on the upper Rio Orinoco was broken during the attack on the camp. His mother tries to gather the scattered ashes and wrap them in banana leaves. In her hurry to escape she drops several of these on the forest floor.

The Exodus

Thus begins an exodus lasting several weeks. The hunted group stays off the trails, often walking at night and without eating, carrying the three wounded little girls. After eight days, the fugitives stop for the night in an allied community at $T^h omokoxipi\ u$. The ten-year-old girl with the head wound dies shortly before dawn. Her mourning parents carry her body throughout the entire next day until they arrive at a new camp in the forest where she too can be cremated. After that, the flight through the forest continues without reprieve for more than two weeks. The group crosses the network of paths belonging to two other communities, *Waraka u* and *Ayaopë*, but they do not take time to stop. They won't rest until after they've crossed the Rio Orinoco, heading south, and reach *Maamapi*, a fourth house.

The people of the $H^w axima\ u$ River are now close to the Brazilian border, which they cross to walk downstream along the Rio Toototobi in the state of Amazonas. They finally arrive at a new communal dwelling called "Marcos's house," where they decide to hide out. It is August 24, 1993. Almost a month has passed since the massacre. The sixty-nine survivors of the community have chosen this area of Brazil for three reasons. There are still no gold prospectors in the area, the inhabitants are longtime allies, and there is a health post they have visited many times since the late 1980s to get treatment for malaria introduced by the *garimpeiros*.

When they stop at the houses in T^h*omokoxipi u* and *Maamapi* and once they arrive at "Marcos's house," the survivors of the massacre pound the charred bones they have carried with them and fill funeral gourds, kept in small open-weave baskets. They intend to organize intervillage funeral ceremonies to mourn their deceased relatives with the help of the allies who welcomed them in Brazil. Potential in-laws will bury the ashes of the adults near the hearths of their own deceased next of kin. They will ingest the ashes of the children mixed in a banana soup. After that, the baskets and gourds that contained the bones and their ashes will be burned. Everything that remains of the physical and social existence of the dead must be destroyed and obliterated—all traces of them, their possessions, any mention of their names, and, finally, the ashes of their bones. The purpose of this ritual is to prevent the ghosts from returning from the sky's back where they belong. It aims to guarantee a permanent separation between the world of the dead and that of the living, and does not end until the buried or ingested bone ashes have been put in oblivion, after long funeral lamentations in remembrance of the deceased and in praise of their virtues.

This explains why the people of the H^w*axima u* River always put these funerary rituals before their own safety, despite the fact that they were in great danger and risked unbridled violence from the gold prospectors. To renege on funerary obligations would mean to condemn the ghosts of their dead to wander between two worlds and to condemn the living to suffer a melancholic torment far worse than death itself. The survivors of the massacre now try to reconstruct an existence that has been laid waste by the incredible savagery of the *garimpeiros*. They prepare to build a new communal house and to plant new gardens. During much of the year ahead, their existence will still revolve around funeral ceremonies to mourn their murdered relatives and others who recently died of malaria. Their immense grief will not find closure until they have emptied the contents of the last funeral gourd. Then, a normal life can resume.

Yet they will never forget that white people are capable of slaughtering women and children in the most bloody, barbaric way. So far they had believed that only man-eater evil beings could be capable of such ferocity. They now renounce vengeance against the gold prospectors. They would have taken revenge on them if they had continued to view these outsiders as worthy of the name *enemy,* as human beings partaking of the same warriors' code of honor. Such is not the case. They hope for just one thing now—that these inhuman creatures are kept captive in the cities they come from and that they never again return to the forest.

In the end, however, even this hope was dashed. The people of the H^waxima u River have returned to the upper Rio Orinoco, where they expect nothing more of white people. In the aftermath of the barbarous slaughter, twenty-three garimpeiros *were nominally charged with supporting evidence. It was not until December 1996 that five of them were finally sentenced to a total of ninety-eight years in prison, although not a single one was imprisoned at the time. Only two of those found guilty were finally imprisoned later. The Haximu Massacre was legally characterized and judged as an attempt at genocide. This ruling was an unprecedented act in the history of Brazilian jurisprudence regarding the massacre of an Indian group.*

—B. A.

Notes

Setting the Scene

1. The notorious Indian Protection Service, or SPI, was founded by Marshal Rondon in 1910. Its field agents are called *sertanistas*.

2. "Yanomami" is a simplification of the ethnonym *Yanōmami*, a term that when followed by the plural *tĕpĕ* means "human beings" in western Yanomami language (for more information, see Appendix A).

3. The Yanomami also occupy the Rio Casiquiare basin, a natural canal between the Orinoco and upper Rio Negro.

4. The population numbers approximately 17,600 in Venezuela (according to the 2001 *Censo de comunidades indígenas*) and 15,500 in Brazil (according to the 2006 census by FUNASA (Brazil's former National Health Foundation).

5. See Albert, 1985, and Duarte do Pateo, 2005. The *Sanima* northern Yanomami are an exception to these general characteristics in their social organization as well as their settlement pattern (Ramos, 1995).

6. The National Indian Foundation, Brazilian administration of indigenous affairs, is today under the auspices of the Ministry of Justice.

7. See Albert, 1993.

8. *Hutukara* is the shamanic name for the ancient sky that fell at the time of the origins, creating today's "land-forest" *(urihi a)*. For the founders of this Yanomami organization, it is "a name that defends the land-forest" *(urihi noamatima a wããha)*.

9. CCPY was co-founded with Claudia Andujar, an exceptional photographer (Andujar, 2007), and Carlo Zacquini, an uncommon Catholic friar, both of whom were similarly captivated by their encounter with the Yanomami in Brazil.

10. Since 2009, CCPY's programs have been integrated into the activities of the largest Brazilian NGO devoted to indigenous and environmental issues, the Instituto Socioambiental of São Paulo (www.socioambiental.org), with which I continue to collaborate closely.

11. There are several other references to cases of young women captured by the

Yanomami on the upper tributaries of the Rio Negro after 1925 (Albert, 1985, 53–56). Moreover, a new version of Helena Valero's account was published under her name in 1984 in Venezuela, compiled by R. Agagliate and edited by E. Fuentes (Valero, 1984; see Lizot, 1987). Born in 1919, Helena Valero died in 2002.

12. See Brumble, 1988.

13. Some examples might include narratives about shamanic calling (see Chapter 3) or accounts of migration trajectories (see Albert, 2009).

14. Davi Kopenawa's shamanic prophecies have a surprising parallel in theories of climate change and the Anthropocene (see Crutzen and Stoermer, 2000).

Words Given

1. To speak "ghost talk" (to be *aka porepë*) means to speak a non-Yanomami language, to express oneself awkwardly, stutter, utter inarticulate sounds, or remain silent.

2. The word *napë* (plur. *pë*) actually means "outsider, enemy."

3. Any existing being has an "image" *(utupë)* from the original times, an image which shamans can "call," "bring down," and "make dance" as an "auxiliary spirit" *(xapiri a)*. These primordial image-beings ("spirits") are described as miniscule humanoids wearing extremely bright, colorful feather ornaments and body paint. Among the eastern Yanomami, the word for "spirits" (plur. *xapiri pë*) also refers to shamans *(xapiri t^{h}ë pë)*. Practicing shamanism is referred to as *xapirimuu*, "to act as a spirit"; to become a shaman is said "to become spirit" *(xapiripruu)*. These expressions refer to the fact that, during the shamanic trance, the shaman identifies with the "auxiliary spirits" he is calling.

4. Here, Davi Kopenawa uses the Portuguese word *histórico*, "historical."

5. The expression *pata t^{h}ë* (plur. *pë*) refers to leaders of local factions or groups ("great men, men of influence") or, generically, to "elders."

6. *Omama* is the demiurge of Yanomami mythology; see Chapter 2.

7. The Yanomami refer to pages of writing and, more generally, printed documents with illustrations (magazines, books, newspapers) as *utupa siki*, "image skins." They use a neologic expression for paper (in Portuguese, *papel*): *papeo siki*, "paper skins." They describe writing by using terms applied to certain patterns of their body painting: *oni* (a series of short dashes), *turu* (a group of dots), and *yãikano* (sinusoidal lines). To write is therefore to "draw dashes," "draw dots," or "draw sinusoids" and writing, *t^{h}ë ã oni*, is a "drawing of words."

8. Audio recordings are usually described by the expression *t^{h}ë ã utupë*, "image, shadow of words."

9. The eastern Yanomami use three generic terms to refer to their elders/ancestors: *pata t^{h}ë pë* (the elders, the "great men"), *xoae kiki* ("the grandfathers," the historical ancestors), and *në pata pë* (the mythical ancestors).

10. *Teosi* comes from the Portuguese *Deus*, "God." These "people of Teosi" are evangelical missionaries of the New Tribes Mission, who first visited the upper Rio Toototobi *(Weyahana u)* in 1958, when Davi Kopenawa was about two.

1. Drawn Words

1. *Yosi* is a first name with Hebrew roots, short for Joseph. Davi Kopenawa associates it with the members of the Brazilian Boundary Commission who explored the upper Rio Toototobi with the SPI in 1958–1959. It was probably given to him by missionaries of the New Tribes Mission who accompanied the first SPI expedition to the upper Rio Toototobi in June 1958. On the village of *Marakana* and the first contacts with white people, see Chapters 10 and 11.

2. By derivation, the word *yoasi pë* refers to a fungal infection (*Pityriasis versicolor*) that produces patches of discolored skin. The myth cycle devoted to these two Yanomami demiurges invariably presents *Yoasi* as a choleric, lecherous, and clumsy being (see M 187, 191, 197, 198 in Wilbert and Simoneau, 1990; subsequent numbered myths are also from this volume).

3. That is, easy to pronounce and not reminiscent of any Yanomami word. When phonetically apt, "white people names" *(napë wããha)* are subjected to endless humorous distortions (for example, "Ivana" becomes *iwa na*, "caiman vagina"). The expression *wããha yahatuaɨ*, "to misuse the name," is an equivalent of "to insult."

4. The sedimentation of "white people names" in Yanomami villages following the passage of successive visitors is worthy of a study. Biblical names, first names of SPI or FUNAI agents, names of doctors, of local politicians, of Brazilian states, of soccer and television stars, of cartoon characters, and even of advertising brands. Considered socially neutral—so long as they are not phonetically close to any Yanomami word—these "white people names" are used not only in contact situations but increasingly between the Yanomami themselves. Yet traditional nicknames, which cannot be pronounced in the presence of those they refer to or their close kin, retain their more confidential circulation.

5. On the kinship terminology of eastern Yanomami, see Albert and Gomez, 1997: 289–298. Note that the vocative noun *õse!* is also applied to brothers and sisters, and even to nephews and nieces when they are still children.

6. Davi Kopenawa here used the Portuguese word for "family" *(família)*, which does not exist in Yanomami. However, "uncle," "aunt," and "grandparents" refer to Yanomami kinship terms: *xoae a*, "mother's brother" and "grandfather"; *yae a*, "father's sister" and "grandmother."

7. The Yanomami refer to these "childhood names" with the expression *wããha oxe kuowi*.

8. Aside from physical, psychological, and behavioral particularities ("Big Legs," "Hostile Face," "Cries All the Time"), Yanomami names sometimes refer to events associated with the bearer's birth (*Waikama*, born after an attack by a group called *Waika*) or place of birth (*Yokoto*, which means "Lake").

9. In this case, the "people from afar" *(praha tʰëri tʰë pë)* and "other people" *(yayo tʰë pë)* are nonrelatives and inhabitants of other communal houses. The Yanomami contrast childhood names with the pejorative adult nicknames referred to by the expression *wããha yahatuaɨwi tʰë ã*, "words to misuse the name."

10. Publicly connecting a person (or his or her close kin) with his or her name(s) is thus proscribed, and to pronounce the name of a dead person before his or her kin is an even greater offense. Abruptly asking a Yanomami his name, as white people invariably do, will greatly embarrass him and generally lead to the answer: "I don't have a name" or "I don't know; ask someone else."

11. Regarding the Yanomami linguistic family, see Appendix A at the end of this volume. "Xiriana" is an appellation taken from the Ye'kuana Indians, the Yanomami's northern neighbors (Arvello-Jiménez, 1981: 22n2). This term was also formerly used to refer both to the Yanomami of the Rio Toototobi (where Davi Kopenawa comes from) and, farther downstream, the Arawak (Bahuana) Indians of the Rio Demini (Ramirez, 1992: 4).

12. A birth certificate in this name was issued for Davi Kopenawa by FUNAI in January 1974 and an ID card in July 1975. The (estimated) birth date on these documents is February 15, 1956.

13. This shamanic self-denomination was finally ratified by the Brazilian courts in March 2008. "Davi Xiriana" was thus able to become "Davi Kopenawa Yanomami," the name he has brought to prominence since the 1980s, in Brazil and beyond, through his fight to defend his people.

14. These murders took place in August 1987, in the area of FUNAI's Paapiú Outpost, on the upper Rio Couto de Magalhães *(Hero u)*. See Chapter 16.

15. Conducting shamanism sessions is also known as *yãkoanamuu,* "to act under the *yãkoana* powder." Davi Kopenawa was initiated into shamanism in the early 1980s by his wife's father, the leader of the community where he now resides with his family, *Watoriki.* Though the verb used is "to drink" *(koai),* *yãkoana* snuff is inhaled. It is prepared from resin drawn from the deep part of the *Virola elongata* tree's bark, which contains a powerful hallucinogenic alkaloid, dimethyltryptamine (DMT). DMT has a chemical structure close to that of the neurotransmitter serotonin and takes effect by attaching itself to certain of its receptors. Its psychic effects are close to those of lysergic acid (LSD). *Yãkoana* also contains various ingredients that probably reinforce its effects: dried and pulverized *maxara hana* leaves, ashes of *ama hi* and *amat^ha hi* tree bark (see Albert and Milliken, 2009: 114–116).

16. It is said that by blowing *yãkoana* snuff into a novice's nostrils, the shaman initiating him transmits the *xapiri* spirits with his "breath of life" *(wixia or wixi aka);* see Chapter 5. Here, Davi Kopenawa translates *wixia* into Portuguese as "strength, richness." As a component of the person, he also refers to it by terms such as "life" or "energy." The *wixia* breath is associated with the abundance of blood and the heartbeat, and thus to the body image or vital essence of the person *(utupë).*

17. The presentation dance *(praiai)* of these image-beings *(xapiri* "spirits") reproduces that of the *yarori* (plur. *pë)* human/animal ancestors (see Chapter 2, note 1) in the origin myth of fire (M 50). It constitutes the superlative prototype for the dancing of guests *(h^wama pë)* at the opening of the intercommunal feasts *(reahu).* The latter dance takes place around the house's central plaza, first individually and then in a group. Pounding their feet on the ground, the men spin and brandish

their weapons or trade goods. The women shake young palm branches while moving backward and forward.

18. The "mark of teaching" that refers to the mythical origin of a custom is known as *hiramano*, from *hira-*, "to name, teach, create"; *-ma* (passive); *-no*, "mark, trace."

19. For the Yanomami of the lowlands, the epicenter of war and warrior's ardor resides in the people of the highlands, in the historical heart of Yanomami territory, in the Serra Parima. For more about these "War People" *(Niyayopa tʰëri)*, see Chapter 21.

20. Ambivalently, the term *waitʰiri* means both "aggressive" and "valiant."

21. A warrior who has killed an enemy must follow a ritual of homicide *(ōnokaemuu)*. He is then said to have "a greasy forehead" because he is supposed to ritually digest his victim's flesh and fat (see Chapter 7, note 40; and Albert, 1985: chapter 11).

22. The *-wë* ending added to this character's name is borrowed from the western Yanomami *(Xamatʰari)* language. See another version of the *Aro* myth told by Davi Kopenawa's father-in-law: M 288. It is associated with an origin myth of war which features *Õeoeri*, the warrior child: M 47.

23. These evil beings of the forest are generically referred to by the expression *në wãri pë*: *në*, "value of"; *wãri*, "bad, evil"; *pë* (plur.).

24. The Yanomami think the flesh and blood of the fetus are gradually formed from the sperm of his genitors accumulated in the womb during successive acts of intercourse throughout the pregnancy.

25. The *Titi kiki* ("night beings") are described in the origin myth of night (M 80) as large Cracidae birds perched in a tree from which they chant the names of the rivers in a plaintive voice. They project a spot of darkness below them, which a hunter will spread by killing them.

26. On the fall of the sky, see M 7. Yanomami mythology mainly consists of two series of narratives. One relates the erroneous ways of the first human/animal ancestors *(yarori)*, a reverse image of the norms of current sociality that induced their metamorphosis into game *(yaro)* and that of their "images" *(utupë)* into shamanic entities *(xapiri)*. The other develops the epic tale of the demiurge *Omama* and his brother, the trickster *Yoasi*, creators of the current world and human (Yanomami) society.

27. The Yanomami cultivate about one hundred varieties of some forty plant species (see Albert and Milliken, 2009: 32–41).

28. For the Yanomami epidemics spread in the form of fumes, hence the expression *xawara a wakëxi*, "epidemic smoke," see Albert, 1988, 1993, and Albert and Gomez, 1997: 48, 112–115. *Xawara* refers to all highly contagious infectious diseases of foreign origin. The eastern Yanomami distinguish eighteen types of *xawara* (Albert and Gomez, 1997: 112–115).

29. The *reahu*, a great intercommunal feast, is both a ceremony of political alliance and a funeral rite (see Albert, 1985).

30. Davi Kopenawa opposes *Omama* to *Teosi*, orality to writing, as well as sha-

manism to Christianity in order to "reverse" the content of the evangelical sermons he heard as a child (see Chapter 11).

31. An allusion to Napoleon Chagnon's famous 1968 monograph about the Venezuelan Yanomami: *Yanomamö: The Fierce People*. This volume (reissued under the same title until the 1980s) and other writings by the same author on alleged Yanomami "violence" (Chagnon, 1988) contributed to spreading pejorative stereotypes about these Indians over several decades (see Borofsky, 2005).

32. Davi Kopenawa's parents' generation was wiped out by two successive infectious disease epidemics in the 1950s and 1960s. His wife's father's group was decimated in the same way in 1973, then again in 1977. See Chapters 11 and 13.

33. This question and the insistence upon it is all the more perplexing to the Yanomami given that this ethnic name is an external adaptation of an expression meaning "human beings" (see Appendix A).

34. Among the western Yanomami, the bone ashes of the deceased are mixed with plantain soup and ingested. Among the eastern Yanomami, only children's ashes are consumed in this manner, while those of adults are buried by the hearth of their closest kin. In both cases, the funeral service is carried out by potential affines of the deceased. The expression "put the ashes in oblivion" *(uxi pë nëhë mohotiamãi)* refers to this ingestion or burial process (see Albert, 1985).

2. The First Shaman

1. This human/animal ancestor's name comes from *yaro*, game, followed by the suffix -*ri* (plur. *pë*), which denotes the original times, the nonhuman, superlative, monstrous, or something of extreme intensity. These first ancestors transformed *(xi wãri-)* into game while their "images" became *xapiri* spirits (see Chapter 1, note 26).

2. On the fall of the sky and the pushing of the first ancestors into the underworld, where they were transformed into chthonian voracious predators—*aõpatari* (plur. *pë*)—see M 7 and Chapters 6 and 7.

3. Vermilion dye obtained from oily seeds found in the fruit of a small cultivated tree *(nara xihi)*.

4. The Yanomami think the celestial level *(hutu mosi)* curves at its ends to get closer to the terrestrial level *(warõ patarima mosi)*, where it rests upon it with huge "feet" (stakes).

5. On the pathogenic power of the metal *Omama* hid under the ground, see Chapter 16.

6. On *Omama* and the origin of rivers, see M 202. On *Omama* and the origin of metal, see Chapter 9.

7. On the aquatic monster *Tëpërësiki* (sometimes associated with the anaconda), his daughter's coupling with *Omama*, and the origin of cultivated plants, see M 197 and 198.

8. On the birth of *Omama*'s son, see M 22. Davi Kopenawa sometimes calls him *Pirimari a*, which is also the name of the star that the Yanomami refer to as "the moon's son-in-law," Venus.

9. This name is based on a feminine redundancy: *t^huë*, "woman, wife"; -*yoma*,

feminine name suffix (like in *napëyoma,* "white woman"). This emphasizes the fact that this is the first woman. It is also a "fish being–woman," with which Davi Kopenawa freely associates our image of the siren (see Chapter 20).

10. These evil forest beings (see Chapter 1, note 23) are also described by the expressions *yanomae tʰë pë rããmomãiwi,* "those who make humans sick," and *yanomae watima tʰë pë,* "eaters of humans."

11. The *Watoriki* shamans say that the ghostly form (his "value of ghost," *në porepë*) of *Omama* (equivalent to his "image," *utupë*) "has many names" *(tʰë ã waroho),* such as the sun being *Motʰokari,* the jaguar being *ɨramari,* and the evil being *Omamari.*

12. The "ordinary people," *kuapora tʰë pë* (literally, "the people who exist simply"), are opposed here to shamans, *xapiri tʰë pë* (literally, "the spirit people"). The latter consider that the former have "ghost eyes" that can only see the deceptive appearance of beings and phenomena. Conversely, shamans are said to be able to see the image-essence *(utupë)* of existing beings at the time of their mythical creation. This primordial image-form is denoted by the suffix *-ri* (plur. *-ri pë*; see above, note 1).

13. For another version of this origin myth of short life, see M 191. Western Yanomami mothers tie their newborn's umbilical cord to this tree and carry the baby around in it in order to assure him a long life (Lizot, 2004: 321).

14. The Yanomami consider the toucan's hoarse and piercing call particularly sad. They associate it with mourning and nostalgia. Hearing the toucans "cry" in the forest announces a death in a distant home; listening to their call in the house at dusk gives rise to amorous longing.

15. Here, Davi Kopenawa refers to *pihi,* a term that denotes conscious thought and volition (but also the gaze), and to the "breath of life" *(wixia).* Death is referred to as *noma a.*

16. The shamanic cure's actions against pathogenic entities that devour the sick person's body image/vital essence *(utupë)* are described by warlike expressions such as *në yuai,* "to take revenge"; *nëhë rëai,* "to ambush"; *nëhë yaxuu,* "to scare away" (see Chapter 6). The Yanomami conceive all relationships of otherness/hostility in terms of predation and vengeance, be it between humans or between humans and nonhumans.

17. This tube, between fifty and ninety centimeters long, is generally made from the hollowed-out stalk of a small *horoma a* palm or from a cultivated *xaraka si* bitter cane.

18. The greatest Yanomami shamans are considered able to regurgitate *(kahikɨ ho-)* the pathogenic objects affecting the sick person's body image/vital essence *(utupë)* or animal double *(rixi).* See Chapter 7.

19. The expressions "act in/enter a ghost state" *(poremuu)* or "to become other" *(në aipëi)* both refer to states of altered consciousness induced by hallucinogens, pain, or sickness in which the "ghost" *(pore)* enclosed in each living being (similar to our "unconscious") takes precedence over consciousness *(pihi).*

20. These bunches consist of feathers torn horizontally and attached to a small stick. Green *werehe* parrot feathers and black and white *maraxi* piping-guan feath-

ers are most commonly used. They can also be made with white ventral feathers from the *paari* wild turkey and the *wakoa* hawk.

21. An ornament made of down feathers from the *watupa aurima* vulture and the *wakoa* and *kãokãoma* birds of prey.

22. This monkey's tail is covered in thick and shiny black fur. This ornament and those described in the previous notes are worn by shamans during their sessions and, more generally, by all men during *reahu* feasts.

23. *Uuxi* refers to "the inner part" of the person, as opposed to his or her body ("the skin"), *siki*. The expression *xi wãri-* (literally, "turn bad") refers to mythical transformations and any kind of ontological change of form or identity. In this case it is synonymous with *në aipëi*, "to take the value of/become other." It can also mean "to lose one's mind, to be out of one's self," and, more concretely, "to get tangled up, become inextricable, no longer cease (state or action), remain stuck."

It should be noted here that nocturnal shamanism associated with the dream time is a fundamental part of Yanomami shamanism. Shamanistic initiation and work colonize the dreams of shamans with hallucinatory remnants of their daytime sessions (see Chapter 22). Both *yãkoana* snuff and dreaming allow shamans to access the mythical times of origin which, for them, continue to unfold immutably in an eternal present, as a parallel dimension to historical time (that of migrations and wars).

24. The Yanomami consider any form of lethal aggression, whether human (war, sorcery) or nonhuman (evil beings, enemy shamanic spirits), visible or invisible, as a form of cannibalism (see Albert, 1985).

25. Shamanic activity is referred to by the verb *kiãi*, "to move, to work."

3. The *Xapiri*'s Gaze

1. *Yai t^hë* (plur. *pë*) is a generic term referring to unknown and dangerous invisible (at least to the eyes of "ordinary people") entities (but also to unnamed, unknown, or inedible visible beings). The *yai t^hë pë* category includes, among others, ghosts *(pore pë)*, evil beings of the forest *(në wãri pë)*, and shamanic spirits *(xapiri pë)*. It is contrasted with *yanomae t^hë pë*, "human beings," and *yaro pë*, "(edible) animals, game" (domestic animals are called *hiima pë*).

2. Davi Kopenawa lived in *Marakana* in the second half of the 1950s.

3. The dream *(mari)* is considered a state of temporary wandering of the person's inner image/vital essence *(utupë)* outside of his body *(siki)*. During shamanic dreams the dreamer's *utupë* image is taken far away by his *xapiri* spirits; it then has the "dream value of the spirits," *xapiri pë në mari*.

4. Davi Kopenawa began his shamanic initiation in the early 1980s, in his late twenties (see Chapter 5).

5. Regarding *Omama*'s flight and the creation of the mountains, see Chapter 4.

6. The *xapiri* are mostly images *(utupë)* of the human/animal ancestors *yarori*; see note 26, Chapter 1.

7. The warriors' black body paint is made of crushed coal and latex from the *operema axihi* tree (see Albert and Milliken, 2009: 111–112).

8. These women are also called *tʰuëyoma* (plur. *pë*), like *Omama*'s wife (see Chapter 2, note 9). They are the daughters of the *yawarioma* (plur. *pë*) water-being people associated with his aquatic father-in-law, *Tëpërësikɨ* (M 198, M 197).

9. Davi Kopenawa's stepfather was his mother's second husband, a highly reputed shaman and warrior who died in 1997. Davi Kopenawa's father died in *Marakana* when he was a young child (see Chapter 10).

10. The upstream of the sky *(hutu mosi ora)* is the west, the downstream of the sky *(hutu mosi koro)* the east.

11. This spirits' house is said to be initially built in the shaman apprentice's chest. Later its peak is said to be fixed to the "sky's chest" and the house to remain suspended there. See Chapter 6.

12. Davi Kopenawa's eldest son (born 1982) was first a schoolteacher at *Watorikɨ*. In 2004 he joined the staff of the Yanomami *Hutukara* organization and later became a university student.

13. Davi Kopenawa's stepfather separated from the rest of the group that remained on the upper Rio Toototobi in 1961 (private correspondence, B. Hartman, New Tribes Mission).

14. It is disgraceful for a Yanomami hunter to "eat his leftovers" *(kanasi wamuu)* or to "bring back to himself" *(koãmuu)*, i.e., eat his own prey. He would fear becoming drowsy, visually impaired, and being condemned to always returning empty-handed *(sira)*. Hunters most often exchange their game in the forest before returning to their home. A solitary hunter will give the game he brings home to his wife, children, and in-laws. See Chapter 23.

15. *Na waɨ*, "eating the vulva," is the common expression used to refer to copulation.

16. A favorite activity of teenagers, which proves perilous when fathers and husbands are light sleepers.

17. Due to these concerns, Davi Kopenawa and his father-in-law have turned *Watorikɨ* into a veritable shamanic center: in 2010, there were sixteen shamans out of a population of 174 people.

18. A strip of beaten *yaremaxi hi* tree bark in which Yanomami women carry children slung over their shoulders.

19. This shamanic vision originates in the fact that tapirs habitually remain in the water (small lakes and watercourses) for long periods in order to get rid of parasites or to escape from their predators.

20. Precious goods, funeral ashes, and feather adornments were referred to as *matihi* (plur. *pë*), which today most often refers to the white people's merchandise. See Chapter 19.

21. Tapir hunters enjoy a particular prestige. Indeed, few men are able to follow this animal's inextricable trail, which can lead the hunt to last several days.

22. Young women sometimes participate in shamanic sessions with men by inhaling *yãkoana* snuff.

23. Diluted wild honey *(puu upë eherexi)* is the favored drink for shamanic initiation. See Chapter 5. The *xapiri*, called "drinkers of flower nectar" *(horehore u koatima pë)*, are said to come down easily to those who like sweet food.

24. Davi Kopenawa takes the example of an eight-year-old boy in *Watoriki* to determine his age when he had his first experiences with *yãkoana*. It is not unusual for particularly intrepid young boys to approach shamans at the beginning of their session and ask for a little *yãkoana* snuff.

25. On the last day of the *reahu* feast, a clay plate *(mahe)* is set up with a small pile of *yãkoana* snuff, which all the men assembled inhale, whether or not they are shamans. Young boys often join them to snort a few pinches. This collective taking of *yãkoana* is followed by ceremonial dialogues *(yãimuu)* and the burial (or ingestion) of the bone ashes of the deceased person mourned during the feast. These sung dialogues in which hosts and guests squat in pairs are conducted to settle material or matrimonial exchanges, intercommunal political conflicts, and to share important news. At the end of this dialogue session, the funerary ashes are buried (or ingested) by a small circle of potential affines of the deceased. This is described by the expression *uxipë yãimumãi:* "submit the funerary ashes to an exchange dialogue."

26. *Yãkoana*'s effect is always described by the verb *nomãi*, "to die."

27. All the nonshaman young people who inhale the *yãkoana* at the end of the *reahu* wind up in the same state. Only shamans who have gone through a long initiation can master the extremely powerful effect of this hallucinogen.

28. It is said that by overindulging in certain foods, one makes those foods "show their own images" *(utupë taamamuu)*. Thus the dangerous power of the *rasa si* peach palm fruit juice *(raxa a wai)* brings down the images of the "fruit women of the peach palm" *(raxayoma)*. This juice is a ceremonial food much appreciated at *reahu* feasts, as is boiled ripe plantain soup *(koraha u)*.

29. Here, Davi Kopenawa uses a thirteen-year-old boy as an example.

30. This shotgun was acquired from SPI agents in exchange for jaguar, deer, and peccary hides.

31. This loop *(hʷaraka a)* made of *masi* vine fits tightly around both feet to ensure they have a firm grip on the tree trunk.

32. *Xoape* is the vocative form of the kinship term *xoae a*, which refers to the mother's brother, the father-in-law, or the grandfather. Davi Kopenawa used to call his stepfather "father-in-law" because his own father referred to him as a "brother-in-law" *(xori a)*.

33. The act of initiating a young shaman is referred to as *huka horai*, "blow (in) the nose." The initiator is referred to as *topuwi a*, "the one who gives."

34. They are called "spirit's sperm" *(xapiri mõ upë)* or "spirit down feathers" *(xapiri hõromae pë)*.

35. The Yanomami contrast good hunters who spend their time in the forest ("inhabitants of the forest," *urihi tʰëri pë*) with lazy young people who are more interested in women than hunting ("inhabitants of houses," *yahi tʰëri pë*).

36. This enumeration of birds suggests a great skill at hunting. Arrowing these

birds is "to arrow finery/precious goods" *(matihi pë niyãi)*, an exercise restricted to the finest young archers, who will flaunt them during *reahu* feasts in order to impress young women (and potential fathers-in-law).

37. The greatest and oldest shaman in *Watoriki*.

38. In this shamanic experience, animals *(yaro pë)* resume their mythological condition as human/animal ancestors *(yarori pë)*.

39. Davi Kopenawa was between twelve and fourteen at the time. White-lipped peccaries are among the Yanomami hunters' most important prey. Their herds can include from fifty to over three hundred individuals, weighing around thirty kilograms each. They have a very strong smell and the smacking of their mandibles can be heard from hundreds of meters away.

40. A specific term refers to the hunger for meat in Yanomami: *naiki*.

41. Shamans consider white-lipped peccaries *pata t^{h}ë pë yai*, "real ancestors," due to their mythical metamorphosis (M 148 and 149).

42. This group of women constitutes a gathering of "mourners" typical of Yanomami funeral laments.

43. Here, Davi Kopenawa's tale has been completed by a direct recording of his stepfather's words. This account follows a traditional pattern of shamanic initiation narratives recounting an encounter with a seductive water-being woman.

44. A matrimonial alliance is thus established between water beings, reproducing—this time at their initiative—the one *Omama* entered into in the beginning of time by "fishing" *Tëpërësiki*'s daughter (see M 197).

45. *Xapiri* who are images of plants are "too close" and only have a "ghost's tongue" and are thus the least powerful. They are the first the apprentice shaman sees, and they prepare the arrival of the animal ancestors' *xapiri* during the initiation. See Chapter 5.

46. The dismembering and recomposing of the initiate body by the *xapiri* is the basis of his initiation. See Chapter 5.

47. There is among the Yanomami a strict avoidance between mother-in-law and son-in-law. Their principal incest origin myth deals precisely with this relationship (M 42).

48. Here the bride-service owed by a man to his in-laws is reversed (from a mother-in-law to a son-in-law). Outside of shamanic reality, this reversed bride-service only happens when a bride is too young and inexperienced, or temporarily disabled. In this case a mother-in-law can cook and provide firewood for her son-in-law.

49. The drowned are said to have been "swallowed" by *Tëpërësiki*. The *xapiri* spirits issued from the mythical image *(utupë)* of *Omama*'s aquatic father-in-law, *Tëpërari pë*, are described as having an enormous mouth, which swallows evil beings and then spits out their bones. *Tëpërësiki* is also associated with the anaconda *õkarima t^{h}oki*.

50. The "goods" *(matihi pë)* of shamanic spirits and evil beings of the forest are the pathogenic objects or weapons with which they affect their victims' inner image *(utupë)*.

51. The expressions "to make weak" *(utitimãĩ)* and "to make new/young" *(oxe-pramãĩ)* refer to the preparatory work on the initiate carried out by his initiator's shamanic spirits. See Chapter 5.

4. The Animal Ancestors

1. In an interview with a representative of the American Anthropological Association, Davi Kopenawa explained in Portuguese his use of the term "spirit": "Now "spirit" *(espirito)* is not a word in my language. I have learned this word "spirit" and I use it in the mixed language I invented (to speak to whites about these things)" (Turner and Kopenawa, 1991: 63).

2. Literally: *napë pë pore pë,* "ghost outsiders." The *xapiri* see humans as ghosts and nonshamans are said to have "ghost eyes."

3. Storms and felled trees are often attributed to the *xapiri*. At the death of the "father" (the shaman), these bereaved and furious spirits cut up the sky with machetes. See Chapters 6 and 24.

4. Adornments the shamans refer to as *topëraki*. For the Yanomami, the regularity of the glass beads' perforations is particularly important in assessing their quality.

5. This description of the *xapiri* emphasizes their beauty and ideal youth. It associates their choreography with that of the *reahu* feasts' presentation dance, which provides a special opportunity for guests to present themselves at their best by "playing young" *(hiyamuu)* and by "exhibiting adornments" *(matihimuu)*.

6. *Yõrixiamari,* the mythological ancestor of the *yõrixiama* cocoa thrush, is the creator of the *heri* songs sung at night during *reahu* feasts (M 41). *Amoa hiki* "song trees" are sometimes also referred to as *yõrixiama hiki,* "*yõrixiama* thrush trees." The *ayokora* cacique bird is known for its ability to imitate other animals' calls and for its songs, as are, to a lesser degree, the *sitipari si* and *taritari axi* birds (on "polyglot" birds, see Dorst, 1996: 61–65).

7. A simple cylindrical basket loosely woven in a hexagonal weave and made with split stems of arouman reed *(Ichnosiphon arouma)*.

8. Tape recorders are called *amoa hiki* ("song trees") but are also referred to by the expressions *amoatoatima hiki* ("song-taking trees") or (by shaman elders) *yõrixia kiki* ("*yõrixiama* thrush-things"). The term *amoa hiki* initially designated harmonicas often given by white people at the time of first contacts.

9. These are very short songs, generally consisting of a single phrase, repeated by a principal singer and taken up by a choir of followers. Each night of the *reahu,* men and women sing in alternating groups. The women sing while dancing side by side, moving back and forth in the central plaza. The men form a single line and circle the plaza. The term *heri* also refers to the calls of groups of batrachians and troops of howler monkeys.

10. The term *Horepë tʰëri* refers to the highland Yanomami to the north of the Rio Toototobi, and *Xamatʰari* to the Yanomami living to the west of this river.

11. On the Yanomami linguistic family, see Appendix A.

12. The "white people's land" *(napë pë urihipë)* and the downstream of the rivers

correspond here to the areas south and east of Yanomami territory *(yanomae t^{h}ë pë urihipë)* in Brazil.

13. Here, Davi Kopenawa contrasts *amoa pë ã siki oni,* "song-drawing skin" (or *amoa kiki ã oni,* "song drawing") with *amoa wãã* ("sound of song").

14. Photographs are called *utupë,* a term which refers, as we have seen, either to an inner component of the person (body image, vital essence) or to the primordial mythical form of all beings. It also means "reflection, shadow, echo," and is used to describe any form of reproduction or representation.

15. The word *representante* is part of Davi Kopenawa's common political vocabulary in Portuguese.

16. *Yarori* (plur. *pë*) refers both to the mythological animal ancestors (the "fathers of the animals," *yaro h^{w}ïïe pë*) and their images that turned into *xapiri* shamanic entities. This triangulation of animal ancestors *(yarori pë),* game *(yaro pë),* and shamanic animal images *(also yarori pë)* is a fundamental aspect of Yanomami ontology.

17. The term "skin" *(siki)* refers here to the "corporeal envelope," i.e., the "body," as opposed to the "image" *(utupë),* of these mythological entities.

18. The expression *në porepë,* "value (form) of ghost" is used as a synonym of *utupë* ("image") here, as is often the case.

19. By "dying" *(nomãi)* under the effect of *yãkoana,* a shaman identifies with the *xapiri* image-beings ("spirits") he "brings down," and in doing so, incorporates their gaze. It is thus by becoming a spirit himself that he is able to see other spirits.

20. This onomatopoeia echoes the verbal expression *si ekekai,* "tear the skin, flay."

21. On this mythological character who used to flay hunters and is associated with cotton ornaments, see M 260.

22. On *Omama's* flight and the origin of mountains, see M 210 and M 211.

23. Manufactured mirrors are called *mirena* (*mire* for the western Yanomami) while *xapiri's* "mirrors" are called *mireko* (plur. *pë*) or *mirexi* (plur. *pë*). The word *mirexi* also refers to the mica-laced sandbanks shining in the waters of mountain streams and *xi* means "light, radiance, emanation." Among the western Yanomami, reflections from the first manufactured mirrors, acquired in the 1950s, were feared to make people blind (Cocco, 1987 (1973): 125) and *mireri noku* refers to a dye that makes the spirits shiny (Lizot, 2004: 222). The *xapiri's* "mirrors" are therefore principally conceived as surfaces reflecting light and, unlike white people's mirrors, not surfaces reflecting images.

24. Geometric patterns of Yanomami body paintings.

25. These are the savannas *(purusi)* of the upper Orinoco (see Huber et al., 1984).

26. The western Yanomami *(Xamat^{h}ari)* refer to the eastern Yanomami as *Waika.*

27. The *xapiri* have their houses in high mountains and move on mirrors in the forest. But once they become spirit helpers of a shaman "father" they live in dwellings whose tops are fixed to the "sky's chest" and whose central plazas are also mirrors. See Chapter 6.

28. On the *xapiri* living in the land of the white people's ancestors, see Chapters 18 and 20.

29. Shamans "call," "bring down," and make "dance" as *xapiri* spirit helpers the primordial "images" *(utupë)* of a highly heterogeneous (and potentially infinite) set of beings, entities, and objects. Outside of animal ancestors and other forest spirits, *xapiri* include all the entities of Yanomami mythology and cosmology, domestic spirits (dog, fire, or pottery) or outsider spirits (ancestors of the white people, bull, horse, or sheep).

30. These *xapiri* are referred to as *yai t*ʰ*ë pë xapiri* ("unknown/evil being spirits") and *napë pë xapiri* ("enemy/hostile spirits") while those from the mythological animal ancestors are referred to by the same name as these animal ancestors: *yarori pë*.

31. Fish poisoning is a fishing method used during the dry season. It consists of soaking leaves or crushed sections of lianas that have the property of asphyxiating fish in streams or ponds (see Albert and Milliken, 2009: 69–73). Dead fish in dried up streams in the height of summer (February–March) are said to be *Omoari* leftovers.

32. *Porepatari* is an ancient ghost who is said to haunt the forest and is associated with the mythical origin of curare (see M 124). On curare among the Yanomami, see Albert and Milliken, 2009: 62–63.

33. This evil spirit's name comes from that of the *koikoiyoma* falcon, who is considered his "representative." A derived verb *(koiai)* generically refers to the attack of the spirits of evil beings.

34. For the Yanomami, the universe consists of four superimposed levels *(mosi)*, surrounded by a great void *(wawëwawë a)*: *tukurima mosi* (the "new sky"), *hutu mosi* (the "[current] sky"), *warõ patarima mosi* (the "old sky"), and *pëhëtëhami mosi* (the "world below"). The "new sky" is a kind of embryonic sky destined to replace the celestial vault after its fall. The current earth is a sky fallen in the beginning of time; see M 7.

35. In Yanomami eschatology, humans *(yanomae t*ʰ*ë pë)* die and become ghosts *(pore pë)* on the sky's back *(hutu mosi)*. Then the ghosts also die and finally metamorphose into fly beings *(prõõri pë)* and vulture beings *(watupari pë)* on the new sky *(tukurima mosi)*. *Warusinari* beings (plur. *pë*) are described as enormous *koyo* ants who "became other"; *h*ʷ*akoh*ʷ*akori* beings (plur. *pë*) are described as giant vultures without feathers.

36. This chthonian spirit is associated with the cataclysm that carried away the *Hayowari t*ʰ*ëri* ancestors and led to the creation of the white people (see Chapter 9). The word also refers to the place where this transformation took place. His name comes from the verb *xi wãri-* ("to turn bad, to metamorphose") (see Chapter 2, note 23).

37. The images of *në wãri* evil beings can be used as shamanic spirit helpers to rescue the image of children captured by their analogs in the forest. To this end, for example, the evening spirits *(weyaweyari pë)* are "brought down" to fight their father-in-law, the evil being *Weyaweyari*. Yanomami shamanism often operates through this kind of symbolic homeopathy.

38. These *waikayoma* (plur. *pë*) spirits are the images of foreign women, living

on the banks of great rivers, whose bodies—and even houses—are covered in magnificent ornaments made of glass beads.

39. The term *hapara pë* also refers to posthumous children.

40. This last term also refers to *Omama*'s aquatic wife, as well as to all the water-being women. See Chapter 2.

41. Coati spirit women *(yarixiyoma pë)* are admired for the beauty of their eyes, and *kumi* vine spirit women *(kumirayoma pë)* for their intoxicating scent. The *kumi* vine has a sweet-scented phloem that men use to make love charms.

42. These feminine ornaments, consisting of sections of the stem of a foxtail grass *(pirima hi)*, are worn at the corner of the lips and under the lower lip.

43. On the *në rope* concept, see Chapter 8.

44. Evil being spirits are also called up in a warlike shamanism exclusively directed against distant villages assumed to be enemies. See Chapter 7.

45. Yanomami tobacco plugs consist of a roll of dried tobacco leaves moistened with a little water and rubbed with fireplace ashes. They are generally placed under the lower lip. Craving tobacco is referred to as "having a concupiscent lip" *(kasi pexi; pexi* refers to sexual desire).

46. *Yoropori*, the mythological caterpillar ancestor, was the first to possess tobacco (M 76). His name is derived from that of the *yoropori* caterpillar.

47. The Yanomami say that this large forest gastropod's "tongue" (crawling sole) leaves "saliva" tracks on the ground.

48. *Sipara* is a term borrowed from the western Yanomami language. It refers to the spirits' machetes and sabers, which are sometimes also called *sipara mireri,* "mirror sabers" (Lizot, 2004: 367).

49. The weapons of the animal ancestor *xapiri* are most often associated with the bodily and behavioral characteristics of the corresponding forest animals. See Taylor, 1974.

50. These are balls of *mai kohi* tree resin used for rope making.

51. These "marks" *(õno)* are left on sick people's "images" *(utupë)* by the pathogenic weapons/objects/substances of evil entities and human aggressors who are held responsible for disease and death: see Chapter 7.

52. Here, Davi Kopenawa names several species of bees—*xaki na* (plur. *ki*), *õi na, pari na, maxopoma na, puu xapiri na, tima na,* and *puu axi na*—and of ants—*kona* (plur. *pë*), *kaxi, koyo,* and *pirikona*.

53. Regarding the *Waika/Xamat^hari* ethnic designations see notes 10 and 26 in this chapter. *Parahori* refers here to the highland Yanomami of the Rio Parima.

54. These fragments of sky are known as *xitikari kiki* or *pirimari pë*, terms which refer to the stars in shamanic language.

5. The Initiation

1. Davi Kopenawa was hired as an interpreter for FUNAI in 1976. He was twenty years old. The Wind Mountain *(Watoriki)* is called the Serra do Demini in Portuguese.

2. FUNAI's Demini Outpost was opened at the foot of the Serra do Demini in

1977. The group whose leader would soon become Davi Kopenawa's father-in-law gradually drew closer to it beginning in 1978 (see Chapter 14 and Appendix C).

3. Here, the initiating shaman seeks to determine the nature of the future initiate's visionary state to adjust his initiation. In this regard, intense nocturnal dream activity is a less "advanced" stage than daytime hallucinatory episodes.

4. Davi Kopenawa's initiation took place during the dry season (October to March), when his oldest son, born in August 1982, was starting "to barely stand up," so in late 1983. Davi Kopenawa was twenty-seven at the time.

5. *Yãkoana* snuff that has been stored a long time is known as "soft" *(nosi)*, like an overused tobacco wad. It is then said that the spirits have taken away its "dangerous power" *(wai)*.

6. We have here a process of identification between shaman and spirits. The initiate inhales the *yãkoana,* which is drunk through him *(he tore)* by the *xapiri* who, like him and at the same time as him, "die," "become ghosts," while in return he imitates *(uĕmãi)* their songs and dances. Regarding the association between the shaman's chest and his spirits' house during the initiation, see Chapter 6.

7. On this variety of *Virola sp.* see Albert and Milliken, 2009: 114–116.

8. These spirits are those of the *yãkoana*'s hallucinogenic power *(wai)*. The second term is borrowed from the Ye'kuana (Carib-speaking neighbors of the northern Yanomami), among whom *aiyuuku* also refers to a shamanic hallucinogenic (K. Vieira Andrade, private correspondence).

9. This sentence translates the expression *pariki kõapë* or *pariki kõakõapë* (literally, "chest in a state of return"), associated with the verb *kõamuu,* "to eat one's own prey" (see Chapter 3, note 14). The images of the leftovers of such game, particularly repulsive to the *xapiri,* are called *yaro pë kõakõari pë.*

10. Female capuchin monkeys and otters are said to breastfeed and carry their young with particular care.

11. *Wari mahi* tree seeds are surrounded by a fluffy whitish fiber, and the *mohuma* eagle's breast and belly are covered in white down.

12. The "upstream" of a speech *(t^{h}ë ã ora)* or a song *(amoa wãã ora)* refers to its last part (as opposed to its full content) or to its superficial content (as opposed to its full meaning).

13. The intransitive verb *reãmuu,* which refers to the *xapiri*'s song, includes two types of sound utterance: *areremuu* (from the onomatopoetic *"arererere!"*), a kind of stridulation, and *krititimuu* (from the onomatopoetic *"kriii! kriii! kriii!"*), a kind of squeak. Among the western Yanomami, *reãã t^{h}eri* refers to a shamanic spirit "who makes his tongue vibrate like the novice shamans" (Lizot, 2004: 352).

14. This "funny dancing" refers to the *yarori* human/animal ancestors' presentation dance in the fire origin myth (M 50).

15. These are large, solitary bees found on sandbanks. Their mythological ancestor *Remori* gave the white people their language, a "ghost language" associated with his inarticulate roar (see M 33).

16. This "thin tongue" *(aka si yahate)* is contrasted with the "thick tongue" *(aka si t^{h}et^h e)* of impeded elocution.

6. Spirits' Houses

1. A large, bell-shaped, tightly woven basket carried by the women on their backs with a frontal bark head sling *(wii a)*.

2. These are little flutelike whistles *(purunama usi, xere a)* made of *Olyra latifo-lia* bamboo and three-hole flutes made of deer bone *(përa a)*.

3. The cock-of-the-rock, called the "jaguar's son-in-law," is a magnificent bird with a bright orange crest (male). Its mating displays are spectacular. The white-tipped dove has a deep and inquiring call repeated at intervals. *Tãrakoma* manakin birds are remarkable for their dances and displays and for their call, which is very loud for their size.

4. A sung ceremonial dialogue performed to invite the elders of an allied house to a *reahu* feast or to join a war raid, or to ask a potential in-law to perform the burial of funerary ashes of one's kin.

5. Guests at a *reahu* feast must spend a final night camping near their hosts' house *(mat^{h}ot^{h}o yërë-)* before performing their presentation dance there. During this night, the hosts intone joyful *heri* songs, which their guests hear from a distance.

6. The first house is the *miamo nahi a*, the "central dwelling," and its annexes, *sipohami nahi pë*, the "exterior houses." Thus, as the shamanic experience progresses and new spirit helpers are acquired, the number of secondary houses increases accordingly. Davi Kopenawa describes these annex homes as "apartments" *(apartamentos)* in Portuguese.

7. The terms *nahi*, "house," and *mireko*, "mirror," are used here as synonyms. For instance, *Tihiri mireko* is the "jaguar spirit's mirror" and *Tihiri nahi* the "jaguar spirit's house." Davi Kopenawa refers to the names of these spirit houses/mirrors as their "bead finery" *(pei a në topëpë)*.

8. In this case, Davi Kopenawa refers to the western Yanomami of the upper Rio Demini, neighbors and allies of the people of his native region.

9. These evil *xapiri* (see Chapter 4, note 30) are sent by shamans to devour the image *(utupë)* of children in distant enemy villages.

10. These pets of the dry season spirit *(Omoari a në hiimari pë)* are, among others, the *aputuma, kraya, maya, raema,* and *wayawaya* caterpillar evil beings.

11. This correspondence between spirit house and initiate's chest is expressed by the word *nõreme*, which denotes appearance, analogy, or simulation. This term is also used as a synonym of *utupë*, "body image/vital essence" and sometimes of *nohi*, "friend, alter ego."

12. The mountains are the homes of the "free" *xapiri* before they become spirit helpers of the shamans. In western Yanomami *(Xamat^{h}ari)* shamanism, these "house-mountains" *(pei maki)* are materialized during the initiation in the form of a pole covered in annatto and white down feathers.

13. Davi Kopenawa visited the United Nations building in New York in April 1991; see Chapter 20.

14. The elder shamans are said to "project their vital breath" *(wixia horamuu)* or

"give their vital breath" *(wixia hipɨamuu)*. It enters the initiate's nostrils and chest with the *yãkoana* and the *xapiri,* giving him his initiator's personal qualities (bravery, wisdom, courage at work, and so on).

7. Image and Skin

1. The expression "harmful things" is intended to translate the term *wai,* which refers to a disease's pathogenic principle or the strong or dangerous effect of a substance (tobacco, pepper, hallucinogen, venom). The plural of *wai (wai pë)* also designates warriors on a raid.

2. The Yanomami say *wai xëɨ* or *wai xurukuu,* "hit, attack a disease," *wai nëhë rëmãɨ,* "to lie in ambush for a disease" (see also Chapter 2, note 16, on this warlike curing vocabulary).

3. In this case, the expression *në wãri kiki* refers to all the evil beings and the diseases they cause.

4. Literally, "plants for healing" *(haro kiki).* Recent research revealed that at least 203 of such medicinal plants are used by the eastern Yanomami in Brazil (see Milliken and Albert, 1996, 1997a; Albert and Milliken, 2009).

5. The sick person's body *(siki)* is described here by the expression *në wãri kanasi:* "leftover of an evil being." In this case, disease consists of a process of capturing and devouring the victim's image *(utupë).* After the shamanic treatment, it was customary for experienced women to apply different remedies to the patient's body, generally ones based on medicinal plants. On the Yanomami interpretation of disease and therapeutics, see Albert and Gomez, 1997, and Albert and Milliken, 2009.

6. The former kind of sorcery is said to cause the victim's lower limbs to putrefy, the latter to make him perish of chronic dysentery. On $h^w ëri$ sorcery substances, see note 17 below. On *paxo uku* magic poison, see Mattei-Müller, 2007: 227 *(pasho ishiki* among the western Yanomami).

7. Conversely, the mass asphyxiation of fish through fish poisoning is often compared to the effect of an epidemic.

8. The epidemics *(xawara,* plur. *pë)* are said to spread in the form of smoke or fumes *(xawara a wakëxi).* In the shamans' view, it takes the form of a cohort of cannibal evil beings *(xawarari,* plur. *pë)* looking like white people. See Chapter 16.

9. The animal ancestor spirit helpers' role in the cure is determined by the morphological and/or ethological characteristics of their forest animal "representative" (see Chapter 4, note 49).

10. The attribution of sicknesses and deaths to human malevolence (sorcery, shamanic warfare, animal double killing) is the dominant political idiom of intervillage Yanomami relations. Their attribution to evil beings of the forest is an indication of the circumstances' political neutrality. See Albert, 1985.

11. *Omoari* is said to roast the image of his human victims on a wooden rack (fevers). All his sons-in-law are spirits of animals and insects associated with the dry season.

12. The shamanic image of *Porepatari,* the ancient ghost guardian of the forest, gives the other *xapiri* their curare arrow points and protects them when they fight powerful enemies.

13. Here again, the *xapiri*'s weapons match those of the animals they refer to: the wasps' arrows/stingers, the swallow-tailed kite's blades, the coatis' club/long tail, the giant anteater's claws (see above, note 9).

14. *Moxari* is said to make the fruits on the forest's trees grow. He protects them by stinging or arrowing humans who try to eat his leftovers—i.e., ripe fruits with maggots *(moxa pë).* He makes their mouths and throats rot, devoured by his pets *(Moxari a në hiimari pë).*

15. Every Yanomami is linked to an animal double with whom he keeps a consubstantial relationship, passed down from mother to daughter and from father to son. This system, reminiscent of Australian "sexual totemism," generally associates a species (terrestrial/aerial) with a gender (female/male). The animal doubles *(rixi)* of the members of a community are said to live at the edges of their social world, close to faraway enemy groups ("unknown people," *tanomai t*ʰ*ë pë).* Attributing a death to the killing of a *rixi* by distant hunters means resorting to the outermost possible degree of social hostility in the Yanomami disease interpretation system. See Albert, 1985.

16. Lanceolate bamboo arrow points *(rahaka)* are used for terrestrial game; harpoon points, made of a monkey bone attached to a wooden shaft *(atari hi),* for birds.

17. *H*ʷ*ëri* substances are principally (but not exclusively) made from plants (often Cyperaceae and Araceae) (see Albert and Gomez, 1997: 95–100). They are rubbed, thrown, or blown on their victims. Their use is not restricted to specialists, and men and women have specific arsenals of them. This common sorcery is used between allied groups, never between members of a community. It is not considered lethal, so long as the victim undergoes a shamanic treatment. See Albert, 1985.

18. The expression *imino në mot*ʰ*a* literally means: "hand mark" *(imino)* "with value" *(në)* "of anger" *(mot*ʰ*a,* which also refers to weariness).

19. This kind of ceremonial dialogue is performed to exchange news on the first night of an intercommunal *reahu* feast.

20. Davi Kopenawa refers here to the sorcery plant or substance's pathogenic principle *(wai);* see above note 1.

21. This kinship term refers to a brother or sister, independently of the speaker's gender. In the vocative *(õse!),* it can apply to the speaker's son and daughter. See Chapter 1, note 5.

22. *Repoma* bees nest in the ground.

23. The verb *hereamuu* describes the speeches of leaders of factions or local groups, the *pata t*ʰ*ë pë* (the "great men," "the elders").

24. The term *mae* refers here to a footprint on the ground, but also means "path." On the substances used in this earth footprint sorcery, see Albert and Gomez, 1997: 99–100.

25. This footprint sorcery is attributed to bad allies seen as malevolent collectors

of footprints while distant enemies carry out the lethal sorcery on them. On this indirect war sorcery, see Albert, 1985. This kind of sorcery can also be attributed to *oka* (plur. *pë*) distant enemy sorcerers directly taking footprints of isolated individuals in their gardens or the forest.

26. These darts *(ruhu masi)* are made from the petiole fiber leaf rachides of *kõanari si* and *õkarasi si* palms.

27. A death "by blowpipe" *(horomani)* blamed on enemy sorcerers is considered equivalent to a killing "by arrow" *(xarakani)*. The attribution of sudden and unexplainable deaths (corpses found in a garden or the forest) to *oka* sorcerers generally concerns socially valued elders *(pata t^{h}ë pë)* and therefore most often leads to revenge through war raids.

28. *Kamakari* is a celestial evil being *(hutukarari a në kamakaripë)* held responsible for toothache, earache, and internal eye ache, as well as articular pain (he is said to devour bone marrow with his sharp teeth). He is also associated with death and funerary ashes, notably those of shamans.

29. The caciques weave pouchlike nests with large and smooth entrances. The tapir is known for its imposing windpipe. The *xapiri* spirit of whom it is the "representative" is said to possess a "throat" *(Xamari a në t^{h}orapë or purunaki)*, which can be placed in a sick person's throat to help her eat or in that of an initiate to make him fit for singing.

30. The toucans swallow palm fruits *(maima si, hoko si, kõanari si)* and spit up their pits. The *wayohoma* nightjar birds have an extremely wide mouth. *Ixaro* and *napore* cacique birds, *taritari axi* birds, macaws, and peccaries are also mentioned in this context.

31. This episode of facial paralysis took place in March 1986, when Davi Kopenawa was organizing the first-ever assembly for the defense of Yanomami territorial rights in *Watoriki*, which was attended by leaders of several Amazonian groups and a delegation of Brazilian parliamentarians.

32. The *ayokora* cacique birds place their nests in high branches near those of *kurira* paper wasps, known for their aggressiveness.

33. The expression used here is *xapiri huu*, "travel in the form of *xapiri* spirit." The sudden deaths of children are mostly attributed to such shamanic attacks.

34. These western Yanomami groups live at a great distance from *Watoriki*. Their names, and those of their great shamans—Õina, Xereroi—circulate in the vast network of intercommunal rumors.

35. Children weaned prematurely following a new birth often suffer severe malnutrition. They are called *totixi pë* and considered far more vulnerable than others.

36. These shamanic incursions are referred to by the same expression as war raids *(wai huu)*.

37. Pathogenic objects belonging to these hostile *xapiri* (weapons and goods) are referred to by the term *matihi* (plur. *pë*), "precious goods, merchandise," like those of the *në wãri* evil beings, of whom they are the image. Note that in the initial period of contact, white people's manufactured objects were feared for their pathogenic

powers (Albert, 1988) and that the northern Yanomami *(Sanima)* still refer to them as *wani de,* "evil things" (Guimarães, 2005: 108).

38. *Herona* is an evil being described as a giant sloth who is said to burn houses with his urine (compared to curare and gasoline) when people grill game in them at night.

39. The Yanomami often interpret malarial convulsions and coma as attacks from evil shamanic spirits. The *waka moxi* sorcery plant is said to cause major convulsions, leading its victim to fall into his hearth fire while sleeping.

40. Any man having killed ("eaten") an enemy during a raid or by invisible means (sorcery, shamanism, or animal double hunting) enters into a ritual homicide state *(õnokae)* during which he is supposed to digest his victim's body. He is thus compelled to follow a seclusion rite *(õnokaemuu)* and observe a series of behavior and food prohibitions. "Having a greasy forehead" refers to exuding the fat of the body subjected to this ritual digestion process (see Chapter 1, note 21).

41. In the beginning of time, the dead instantly came back to take their place among the living until they were sent into flight and their path between sky and earth was interrupted (see M 35).

42. Here the *xapiri* speak in the name of the humans they protect.

43. On Yanomami eschatology and cosmological layers, see Chapter 4, note 35.

8. The Sky and the Forest

1. This episode took place in 1974, when Davi Kopenawa was eighteen years old.

2. The "land-forest of human beings" *(yanomae t^hë pë urihipë),* the Yanomami territory, is said to occupy the center *(miamo)* of the terrestrial layer *(wãro patarima mosi),* a center from where the demiurge *Omama* made rivers spring forth and created mountains (M 202, 210, 211).

3. On the fall of the sky and the death of the shamans, see Chapter 24.

4. The first metal tools (axe head or machete blade fragments) to enter Yanomami territory via interethnic trade before direct contact with white people were attributed to *Omama.* See Chapter 9.

5. The highlands *(horepë a)* of the Orinoco-Amazon interfluve (Rio Branco and Rio Negro) form the historic center and demographic heart of the Yanomami territory. The rugged landscape is attributed to the sky's fall on the canopy of the mythical cacao tree. For another version of the myth of the sky's fall, see M 7.

6. Allusion to the origin myth of the other Amerindian groups, then of white people (M 33), as well as to the great forest fire described in Yanomami oral history (see later in this chapter).

7. These meteorological events are considered the sound signal, the "heralding call" *(heã)* of a great shaman's death.

8. On the risk of the sky collapsing again following a great shaman's death, see M 13.

9. Fishing is very difficult when the rivers are overflowing and the fish are look-ing for food in the flooded forest.

10. Yanomami men tie their foreskin to a cotton string around their waist.

11. The "representative" (visible correspondent) of the spirit *Tʰorumari* is *tʰoru wakë,* a term which refers, according to Davi Kopenawa, to "a great star fallen from the sky," perhaps a comet.

12. Joking relationships are constant between two potential brothers-in-law: os-tentatious joviality, close physical contact, sexual mockery.

13. Though they evoke *Yãri,* "the" thunder, the Yanomami consider many thun-ders to live on the "sky's back." For another version of this myth of the thunders' origin, see M 4.

14. *Yariporari* literally means "waterfall of wind (supernatural) being."

15. The spirit *Toorori*'s "representative" in the visible world is the *tooro* toad, whose song can be heard in the rainy season.

16. Allusion to the great drought and fires caused in the forest by El Niño during the summer of 1998 (December to March).

17. The radio of the local FUNAI outpost at Demini.

18. This oral tradition confirms the importance of climate factors (and not only anthropic ones) in the origins of the savannas of the highlands in Yanomami terri-tory in Brazil (see Alès, 2003).

19. This sorcery plant is said to cause states of unconsciousness and intense agi-tation during which the victim runs frenetically through the forest.

20. The white-lipped peccary disappeared from Yanomami territory for many years in the late 1980s following this territory's massive invasion by gold prospec-tors. See Chapter 15.

21. *Xotokoma* birds are considered the "peccaries' brothers-in-law," and their call is said to be these animals' "heralding call" *(heã)* in the forest.

22. Their calls are also considered tapir's "heralding calls." The *herama* falcons feed on ticks on the tapirs' back.

23. On tapir hunters, also see Chapter 3. Another expression—this one pejora-tive—is constructed in the same way: *napë xio,* "white people's bottom," refers to Yanomami too interested in chasing after white people and their merchandise.

24. The terms used about tapir hunting come from the psychology of love: *pihi kuo* ("to be in love, to have your mind set on someone"), *piri wariprao* ("to long for someone").

25. The expression *në rope* (or *në ropeyoma*) literally means "value of quickness (of growth)," "value of fertility." This "value of growth of the forest" *(urihi a në rope)* is opposed to its "value of hunger" when devoid of fruit or game *(urihi a në ohi).* The expression *në rope* is synonymous with *në wamotimapë,* "value of food." Both the images of the forest's "value of growth" and "value of hunger" can be used by shamans as spirit helpers *(xapiri),* respectively becoming *Në roperi,* the "spirit of growth," and *Ohiri,* the "spirit of hunger." *Urihi a,* the "forest," also means "earth" and "territory."

26. Fruit of the following trees is mentioned here: *oruxi hi, xaraka ahi, wapo kohi, krepu uhi, pooko hi, apia hi, õema ahi, horomona hi, hoko mahi, poroa unahi, himara amohi, hawari hi.* Fruit of the following palms is also mentioned: *õkorasi si, rio kosi, hoko si,* and *mai masi.*

27. Here, *hutuhami t^hẽ pë nẽ rope,* "the value of the growth of garden things," is contrasted with *urihi a nẽ rope,* "the value of growth of the forest."

28. On the abundance of forest resources available in *Watoriki,* see Albert and Le Tourneau, 2007.

29. These two very tall forest trees (up to forty meters for the first, more than fifty meters for the second) have compound leaves: the jatoba *(aro kohi),* with double leaflets, and the kapok tree *(wari mahi),* with eight leaflets.

30. *Koyori* is the mythological "father" of the *koyo* leaf-cutting ant, major scourges of manioc plantations. *Koyori* is associated with the earth's fertility and gardens' opulence. He is an indefatigable land-clearer and the owner of vast maize gardens (M 86). Note that maize is the only plant that leaf-cutting ants do not attack in plantations (Gourou, 1982: 83).

31. On the Yanomami theory of conception, see Chapter 1, note 24.

32. The Yanomami associate valued personal qualities (generosity, courage, eloquence, skill at hunting, zeal at agricultural work) with the incorporation and transmission (from father to son) of certain animal ancestor images *(utupë)* who constitute their mythological archetype (see Albert, 1985: 157–163). They must not be confused with the *xapiri* spirit helpers, which settle in a shaman's spirit house.

33. Banana plant shoots are known as *moko si,* "pubescent girl-plants."

34. The expression used, *hewëri/paxori a nẽ roperipë* ("the supernatural value of growth of the bat/spider monkey spirit"), suggests that *nẽ roperi* refers here to a generic insemination principle attributed to a group of shamanic spirits in charge of the growth of cultivated plants.

35. These are embryonic leafy shoots, still curled into themselves *(ako).*

36. The giant *waka* armadillo is a nocturnal animal fond of manioc tubers.

37. The *marokoaxirioma* is a small bird that often nests in *rasa si* peach palms in gardens. Its ancestor is the central character of the *rasa si* peach palm's origin myth (M 102).

38. The green cacique bird is often seen in gardens. The bark slings with which Yanomami mothers carry their babies are generally made either of the interior bark of *rai natihi* trees or of pounded bark from *yaremaxi hi* trees.

39. The Yanomami cultivate the cush-cush yam *(wãha a),* which has a separate origin myth (M 92). In this myth it enters a house in answer to a starving old woman's call. They also collect several kinds of wild yam (including *Dioscorea piperifolia* and *Dioscorea triphylla*).

40. The *reahu* feast ceremonial foods most often mentioned in Yanomami myths are maize and the fruit of the *momo hi* highland tree.

41. The *reahu*'s ceremonial etiquette requires that guests are given the maximum possible amount of plantain soup (or of the juice of the *rasa si* peach palm

fruit). The resulting indigestion puts them into an altered state of consciousness attributed to the "dangerous power *(wai)* of plantain soup," *koraha u wai* (or "of the juice of the *rasa si* palm fruit," *raxa u wai*).

42. The following fruit trees are mentioned here: *oruxi hi, hai hi, xopa hi, makoa hi, õema ahi, ixoa hi, aro kohi,* and *okoraxi hi.*

43. Davi Kopenawa refers here to the "value of growth's ripeness" *(në rope a tʰa-tʰe)* and to its sweet taste *(në rope a ketete).*

44. The *hutuma* bird's call is considered the "heralding call" *(heã)* of the "time of the fatty monkeys" *(paxo pë wite tëhë)*, which corresponds to the months of June through August.

45. This unknown/invisible being *(yai tʰë)* is referred to by the expression *hutu-karari paxori a në witepë*, literally the "spider monkey being of the sky with the value of fat."

46. The *xapiri* are said to expel the concretions *(xapo kiki)* obstructing sterile women's uteruses. The spider monkey spirit then copulates with their image, allowing their husbands to fertilize them afterwards. A shaman can also be asked to determine the child's gender by putting a male or female baby sling on the mother's image.

9. Outsider Images

1. The Yanomami territory was considered impenetrable until the twentieth century: "It is my conviction—and that of the people who know these distant, mysterious, and deserted regions—that so long as they remain in their current state, which is to say deprived of resources and dominated by the fierce hordes of Maracanas, Kirishanas [Yanomami], and the many others that infest them, the secluded Parima will remain inaccessible to civilized men, surrounded by the mysteries in which it has been cloaked to the present day" (Lopes de Araujo, 1884 [Report of the Brazilian Boundary Commission]).

2. The ancient Yanomami used machetes and digging sticks made of palm wood. They probably also had stone and giant armadillo shell *(waka husi)* hatchets (Albert and Milliken, 2009: 32–34, 101–102).

3. The Brazilian Boundary Commission (CBDL), which came into contact with isolated Yanomami groups on the upper Rio Mucajaí in 1943, observed: "We confirmed that they had heavily worn metallic tools. . . . Among these objects, we noted a curious axe head cleverly bound to a piece of wood with fibers covered in wax: the ring which once served to hold the handle was no longer present and the blade was reduced to a few centimeters of its original size" (Aguiar, 1946).

4. Literally, *Omama poo e xiki*, from *poo*, "metal," and *xiki* (plur.), "material," as in *hapaka xi*, "clay for pottery-*hapaka*." The western Yanomami use the term *poo* to refer to stone axe heads (Lizot, 2004: 319; Mattei-Müller, 2007: 239), whereas the eastern Yanomami use it to refer to metal blades. It is therefore likely that, among the latter, iron was initially referred to as "*Omama's* (stone) tool" and that the term *poo* was then gradually associated with metal.

5. In-laws are subject to strict avoidance among the Yanomami. For another version of this myth, see M 198 and a variant, M 128.

6. A digging stick *(sihe enama)* is made of *rasa si, hoko si,* or *kōanari si* palms.

7. A 1958 SPI report states: "manioc used to make cassava bread is grated on rough tree bark, which demonstrates their primitivism" (Andrade Gomes, 1958). Around the same time, the German ethnologist Hans Becher also noted in the upper Demini: "They do not even have manioc graters. They use rough granite stones" (Becher, n.d.).

8. The hammocks were made from the bark of various trees: *rai natihi, ara usihi, hokoto uhi,* and *hotorea kosihi.*

9. Malaria in its epidemic form is referred to by the expression *hura a wai,* "dangerous disease of the spleen" (malaria causes painful splenomegaly). Malaria was never detected in the Yanomami highlands until they were invaded by *garimpeiros* in the late 1980s. However, it had been reported in the lowlands since the early twentieth century (see Albert and Milliken, 2009: 135–137).

10. These impromptu *reahu* feast couples were formed by men and women of allied houses, both guests and hosts, in a kinship relationship of potential spouses (see Albert, 1985: 463–470).

11. Medical evacuations to the small city of Boa Vista, capital of the state of Roraima, are frequent. Since the 1980s, Davi Kopenawa—followed by other representatives of the Yanomami—has participated in many national and international meetings in order to defend their territorial rights.

12. These are references to the Perimetral Norte road, which opened in the south of Yanomami territory from 1973 to 1976, to settlement and cattle ranching projects bordering Yanomami territory, and to the fact that more than half of Yanomami territory in Brazil has been requested for mining concessions (Ricardo and Rolla, 2005: 50).

13. *Cachaça* is Brazilian sugarcane alcohol.

14. From *napë,* "enemy, outsider" (later "white person"), *-ri,* a suffix indicating supernaturality, monstrosity, or extreme intensity. The plural of this word is *napënapëri pë.*

15. Literally, *napë* ("outsider/enemy") *kraiwa* (plur. *pë*) as opposed to *yanomae t^hë pë napë* ("outsider/enemy human beings") or *napë pë yai* ("real outsiders/enemies"). These latter two expressions denote other Amerindian groups. *Kraiwa* probably comes from *karai'wa,* a term from Old Tupi referring to white people. It is also used by the Yanomami's Carib-speaking neighbors, the Ye'kuana (Heinen, 1983–1984: 4). Following the disappearance of neighboring Amerindian groups from the beginning of the twentieth century (with the exception of the Ye'kuana), the term *napë pë* is now exclusively applied to white people (Albert, 1988). Similarly, the "ancient outsiders' spirits" *(napënapëri pë)* are now only "white people ancestors' spirits."

16. In 1946, a Yanomami group from the upper Rio Parima presented a team of the CBDL with "interesting objects" (knives, machetes, fabrics, feather ornaments) "received from civilizeds from the Rio Uatátas (Uatataxi)" "who traveled in dugout canoes" and "came from the North" (Aguiar, 1946). The name *Watata si* (plur. *pë*),

also used for the Rio Parima's upper course, was probably applied to the Maku, a group which still had a house downstream of the river at the time (Albert and Kopenawa, 2003: 167). The Maku were already living in the area in 1912. They were in close contact with a Yanomami group to whom they provided manufactured objects, which they obtained by taking long trips among Carib groups in the east (Makuxi, Taurepang) (Koch-Grünberg [1924] 1982, vol. III: 28, 266).

17. The *Mait^ha* (plur. *pë*) were another Amerindian group once located in the area of *Takai maki* mountain (the Serra do Melo Nunes, between the upper Rio Uraricoera and the upper Rio Mucajaí). Oral history attributes the first malaria contamination to this group (see Albert and Milliken, 2009: 136). A myth also attributes the origin of lice to the hammock of *Sutu,* a *Mait^ha* shaman (M 178).

18. For Davi Kopenawa, the swords, helmets, and armor of the spirits of the white people's ancestors (from "the time when David killed Goliath") echo the biblical placards ("Biblavision" flashcards) used by the New Tribes Mission's missionaries to teach Bible stories. But as he heard about these visions from shaman elders in his youth, these images may have also had their roots in interethnic rumors from colonial times. The mirror-eyeglasses and white uniforms (medical and/or military) are obviously of a more recent vintage.

19. These are allusions to the outsiders' origin myth (M 33); see further in this chapter. Like the other *xapiri,* the *napënapëri* spirits are the "images" *(utupë)* or "ghost forms" *(në porepë)* of mythical ancestors.

20. In Brazilian Portuguese, *rezadores* means "those who pray." Davi Kopenawa has occasionally met some of these healers in the cities of Manaus and Boa Vista. See Chapter 15.

21. On *Omama* and the origin of the rivers, see Chapter 2. Davi Kopenawa refers to this mythical point of origin of the rivers by the expression *mãu upë monapë,* the "lever of waters." The term *monapë* is used to refer to the notch of an arrow and the trigger of a gun, as well as faucets, light switches, and keys.

22. For another version of the outsiders' origin myth, see M 33.

23. The principal host of a *reahu* feast is the holder of the funerary cinerary gourd *(pora axi)* containing a close kin's bone ashes and of the smoked game *(uxipë h^weni,* the "ashes' game") hunted by the villagers for the ceremony (see Albert, 1985: 440).

24. This *reahu* collective taking of hallucinogens and *yãimuu* ceremonial dialogues (see Chapter 3, note 25) often lead to ritual duels during which partners take turns punching each other in the chest *(pariki xëyuu)* or hitting each other on the head with sticks *(he xëyuu).*

25. When young girls are married before puberty, their husbands must follow the first menstruation seclusion with them.

26. Davi Kopenawa draws a curve in the air from west to east, roughly describing the course of the Rio Branco (known as the Rio Parima, then as the Rio Uraricoera on its upper course).

27. Whirlpools in rivers are considered indications of the presence of these be-

ings. Their father/mythological prototype is *Tëpërëriki*, Omama's underwater father-in-law. See Chapter 3, note 49.

28. Recent shamanic incorporation of a classic of ancient whaling iconography.

29. The Pauxiana, a Carib-speaking group, lived along the middle course of the Rio Catrimani and Rio Mucajaí until the nineteenth century. The Amerindians of the mid-Demini are Bahuana, an Arawak group. Regarding the *Watata si*, see note 16 above. The last villages of all these groups became extinct during the first decades of the twentieth century.

30. These groups are ones that Davi Kopenawa knows well, either because they are close by (Ye'kuana, Makuxi) or because he worked among them for FUNAI (Tukano) or because he visited or met them during interethnic political assemblies (Wayãpi, Kayapo).

31. Large, solitary *remoremo moxi* bees are often found on the sandbanks of rivers when the water level is going down.

32. On the dreams following daytime shamanic sessions, see Chapter 2, note 23, and Chapter 22.

10. First Contacts

1. For the Yanomami, avenging a dead person consists of "killing/eating" his murderer, who is said to be "in *õnokae* homicide ritual state." See Chapter 7, note 40.

2. The prohibition regarding the use of names (actually nicknames, see Chapter 1) is even more inflexible in the case of a deceased person. Even mentioning his death and referring to anything potentially related to it are also strictly avoided through the use of a set of highly codified circumlocutions (see Albert and Gomez, 1997: 166–170, 240).

3. This stepfather lived over ten days on foot from the *Watoriki* house where Davi Kopenawa lives (on a tributary of the upper Rio Demini, the *Wanapi u* River). He died of old age in the late 1990s.

4. Regarding these temporary triangular shelters *(naa nahi pë)* covered in large leaves *(irokoma si, ruru asi)*, see Albert and Milliken, 2009: 73.

5. This communal house was first visited by an SPI expedition accompanied by missionaries from the New Tribes Mission in June 1958. One of them described it as a circular shelter that was fifty-eight by forty meters and inhabited by some two hundred people (McKnight, 1958). The people of the upper Rio Mapulaú referred to here (then known as *Mai koko* or *Mai koxi*) now live on the Rio Jundiá (a tributary of the Rio Catrimani).

6. One communal house was occupied by "the people of *Yoyo roopë*," another by "the people of *Sina t^{h}a*," and the third by the group of a leader called "Paulino," who had remained separate from the temporary grouping at *Marakana*. Davi Kopenawa's parents and stepfather belonged to the first group.

7. In January 1959, a Brazilian Boundary Commission (CBDL) report described

the site as follows: "On the upper Toototobi, a little downriver from its tributary the Cunha Vilar, is a site consisting of barely two houses and inhabited in a transient fashion by Xirianãs Indians [Yanomami]. These Indians are principally concentrated in their Marakana house, which is situated upriver from the divisor" (L. P. de Oliveira, 1959: 16).

8. The Yanomami continued to pick the fruit of cultivated *rasa si* peach palms in *Marakana*'s former gardens until the early 1980s (a six-hour journey on foot from the Toototobi Mission).

9. "Isolated" Yanomami spent between one-third and half of the year in such forest camps (Lizot, 1986: 38–39; Good, 1989: 89; 1995: 115).

10. During this ritual (*watupamuu*, "acting as vulture"), the warriors imitate the image *(utupë)* of various scavengers (in addition to the vulture) and carnivores (including the jaguar). In so doing, they represent the devouring of the corpses of the enemies they will attack. See Chapter 21.

11. Reference to the ceremonial smoked game distributed to guests at the end of *reahu* feasts (see Chapter 9, note 23).

12. Raiding the people of the Rio Mapulaú (and the upper Rio Catrimani) was at its height during the 1950s. Incursions carried out against the *Xamatʰari* of the upper Rio Demini were much less frequent. See Chapter 21.

13. The father of these two women, a widower, later came to join the group of their abductor and husband (Davi Kopenawa's stepfather) and was given there a second wife. The abduction of women and children is only a "fringe benefit" of war raids, appreciated but uncertain.

14. Regarding this ritual of the departure to war, also see Chapter 21 and Albert, 1985: chapter 11.

15. This site located at the headwaters of the Rio Toototobi was inhabited from the second half of the 1930s to the beginning of the 1940s. Davi Kopenawa's stepfather, who was born in the late 1920s and spent most of his childhood there, still referred to himself as "an inhabitant of *Yoyo roopë*" when he was over seventy years old.

16. The old network of access to metal tools, which was centered in the north at the beginning of the twentieth century (Rio Parima), gradually shifted to the south in the very beginning of the 1940s (Rio Demini).

17. Balata is a natural latex extracted from the *Manilkara bidentata* tree, formerly used to manufacture boots, cable insulating material, and golf balls. Piassava is a textile fiber extracted from the *Leopoldina piassaba* palm, from which brushes, brooms, doormats, and cords are made.

18. The *Xamatʰari* of the *Kapirota u* River began maintaining peaceful relations with the white people of the Rio Aracá in the early 1940s, putting an end to over a decade of conflict (skirmishes with the regional population in the 1930s, attack on a CBDL camp in January 1941). Hans Becher, who visited the region in 1955–1956, recounted that during the rainy season they lived halfway down the Rio Aracá, in the place known as Cachoeira dos Índios ("the Indians' Waterfall"), seventy kilometers from their communal house, in order to work for forest product collectors or

directly for their "boss," a traveling Portuguese salesman, who relentlessly exploited them: "there was not enough food given the heavy work they had to do and which they were not used to. Six Indians died because of this. Payment consisted of a few knives, axes, pots, etc., cheap things with no real value. To him, the Indians are nothing more than slaves" (Becher, 1957). Their sanitary conditions were disastrous: anemia, flu, verminosis, chronic malaria (Avila and Campos, 1959). Hans Becher also describes the visit of some twenty "Xirianá" warriors (the Yanomami elders referred to above by Davi Kopenawa) who came to get manufactured objects from the $Xamat^hari$, who "after such a visit, literally found themselves stripped of everything" and claimed they were "forced to trade with the piassava gatherers because they have to give all the objects they obtain to the Xirianá, who do not have any contact of their own with the white people."

19. The Inspetoria refers to the Primeira Inspetoria of the SPI (Amazonas state). The Yanomami referred to SPI agents as $Espeteria\ t^hëri\ pë$, "the Inspetoria's people." The members of the CBDL were referred to by the name $Komisõ\ pë$, from the Portuguese $comissão$. After its first visit to $Marakana$ in June 1958, the SPI returned there with members of the CBDL in October 1958 (SPI Reports, Ajuricaba Outpost). At the time, Davi Kopenawa was about three or four years old (his official birth date, February 1956, is approximate).

20. The CBDL finally carried out its work in the Rio Tootobi headwaters in the first half of 1959. Its team reached $Marakana$ in January and had a good first impression: "The Xirianãs [Yanomami] are Indians of robust complexion and based on the expanse of gardens we saw, covered in manioc and banana plants, they are also good workers. However, the long summer delayed the maturing of the bananas and the growth of the tubers" (L. P. de Oliveira, 1959: 16). Yet the Yanomami were initially not very cooperative: "The Indians, starving and in larger numbers than us, helped themselves to all our open supplies and, to top it off, did not want the company to continue on its path" (Oliveira, 1959: 7). Finally, by being generous with its food and trading goods, the CBDL team won everyone's support and work continued without any trouble: "All in all, we have no complaints about the Indians of the Rio Tootobi for, while they proved to be irritable and demanding at the beginning, they gradually became more docile and eventually helped us with some work, including transporting loads and collecting lianas for the carrying baskets" (Oliveira, 1959: 17).

21. This part of the house *(yano a xikã)*, located behind the family hearth, is an essentially feminine space where cooking implements, basketry, and firewood are kept (see Albert and Milliken, 2009: 76).

22. The CBDL's first visit to the Rio Mapulaú took place in the period 1941–1943: "On a small tributary of the Mapulaú we surprised a group of Uaicá [Yanomami] in their house, located near a vast clearing where there was a garden. The Indians did not have time to take their weapons and, while some escaped into the garden or by climbing up trees, others, perhaps the more courageous, stayed in the house, constantly talking and gesticulating. When the members of our team started taking them into their arms, they were shaking, but once this first fright had passed, they

calmed down and the runaways also drew close . . . The natives then reciprocated our team's gifts with produce from their garden: bananas, papayas, sugarcane . . . and cassava bread baked by their wives" (Jovita, 1948: 64).

23. This group was the target of the upper Rio Catrimani's inhabitants until the early 1980s. These two children were raised in a small regional village on the lower Rio Demini (upriver from Barcelos).

24. These were planes taking aerial photographs during the CBDL's work. See Le Tourneau, 2010: chapter 1.

25. The CBDL's teams transported huge amounts of metal tools, red cotton fabric, and other trade goods, generously giving them away in order to "pacify" the Indians (see the photo documentation of this trading in Albert and Kopenawa, 2003: 168–169): "Over one hundred forest dwellers, men, women, and children, remained in our camp for twenty days, requiring food, machetes, knives, tobacco, fishing hooks, and other objects. When our boats arrived, bringing a new stock of supplies and trade goods which we had sent for, the presents were distributed, and, in exchange, we received bows, arrows, and bananas" (L. P. de Oliveira, 1959: 16).

26. The machetes were greased and wrapped in paper probably to avoid oxidation.

27. Sweet odors *(ria rieri)* are considered dangerous *(wai)* because they can make people "become other."

28. The "metal (tools) smoke" *(poo pë wakëxi* or *poo xiki wakëxi)* is also called the "metal epidemic" *(poo xiki xawara)*.

29. Conjunctivitis is a usual complication of flu.

30. These "epidemic trees" are said to include *t^{h}oko hi* ("cough tree"), *hipëri hi* ("blindness tree"), or *mamo wai hi* ("conjunctivitis tree"), and *xuu hi* ("dysentery tree"). Davi Kopenawa's father-in-law explained to us that at that time pieces of "cough tree" bark were boiled by the white people to coat the red cloth distributed to Yanomami.

31. "Clothing" is *kapixa,* from the Portuguese *camisa,* "shirt."

32. Red cloth has long ago disappeared from the common trade goods. But when it reappears, old fears from the first contacts can immediately be rekindled among the elders (see Albert, 1988: 168).

33. "On the Rio Demini in 1942, certain members of the team lit cigarettes, unaware that they would set off a genuine panic among the savages. As soon as they saw the flames of the matches and lighters they were terrorized, gesticulating convulsively, crying out in horror, and they started to run away" (Jovita, 1948: 112).

34. To this day, the Yanomami remain hostile to white people burning debris of any manufactured objects or industrially produced waste of any kind in their presence (particularly paper and objects made of plastic and fabric), fearing that their combustion will spread "epidemic fumes."

35. The flu virus can be transmitted through a wide variety of contaminated manufactured objects (Lacorte and Veronesi, 1976: 17).

36. Oswaldo de Souza Leal, then forty years old, was an agent of the SPI outpost at Ajuricaba (upper Rio Demini). "The people of *Sina t^{h}a*" were one of the two communities living in *Marakana* in the late 1950s.

37. The distance from Manaus, capital of the state of Amazonas, to Toototobi as the crow flies is 680 kilometers. The trip upriver by boat and motorboat could take up to three weeks during the period when the waters were low.

38. Anger due to sexual frustration *(pexi hixio)* is subject to contempt and scorn.

39. This was probably dynamite or a homemade, powder-based explosive. Note that at the time, the SPI recommended using firework rockets or explosives and firing weapons in the air as intimidation tactics for its "pacification" expeditions among Amerindians (Magalhães, 1943). The Yanomami still have a word for firework rockets: *hukrixi a.*

40. This oblong funerary bag *(paxara ãhu)*, woven with palm leaves *(mai masi, hoko si,* or *kõanari si)*, can be carried on a person's back with a frontal sling. Once carried into the forest, it is wrapped in a lattice *(yorohiki)* and attached to a wood structure halfway up a young tree or on a platform. Once the body has decomposed, the bones are gathered and cleaned before being kept for some time in a basket above the hearth of the deceased's closest kin. These dried bones will later be burned and ground to fill funerary gourds *(pora axi)* whose contents will be ingested or buried during one or several *reahu* feasts.

41. The symptoms described are reminiscent of an epidemic of exanthematic disease (such as measles, rubella, or scarlet fever). This would have taken place in 1959, probably following the intense circulation of SPI and CBDL agents in the area (1958–1959). A letter from the head of the SPI outpost in Ajuricaba dated June 7, 1959, reports the unauthorized return to Manaus of five workers from the outpost, including Oswaldo Leal, on May 30, 1959. Other reports and telegrams from the outpost establish that Leal visited the upper Toototobi in January, February, and April 1959.

42. The concurrence of the epidemic and Oswaldo's pyrotechnic intimidation, followed by his hasty departure, reinforced the traditional association the Yanomami draw between "epidemic smoke" and white people sorcery (see Albert, 1988). On the *õnokae* homicide state, see Chapter 7, note 40.

43. This is a reference to the SPI outpost established far downstream of the mouth of the Rio Toototobi on the Rio Demini, in the place known as Genipapo (see Chapter 11).

44. In October 1958, the SPI estimated that the number of Yanomami on the upper Rio Toototobi was 335 (Andrade Gomes, 1958). In 1981, the region's population was still only 230 (B. Albert census).

45. Davi Kopenawa refers to the dynamiting of rocks during the opening of the Perimetral Norte in 1973–1975, in the southern part of the Yanomami territory in Brazil.

46. The return of ghosts is a major theme in Yanomami mythology and ritual system (see M 36 and Albert, 1985: chapters 12 and 14). The evangelist missionaries of the Rio Toototobi where Davi Kopenawa grew up describe a visit of some of their numbers to an isolated village on the upper Orinoco in 1968: "One village, Bocalahudumteri, did not receive them well at all. They indicated to R. [Davi Kopenawa's stepfather] that the foreigners were actually the spirits of dead yanomama that had

come back again. B. H. was formerly a yanomama that was shot with an arrow and God rubbed his flesh in some mysterious manner and he came back to life again. I am also a spirit. They say that it is quite obvious because I don't have any hair on my forehead. F., too, is a spirit. This also is quite evident to them because his teeth are loose. (They have never seen anyone with false teeth.)" (K. and M. Wardlaw, 1968a). Other evangelists living in the neighboring region (Serra das Surucucus, upper Rio Parima) in the early 1960s report the same interpretation: "The missionaries . . . [were] believed at first to be spirits of dead ancestors, and for a while the Indians would not touch anything from the mission. Then they suddenly decided we were real, and then the problem was to stop them stealing everything" (Brooks et al., 1973).

47. This expectation is often expressed in the origin myth of outsiders. See, for example, the version recorded in 1979 from a man who had experienced the first contacts with the CBDL in 1941–1943 on the Rio Mapulaú (M 33). The demiurge (*Remori* in this version) says to the outsiders he just created: "You shall go back where you came from and then you will give presents to your people who remained there! . . . You shall go back and give metal tools to all of them."

48. Davi Kopenawa sometimes uses the Portuguese term *cultura,* which he defines as: "the things *Omama* taught us and which we continue to do," or the expression "have a culture," which to him means "to continue to be as our ancestors were."

49. Diverse sources are interwoven here. Davi Kopenawa mixes what he has learned from the white people's recounting of the Conquest with the Yanomami history of contact and also, perhaps, ancient shamanic interethnic rumors.

50. Brazilian textbooks teach that Pedro Álvarez Cabral "discovered Brazil" in 1500.

51. $H^w ara\ u$ is the name of the watercourse at the origin of all rivers that sprang out when *Omama* pierced the terrestrial layer (M 202). It is also the name given by the Yanomami of Brazil to the headwaters of the Orinoco. The waters of the underworld are known as *Motu uri u.*

11. The Mission

1. *Teosi* the Yanomami word for God comes from *Deus* in Portuguese. Nine Yanomami men from the upper Rio Toototobi made an initial visit to the SPI's Ajuricaba Outpost in May 1958; on this visit, they worked in exchange for metal tools (Andrade Gomes, 1959). The *Xamat^hari* mentioned here (known to the SPI as "Paquidai" or "Paquidari") took refuge at the SPI outpost in the beginning of 1943 following a devastating epidemic that had taken place shortly before (Jovita, 1948: 313). This group came from the *Xamat^hari* of the *Kapirota u* River mentioned in the previous chapter. However, the Ajuricaba Outpost practically lost contact with the Indians starting in 1949, and was only occasionally visited by two families living two days away. A report to the SPI by Hans Becher (n.d.) describes the precarious situation of this outpost in 1955, infested with mosquitoes and black flies, without medication, radios, or canoe motors, inhabited by twelve agents who depended on

a few gardens, hunting, and fishing, and who received more visits from traveling salesmen (turtle trade) than from the SPI (twice a year). The Yanomami only came to clear a garden and settle near Ajuricaba again in April 1958, after the outpost was reactivated in preparation for the arrival of a new Brazilian Boundary Commision (CBDL) team. A 1959 report recorded the reactivation of the Ajuricaba Outpost and the presence of missionaries from the New Tribes Mission in its surroundings: "The largest site inhabited by civilized people on the Rio Demini is in the place called Genipapo. The SPI has an attraction outpost there and the Indians who appear, coming from upriver, are employed to cultivate the land on a manioc, sugarcane, maize etc. plantation. A little downriver from the outpost is a rapidly growing American mission whose objective is to teach the Indians religion" (L. P. de Oliveira, 1959: 15).

2. These were allies of the Rio Mapulaú enemy group mentioned in the previous chapter.

3. Leaving the communal house to trek to a distant forest camp is a standard Yanomami defensive strategy after an incursion. Before the establishment of the Toototobi New Tribes Mission outpost in 1963, raiding activities were quite usual. For instance, while visiting the region in March–April 1962, the Swiss anthropologist René Fuerst (1967: 103) witnessed the departure of a war party from Davi Kopenawa's stepfather's house against a $Xamat^hari$ group from the upper Rio Demini. Incursions significantly decreased after that time (however, see note 56 below).

4. This initial visit of the New Tribes missionaries to *Marakana* with an SPI team from the Ajuricaba Outpost took place in June 1958 (McKnight, 1958).

5. The installation of border markers on the upper Rio Toototobi lasted from November 1958 to mid-1959.

6. After their brief visit in June 1958, the New Tribes missionaries made their first stay at *Marakana* (one month) in early 1960 (Zimmerman, 1960).

7. "K. has tried to explain the Gospel messages to them by using the Guaica [Yanomami] Bible Stories translated by Jim Barker in Venezuela. We have also used some Gospel records in the Guaica language that were made in Venezuela by a representative of Gospel Recordings Incorporated. But this Message is all new to them whereas, on the other hand, these practices ("singing for the demons") have been a vital part of their culture for generations. Only the Lord God Almighty is able to liberate them from this bondage of superstition and fear" (Zimmerman, 1960).

8. The Ajuricaba Outpost was part of the SPI's first regional Inspectorate (Inspetoria). The New Tribes missionaries' first contact with the Ajuricaba Outpost took place in 1956.

9. A letter from the SPI-1[a] Inspetoria Regional archives (December 1958) states that "the head of the Ajuricaba Outpost, agent A. de Andrade Gomes, . . . vehemently condemns the foreign influence on Indian elements, whether in the form of exploiters, or under the appearance of missionaries, which they always use to conceal themselves."

10. Regarding this epidemic, see the previous chapter. Curiously, neither the New Tribes Mission journal *(Brown Gold)* nor the SPI archives refer to the "Os-

waldo epidemic," which is therefore difficult to date accurately. As we have seen in Chapter 10, note 41, the SPI archives suggest it may have taken place in mid-1959. Davi Kopenawa's narrative also leads us to believe it took place before the missionaries' second visit in 1960.

11. An allied community from the sources of the Rio Toototobi.

12. The missionaries began settling with their families on the site that would become the Toototobi Mission in June 1963. Their first extended stay here lasted from June 1963 to March 1964 (K. Wardlaw, 1964).

13. "Jim Barker [the first missionary to settle among the Yanomami in 1950 in Venezuela] has prepared some phrases in the Guaica language [Yanomami] for us and through these we were able to get some of the simple truths of the Gospel to this group such as: 'God loves us. He hates sin. God's Son died for our sins and has a place in Heaven for those who love and obey Him'" (McKnight, 1958).

14. "Progress has been made in learning the language and we have made some simple Gospel presentations. We are endeavoring to explain God's love for mankind, His hatred for sin, His knowledge of all we do and say, and, most of all, our need of a Saviour . . . A few Bible stories and some songs have also been written. While their musical ability is far from good, they at least seem to be understanding some of the message" (K. Wardlaw, 1964).

15. *Xupari* is from *shopari wakë*, the fire of the celestial world, where according to the western Yanomami the greedy burn after their death (Lizot, 2004: 401; Mattei-Müller, 2007: 305). This shamanic notion was appropriated by the missionaries to serve as "hell."

16. Satan, from the Portuguese *Satanas*.

17. Jesus, from the Portuguese *Jesus*.

18. "We meet each morning about 6:15 A.M. for prayer, preceded by about five minutes of instruction in some Christian truth pertinent to their needs. Yesterday it concerned taking a wife, disciplining children, and abstaining from committing adultery. . . . Everyone prays at the prayer meeting. We will be dividing into two groups tomorrow, since it is taking too long and some of the older believers get impatient when the new ones attempt to pray" (K. and M. Wardlaw, 1968a).

19. On the re-elaboration of Christian myths by the Yanomami of the Toototobi region, see Smiljanic, 2003.

20. "B. H. has used his little projector here very effectively. The story of Noah and the ark was a real blessing the other night and four people indicated that they wanted to accept the Lord the following day. Most of this village have now professed faith in the Lord, twenty-six in all. There were four more that came to the prayer meeting this morning, but we are a little dubious of their motives. No doubt there are some that are following the crowd, as it were, but there has been a real transformation in the lives of most" (K. and M. Wardlaw, 1968a).

21. "Our daily 6 A.M. prayer meetings have been a real blessing. First, B., K. or C. [the missionaries] shares something from the Word along a very practical line . . . Then we go to prayer with each one, in turn, taking to the Lord the burdens of his heart. One may not have slept well because of some anxiety, and he asks the Lord to

throw his worry away. Another tells the Lord he sinned a little bit yesterday by being angry, and asks forgiveness. Those with sick loved ones admit that their Father God alone knows how to make them well. Many pray in length for fellow Christians away at festivals, naming each one and the weaknesses this one has, and asks for their protection" (T. and M. Poulson, 1968).

22. Regarding hunting prayers, see this anecdote told by the missionaries about Davi Kopenawa's stepfather (R.): "The other day R. was going hunting and his experience and testimony was a real blessing to us. He said he told God that he wanted to get a tapir, but didn't want to take a lot of time and go far. He asked God to send one close by and, sure enough, God sent the tapir across his trail! He shot it and he stopped to say thank you to God. The animal was so near that we heard the shot from here. We have never known a tapir to come this close before. The Uaicas [Yanomami] do not have a word for 'thank you' in their language, so we have been teaching them the Portuguese word [obrigado!] and how to use it. Some have learned to use it quite well and it was that word R. used when he said thank you to God" (K. and M. Wardlaw, 1968a).

23. The "Americans" (*Merikano pë*, from the Portuguese *Americanos*) of the Toototobi Mission (as opposed to the Brazilians, *Prasirero pë*) were actually of different English-speaking backgrounds: Americans, Canadians, and English.

24. The construction of this airstrip (nearly 600 meters) in Toototobi took place in 1964–1965. It was opened by the New Tribes missionaries under the control of the Brazilian air force (FAB): "the FAB would like an airstrip in that area. The airstrip would be completely our responsibility, but would actually belong to the Air Force" (Mrs. K. Wardlaw, 1965a). Regarding the FAB's policy on American missions, see Le Tourneau, 2010: chapter 2.

25. There were several flyovers of the landing strip in 1964 and 1965, including a memorable drop of supplies, mail, and trade goods on February 23, 1965: "The knives and scissors which were included in the drop was to be a gift to the Indians, as previously arranged. They have had some fears already about planes, when M. H. went with E. on the survey last summer. Before we left the Toototobi last November, the story came to us about a village of Uaicas on the upper Demini who attributed a severe illness and many deaths among them to the plane when it went over their village. We felt it wouldn't do any harm to try to encourage the Indians with the gift, and that way attribute something good to the plane" (Mrs. K. Wardlaw, 1965b).

26. This popular association in Brazil originates with the one drawn between God and the Thunder divinity of the ancient Guarani *(Tupã)* by sixteenth-century Jesuits.

27. Regarding the myth of Thunder, see Chapter 8.

28. The threat of traps and poison could worry Yanomami concerned about their children's safety, but this cannibal boasting placed the missionary's threats in the familiar vocabulary of war rituals (the Yanomami homicide ritual *õnokaemuu* specifically deals with the symbolic ingestion of arrowed enemies' bloody flesh; see Chapter 7, note 40).

29. This measles epidemic took place in September 1967.

30. This mission, which was established on the upper Rio Parima in 1962–1963 (Migliazza, 1972: 390), was a thirty-minute flight from Toototobi. It belonged to another evangelical organization, the Unevangelized Fields Mission. Ten years after it was founded, the missionaries' situation there was far from comfortable: "On our first day here we suffered a break-in in our storeroom and several threats on the post. There has been some shooting near the edge of the post on two different occasions between the Indians. But on June 27, 1973, we had a 3-hour war in our front and side yards between two Uaicá [Yanomami] groups" (Moore, 1973). This mission was also attacked by Indians seeking to steal shotguns during a confrontation with illegal gold prospectors *(garimpeiros)* in 1975 (see Albert and Le Tourneau, 2005). It was finally closed by FUNAI shortly thereafter.

31. In fact, Kixi was a Canadian missionary, one of the founders of the Toototobi Mission.

32. This child, whose name is Tomé, was born in 1961, and was therefore five or six years old at the time.

33. Darts carved from the petiole fiber of *õkarasi si* and *kõanari si* palm leaves.

34. Missionary Keith Wardlaw's account of this epidemic is available online (www.sil.org/~headlandt/measles3.htm#measles3). He reports that he left the mission for Canada in June 1967. His wife and two children joined him in Manaus upon his return in late July, and they returned to Toototobi shortly thereafter. His two-year-old daughter was incubating measles (duration ten to twelve days).

35. The Yanomami interpretation of epidemics as sorcery smoke produced by white people's desire for revenge after a conflict (see Albert, 1988 and 1993) is so powerful that Davi Kopenawa continues to refer to the missionaries' alleged sorcery guilt, despite knowing the epidemic's viral etiology. These two hypotheses must have been heavily circulated among the Yanomami of the Toototobi at the time.

36. The ceremonial foods distributed to guests at the end of an intercommunal *reahu* feast generally consist of piles of cassava bread (or a bundle of boiled plantain bananas) and pieces of smoked game.

37. *Sarapo a wai* is from the Portuguese *sarampo*. The term *wai* ("dangerous, powerful," see Chapter 7, note 1) is included in the composition of all epidemic names (see Albert and Gomez, 1997: 112–115).

38. As we have seen in Chapter 8, the theme of the falling sky is a central one in Yanomami cosmology. It is also a key motif of their shamanic prophecy (Chapter 24). It is reproduced here on an individual level in a dream (also see Chapter 20).

39. The epidemic manifested itself in early September 1967: "a measles epidemic broke out in the tribe and three different groups were exposed. After an urgent message from K. W. . . . . B. left for the tribe on the twelfth of September. Since that time until about the first week of November, the missionaries have been battling with this disease. All in all, I guess they treated around 130 cases and had about twenty deaths, counting babies. They worked night and day over a period of weeks treating, giving shots, getting food, cutting wood, bringing water, and anything else that needed to be done" (Mrs. B. Hartman, 1968). The missionaries were

assisted by a pilot and a doctor from the Missionary Aviation Fellowship (MAF). They received a donation of 600 ampoules of injectable penicillin (against secondary infections) from a Brazilian pharmacist in Boa Vista, capital of the state of Roraima. The reported number of victims of this epidemic is different (twelve dead, 150 to 200 cases) in Neel et al., 1970: 421, 425 (according to the MAF's doctor, C. Patton) and in the Toototobi Mission's archives consulted by Bruce Albert in the 1980s (seventeen dead and 165 cases). This latter figure is confirmed by the missionary K. Wardlaw's account (see note 34 above).

40. On this first phase of the Yanomami funerary ritual see Chapter 10, note 40.

41. According to Yanomami therapeutic logic, shamanic spirits and the effects of industrial medicine are perfectly compatible, the former acting upon the diseases' etiology, the latter on the treatment of symptoms (see Albert and Gomez, 1997: 51).

42. Davi Kopenawa uses here the term *xoae a,* which means both "mother's brother" and future "wife's father," the ideal Yanomami uncle/father-in-law.

43. As a last resort, a great man's death is nearly always blamed on *oka* (plur. *pë*) distant enemy sorcerers; see Chapter 7. At the time, the people of *Amikoapë* lived in the highlands of the Rio Mucajaí headwaters. Their leader, *Naanahi,* was regularly accused of sorcery attacks on inhabitants of Toototobi.

44. The Christian burial ritual is considered by the Yanomami to be a revolting practice in that it prevents "putting in oblivion" the deceased's bone ashes, which is the only way to put an end to the mourning process, by permitting the dead ghost's final separation from the world of the living (see Albert, 1985; Smiljanic, 2002).

45. According to the Yanomami, one feels both anxiety *(xuhurumuu)* and anger *(hixio)* during the mourning period.

46. In early 1968, the Yanomami at Toototobi held a *reahu* feast for those who died in the epidemic. The missionaries sang religious chants: "The second evening we introduced a new song in Uaica [Yanomami] style and music . . . It imparts the truth of God's existence, that God is alive, He doesn't lie, He tells the truth, and ends with God is good" (K. Wardlaw, 1968). The consequences of this deadly epidemic, referred to as a "crisis," do not seem to have unduly worried them. Instead, they saw it as an encouraging sign for their evangelizing work: "It was hard to realize that many of our friends had passed on to eternity without knowing Christ. Yet, we know that God never makes a mistake and now that the crisis is passed we can see how the Lord is working in hearts through the things that have happened" (Mrs. B. Hartman, 1968).

47. "The reality of Hell and the escape through Jesus Christ is the basic message that has reached these people" (K. and M. Wardlaw, 1968a).

48. "Near Christmas time [1967] the chief R. took Christ as his Saviour as well. We have hardly been able to keep up with what has been taking place since then. God has given R. a very open heart and a very live and real testimony . . . The fellows started teaching about baptism, sensing that both R. and his stepson C. were ready for this step. Both of them agreed that they wanted to be baptized, so January 14 [1968] was the first baptism here on the Toototobi River!" (K. and M. Wardlaw,

1968a). Then, a few months later: "R., our chief, was one of the first to turn to the Lord, and soon after, all his family . . . Since the first decisions at the end of last year, one after another has come to the Lord to have his 'insides made clean'" (T. and M. Poulson, 1968).

49. In Portuguese the evangelists are *crentes*, "believers." It appears that trauma due to the epidemic and the return to the missionary credo of Davi Kopenawa's stepfather, a highly regarded leader and shaman, did indeed lead to a wave of conversions in early 1968 at Toototobi. The mission had twenty converts in January 1968 and fifty-two in June (K. and M. Wardlaw, 1968a and b). The missionaries were euphoric: "Some are won to the Lord through the Indians themselves, others just start coming to the daily prayer meeting all on their own and there make their stand in public, while still others come to one of us missionaries to let us know that they, too, want to come to the Lord. We have just never seen anything quite like this and it surely is an evidence of what God can still do in hearts . . . The power of God is at work and it is a great and marvelous thing to behold—after about ten years of labor in this field by various missionaries and a host of prayer warriors. Praise him!" Yet a hint of puzzlement showed through this enthusiasm: "The events of the past weeks have certainly been unusual, to say the least. It is sometimes very difficult to accurately appraise what is happening" (K. and M. Wardlaw, 1968a).

50. Chico was back at Toototobi beginning in the first months of 1968. He played a significant part in the strategy to evangelize the region's communities: "Chico, our Brazilian missionary, has been visiting NEW villages. He is presently on his third trek with the Indians. He has been a real help and blessing to the NEW believers whom he has accompanied on these trips" (K. and M. Wardlaw, 1968b).

51. Small presents and gifts of food are the foundations and signs of a new love affair.

52. After being dismissed from the Toototobi Mission, Chico was recruited by FUNAI. In a letter addressed to this administration on January 23, 1969, he wrote: "Unfortunately the mission dismissed me from my job under the pretext that I had accepted a young Indian girl of fourteen–fifteen as a wife." Assuring FUNAI that he had the agreement of the girl's father and the community's headman, he asked his correspondent both for authorization to keep the young girl ("because according to the chief in his culture it is legal") and a job (FUNAI Archives, Brasília).

53. In an evocatively titled article from early 1970 ("Satan's Counter-attack"), the missionaries expressed their concern about the decline of evangelical belief in Toototobi: "Good times and an abundance of food, plus other sneaky factors, seem to be detrimental to the growth of the Church and the propagation of the Gospel of Christ. Spiritual sluggishness, unthankfulness, and a rejection of our wonderful God and Savior are the principal earmarks of this hour on the Toototobi! . . . Pray! F. has taken a second wife and has turned his back on the Lord . . . G. has identified himself clearly as a rejecter of God and has gone back to his witchcraft" (Toototobi Gang, 1970a).

54. The Yanomami use the fishing lines' lead to make large, bullet-shaped pro-

jectiles, which they use in their cartridges for hunting large game (tapir, deer) and when going on a war raid.

55. For the Yanomami, the affection one has for a person—or the nostalgia one has for him—is linked to the generosity which that person has displayed in exchanges one has had with him. On this subject, see Chapter 19.

56. The missionaries reported an incident with Davi Kopenawa's stepfather in early 1970: "R. came to prayer meeting one morning and announced that he has happily come to the conclusion that witchcraft is good, God doesn't exist, Americanos are liars, and he warned us that he is going back to his old ways and is going to become real fierce again. He concluded by saying that he thinks we will all be moving away now. He then left the meeting, called the witch doctors back to active status again, and announced to some of the others that he wants to shoot the foreigners . . . A number of the young men and women have also stopped coming to the meetings" (Toototobi gang, 1970a). Despite another diplomatic rapprochement between R. and the mission later in 1970 (K. Wardlaw, 1970a), the decline of evangelical belief was patent: in April, warriors armed with shotguns launched an incursion against a neighboring group, causing eight casualties (K. Wardlaw, 1970b), and by June there were only "12–18 believers" who "fail to take a strong stand and speak out against evil" (Toototobi gang, 1970b).

12. Becoming a White Man?

1. The Yanomami distinguish the upstream $(t^h\ddot{e}\,\tilde{a}\,ora)$ and the downstream $(t^h\ddot{e}\,\tilde{a}\,koro)$ of a speech, narrative, or conversation to contrast its superficial and deep meanings (see also Chapter 5, note 12).

2. Davi uses the Portuguese words *padre* (father) and *madre* (mother) here. This dream seems to mix memories of the Eucharist iconography and reminiscences of Catholic clergies he may have come into contact with in Boa Vista. (Catholics are traditionally strong supporters of the Indian movement in Brazil.)

3. The name of this being comes from a root also found in the verb *wãiwãimuu*, which could be translated as "to be shaken by a soft pulsation."

4. On *Omama*'s flight and the creation of the white people, see, respectively, Chapters 4 and 9.

5. See Chapter 2. It is no surprise that *Teosi* (God) is associated with the white people's epidemic diseases, which appeared in the forest at the same time as his words (*Teosi* $t^h\ddot{e}\,\tilde{a}$, the "words of God"). Among the highland Yanomami in Brazil, *Teosi a wai* (the "disease-epidemic of God") is sometimes a synonym for *xawara a wai* ("epidemic disease"). *Teosi* is thus opposed to *Omama*, the Yanomami demiurge, and associated with his brother *Yoasi*, the trickster responsible for the loss of human immortality (M 191).

6. An investigative mission from the Aborigines Protection Society visited Toototobi on August 28, 1972. Its report summarizes a conversation with the missionaries: "We were told however that the local Indians had a strong culture which

was 'difficult to break', and it seemed that in 9 years only one convert had been made. This man [Davi Kopenawa's stepfather], proved to be the Chief, who met us clad in a mustard-coloured Playboy Club shirt; but we were told that even he showed signs of losing his faith. Another complaint of this missionary concerned the Indians' unwillingness to accumulate material possessions by work and saving . . . In short, the Yanomami seemed to be content with their culture, and had proved strong enough to resist the converting zeal of the missionaries. The lesson however has been learned: 'We were too aware of the urgency of bringing the message of Jesus-Christ to these people . . . It is a great mistake to belittle the spirit world. It is very real to the Yanomami'" (Brooks et al., 1973).

7. The missionaries settled at Toototobi in 1963, and the epidemic that killed Davi Kopenawa's mother took place in 1967. This period, during which Davi Kopenawa was seven to twelve years old, can be considered the conversion period he refers to here.

8. The local New Tribes Mission used literacy education and the translation of the Bible into vernacular languages as the means for their proselytizing. In 1965–1966, when Davi Kopenawa was about ten years old, the Toototobi Mission school had an average of fifteen to twenty-five students (Smiljanic, 1999: 39–41). Davi Kopenawa continued learning to be literate until the beginning of the 1970s: "Davi still has his problems but continues to show some growth and is doing well in his reading" (Toototobi Gang, 1970b: 3).

9. When preaching, the missionaries used a series of placards depicting biblical stories displayed on a tripod ("Biblavision" flashcards).

10. These were the first Yanomami to be baptized in the waters of the Rio Toototobi on January 14, 1968, after the measles epidemic in late 1967 (see the previous chapter).

11. These visits, directed towards the upper Rio Demini and the upper Orinoco, were quite frequent in the late 1960s: "According to F. [Chico], R. [Davi Kopenawa's stepfather] was faithful to transmit the message of salvation during his 'news chant' [wayamuu ceremonial dialogue] with the Indians in the other villages. In some places they were well received and in other places they were not. One group is coming to Plinio's village just north of here in a few days. They said they were interested and wanted to hear more about God. One village, Bocalahudumteri, did not receive them well at all. They indicated to R. that the foreigners were actually the spirits of dead yanomama that had come back again" (K. and M. Wardlaw, 1968a). This evangelical campaign also served the political goals of Davi Kopenawa's stepfather, R., leader of the mission's group. It allowed him to achieve an unprecedented regional political influence by giving him direct control over the upper Demini and Orinoco groups' access to the manufactured goods dispensed by the mission (notably by trading basketry on behalf of the missionaries): "Recently our Indians contacted an old enemy group and last week about 35 men, women, and children came here to visit at R.'s request. These people show more interest in the Gospel than previous groups who have been here" (Toototobi Gang, 1970b).

12. Six children born between 1961 and 1972.

13. It is usual mourning behavior for young Yanomami to escape their native community after their parents' death.

14. SPI, then FUNAI (which replaced it in 1967) considered these exchanges both a means of integrating the Indians and an economic support for its constantly impoverished outposts. The Ajuricaba Outpost's monthly reports include meticulous accounts of this trading with the Indians (FUNAI Archives, Brasília).

15. For example, a radio message sent from the Ajuricaba Outpost on June 16, 1965, reports a trip to Toototobi from which twenty-five peccary skins, nine deer skins, two ocelot skins, and one otter skin were brought back.

16. A term used in Portuguese *(costume)*, to which Davi Kopenawa gives the meaning "way of life" ("hunting, fish-poisoning, preparing our food").

17. Once they reach adolescence, young Yanomami spend long periods traveling from community to community and from feast to feast, as far from their native village as they can. Today this period's initiatory journeys increasingly include travel to the white people's cities.

18. Davi Kopenawa probably left in 1971–1972, when he was about fifteen or sixteen years old.

19. SPI and then FUNAI subaltern agents used to hire Indians for personal purposes on the fringes of their institution. Yet this was frowned upon by the "outpost chiefs," who were considered the only legitimate operators of the local paternalist economy, which was based on the remuneration of Indian labor with "gifts" *(brindes)* and food.

20. To this day, broad and brightly colored manufactured hammocks are particularly valued by Yanomami adults. One can imagine how precious such a hammock could have been for a teenager in the early 1970s.

21. Freshwater turtles (*Podocnemis* sp.) are still illegally hunted and traded on a large scale in various regions of the Brazilian Amazon, to the point that the species is threatened with extinction.

22. Davi Kopenawa's stepfather told him that this young man suffered from an unsightly skin disorder, probably caused by onchocerciasis, a filariasis endemic to the Toototobi region.

23. Vocative of "father-in-law."

24. Ajuricaba is about six days from Manaus by motorboat during the rainy season and fifteen days during the dry season.

25. Manaus's former airport (Ponta Pelada, now Manaus Air Force Base) was on the banks of the Rio Negro.

26. In Portuguese: *jato* (from the English "jet").

27. The pilots of the Missionary Aviation Fellowship, which served Toototobi, flew small Cessnas. Flights operated by national airlines flying medium-haul Boeings regularly landed at the old Manaus airport.

28. The Volkswagen "Beetle" *(fusca)* was the most common car in Brazil in the 1960s and 1970s.

29. Yo Matsumoto, a student with the University of Kansai's exploration club (Osaka, Japan), stayed at Ajuricaba during his two trips on the Demini, Mapulaú,

and Toototobi rivers in June and October 1972 (Matsumoto et al., 1974). Davi Kopenawa would have been sixteen years old at that time.

30. This was the Adriano Jorge sanatorium, opened in Manaus in June 1953 by
the Brazilian Ministry of Health's Companhia Nacional Contra a Tuberculose.

31. This is what most Yanomami do when they are hospitalized for such a long
time for tuberculosis treatment.

13. The Road

1. Translation of a Yanomami verb which means "to establish (or re-establish)
peaceful relations with an unknown group (or with former enemies)": *rimiai* (tran-
sitive), *rimimuu* (intransitive).

2. This trip was carried out as part of the campaigns organized by FUNAI in
1973–1974 to "pacify" the Indians along the future path of the Perimetral Norte
highway. This northern parallel to the Trans-Amazonian was supposed to connect
the Atlantic (Macapá) to southern Colombia (Mitú), nearly 2,500 kilometers. Work
on the Caracaraí-Padauari section, which cut through the southeast of Yanomami
territory, began in September 1973.

3. This name comes from the way in which the men of this group keep the
foreskin of their penis *(moxi)* stuck *(hatëtë-)* under a cotton string worn around the
waist. They are probably descendants of an older southern migration wave of the
eastern Yanomami *(Ninam/Yanam)*, who had remained isolated in the headwaters
region of the Rio Apiaú. Their location was only discovered by a FUNAI reconnaissance flight in July 2011. They continue to this day to refuse any peaceful contact
with the other Yanomami and white people (see Albert and Oliveira, 2011).

4. Following a raid on these Indians in 1975, the warriors from a village on the
upper Rio Catrimani brought back a stone hatchet blade which was clearly still
in use.

5. This small group of $Xamat^hari$ is related to those of the Ajuricaba Outpost
and of the Rio Jutaí *(Kapirota u)*. They had split from the other groups and settled
within reach of the Toototobi Mission. This expedition probably took place in September 1973. The report of the FUNAI agent who described it (F. Bezerra de Lima)
was dated October 1973.

6. Respectively, these were guests from the upper Rio Toototobi and the Rio
Couto de Magalhães (an affluent of the upper Rio Mucajaí).

7. The cremation of the deceased's bones generally takes place a few weeks after the body is hung to decompose in the forest. See Chapter 10, note 40.

8. The official technique to "attract" and "pacify" "isolated Indians" used by
agents of SPI, later FUNAI (the *sertanistas*), was to hang manufactured objects in
the forest to establish a silent barter with Indians, referred to in Portuguese by the
term *namoro* ("flirt").

9. This first contact probably dates from early 1970 (or late 1969): "A group of
visitors from the Mãykoko [*Mai koxi* or *Mai koko*, Davi Kopenawa's father-in-law's
old group] whom we tried to contact over a year ago, arrived here about two weeks

ago. Apparently they are interested in forgetting old feuds and want to establish friendly relations once again with R.'s people [Davi Kopenawa's stepfather in Toototobi]. We tried to tell them about God, but they were not interested and we got nowhere" (Toototobi Gang, 1970a). It was following this first contact that the group settled on the *Werihi sihipi u* River, first upstream, then close to its mouth on the Rio Mapulaú. It migrated there from the headwaters of the Rio Lobo d'Almada, a tributary of the upper Rio Catrimani.

10. Now that Chico was a FUNAI *sertanista,* he saw Davi Kopenawa as an ideal middleman for his expeditions to allegedly "pacify" the Yanomami along the path of the Perimetral Norte highway.

11. Chico took the initiative (as "declarer") of having a birth certificate made out for Davi Kopenawa in January 1974 in Manaus ("Davi Xiriana, born on February 15, 1956, municipality of Barcelos"). Davi Kopenawa was possibly employed by FUNAI 10th Delegacia in Manaus during this period under the terms of a casual employment system, the "service provided against receipt" *(serviço prestado contra recibo).*

12. As usual, the Yanomami were unscrupulously exploited by the local white population: "The Indians I met complained that the trader Chiquinho and the hunters who come up there are not honest with them . . . These people who negotiate with the natives are adventurers who illegally fish and hunt in Indian territory, . . . the trader Chiquinho . . . was the person most frequently accused of taking animal skins [giant otter and ocelot hides] and other forest products, promising to pay the Indians on a future trip, which he never did" (Bezerra de Lima, 1974).

13. These unexpected visitors were probably one of the geological prospecting teams that crisscrossed the region at the time the Perimetral Norte highway was cleared. Several helicopter flights were carried out from the Catrimani Mission in March 1973. Explosives or fuel were possibly left on site and accidentally exploded after prolonged exposure to the sun.

14. The flu virus can be transmitted via infected objects; see Chapter 10, note 35. Yet this epidemic's origin was more likely the result of a visit to FUNAI's Ajuricaba Outpost by the people of the *Werihi sihipi u* River, their guests, or travelers passing through. At the time, the area around this outpost was regularly visited by white hunters. Indeed, some of the survivors also mentioned the passage through their village of visitors from the Rio Jundiá who had allegedly stolen shotguns from white people living downriver of the Ajuricaba Outpost and attribute the onset of the epidemic to their revenge by sorcery.

15. According to the Japanese expedition that visited the area in 1972 (see Chapter 12, note 29), the people of the *Werihi sihipi u* River lived in a communal house "numbering more than 50 people" (Matsumoto, 1974: 25). This epidemic killed at least twenty-four adults (including sixteen women) and probably between ten and fifteen young children (genealogies gathered by Bruce Albert). A FUNAI report mentions it, but underestimates the number of its victims: "One of the groups we encountered is the one that formerly lived on the Mapulaú and ran away, frightened by a helicopter landing near its communal house last year . . . Of this group, 15 people died of 'xawala', a dangerous disease ['epidemic' in Yanomami] which, ac-

cording to their information, the helicopter brought; this group continues to fear meeting white people and shows no trust for it believes they are responsible for this 'xawala'. I tried to convince them again that white people do not possess this 'xawala' and do not know how to transport it since they do not know it at all; they still have a lot of doubts, but I think this 'xawala' business will be cleared up with time" (Bezerra de Lima, 1974).

16. Reference to the traditional Yanomami association of sorcery fumes and epidemic contagion (see Albert, 1988 and 1993).

17. After their departure from the Catrimani Mission (on February 6, 1974), the members of the expedition walked nine days to reach the Rio Mapulaú where they "cleared a garden to found an outpost, for this was the river-accessible place closest to the six villages we had encountered on the trip" (Bezerra de Lima, 1974). FUNAI named this new outpost the "Subposto Mapulaú."

18. At the time, the Ajuricaba Outpost was used "as a support base for attraction work on the Mapulaú and on the work fronts on the Perimetral Norte" (Mont'alverne Pires, 1974).

19. In the early 1940s.

20. A *Boletim Informativo da FUNAI* (1973–4, 9/10: 41) reported that this "native attraction outpost" was established in April 1974.

21. The *delegado* was the head of the Delegacia Regional of FUNAI (local administration, one per state of the Brazilian federation), a kind of regional inspector.

22. Cassiterite is the name of natural tin dioxide, the principal ore of tin. Regarding the invasion of the Surucucus region by illegal cassiterite miners in 1975–1976, see Taylor, 1979, and Albert and Le Tourneau, 2005. Chico arrived in Surucucus in January 1976 in very difficult conditions: "This government agent was thrown into the region without any means of subsistence. He lives in a shelter made of a few posts covered in a plastic tarp loaned by policemen currently here [to expel the miners] and does not even have a radio" (Pacheco Rogedo, 1976).

23. A FUNAI report from July 1977 (Horst, 1977) confirmed the burning of the Mapulaú Outpost by the Yanomami. A later report (Melo, 1982) commented on the incident: "The Indians did burn down the outpost [Mapulaú], appalled by the deaths from a new epidemic."

24. The area is mostly inhabited by speakers of different languages of the western Tukano family. Davi Kopenawa thus uses the name *Tukano* generically to refer to all the area's inhabitants. He worked there briefly from late 1974 to early 1975.

25. The whole region of the Northwest Brazilian Amazon is a multiethnic mosaic of twenty-two peoples (almost 32,000 individuals in 2000, according to the Instituto Socioambiental of São Paulo) who speak languages belonging to three linguistic families: Arawak, Tukano, and Maku.

26. Davi Kopenawa probably began his studies in November 1975. A document from October 1975 establishes that it was the anthropologist K. I. Taylor (University of Brasília), then coordinator of FUNAI's Plano Yanoama, who recommended that Davi Kopenawa take this course: "We would like to suggest as a candidate for this year's course for indigenous health workers (the one in Boa Vista or Manaus), the

Indian known as David Xiriana who, according to Dr. Silverwood-Cope's information, is at the Iauaretê Outpost. If he is approved, we would like him to be put to use as one of the monitors mentioned . . . to work in an indigenous Yanoama [Yanomami] community in the area of the Perimetral Norte" (Taylor, 1975c). Regarding the Plano Yanoama project, see Taylor, 1975a; Ramos and Taylor, eds., 1979; Bigio, 2007: chapter 4; and the final chapter, "How This Book Was Written," in this volume.

27. "Davi Xiriana Yanomami . . . speaks and understands Portuguese, but does not read or write it. Davi is a native of the Rio Demini and he was taught to read and write in his tribal language. He has health knowledge for he was sent to Manaus to follow a health worker course . . . but he was not employed as such because he could not read the instructions for the medicine. Davi wants to study Portuguese and perfect his paramedical knowledge, in order, he has told us, to be able to help his group, but to do so we will have to remove him from the region and send him to a training center and his development will have to be constantly monitored" (Paixão, 1977).

28. A team from SUCAM, the former Brazilian department of malariology. Toototobi is the first source of onchocerciasis detected in Brazil. This is a filariasis transmitted by small hematophagous gnats. This expedition, led by Dr. A. J. Shelley, reached Toototobi in December 1975 (see Shelley, 1976).

29. Boa Vista (capital of the state of Roraima) appears here for the first time as an urban reference in Davi Kopenawa's account. He arrived here in late 1975, the period of the Perimetral Norte highway's construction. Previously, he had had relationships only with the Manaus office of the SPI, then FUNAI (in the state of Amazonas), via the Rio Demini and Rio Negro. In the early 1970s the traditional Amazonia of rivers begins to shift towards an increasingly urbanized Amazonia of roads.

30. The section of the Perimetral Norte that crossed the Yanomami territory in 1973 (see note 2 above) was permanently abandoned in early 1976. It ran west from the Manaus–Boa Vista BR-174 road for 225 kilometers, ending in the heart of the forest, shortly after the Serra do Demini. Only the first fifty kilometers of this road are now relatively usable and are occupied by cattle ranches and settlers.

31. A colonial reminder of Amerindians' status as minors under FUNAI's guardianship following the Brazilian Indian Statute of 1973 (see Albert, 1997b and 2004).

32. The Yanomami, located both in the state of Amazonas and in the state of Roraima, were at the time under the authority of two different FUNAI *delegacias* (the 1st Delegacia, located in Manaus, and the 10th Delegacia, located in Boa Vista).

33. Amâncio had just won responsibility for the Yanomami area after the Brazilian military revoked the Plano Yanoama, an assistance project directed in 1975–1976 by K. I. Taylor (University of Brasília) (see Taylor, 1975b, and the last chapter of this volume, "How This Book Was Written"). Amâncio had Davi Kopenawa hired as an interpreter in December 1976. He truly needed his help to assert himself in the field because he knew nothing about the Yanomami.

34. The Ajarani FUNAI "control outpost" was opened at kilometer 50 of the Perimetral Norte in 1974: "The Ajarani control outpost was temporarily opened here because the major indigenous concentrations are located beginning at the Rio Catrimani, nearly 100 kilometers from the Rio Ajarani. Pending definition of the indigenous area to be delimited, the outpost was built with local materials and consists of a plain palm hut, because the intention was to move the three small [Yanomami] groups in the region from the Rio Ajarani towards the Rio Catrimani" (Costa, 1976b).

35. On this illegal tin ore miners' invasion, which began in March 1975, see note 22 above. They were expelled from Surucucus in September 1976, to be replaced by the official prospecting of a state mining company, the Companhia Vale do Rio Doce.

36. This trip was carried out in August and September 1977. The expedition's arrival threw this isolated group into turmoil: "We really startled this tribal group because they were on the move; most of the group ran off, with women and children crying and yelling; others lay down on the ground and spoke incessantly; then our team stopped and waited, our guides Ueicoá-Teli [*Weyuku t^{h}ëri*] did not take any initiative; . . . it was left to the chief of the Xirroma-Teli [*Xihoma t^{h}ëri*] to do so, he came towards us, bow and arrows in hand, with an expression of despair, ready to defend himself, at that moment our interpreters took action and very slowly the situation returned to normal, but those who panicked and ran off did not come back until the next day" (Costa, 1977).

37. Reference to a project that FUNAI and the military tried to promulgate in 1977–1978 and that aimed to reduce and break up Yanomami land into an archipelago of twenty-one "islands" surrounded by colonization corridors (see CCPY, 1979). Amâncio participated in all field reconnaissance operations to prepare for this project.

38. Davi Kopenawa had this identity card made in July 1975 during a stay in Manaus.

39. The departure for the Ajarani Outpost took place early in 1977. After having been an "attraction base" in 1973, the year the road works began, Ajarani became a permanent "indigenous outpost" in August 1974 (Costa, 1976b). The usefulness of this outpost on the edge of the road was questioned, including at FUNAI: "[its creation] greatly surprised us, for according to what we have observed in the indigenous region of the Ajarani Outpost, the difficult thing is to keep the Indians away from the white people and not to attract them [to the road]" (Pacheco Rogedo, 1976).

40. These Indians' territory is at the extreme southeast of Yanomami land (Ajarani, Repartimento, and Apiaú rivers). They are probably related to the *Moxi hatëtëma* (see above, note 3). The other Yanomami call them *Yawari*, a term by which the western Yanomami refer to underwater supernatural beings (Mattei-Müller, 2007: 385). Their territory had already been visited by missionaries and loggers since the 1950s. In 1967, they suffered a severe measles epidemic (Figueiredo Costa, 1967). Their population was estimated at 102 individuals just before the

opening of the road (thus after the measles epidemic of 1967), but they were down to seventy-nine in March 1975 (Ramos, 1979).

41. The road construction site reached the Yanomami of the Rio Ajarani in November 1973, at kilometer 32 of the Perimetral Norte: "A group of about fifty Indians, naked, constantly talking and gesticulating but acting friendly, were encountered by the workers building the Perimetral Norte road near Caracarai. The Indians offered them arrows and necklaces and received hammocks. The group of workers was led to the chief of the community—who lives right on the road's path—but did not succeed in understanding anything that he said to them. Yet they did understand that the Indians do not want violence, though they are big and strong" (Estado de São Paulo, November 29, 1973). See Bigio, 2007, for a synthesis of eyewitness accounts of the tragic situation of these Indians after the road reached their territory.

42. The clearing of the road's path took place in sections of 50 by 500 meters, carried out by workers (two to four men per kilometer) employed by a subcontractor (Nordeste Desmatamentos) of the major public works company in charge of the road's construction (Camargo Corrêa).

43. In the early 1970s, the territory of the Waimiri-Atroari Indians, who were resisting contact, was occupied by a FUNAI "attraction front" backed by the army in order to allow the passage of the Manaus–Boa Vista BR-174 road (see Baines, 1991). The local army's intimidation measures are described in a book by a former FUNAI delegate in Manaus (Carvalho, 1982: 74–79, 97–98), who quotes this statement by General Gentil Nogueira Paes: "The road must be finished, even if we have to open fire on these murderous Indians to do so. They have already greatly defied us and they are getting in the way of construction."

44. For their part, the road workers were surprised and perplexed by the Yanomami's lack of hostility: "The Indians contacted thus far are exceptionally docile, but there is no guarantee that they will all continue to behave this way" (Marcos, 1976). This lack of hostility would last throughout the construction: "Along the route . . . there is no indication of problems with the Indians, aside from the occasional invasion of our camps . . . which only amount to requests for food and merchandise" (Yssao, 1975).

45. The Yanomami health situation along the road rapidly became critical: "As soon as construction on the road began, the entry of unauthorized individuals and crews involved in the construction on indigenous territory, as well as the lack of minimal precautions to avoid the transmission of disease, put the region's indigenous population at risk from successive epidemics of flu, measles and other illnesses, and more than 11% of its population has died needlessly" (FUNAI, 1975).

46. Rivers crossed by the path of the 412-kilometer Caracaraí–Padauari section of the Perimetral Norte. On the Rio Aracá, the house of the Yanomami of the Rio Jutaí (Kapirota u) was one kilometer away from the road's path (kilometer 307). The FUNAI agent responsible for their "attraction" commented: "The problems encountered are the following: 1) I found all the Indians sick with flu and malaria, 2) the Indians did not want the road to go through this place" (Arantes, 1974).

47. The FUNAI teams accompanying the topographers received regular aerial drops of presents and food from the Camargo Corrêa construction company (Arantes, 1974). A missionary established on the Rio Ajarani also reported: "The company's personnel was so nice to them, they gave a lot of food because they were worried they would attack them when they were working on the road, so that as soon as the Indians arrived on a construction site, the workers gave them large quantities of sugar, salt, rice, and boxes of sweets" (interview, Elim Mission, Rio Ajarani, March 1984, Bruce Albert archives).

48. This *Hewë nahipi* group was in regular contact with the Catrimani Mission located at kilometer 145 of the road. The topographers arrived in the region in January 1974. The clearing of the road's path took place in March–April and the actual earthwork began in October of the same year: "hordes of workers showed up on the road by plane or appearing around a bend in the river. Fascinated, the Indians would go to the construction site, where the axes, tractors, weapons, and cachaça were found" (*Veja* magazine, August 10, 1977).

49. See Campbell (1989: 39–42) regarding the construction of a Perimetral Norte section on the Waiãpi Indians' land in the state of Amapá: "From miles away in the still night we could hear the road coming towards us. The gigantic machines which flattened ridges and filled up valleys worked twenty-four hours a day."

50. "In December 1976, without knowing that a Yanomami child committed to a Boa Vista hospital . . . had contracted measles, the missionaries brought him back to his group . . . near the mission led by Father S. Despite the vaccinations carried out, the mission was soon turned into a hospital. In February 1977, the epidemic appeared to be under control, but other groups who had visited the village at the mission and contracted the disease arrived looking for help. Entire families were decimated. Including adults, elders, and children, 68 dead were recorded, a terrible loss for this small tribe . . . In one of Father S.'s last drawings, he represented himself, tormented, addressing a young Yanomami appearing among the dead: 'Why didn't you come to call me before everyone died?' The young Indian answers: 'What good would that do? You're the one who brought the disease. It would have been better if you never came to us!'" (*Veja* magazine, August 10, 1977).

51. This measles epidemic killed sixty-eight people, 55 percent of the inhabitants of the four houses established on the Rio Lobo d'Almada at the time. Ten to fifteen of these sixty-eight victims were survivors of the 1973 epidemic on the *Werihi sihipi u* River, who had left there to settle again on the upper Rio Catrimani at their former *Hapakara hi* site (genealogies established by Bruce Albert).

52. Father-in-law/son-in-law relationships express subordination. Relations between closely related but differently sized animal species, Moon and Venus as they appear in the sky, a master and his dog, and bosses and workers (road workers and gold prospectors) are conceived according to this kinship model. On the *xawarari* (plur. *pë*) epidemic beings, see Chapter 16.

53. The Yanomami lowland groups were decimated by epidemics through direct contact with the white people at different times from the 1940s to the 1970s. The groups of the highlands, historical and demographic heart of Yanomami territory,

remained isolated until the massive invasion of their territory by gold prospectors in the late 1980s, a period during which they experienced a deadly epidemiological shock in their turn.

14. Dreaming the Forest

1. The Serra do Demini, at the foot of which this FUNAI outpost is located, rises to over 900 meters. The Demini Outpost is located at kilometer 211 of the Perimetral Norte highway. This former camp of the Camargo Corrêa company's roadwork then became the base of a FUNAI "Rio Demini attraction front" beginning in December 1976.

2. "On January 12, 1977 work began to clear 20 hectares for farming in order to provide for the attraction team and contacted tribal groups who will come to visit" (Costa, 1977). Amâncio planned to turn the Demini Outpost into an "indigenous agricultural colony" whose products could be exported to the city of Boa Vista.

3. This was a community located within the Catrimani Mission's sphere of influence (at kilometer 145 of the road), but which also came to visit the FUNAI's Ajarani Outpost (at kilometer 50). At the time, Amâncio's strategy was to weaken the Catrimani Mission and replace or duplicate it with a FUNAI outpost.

4. This aborted raid took place in 1978. The Hewë nahipi's headman had invited the Opiki tʰëri because his son-in-law belonged to this group. About the paxo uku sorcery substance, see Chapter 7, note 6.

5. This great man and shaman of the Opiki tʰëri was over seventy-five years old at the time and had three wives. The youngest was about thirty and came from a Yawari group of the Rio Ajarani. He died in 1988.

6. This first visit to the Demini Outpost took place in early 1978.

7. Banana plants take about a year to give mature fruit.

8. In 2012, the Watoriki tʰëri near the Demini Outpost included 174 people, living in two houses located two and a half kilometers from the Demini Outpost. The group took fifteen years (from 1978 to 1993) to move the twenty kilometers separating Haranari u from Watoriki, moving from site to site (see Albert and Kopenawa, 2003; Albert and Le Tourneau, 2007).

9. In Yanomami kinship terminology, maternal aunts (mother's sisters) are also called "mothers."

10. Regarding this dance by couples that mixes hosts with guests and brings together people who are potential spouses, see Chapter 9, note 10.

11. Davi Kopenawa's future wife, Fatima, born in 1963, was fifteen at the time and Davi Kopenawa was twenty-two.

12. Footprint (mae) sorcery (see Chapter 7, note 25) is frequently used against sons-in-law coming from allied houses, considered as rivals and intruders by young local suitors.

13. As a sign of deference, a father-in-law is generally directly addressed with a third person plural personal pronoun (wamaki).

14. Thirty-five years later, Davi Kopenawa is still married to the same woman,

with whom he has had five children: a son, Dario, born in 1982, followed by three daughters—Guiomar, Denise, and Tuira—born in 1985, 1987, and 1993, respectively. His youngest son, Vitório, was born in 2001.

15. During this period (late 1970s) Davi Kopenawa accompanied many FUNAI reconnaissance expeditions (outpost inspection, contact with new groups, expulsion of illegal miners), as well as medical teams, the federal police, and inspectors from the forest departments of the IBDF (Instituto Brasileiro de Desenvolvimento Florestal).

16. Yanomami men owe their wife's father a bride-service *(turahamuu)*, which consists mainly of agricultural work (clearing new gardens) and regular gifts of forest products (game, palm fruits, honey). Today, trade goods *(matihi pë)* are increasingly part of *turahamuu*. These material requirements are coupled with a duty of political solidarity (in quarrels, ritual duels, and war raids). A father-in-law is the only person from whom orders for work to be done *(nosiamuu)* are tolerated; see also Chapter 13, note 52.

17. This house, which was bigger than the previous ones, was built in the early 1980s about seven or eight kilometers from the Demini Outpost.

18. Amâncio left the Yanomami area in June 1978 following an article accusing him of making Makuxi Indians deported by FUNAI as "agitators" work at Demini in near-slavery conditions. See: "O Erro Histórico de Krenak Se Repete: Prisão Indígena Clandestina," *Jornal de Brasília,* April 2, 1978.

19. Regarding this invasion of illegal tin ore (cassiterite) miners in Surucucus in 1975–1976, see Chapter 13, note 22.

20. "The majority of the employees of the indigenous outposts are recruited from the regional population. These are generally people with a limited education who are drawn to work with Indians by the salary offered by FUNAI . . ., a 'good salary' in comparison with other administrations and local companies" (Melo, 1985)

21. In the 1990s, this *delegado* efficiently ran the operations to expel illegal gold prospectors from Yanomami territory. He later became president of FUNAI.

22. Yanomami have been murdered by armed FUNAI employees on two occasions in the 1980s and 1990s (at Ericó and Ajarani Outposts in the state of Roraima).

23. Amâncio came back to Roraima state to start as head of the FUNAI *delegacia* in Boa Vista in October 1984.

24. Davi Kopenawa took charge of the Demini Outpost in October 1984. He kept his duties as an interpreter and was initially only appointed as substitute outpost chief. His full appointment to both positions was made official in May 1986. He was relieved of his duties in January 1989 following his protesting a new government attempt to dismember Yanomami land. However, he was reinstated the following month under pro-Indian media pressure during the gold rush in the Yanomami territory.

25. *Demarcação* is a Brazilian administrative term referring to the process of delimiting and legally recognizing an Indian territory.

26. After the Surucucus invasion of 1975–1976 (see above note 19), Davi Ko-

penawa had been sent to the Rio Uraricaá with a FUNAI team in December 1977 to investigate a new invasion of illegal miners. Additionally, settlers and ranchers were starting to rush into Yanomami lands along the beginning of the Perimetral Norte highway.

27. A traffic barrier was actually established in 1975 at kilometer 50 of the road, on the edge of Yanomami territory. There, the FUNAI Ajarani Outpost agents prevented unauthorized people from entering the Yanomami territory.

28. Davi Kopenawa is referring here to the official project of reducing Yanomami territory which began in June–July 1977, when FUNAI agents in the field secretly studied dismembering it into an archipelago of twenty-one separate reservations (see Chapter 13, note 37).

29. On the demographic decline of the Waimiri-Atroari and the spoliation of their land during the 1970s, see Baines, 1991 and 1994.

30. CCPY (Comissão Pró-Yanomami) is a Brazilian NGO that obtained the legal recognition of a continuous Yanomami territory covering over 60,000 square kilometers in 1992. CCPY was founded in 1978 in São Paulo by Claudia Andujar (photographer), Carlo Zacquini (Catholic brother), and Bruce Albert (anthropologist). After the creation of a Yanomami association (*Hutukara*) in 2004, CCPY dissolved and transferred its activities to a partnership between *Hutukara* and the Instituto Socioambiental of São Paulo—www.socioambiental.org.

31. In 1978–1979, CCPY had just launched a campaign against the Brazilian government project to dismember Yanomami land (see note 28 above). The CCPY founding members quoted are of European background or nationality. Here, the head of the Demini Outpost was repeating the Brazilian military's traditional xenophobic fantasy, which assimilates defense of Indian rights and environmental protests with foreign maneuvers intended to "internationalize Amazonia."

32. Davi Kopenawa briefly recalls this meeting in a book by the photographer Claudia Andujar published in France (2007: 167): "I met her [C. A.] when she was talking with people who wanted to help us: [the anthropologists] Alcida [Ramos], Bruce [Albert] and a few others . . . She told me she wanted to fight for my people . . . I believed her and I became friends with her, with Carlo, and with Bruce—who already spoke the Yanomami language."

33. CCPY's campaign for the recognition of Yanomami territorial rights began in 1978. Davi Kopenawa's joining of the campaign in 1983 was obviously a decisive step towards its success. See Chapter 17.

34. After the initial episodes of Surucucus in 1975–1976 and of the Rio Uraricaá in 1977, a second wave of invasions of Yanomami territory by illegal miners began in the 1980s (see Chapters 15 and 16).

35. Davi Kopenawa was initiated in 1983 at the age of twenty-seven (see Chapter 5). The same year, he went to Manaus to participate in his first assembly of Amerindian leaders.

36. In this context the words "drawings" and "to draw" translate the terms *turu* ("drawing of a dot") and *turumãi* ("to draw dots"), which refer to a body paint motif *(turumano)*.

37. The Yanomami territory is located on both sides of the Serra Parima mountain range, from which the Orinoco and most of the tributaries of the Rio Branco's right bank and Rio Negro's left bank flow. These highlands and their piedmont are considered the "forest of the human beings" *(yanomae tʰë pë urihipë)* and the "center" *(miamo)* of the terrestrial layer. The Rio Branco's savannas that surround this region are considered its "margins" *(kasiki)* and the "land of the outsiders" *(napë pë urihipë)*.

38. This is a reference to the myth which describes the sky's fall at the beginning of time (M 7); see Chapter 8.

39. There is a change of perspective here: initially, the shaman "calls" his spirit helpers *(xapiri, plur. pë)*, "brings them down," and "makes them dance." Then, once they have carried away his image *(utupë)*, he himself becomes a "spirit person" *(xapiri tʰë)*, acting *(xapirimuu)* and moving *(xapiri huu)* as a spirit, seeing what the spirits see.

40. Regarding these stone houses, see Chapter 18.

41. Here, Davi Kopenawa evokes *Omama*, the Yanomami demiurge, in his "image form" *(a në utupë)* or "ghost form" *(a në porepë)*, i.e., as *xapiri* spirit helper.

42. On Yanomami shamanism's logic of symbolic homeopathy, see Chapter 4, note 37.

43. Shamans identifying with their spirit helpers (see above, note 39) are able to see the images *(utupë)* of the mythical human/animal ancestors *(yarori, plur. pë)*, which are themselves potential spirit helpers.

44. These are collective hunts to accumulate smoked game to prepare *reahu* feasts *(hʷenimuu)* and long village expeditions for hunting and gathering valued fruit *(waima huu)*.

15. Earth Eaters

1. Davi Kopenawa uses the term *maquinário*, which in Brazilian Portuguese refers to the motor-pumps used in placers both to break up the banks of watercourses with high-pressure water hoses *(bico jato)* and to suck up auriferous sand and gravel *(chupadeira)*.

2. In Brazilian Portuguese, the word *azougue* describes a lively and edgy person. It also refers to mercury, like the English expression *quicksilver*. Regarding the contamination of the Yanomami with mercury from gold prospecting in the late 1980s, see Ação pela Cidadania, 1990, and Castro, Albert, and Pfeiffer, 1991.

3. The white-lipped peccary herds disappeared from Yanomami territory in Brazil for at least a decade following the invasion of the gold prospectors in the late 1980s. This disappearance may have been caused by an epidemic related to the introduction of domestic hogs (Fragoso, 1997). Along with tapirs, white-lipped peccaries are Yanomami hunters' most sought-after prey.

4. For myths about the origin and dispersal of the peccaries, see M 148 and 149.

5. The *garimpeiros* are saddled with many other similar expressions: "stone eaters" *(maama pë watima pë)* or "metal eaters" *(poo xi watima pë)*, "destroyers of land-

forest" *(urihi wariatima pë)* or "firebrands" *(wakoxo pë),* "for they destroy the forest like fire." Yet, they are most often simply referred to by the neologism *karipiri pë* whose *-ri* (plur. *pë*) ending connotes excess and supernatural danger.

6. Regarding the *Moxi hatëtëma,* see Chapter 13. The invasion of Yanomami territory began gradually in 1980–1981 in the northeast, along the Rio Uraricaá (Santa Rosa), then in 1982 to the southeast, along the Rio Apiaú (Apiaú Velho). The Apiaú Velho gold placer, which initially remained relatively small (about 400 prospectors), began to expand in 1984. This gold prospectors' invasion of Yanomami lands is related to an abrupt increase in the price of gold on the international market beginning in 1979 (see Albert, 1993).

7. Amâncio had become the head of FUNAI in Boa Vista in October 1984, and the attempts to drive the *garimpeiros* out of the upper Rio Apiaú took place in January and February 1985 (see Albert and Le Tourneau, 2005: 8).

8. This undoubtedly constituted a spectacular scene. This expedition gathered about fifty warriors, their bodies entirely covered in black paint, a blend of crushed wood coal and latex from the *operema axihi* tree (see Albert and Milliken, 2009: 111–112), and armed with bows and arrows over two meters long.

9. "Gold holes" *(oru pëka pë)* is the Yanomami name for placers.

10. Following this federal police operation (February 1985), FUNAI set up a "vigilance outpost" in the area operated by five military policemen from the state of Roraima (Melo, 198: 11–12). However, the expelled *garimpeiros* soon returned, traveling up the Rio Mucajaí and Rio Apiaú in small groups, then bypassing the FUNAI outpost. In July 1985, a new auriferous site was discovered on the Rio Novo, a tributary of the upper Apiaú, and the contingent of gold prospectors grew to 600 people. By the end of the year, it had doubled. Overwhelmed, the FUNAI outpost was abandoned (see Albert and Le Tourneau, 2005: 8).

11. This FUNAI president (now a senator), Romero Jucá Filho, was appointed by the military in the late 1980s to dismember Yanomami land and favor the gold prospectors' settlement in the context of the Calha Norte Project launched to occupy and control the Brazilian northern Amazonian border strip (see Albert 1990a; Albert and Le Tourneau, 2005).

12. Four Yanomami leaders were murdered at the Novo Cruzado placer on August 12, 1987 (see CCPY, 1989b; Geffray, 1995, and MacMillan, 1995).

13. The gold prospectors' passage from the upper Rio Apiaú to the basin of the Rio Couto de Magalhães *(Hero u),* upper Mucajaí tributary, gave rise to the gold rush of 1987 in western Roraima, with nearly 40,000 gold prospectors and more than ninety illegal landing strips (see MacMillan, 1995). The two key auriferous sites of these developments were Cambalacho, on the upper Apiaú/upper Catrimani, opened in 1986, then Novo Cruzado, on the Couto de Magalhães, in 1987.

14. The Yanomami in the area of the Rio Couto de Magalhães had acquired basic gold-panning techniques from the Yanomami of the Rio Mucajaí, who had learned them from the *garimpeiros* on the Rio Uraricaá (see Ramos, Lazarin, and Gomez, 1986). They had sporadically exploited a placer in the region of Paapiú since the early 1980s. From August to December 1986, for example, they obtained 733 grams

[25 ounces] of gold from it (Lazarin and Vessani, 1987: 60). The chief of the local FUNAI outpost sold this gold in Boa Vista and used the return to buy trade goods the Indians ordered from the city.

15. The catastrophic situation of the area in the late 1980s was vividly described by the late Brazilian senator Severo Gomez: "The landing strip at Paapiú looks like it is out of the Vietnam War. A plane lands or takes off every five minutes. A constant patrol of helicopters flies over the tropical forest . . . The FUNAI outpost is abandoned. Syringes and medicine are piled up in disarray, mixed with empty beer cans. The wind leafs through the medical log. The radio has disappeared . . . The Yanomami have been abandoned to the gold prospectors. The roaring of the motors only comes to an end after nightfall. Then—an elder tells me—you hear a sound far worse: that of our children crying of hunger." ("Paapiú-Campo de exter-mínio," *Folha de São Paulo*, June 18, 1989; see also Albert, 1990a; Albert and Mene-gola, 1992, and Ação pela Cidadania, 1989 and 1990, as well as MacMillan, 1995, and O'Connor, 1997). Official medical assistance and food supplies were only brought to the Yanomami beginning in January 1990, when their decimation by the *garimpeiros* became an international media scandal.

16. The events described below took place in 1988.

17. Yanomami women chew pieces of this cultivated plant rhizome and mix them with annatto red dye. They cover sticks with this mixture and throw them in the direction of enemies in order to make them lose their courage.

18. Zeca Diabo is the character of a reformed hitman in an old Brazilian *tele-novela: O bem-amado* (The Beloved), rebroadcast by TV Globo as a series from 1980 to 1984.

19. This black dye is obtained with the soot from the resin of *aro kohi* and *warapa kohi* trees.

20. From 1987 to 1990, about 13 percent of the Yanomami population in Brazil died, victims of the gold prospectors' violence and, especially, their diseases (Albert and Le Tourneau, 2005: 11).

21. Regarding the *rezadores*, Manaus healers, see Schweickardt, 2002.

22. Another reference to the Brazilian military's projects to dismember Yano-mami territory in the late 1970s (see Chapter 13, note 37, and Albert, 1990a and 1992).

23. Regarding these spirits of the bare lands *(purusi)*, see Chapter 21.

24. *Omama* is considered the first owner of metal (see Chapter 9).

25. Literally, a nose "without taste" *(oke)*, that is, a nose that has not inhaled *yãkoana* powder.

26. Chico Mendes, leader of the rubber tappers *(seringueiros)* union in the state of Acre, had already become an icon of the fight to preserve Amazonia. He was as-sassinated on December 22, 1988, in Xapuri, Acre.

27. This ancient ghost, described in the form of a whitish humanoid, is associ-ated with the deep forest, without paths *(urihi komi)*, where he is said to hunt the humans he encounters at night by shooting them with his curare-poisoned arrows (see Chapter 4, note 32). In 1987–1988, Davi Kopenawa was beginning to be well known in Brazil and abroad for defending Yanomami land and the Amazon for-

est. His growing fame probably dissuaded the *garimpeiros* from carrying out their threats, all the more so after the international scandal provoked by the assassination of Chico Mendes.

28. The assault on murderers' eyes and intestines by the celestial evil being *Kamakari* (see Chapter 7, note 28) is due here to the failure to observe the behavior and food prohibitions of the *õnokaemuu* homicide ritual (see Chapter 7, note 40).

29. The Yanomami expression indicating that someone is ugly or without interest is *pihi wehe*, literally "dry thought."

30. The notion of value in Yanomami is expressed by the word *në* (or *no*), which is used to compose expressions such as: *në tire/në hute*, "of great (high/ heavy) value," *në kohipë*, "of solid value," and *në kõamãi*, "to return (compensate) the value" of an object obtained in a barter.

16. Cannibal Gold

1. Oil is referred to by the neologism *oleo* ("oil" in Portuguese) *upë* ("contained liquid").

2. The Yanomami association of metal (and, in a general manner, manufactured objects) with "epidemic smoke" has been a constant since the first contacts (see Chapter 1, note 28).

3. On the falling sky, see Chapters 8 and 24. *Hutukara* is the shamanic name of the old fallen sky, which became today's earth.

4. Davi Kopenawa added this comment: "The moon that fell with this first sky died. But it was a *yai t^{h}ë* (supernatural) being and now there is another moon in the sky that is its image, its ghost. The same is true of the sun."

5. *Mareaxi* is also the name first used by the Yanomami to refer to aluminum pots and, by derivation, the name of the triangular or round pendants they acquired from their Ye'kuana neighbors, which were made from fragments of these pots' lids. These pendants, which are found among all the Carib groups in the region, were formerly made of silver (see Koch-Grünberg, 1982: 43, for Taurepang/Pemon examples). *Xitikari* refers to the crescent-shaped aluminum pendants of these Carib groups and also means "star" in western Yanomami (Lizot, 2004: 396). Also note that the suffix *-xi* also denotes "radiance, emanation," as in *wakaraxi*, "brightness, light" or *poripoxi*, "lunar radiance."

6. *Poo xiki* is metal; *Hutukara xiki*, the "metal of the old sky."

7. Regarding *Omama* and the origin of metal, see the beginning of Chapter 9.

8. The "children of metal" *(poo ihirupë e xiki)* and the "father of gold" *(oru h^{w}ii e)*.

9. An outcrop of cassiterite was exploited by illegal miners in the Yanomami highlands in 1975–1976 (see Chapter 13, note 22). Geologic prospecting during that period also revealed traces of radioactive minerals in the same area.

10. *Napë wakari pë*, the "giant armadillo spirit outsiders," is a Yanomami expression used to refer to mining companies.

11. See Chapter 9 on the chaos being *Xiwãripo* and the transformation of the *Hayowari t^{h}ëri* ancestors into outsiders.

12. *Mõhere* (plur. *pë*) are probably mica flakes.

13. Here, Davi Kopenawa uses the expression *poo xi tʰaixi*, "metal shaving," or *minerio* (in Portuguese) *tʰaixi*, "mineral shaving."

14. This sorcery substance's name is derived from the root *hipë-*, "blind." It is prepared with fragments of mica *(mõhere pë)* and the powder of an insect found stuck to rocks in streams—an insect sharing its name *(hipëre a)*.

15. Here, Davi Kopenawa uses the expression *wixia a wakëxi*, "breath of life smoke."

16. Here, Davi Kopenawa associates "gold breath of life smoke" *(oru wixia a wakëxi)*, "metal smoke" *(poo xiki wakëxi)*, and "mineral smoke" *(minerio a wakëxi)*; all are considered "epidemic smoke" *(xawara a wakëxi)*. By doing so, he updates and expands the old "metal (tools) smoke" *(poo pë wakëxi)* notion used during the first contacts (see Chapter 10, note 28). On this series of associations, see Albert, 1988 and 1993.

17. The Yanomami refer to the reddish cloud formations of sunset as *xawara a* ("epidemic"). Davi Kopenawa has stayed in São Paulo several times, a city often covered with a thick layer of atmospheric pollution.

18. Gold flakes are mixed with mercury, and this amalgam is then burned to produce gold nuggets.

19. *Sarapo a wakëxi*, "measles smoke" (in Portuguese, *sarampo*). This was one of the infectious diseases that most affected the Yanomami in the first decades of contact with outsiders. Regarding the 1967 measles epidemic at the Toototobi Mission, see Chapter 11.

20. The epidemics caused by indirect contamination that affected the Yanomami before their first encounters with the white people used to be interpreted according to this traditional model of intercommunity sorcery (see Albert, 1988 and 1993). Regarding epidemic sorcery plants and substances, see Albert and Gomez, 1997: 114.

21. The *mãu tʰëri pë*, "inhabitants of the rivers," are the first white people—generally involved in exploiting forest or fishing resources—to enter Yanomami territory (rubber and palm fiber collectors, hunters, and fishermen).

22. Expression used in Portuguese: *doença do minério*.

23. The Yanomami theory about the pathogenic emanations of metal and manufactured objects, rooted in the experience of contamination through the first contacts with white people, was initially extended to mining and oil extraction before being recently combined with the notion of pollution, which was borrowed from the environmental discourse of the 1980s. See above note 16.

24. Coughing is called *tʰoko*, the flu epidemic *tʰoko a wai* ("powerful-dangerous flu") or *tʰokori a wakëxi* ("smoke of the cough-being").

25. *Xuukari* is also an evil being who lets a pathogenic liquid seep from the sky. This liquid to which epidemic dysenteries *(xuu upë)* were attributed in the past is referred to as "the sky spirits' dysentery": *Hutukarari a në xuukari pë* or *hutukara a në xuu upë*, "the dysentery of the sky."

26. These are the *xawarari a në hiimari pë*, "the domestic animals of an epidemic being."

27. These are the *xawarari a në mahe pë*, "the baking plates of an epidemic being."

28. Shamans call these "employees" *(empregados* in Portuguese) *xawarari a në naikiari pë*, "the cannibal *(naikiari)* beings of an epidemic being" or *xawarari a në kamakari pë*, "the devouring *(kamakari)* beings of an epidemic being."

29. The Yanomami keep some of the bones of the game they eat by suspending them in their hearths so the forest animals do not feel mistreated and later flee the hunters.

30. *Xawarari a mae* is an "epidemic being path," *xawarari a periyoka*, "a door to an epidemic being path."

31. "Merchandise has the value of epidemic," *matihi pë në xawarapë.*

32. These "metal rods of the epidemic" that pierce the sick are referred to as *xawara a në pooxi pë.*

33. These spirits *(remori pë)* are associated with the mythical being responsible for the creation of the outsiders' or white people's language (see M 33).

34. A new example of Yanomami shamanism's "homeopathic" logic, which in this case draws on the primordial images of the white people's ancestors *(napënapëri pë)* and that of the epidemic itself *(Xawarari a)* as spirit helpers against the current "epidemic smoke." See Chapter 4, note 37.

35. Here, Davi Kopenawa uses the western Yanomami term *parimi*, "immortal, indestructible" (see Chapter 19, note 11).

36. Davi Kopenawa uses the Portuguese expression *mundo inteiro* to translate the Yanomami expression *urihi a pree* or *urihi a pata*, "the big land-forest," which refers to the entire terrestrial layer.

17. Talking to White People

1. These *pata tʰë pë* ("elders" or "great men," plur.) are merely influential figures without any coercive power; however, like fathers-in-law, their authority over their kindred's many sons-in-law is clearly defined (see Chapter 14, note 16).

2. The root of the verb *(here-)* is also that of the terms used to refer to the lungs and respiratory functions. The *pata tʰë pë*'s harangues are driven by powerful exhalation and punctuated by exclamatory syllables *(kë, yë, xë!).* They "speak with wisdom" *(mõyamu hʷaï),* organizing and commenting on the group's collective activities (economic, social, political, and ceremonial) or passing on their historical and mythological knowledge.

3. Although they are translated in the text here as substantive nouns for the sake of convenience, *hereamuu, wayamuu,* and *yãïmuu* are intransitive verbs. The *wayamuu* essentially conveys community and intercommunity news. The *yãïmuu,* which involves the oldest men, is largely intended for negotiating economic and matrimonial exchanges (or disputes), as well as managing political and ceremonial relations. It is resumed on the last day of the *reahu,* just before the burial or ingestion of the funerary bone ashes of the dead celebrated in the feast. These two types of ceremonial dialogue are characterized by the use of long circumlocutions whose

rhetorical figures and complex prosody have not yet been studied in depth by ethnolinguists.

4. Regarding *Titiri* and the origins of ceremonial dialogues among the western Yanomami, see Lizot, 1994, and Carrera Rubio, 2004.

5. *Xõemari*, the "dawn being," is the son-in-law of *Harikari*, the "dew being," who announces the day before him.

6. What is merely spoken "with the mouth" *(kahini)* consists of informal discourse and rumor and is therefore contrasted with the legitimate public speech of the ceremonial dialogues and formal speeches of the elders *(hereamuu)*.

7. This statement was made in 1993. Davi Kopenawa was about thirty-seven at the time, and his fourth child (a third daughter) had just been born, which was beginning to turn him into an attractive potential father-in-law and consolidate his claim to "act as a great man" *(patamuu)*.

8. *Kãomari* is the primordial image of the *kãokãoma* falcon. Its loud call is considered the "heralding sign" *(heã)* of far-off *hereamuu* speeches. When its image "comes to live" in a man, he becomes good at making exhortations *(herea xio)*, his "discourse is close" *(tʰë ã ahete)*, he knows how to "command with rightness" *(nosiamuu xariruu)*, and people "respond to his words" *(wãã huo)*.

9. Adults in the village of *Watoriki* often complained that youngsters disappear for long periods of time to go adventuring from one *reahu* feast to the next, making endless female conquests, in order to avoid family responsibility, which they have reached the age to take on.

10. The verb used to describe the acquisition of this animal image into a person's chest is *yãmapu* (literally "to make somebody hang his hammock in one's house, to keep him living there.") These images are not acquired as *xapiri* spirit helpers but embodied as prototypes of socially valued personal qualities (see Albert 1985: chapter 5.)

11. A man's wife can join a group of women to harvest food in the garden of another man who has given his agreement *(naremuu)*. The situation, which suggests her husband is lazy, lacking in foresight, or an invalid, is clearly humiliating, except in the case of a refugee family who has not yet had time to clear their own garden.

12. All the local groups of the lowlands *(yari a)* developed from a movement of successive residential fissions and migrations from the highlands *(horepë a)* of the Serra Parima (Orinoco/Rio Parima interfluve), the Yanomami's historical center.

13. Reference to the male puberty ritual, signaled by the change in the voice (when "the throat imitates the curassow," *ureme paaripruu*).

14. Here, Davi Kopenawa uses the term *kanasi* to refer to the catch from a hunt. *Kanasi* primarily refers to leftover food. The word is also used in a general way to refer to what remains from an act of predation. It can thus be applied to a hunter's prey or the victim of an act of sorcery (sick body) or a war raid (wounded body).

15. For a version of this myth about the menarche ritual, see M 305.

16. For a version of this myth on the origin of war, see M 47.

17. For a version of this myth about the flight of the wild honeybees, see M 110.

18. On this myth about the first maize garden, see Chapter 8 and M 86.

19. In mythological narratives and shamanic songs, this call signals the presence or arrival of ghosts. About *Porepatari,* see also Chapter 4, note 32, and Chapter 7, note 12.

20. This meeting of the Union of the Indigenous Nations (UNI) was held in July 1983 at the headquarters of the Indigenous Missionary Council (CIMI) in Manaus (see "Cacique diz que FUNAI está matando os índios," *A Crítica,* July 11, 1983). UNI was founded in 1980 and was active until the early 1990s (see Albert, 1997b: 188). Davi Kopenawa was then invited to another UNI meeting in Brasília, held from November 26 to 28, 1984.

21. This assembly was held in early January 1985 at the Surumu Catholic mission on the territory of the Makuxi. About 150 people attended, principally representatives of six Amerindian groups (Makuxi, Wapixana, Taurepang, Yanomami, Munduruku, and Apurinã), UNI's coordinators (Ailton Krenak and Alvaro Tukano), and a contingent of white observers (church, FUNAI, anthropologists, indigenist NGOs). A transcription of Davi Kopenawa's speech can be found in Portuguese in Albert, 1985: 81.

22. See Chapter 11.

23. This assembly took place in March 1986. It brought together about one hundred Yanomami from fourteen houses representing most areas of the group's territory in Brazil. The non-Indian audience was relatively small (a few representatives of CCPY and FUNAI, the Ministry of Justice's chief of staff, a senator, a representative of a human rights commission, and a journalist from the national press agency).

24. The first Yanomami political assemblies were always framed by the traditional ritual context of the *reahu* feasts' food sharing.

25. In July 1986, Davi Kopenawa was named the UNI and Workers' Party (PT) candidate from Roraima state for the Constituent Assembly, which drafted the 1988 Brazilian Constitution.

26. None of the country's nine Amerindian candidates was elected.

27. This was in 1988–1989, the height of the gold rush in Yanomami territory.

28. On the Yanomami theory of conception, see Chapter 1, note 24.

29. In the same way that today's animals *(yaro pë)* are considered the ghosts *(pore pë)* of the human/animal ancestors from the time of origins *(yarori pë)*.

30. Davi Kopenawa and Macsuara Kaduweu were received by the then-president of the Republic of Brazil, José Sarney, on April 19, 1989, during the most difficult period of the gold rush on Yanomami territory.

31. Davi Kopenawa is referring here to local politicians, most of whom are closely connected with a wide variety of illegal endeavors to plunder indigenous lands (gold prospecting, logging, land-grabbing, cattle ranching).

32. The Yanomami contrast game *(yaro pë)* with domestic animals *(hiima pë)*, which are absolutely inedible. On the *Hayakoari* tapirlike being, associated here with oxen and sheep, see Chapter 8.

33. *Heri* songs are intended to celebrate the joy of abundant food during a *reahu* feast (see Chapter 4, notes 6 and 9).

34. A reference to accusations by the Brazilian military and local politicians opposed to the legal recognition of indigenous lands in border areas on the grounds of an alleged risk of indigenous separatism fomented from abroad.

18. Stone Houses

1. Davi Kopenawa was invited to the United Kingdom by Survival International (SI), a global organization for the defense of tribal peoples' rights whose headquarters is in London. In December 1989, the Right Livelihood Award (RLA), considered the "alternative Nobel Prize," was awarded to SI, which shared the prize with Davi Kopenawa, thus offering him a platform to defend his people, threatened with decimation by the Roraima gold rush: "We have asked Davi Kopenawa Yanomami to stand with us for the Right Livelihood Award ceremonies. This is the first time he has left Brazil. He is spokesman of the 10.000 Yanomami Indians there and has spent many years fighting for the Indians' rights to their traditional lands" (RLA Award, Acceptance Speech, S. Corry, SI, Stockholm, December 9, 1989). On his first trip to Europe (November to December 1989), Davi Kopenawa stayed in London before traveling to Stockholm for the award ceremony. He has few memories of Sweden, save for the extreme cold, which nearly paralyzed him.

2. Regarding the creation myth of the outsiders, see Chapter 9.

3. During this trip, Davi Kopenawa visited the megalithic site of Avebury in the south of England (he later visited Stonehenge in 1991). He brought home a tourist brochure that includes diagrams of vast circular structures similar to Yanomami communal houses.

4. Regarding *Omama*'s flight and the creation of the mountains, see Chapter 4.

5. Regarding *Koyori* and the origin of gardens, see Chapter 8.

6. The thunders' angry vociferations are the "heralding call" *(heã)* of the death of a shaman; see Chapter 8 and Chapter 24.

7. A documentary crew from TV Globo accompanied Davi Kopenawa to the area of FUNAI's Surucucus Outpost, probably in the mid-1980s.

8. Regarding shamanic song trees, see Chapter 4.

9. The land of the white people ancestors is considered a "spirit land" *(xapiri urihipë)*, a "land from which the spirits come down to us" *(xapiri pëni wamare ki napë it^{h}uwi t^{h}ë urihi)*.

10. Literally: *urihi mirekopë*, the "land-forest-mirror."

11. In Yanomami, *mãu u pesi*, "water envelope," refers to bottles of mineral water. During another activist trip to Europe, Davi Kopenawa took a brief excursion to the Italian Alps, and he seems to call upon that memory here.

12. See Chapter 11 on the New Tribes Mission sermons Davi Kopenawa heard during his childhood.

13. For a version of this myth about the honeybees' flight, see M 110. The Yanomami consume more than forty kinds of wild honey ranging widely in flavor from the sweetest to the most acidic.

14. The Portuguese *loja* ("boutique, store") is translated by the expressions

matihi pë tʰari, "merchandise container/shelter," or *matihi pë rurataatima yahi*, "house for acquiring merchandise."

15. A reference, of course, to zoos and natural history museums.

19. Merchandise Love

1. Literally, the "inhabitants/people of merchandise" *(matihi tʰëri pë)* or the "possessors/masters of merchandise" *(matihi pë potima tʰë pë)*.

2. White people's shelters in the Amazon are often covered in corrugated iron, which the Yanomami call *yano siki*, "house skins/leaves." This list of "white people merchandise" *(napë tʰëpë matihi pë)* is characteristic of what can usually be found at a FUNAI outpost or a mission.

3. The verbal expression used here, *xi toai*, refers both to euphoric avidity and sexual climax.

4. A "word from the beginning" is a translation of the Yanomami expression *hapa tʰë ã*.

5. The word *paixi* (plur. *pë* or *ki*), which refers to bunches of feathers placed in armbands, is often used as a synonym of *matihi* (plur. *pë* or *kiki*), a term that could thus be translated by "ornaments, precious objects." Lévi-Strauss (1996: 41) likens the value of Amazonian feather adornments with that of gold in our history.

6. This "word has value of spirits": *tʰë ã në xapiripë;* it "shows the value of beauty of the spirits": *në taamuu xapiripë totihi;* it "makes us think about the spirits": *pihi në xapiripë.*

7. A dead person's cinerary gourds are generally divided among the members of several friendly houses, who will take turns holding *reahu* ceremonies to "put them in oblivion" (see Chapter 1, note 34).

8. The most highly appreciated manufactured objects were those that could be seen as improved versions of objects the Yanomami already possessed (metal machetes versus palm wood machetes, knives versus bamboo blades, aluminum pots versus clay pottery, etc.). Genuinely unknown objects only inspired fear or indifference. Regarding this subject, see Albert, 1988.

9. *Poo pë* (plur.) refers to metal tools, *mareaxi pë* (then, more recently, *rata pë*, from the Portuguese *lata*, "metal box") to aluminum pots, *kapixa pë* to clothing (from the Portuguese *camisa*, "shirt"), *tʰooraa si pë* (then, later, *tʰoutʰou si pë*) to manufactured hammocks, *mirena pë* to mirrors, *tʰaima hipë* (then, later, *moka pë*) to shotguns, etc.

10. In this context, the mark of touch *(hupano)* is also referred to as *imisi* ("skin of the hand/fingers"), *imino* ("mark of the hand/fingers") or, simply, *õno* ("mark, trace"), and it is said that it "takes value of grief" *(në õhotai)*. All these terms are considered synonyms ("close words," *tʰë ã ahete*).

11. Regarding the term *parimi*, "immortal, indestructible," borrowed from the western Yanomami's language, see Lizot, 2004: 296–297, and Mattei-Müller, 2007: 224–225. A great shaman is sometimes referred to by the expressions *xapiri tihi*, "spirit tree," or *parimi tihi*, "tree of eternity."

12. Most Yanomami "exchanges" are made in this relatively blurry deferred

mode. More than bartering, the essential here is to show that one is prepared to give up the requested goods without too much concern for compensation. The verbal roots designating this operation refer essentially to the idea of giving away (*hipi-*, "give"; *topi-*, "offer"; *weyë-*, "distribute"). On the other hand, acquiring a coveted good through defined compensation is known as *rurai* (a term that now also describes a commercial purchase), and a direct exchange is described by the verb *nomihiai*, which denotes immediate reciprocation.

13. The Yanomami ethos closely associates courage, humor, and generosity.

14. A "generous people path" *(xi iheterima t^hë pë mãe)* is also called "a path by which merchandise is brought" *(matihi pë hirapraiwi t^hë mãe)*. In opposite cases, it is referred to as a "greedy people path" *(xi imi t^hë pë mãe)*.

15. Here, a "path of merchandise" translates the expression *matihi pë mãe*.

16. In Portuguese, *namoradas*.

17. Here, the expression "all that is dangerous" is a translation of the term *waiwai a*, from *wai*, "dangerous, powerful, harmful" (see Chapter 7, note 1).

18. Traditionally handmade objects are often referred to as "leftovers" *(kanasi)* of the person who made them.

19. This expression *(imiki yãkete)* refers to the narrow hands of the *hopë* porcupine, the mythological greedy keeper of the *nãi hi* tree's sweet edible flowers (see M 153).

20. Here, *matihi pë mae* ("path for merchandise") or *matihi pë toayuwi yo* ("path for acquiring merchandise") is contrasted with *poriyo në napë* ("path with value of hostility") or *në napëowi t^hë pë mãe* ("path of hostile people").

21. This intransitive verb refers to the act of making friendly contact with an unknown group encountered during a long distance migratory move or re-establishing peaceful relationships with a former enemy group (see Chapter 21).

22. Here, the "mark, trace," *õno* (or "leftovers," *kanasi*, see note 18 above) of a person refers to the objects he has made or, at least, owned for a long time and thus those that would be burned if he were to die.

23. The terms *nõreme* and *utupë* ("body image/vital essence") are used as synonyms (see Chapter 6, note 11). These components of the person are related to the breath *(wixia, wixiaka)* and the blood *(iyë)* as sources of the *animatio corporis* and vital energy.

24. The qualifier used here, *wait^hiri*, does not lack ambivalence in its own right, as it can mean "valiant, courageous, stoic" but also "aggressive, violent, bellicose."

25. This process of ritual erasure is described by the expression *õno ki wãriai*, "to destroy the marks/traces."

26. Sometimes these objects are said to be *hapara pë*, a term that also refers to the spirits of dead shamans and posthumous children.

27. *Osema* is a kinship term that refers to one's siblings (and children in its vocative form; see Chapter 7, note 21).

28. After the co-residents' collective crying *(ikii)*, the deceased's close kin continue their lamentations *(pokoomuu)* each time they nostalgically remember him throughout the stages of his funerary ceremonies, from the hanging of his body

in the forest to the burial of his bone ashes, but also following their dreams or during storms (for it is said that the thunders welcome the ghosts onto the sky's back).

29. Funerary mortars are often made with wood from the *hoko mahi* tree.

30. The main part of the deceased's bone ashes will be kept in gourds *(pora axi)* whose contents will later be drunk or buried during successive *reahu* feasts (see Chapter 1, note 34, and Chapter 3, note 25).

31. The central purpose of Yanomami funerary ceremonies is to "put into oblivion" the ashes of the dead's bones in order to allow their ghost to permanently return to the sky's back (see Chapter 1, note 34, and Chapter 21, note 28). The incorporation of the deceased's personal qualities (such as generosity and courage) through the acquisition of the "image of his breath" *(wixia utupë)* and the "imitation/keeping of his vital principle" *(nõreme uëpu)* is only a secondary, occasional aspect of the ceremony.

32. In 1991, Davi Kopenawa followed the Brazilian TV Globo reports on the Gulf War and was deeply struck by the images of burning oil wells in Kuwait.

33. The expression "people of factories" is a translation of *haprika t^{h}ëri pë*, from the Portuguese *fabrica* and *t^{h}ëri pë* "people of, inhabitants of."

34. These ceremonial troughs are generally carved in the trunks of *oruxi hi, wari mahi, apuru uhi, hoko mahi,* and *ruru hi* trees.

35. The ostentatious generosity displayed during this ritual episode is conceived as a war parody (see Albert, 1985: chapter 12).

36. The Portuguese word *caro,* "expensive," is taken here as an equivalent of the Yanomami expression *në kohipë,* "strong or hard valued." This word *caro* is repeated over and over by local white people (missionaries, health and FUNAI agents, gold prospectors, etc.) as a master argument to justify their refusing food or merchandise to the Yanomami.

20. In the City

1. After his first trip to England and Sweden in 1989 (Chapter 18), Davi Kopenawa participated in a session of the People's Permanent Tribunal about the Brazilian Amazon held in Paris from October 12 to 16, 1990. In Yanomami, *kawëhë* means "unstable, vacillating, shifting" and the verb *kawëkawëmuu* "to walk in a hesitant manner."

2. Allusion to the long moving walkways at the Roissy Airport.

3. In Yanomami cosmology, the "middle" *(miamo)* of the earth—where the "land-forest of the human beings" *(yanomae t^{h}ë pë urihipë)* is found—is the place where the sky is highest. The white people ancestors' land, located "at the edge" *(kasikiha)* of the terrestrial layer, is therefore considered closer to the sky.

4. The Eiffel Tower.

5. The Obelisk on Place de la Concorde.

6. During his stay in Paris, Davi Kopenawa visited the former Musée de l'Homme [Museum of Man] at the Trocadéro.

7. Shamans' ornaments and those worn both by men and women during *reahu* feasts are considered clumsy imitations of those worn by the spirits.

8. The exact expression in the passive voice is *xapiri pë marimãi*: "make dream/ dream the spirits." The Yanomami also say *xapiri pë në mari*, the "value of dream of the spirits," to refer to their dream images.

9. This bamboo is only found in the world of the *xapiri* (see Mattei-Müller, 2007: 267).

10. These glass bead ornaments were the specialty of the Amerindian groups that surrounded the ancient Yanomami who had to acquire them through long-lasting intervillage exchanges or through perilous trading expeditions (see Chapter 9 and Albert, 1985: chapter 1).

11. On these darts, see Chapter 11, note 33.

12. The first term is more common among the eastern Yanomami, the second among the western Yanomami. Regarding the *waikayoma* women spirits see also Chapter 4, note 38.

13. To refer to these mummies, Davi Kopenawa uses the word *matihi*, which is applied to human bones and their funerary ashes (but also, as we have seen in the previous chapter, to feather ornaments and white people's merchandise). During their raids the Yanomami warriors let their enemies safely rescue the bodies of their dead so that they can subject them to the traditional funerary rituals. Throwing a dead body in the river, burying it, or getting rid of it in any other manner is considered an ultimate act of hostility. Keeping it for public display is therefore considered absolutely inhuman.

14. Allusion to the *õnokae* homicide ritual state said to be produced by the murderer's digestion of the fat from his victim's body (see Chapter 1, note 21).

15. On the role of these water beings and superlative hunters in shamanic vocations, see Chapter 3.

16. Davi Kopenawa traveled to New York in April 1991, once again with the support of Survival International. Among those he met with were the secretary general of the United Nations at the time, Javier Pérez de Cuéllar, and various officers of the World Bank, the Organization of American States, and the U.S. State Department. Regarding Davi Kopenawa's visit to New York, see the article by T. Golden (1991) and the book by G. O'Connor (1997: chapter 21).

17. Davi Kopenawa visited the South Bronx and met homeless people on Southern Boulevard (see Golden, 1991: B4).

18. As during his trip to Paris, Davi Kopenawa experienced recurrent attacks of vivax malaria, for which he was then being treated (see O'Connor, 1997: 233–234). At the time, the malaria introduced by gold prospectors had reached epidemic levels in Yanomami territory.

19. Probably the Triborough Bridge over the East River, which connects Manhattan, Queens, and the Bronx and is near where Davi Kopenawa was staying. It attracted his attention as he arrived in New York (see O'Connor, 1997: 236–237).

20. Regarding the *Hayowari tʰëri* ancestors and *Omama's* creation of the white people downstream of all rivers, see Chapter 9.

21. Regarding this sorcery powder *(hipëre a)*, see Chapter 16.

22. Davi Kopenawa looked at Dee Brown's *Bury My Heart at Wounded Knee* with the friend who was hosting him in New York (see O'Connor, 1997: 237–242).

23. These are the Onondaga (People of the Hills) of the Six Nations Confederacy (Haudenosaunee), known as the Iroquois Confederacy and located in the state of New York. From 1788 to 1822, the Onondaga Nation was despoiled of 95 percent of its land. Its current territory has been reduced to a little less than thirty square kilometers south of Syracuse, near Nedrow, New York. During his 1791 visit to the Onondaga, the French romantic writer François-René de Chateaubriand reported that "their first Sachem . . . complained about the Americans, who would soon leave the people whose ancestors had welcomed them with not enough land to cover their bones" (Chateaubriand, 1969: 690).

24. Maple water (*Acer nigrum* and *A. saccharum*).

25. This experience of intense air pollution mostly struck Davi Kopenawa during his visits to São Paulo, Brazil's economic capital.

21. From One War to Another

1. The Yanomami refer to their incursions by a verb, *niyayuu,* which literally means "to arrow each other." The expression *niyayotima t^{h}ë,* which can be translated by the substantive noun "war," refers to the same idea of reciprocal arrowing.

2. Literally, the "*õnokae* people"; that is, the warriors who have killed (or participated in a killing) during a raid and followed the *õnokaemuu* homicide ritual (see Chapter 7, note 40).

3. On Yanomami warfare and social organization, see Albert, 1985, 1989, and 1990b. Yanomami incursions are only launched to avenge a death after the holding of a funerary ceremony (the cremation of the deceased's bones and/or burial/ingestion of his bone ashes). This instigating death must be caused by arrow (or shotgun) during an ambush or raid, or attributed to an attack by *oka* (plur. *pë*) enemy sorcerers (but sometimes also caused accidentally in a head-beating duel; see note 45 below).

4. In the late 1970s the Brazilian military dictatorship extensively used the stereotype of Yanomami fierceness to justify the dismembering of their territory. A particularly racist and outrageous version of this propaganda is found in a 1977 official report written by a FUNAI general (D. de Oliveira, 1977): "We observe . . . that the group lives in strongholds composed of 50 to 200 Indians and that each of these groups is hostile to the others, which leads us to conclude that relations between men and women take place between brothers and sisters, fathers and daughters, mothers and sons, and maybe even between grandmothers and grandsons and grandfathers and granddaughters, constituting genuine incest, which has, over the course of the centuries, caused the physical and intellectual atrophy of this indigenous group."

5. As previously noted, the term *wait^{h}iri* is quite ambivalent, as it can affirm a quality ("valor, courage, endurance") or denounce a rejected behavior ("aggressive, violent"), depending on context.

6. The eastern Yanomami's principal origin myth for war features an orphan

child *(Ōeōeri)* becoming a frantic warrior to avenge his mother killed by enemy sorcerers (M 47). *Arowë* is a very aggressive and invincible warmonger who turns into a jaguar (see Chapter 1; M 288). Davi Kopenawa describes *Aiamori* as the "image of a warrior elder," the "image of valor." The western Yanomami describe *Aiamori* as an evil and insatiable warrior spirit (Lizot, 2004: 6). All these mythic characters highlight the ambivalence of the *wait^hiri* notion (see note 5 above).

7. The ancient *Xamat^hari* (western Yanomami) of the Orinoco's headwaters in the northern highlands are mentioned as prototypical enemies in several myths recorded in the area of the Rio Catrimani and Rio Tootobi: in the origin myth of war (M 47) and in two narratives about *oka* sorcerers (M 141, M 359), and in one about a decapitated messenger (M 362). Regarding the contrast between the *Xamat^hari* and other Yanomami subgroups, see Albert, 1989.

8. This name is derived from an onomatopoeia associated with babies' cries *("Ōe! ōe!")* to which is added the *-ri* suffix, which characterizes mythic characters, shamanic spirits, and evil beings.

9. Savannas located in Venezuela, in the Serra Parima region ("Parima B"), north of the Orinoco headwaters *(H^wara u)*. In the 1970s, a group called "*Niyayoba-teri*" could still be found in this area (see Smole, 1976: chapter 3 about the "Parima Barafiri aerea").

10. These shamanic spirits of the savanna *(purusi)* are considered particularly fierce warriors (see Chapter 15).

11. Yanomami raids only target men and prioritize warriors reputed for their valor and aggressiveness (see note 2 above and Appendix D).

12. See Chapter 19, note 32, on the Gulf War.

13. The "smoke of bombs" is *pōpa pë wakëxi.*

14. "Take back the value of someone's blood" is a translation of the Yanomami expression *iyë në kōamāï;* "making the homicide ritual reciprocal" is *ōnokae nomiha-yuu.*

15. Warriors are referred to by the term *wai pë* (plur.), which, as an adjective (sing., *wai*), means "strong, toxic, venomous, dangerous" (see Chapter 7, note 1). Going on a war raid is *wai it^huu* (literally, "going down dangerous"), *napë it^huu* ("going down enemy"), or *wai huu* ("go dangerous").

16. The imputation of a death to *oka* sorcerers is often subject to triangular political manipulations. If the victim belongs to house A, his allies from house B can claim to have heard the members of a distant group C—with whom their relations have deteriorated—mention their sorcery attack on house A. It is then said that B "indicated, denounced" *(noa waxuu)* C after C "confessed" his misdeed *(noa hekuu),* thereby allowing A to "put right C's path" *(māe xariramāï).* This kind of network of rumors can indirectly set off a cycle of hostility between distant groups without prior interactions.

17. This ritual act is referred to by two expressions: *uxipë wariāï* ("damage the ashes") and *uxipë hiprikaï* ("rub and scatter the ashes"). It is intended to exacerbate the warriors' mourning anger and to make their future victims heedless of danger.

18. Dye made of a mixture of wood coal and *operema axihi* tree sap.

19. Warriors shake these game bone bunches while moving their heads from side to side, then drop the bunches to the ground with a crash. The ritual departure on a war raid is described by the verb *watupamuu*, "act as a vulture." On Yanomami war rituals, see Albert, 1985: chapter 11.

20. The *yarima* monkey is an extremely lively and aggressive monkey, with eyes always on the alert.

21. *Wainama* or *waiyoma* is a shamanic image associated with warriors *(wai pë)*; *õkaranama* or *õkorayoma* with enemy sorcerers *(oka pë)*. War raids *(wai huu)* and sorcery incursions *(õkara huu)* are considered equivalents. Additionally, *õkara huu* also refers to the reconnaissance expeditions of warriors preparing a raid.

22. This falcon eats ticks on tapirs and, occasionally, feeds on animal carcasses and human remains.

23. Warriors identify with these shamanic images *(utupë)* of predators and carrion feeders and, through them, will devour the flesh and fat of their dead enemies during the *õnokaemuu* homicide ritual.

24. The *yorohiyoma* spirits' name refers to the funerary envelope made of slats and vines in which corpses are hung in the forest *(yorohi ki)*. Mention of the *hixãkari* spirits refers to the cleaning of bones removed from putrefied flesh following the hanging of the corpse—this process is described metaphorically by the expression *imiki hixãmuu*, "to wash one's hands" (by rubbing, with a stick or other object). *Õrihia* refers to ill omens (see Lizot, 2004: 288; Mattei-Müller, 2007: 216). Finally, the term *naiki* describes the hunger for game meat.

25. These representations of enemies are called *në uë*, literally "value of imitation."

26. If the victim died of an arrow wound ("arrow leftover," *xaraka kanasi*), some of the ashes can be scattered again and rubbed onto the ground to fuel the anger of revenge (see note 17 above).

27. The Yanomami distinguish the upper *(heaka)*, middle *(miamo)*, and bottom *(komosi)* parts of the ashes *(uxipë)* in a funerary gourd *(pora axi)*. It is rare for Yanomami raids, whose privileged targets are a few reputed enemy warriors, to be successful on their first attempt.

28. As noted, the Yanomami funerary rituals aim to erase any social and physical trace of the dead (of whom bones are the last element) in order to send their ghosts to the sky's back (see M 35 about the primeval coexistence between dead and living). Yanomami war and funerary rituals organize a symbolic division of labor between allies (potential affines) and enemies in the ritual treatment of the dead person's body. The former consume or bury the bone ashes during *reahu* feasts while the latter are supposed to digest the flesh during the *õnokaemuu* homicide ritual (see Albert, 1985).

29. The literal Yanomami expression is: "as long as the hand does not fall" *(imiki keo mão xoao tëhë)*. We have seen that funerary gourds can be entrusted to classificatory relatives (brothers and brothers-in-law) in other houses, who are considered friends of the dead man (Chapter 19). *Reahu* feasts can therefore be held one after the other by several keepers of funerary gourds in a group of allied houses, and

raids can be launched at the end of each one until vengeance for the deceased is considered accomplished.

30. To "carry the grievance of the *pora axi* gourd" is said *pora axi nõwa tʰapu*.

31. This verb also refers to entering into friendly contact with an unknown group; see Chapter 19, note 21.

32. The unexpected death of an elder during this reconciliation process could also be attributed to the *oka* sorcerers of the former enemies and start the cycle of vengeance anew.

33. These great warriors are referred to by several expressions: "the bellicose (valorous) people" *(waitʰirima tʰë pë)*, "the people in homicide state" *(õnokaerima tʰë pë)* or "the sated people (with their enemies' flesh)" *(pitirima tʰë pë)*, and, finally, "the war makers" *(wai tʰë thaiwi tʰë pë—wai tʰë* being here, literally, "the dangerous and warlike thing").

34. Yet certain particularly fierce warriors could still sometimes use women's traditional role as peace emissaries to lure their enemies into an ambush.

35. In the 1950s and 1960s.

36. Regarding raiding activity in the highlands area of Yanomami territory in Brazil, see Duarte do Pateo, 2005. The frequency of incursions is higher in this more isolated region proportionate to its historically high population density. The recent introduction of shotguns among its inhabitants has also contributed to expanding the number of raid casualties, thus further intensifying the cycles of vengeance.

37. The Yanomami oppose "people in a homicide state" *(õnokaerima tʰë pë)* and "innocent (lit. oblivious) people" *(mohoti tʰë pë)* or "dry people" *(weherima tʰë pë)*. This last expression refers to the "greasy forehead" of warriors said to exude the fat of the enemy they have "eaten/killed" (see Chapter 7, note 40).

38. This is a reference to the 1993 Haximu Massacre, during which *garimpeiros* killed sixteen Yanomami (see Appendix D).

39. In the early twentieth century, Davi Kopenawa's forefathers lived on the *Amatʰa u* River, a tributary of the right bank of the Orinoco's headwaters, where they were raided by ancestors of the Rio Catrimani people, who then lived at *Arahai*, on the Rio Mucajaí headwaters. Davi Kopenawa's long-ago elders then moved south, successively occupying several sites on small tributaries of the left bank of the Orinoco headwaters *(Manito u, Kõana u)*, where they were repeatedly raided by the *Hayowa tʰëri*.

40. They thus continued their move south, in the direction of the lowlands of the upper Rio Demini basin.

41. Raids against the people of *Amikoapë* (ancestors and/or allies of the groups currently located on the *Hero u* River) and the *Mai koxi* (current Rio Catrimani groups) were launched from the *Yoyo roopë* and *Mõra mahi araopë* sites on the upper Rio Toototobi in the 1930s and 1940s. The *Mai koxi* are descendants of the people of *Arahai* who moved down in the direction of the Rio Catrimani basin (see note 39 above). The elders of the people of *Watoriki*, the community into which

Davi Kopenawa married and with whom he currently lives, belonged to the *Mai koxi*.

42. See Davi Kopenawa's childhood dream about the $H^w axi$ warriors, Chapter 3. The *Ariwaa $t^h ëri$* later became known as the $H^w aya$ *siki $t^h ëri$*, who in the 1990s settled near the Balawaú health outpost (then run by CCPY) on the upper Rio Demini *(Parawa u)*.

43. The groups on the Rio Toototobi stopped most of their raiding activities in the 1960s, following their contacts with the New Tribes Mission and, above all, the epidemics that decimated their population over that decade. A few raids were still launched by these groups in the 1970s, however, and until the mid-1980s by the upper Rio Catrimani groups (including the *Watoriki* people, then living on the upper Rio Lobo d'Almada).

44. In this head-beating collective duel *(he xëyuu)*, the initial adversaries (husband and lover) are both relayed by a series of relatives (kin and affines).

45. However, these ritual head-beating duels over women held between allied houses (or sometimes even within the same house) could lead to accidental deaths (cranial trauma) and also be the cause of cycles of war raids.

46. A distinction is made here between people of the same historical origin living in close allied houses (*kami yamaki*, "us")—who are considered guests, $h^w ama$ *pë*—and the "other people" *(yayo $t^h ë$ pë)*, "distant people" *(praha $t^h ëri$ $t^h ë$ pë)*, seen as "different people" *(xomi $t^h ë$ pë)*, which include enemy warriors *(wai pë)* and sorcerers *(oka pë)*.

47. The expression used here is: "because of the arrow's value of anger," *xaraka në wãyapëha*. In this case, it is the death itself (rather than the initial conflict whose development provoked it) that will be considered the cause of the raid launched to avenge it. Nevertheless, while conflicts over women do not directly lead to raids, which are always motivated by vengeance for a death, warriors sometimes take captives to marry them. This is considered a "secondary benefit" of a raid—as are instances of captured children or theft—and not its primary motive.

48. This ceremonial dialogue takes place between hosts and guests who are gathered in pairs, squatting face to face holding each other's neck with an arm (see Chapter 17). When tempers flare, angry participants try to squeeze and twist each other's neck as much as possible *(aikayuu)*.

49. Participants and their respective relatives in these chest-pounding duels *(pariki xëyuu* and *si payuu)* follow the same relay system as in the head-beating duels *(he xëyuu)*, but the former are held for less important grievances (insults, pilferage, malicious rumors).

50. The coati is a small procyonid carnivore that lives in noisy packs and is known for its aggressiveness.

51. Here, Davi Kopenawa primarily refers to the lowlands area, where contacts with missions and epidemics have drastically reduced raiding activities since the 1960s. See note 36 above on the highland area situation.

52. Both the memories of old wars and the knowledge of war rituals are still kept

in people's minds. It is necessary to distinguish here between the permanent symbolic "state of war" among clusters of allied houses as a political framework (the "war words," *niyayotima t*^{h}*ë ã*) and actual instances of raiding *(wai it*^{h}*uu)*, whose frequency can vary according to the region and the period, in relation to contingent social and historical factors.

53. In 1993, a war departure ritual *(watupamuu)* was held in *Watoriki*, not to attack an enemy house, as was the custom, but to launch a raid against gold prospectors in solidarity with a previously unknown Yanomami group *(H*^{w}*axima u t*^{h}*ëri)* who had been massacred by them (see Appendix D and Albert and Milliken, 2009: 112).

22. The Flowers of Dream

1. Davi Kopenawa sometimes translates *urihinari* (plur. *pë*) into Portuguese by the expressions *filhos do mato, filhos da natureza, espíritos do mato* ("sons of the forest, sons of nature, spirits of nature").

2. Regarding writing as "word drawing" *(t*^{h}*ë ã oni)*, see the chapter "Words Given," note 7. Lines in writing are more generally described as *onioni kiki*, an expression in which the repetition of the word denoting the short dash body painting motif *(oni)* is completed by a plural denoting a group of inseparable elements *(kiki)*.

3. In Yanomami, the verbs "to see" *(taai)*, "to make see, to teach" *(taamãi)* and "to know" *(tai)*, "to make know, to inform" *(tapramãi)* have the same root.

4. Here, of course, Davi Kopenawa refers to geographic maps.

5. "Paper skin": *papeo* (from the Portuguese *papel) siki* ("skin"); "image skin": *utupa* ("image") *siki;* tree skin: *huu tihi* ("tree") *siki.*

6. On Yanomami plant dyes and fragrances, see Albert and Milliken, 2009: 110–112.

7. Regarding the origin of body painting and the *reahu* presentation dance, see the origin myth of fire, M 50. Skin without body paint is considered "gray" *(krokehe)*, and smudged with ashes from the hearth fire *(yupu uxipë)*.

8. Yanomami body paint consists of graphic geometric patterns (over fifteen motifs) that most often refer to animal characteristics.

9. Reversing missionary attempts to associate *Omama*, the Yanomami demiurge, with the Christian god *(Teosi)*, Davi Kopenawa identifies *Teosi* with his brother the trickster *Yoasi*, a quick-tempered, envious, and muddled character who created death and the ills that afflict humanity.

10. Here, Davi Kopenawa uses a Portuguese expression: *nosso histórico*, "our history."

11. The "words of songs" is a translation of the Yanomami expressions *amoa t*^{h}*ë ã* or *amoa wãã*.

12. Shamanic chthonic fire that Davi Kopenawa associates with volcanoes.

13. This statement was recorded before the CCPY began a literacy program in the Yanomami language in *Watoriki* in 1996. Despite his shamanic criticism of

written knowledge, Davi Kopenawa founded this project in order to allow the young people in his community to master white people's writing so they could better defend their rights.

14. The verb used here is *ira-*, which is a component of expressions such as *wai ira-*, "contaminate (sickness)"; *tʰë ã ira-*, "assimilate (language)"; *pihi ira-*, "to fall in love."

15. "Thought" is a translation of the word *pihi*, which refers to reflexive consciousness and volition as well as the expression in a person's gaze. This term is a component of all verbs relating to cognitive activities and the expression of sensations as well as emotions in the Yanomami language.

16. See Chapter 6 on the relationship between initiates' chests and spirit houses.

17. The dreaming induced by the *xapiri* who carry away the shamans' image during his sleep is referred to as *xapiri pë në mari*, literally, the "the spirits' value of dream."

18. Possessing the "spirits' value of dream," a prerogative of the shamans *(xapiri tʰë pë)*, is contrasted with the "simple dreaming" *(mari pio)* of "ordinary people" *(kuapora tʰë pë)*. During the oneiric activity, the person's image/vital essence *(utupë)* is said to extract itself from his body ("the skin," *pei siki*) to travel *(mari huu)*, alone in the case of "ordinary people" or flying away with the *xapiri* in the case of shamans. The dreamer's conscious thought *(pihi)* having been turned off, he is also said to have "value of ghost" *(a në porepë)*.

19. Male love magic consists in covertly making desired women inhale aromatic plant charms during their sleep (see Albert and Milliken, 2009: 138–144).

20. Here, "dreamer" is a translation of the expression *maritima a*, which refers to a person whose dream activity is particularly intense.

21. The "flowers of dream" are called *mari kiki hore*.

22. Two birds of prey that hunt birds and reptiles (the former also hunts small mammals).

23. See Chapter 6 on spirit houses and initiation.

24. In this paragraph, Davi Kopenawa refers to various episodes of the saga of the demiurge *Omama* (M 202, M 197, M 198), then to myths telling of the animal ancestors' transformations at the beginning of time (M 80, M 50, M 86).

25. A reference to cars, which some of Davi Kopenawa's elders had seen for the first time when accompanying SPI agents to Manaus in the 1950s.

23. The Spirit of the Forest

1. Regarding the concept of *në rope*, see Chapter 8, note 25. It can be compared to the Maori notion of the forest's *hau*, perceptively revisited by Geffray, 2001: 149–154.

2. Regarding the *koyo* leaf-cutting ant and the origin myth of maize plantations, see Chapter 8, note 30. Small lizards *(waima aka)* are usually found in gardens. Regarding animal images and personal qualities, see Chapter 8, note 32.

3. According to the Brazilian Constitution of 1988, Amerindians have exclusive

usufruct of their lands, but they are considered public lands (*terras da União*; see Albert, 2004).

4. This Yanomami group *(Yawari)* was contacted by the roadwork crews of the Perimetral Norte road in 1973 (see Chapter 13). Its land is now largely cleared and has been invaded by cattle ranches *(fazendas)*; see Albert and Le Tourneau, 2004.

5. Regarding the environmental impact of gold prospecting in the highlands of Yanomami territory in Brazil, see Milliken and Albert, 2002, and Le Tourneau and Albert, 2010.

6. When the presence of specific plant species indicates an optimal area for cultivation in the forest, it is customary to use the expression *hutu a praa*, "a garden lies down on the ground" (see Albert and Milliken, 2009: 32–37). Thus the terms *hutu a* or *hutu kana a* ("garden," sing.) refer both to cultivated plots and to forest patches suitable for cultivation.

7. *Wahari a* refers to the cold, damp emanation from the forest soil: it is *urihi wixia*, the "forest's breath of life," *Xiwãripo wixia*, the "chaos being's breath of life," or *Motu uri u wixia*, the "underground river's breath of life."

8. Regarding swidden cultivation and the forest's value of growth (and the leaf-cutting ant ancestor, the bat and giant armadillo spirits), also see Chapter 8.

9. There is an interesting connection here with a recent theory that reveals the important effect that the tropical forest's "pumping" of atmospheric humidity has on the climate (Pearce, 2009).

10. Reference to the fall of the first sky, which came to form the current terrestrial layer in the beginning of time (see Chapter 8).

11. The word *urihi a* refers to the forest, to the land that supports it, and to a notion of territory, whereas *maxita a* refers to the soil, the earth (see Albert, 2009).

12. On the animal transformation of the *yarori* ancestors and their acquisition of game skin/feather "paintings," see M 130. Human body paintings are considered "marks of the human/animal ancestors" *(yarori pë õno pë)*.

13. This relationship of similarity is expressed by the expression *ai yamaki h^wëtu*, literally: "(we are) others similar." See also Chapter 4 about the relationship between human/animal ancestors, game, and *xapiri* spirit helpers.

14. Literally: *"yanomae t^hë pë, yaro yahi t^hëri t^hë pë!"* Here, Davi Kopenawa opposes *yahi t^hëri yaro pë*, "game living in houses" (humans), and *urihi t^hëri yaro pë*, "game living in the forest" (animals).

15. Therefore the "savage" cannibalism of the human/animal ancestors of the beginning of time has been replaced in today's human cultural world on the one hand by the ritualized "endocannibalism" of funerary ceremonies and on the other by the culturally controlled hunting and eating of game.

16. Breaches of the ideal rule of exchanging prey followed by good hunters are denoted by two expressions: *kanasi wamuu*, "eating one's own leftovers," and *kõamuu*, "bringing back to oneself." It is highly probable that the term *kõaa pë* (plur.) is derived from the same root as the latter expression (from the verb *kõai*, "bring back"). Regarding this term, Davi Kopenawa commented: "I don't know how

to say that in the white people's language. *Kõaa pë*, that comes from the fact that a hunter who kills a piece of game cannot eat it himself."

17. This smell is also attributed to egg whites and raw fish.

18. *Kãomari* is the image of the *kãokãoma* falcon, a reputed hunter. The *yawari-oma* water beings are also considered outstanding hunters (see Chapter 5). The *Uri-hinamari* forest spirit (the nocturnal twin of the *Urihinari* forest spirit) also con-notes excellence at hunting (it is associated with men who sleep little and go hunting before dawn or at the beginning of the night). Images of these entities are said to accompany the great hunters who spend their time roaming the forest in search of game.

19. This large forest gastropod is considered particularly repugnant.

20. The wind raised by the spirits' movement in the forest is described by the expression *xapiri pë në watoripë*.

21. The "hot season" *(t^{h}ë mo yopi)* is said to have "value of epidemic" *(t^{h}ë mo në xawarapë)*.

22. Here too, Davi Kopenawa appears to refer to São Paulo, with an airport in the city center (Congonhas) and hundreds of buildings with rooftop heliports.

23. See Chapter 4 and M 210 and 211 on *Omama*'s flight and his creation of the mountains.

24. *Omoari* is said to also occasionally capture the images of human beings to roast them before devouring them.

25. *Toorori* is also an evil being said to catch young children's images like fish in a loosely woven *sakosi* basket and roast them on a clay plate. On the alternation be-tween *Omoari* and *Toorori,* see Chapter 8.

26. On *Omama* and the origin of metal, see Chapters 9 and 16.

27. "*Omama*'s metal" is *Omama poo e xiki*; "nature's metal" is *natureza poo e xiki*; and the "sky's metal" is *hutukara poo e xiki* (see Chapter 16).

28. Shamanic weapons that refer to the strong beaks and powerful tails of the corresponding animals.

29. In Portuguese, *o poder da natureza*.

30. Here, "sickness" is a translation of the term *waiwai a*; see Chapter 19, note 17.

31. Here, Davi Kopenawa specifically mentions dots *(turu)* and short lines *(oni)*, which are basic motifs of the graphic repertoire of body painting.

32. Regarding *Omama* and the acquisition of cultivated plants, see M 198 and Chapter 9.

33. In Yanomami, the "words of ecology" are *ekoroxia t^{h}ë ã*.

34. Here, "words to defend the forest" is a translation of the Yanomami expres-sion *urihi noamatima t^{h}ë ã*.

35. The shamanic spirits are said to be "defenders of the forest": *urihi noama-tima pë*.

36. In Yanomami, ecologists, the "people of the ecology," are called *ekoroxia t^{h}ëri pë*.

37. Celebrated Brazilian leader of the *seringueiros* (rubber tappers) who fought

the devastation of the Amazon forest by large-scale cattle ranchers (see Mendes, 1990) and was assassinated on December 22, 1988, in Xapuri, state of Acre. Chico Mendes received the Global 500 Prize from the United Nations Environment Program (UNEP) in 1987, two years before Davi Kopenawa.

38. Early in April 1977, Davi Kopenawa joined an expedition of FUNAI, the IBDF environmental agency (which became the Brazilian Institute of the Environment–IBAMA in 1989), and the federal police against illegal hunting on the Rio Catrimani. During this trip, five hundred turtles were thrown back in the river and some fifty giant otter skins were destroyed (Monteiro Caltaneão, 1977). A FUNAI report mentions that a Yanomami community which had been attracted to the lower Rio Catrimani by white settlers "barter latex rubber, Para nuts, and forest animal skins," and that "they are shamelessly exploited by local tradesmen. As for this group's health, one finds tuberculosis, measles, flu, dysentery, and many cases of malaria" (Costa, 1977).

39. See Chapters 13 and 14.

40. The river turtle and the Amazon dolphin are protected species.

41. In Yanomami "game meat hunger" *(naiki)* is distinguished from "plant food hunger" *(ohi)*.

42. For the Yanomami, this forest without human trails—*urihi komi*, literally the "closed forest"—is the favored domain of the *në wãri* evil beings.

43. Synonyms of this widest sense of "land-forest" *(urihi a)* are *urihi a pata* (the "great land-forest") or *urihi a prauku* (the "vast land-forest"). On the multiple meanings of the *urihi a* concept, see Albert, 2009.

44. Here, Davi Kopenawa uses the Portuguese expression *meio ambiente*, equivalent to the concept of "environment" for English speakers.

45. Literally, *urihi a xee hëaiwi*, the "rest of the forest" *(urihi a xee)* "which still remains" *(hëaiwi)*.

46. Here, Davi Kopenawa plays on the double meaning of the Portuguese word *meio* (environment/middle) to state that the earth must not be cut down the middle, that is, divided into a center—the white people's world—and a subordinate periphery—the tropical forest conceived as a residual surrounding ("environment") of that center. On this subject, see Albert, 1993.

47. Reference to the United Nations Environment Program (UNEP) Global 500 prize awarded to Davi Kopenawa in 1989. The award ceremony was held on February 1, 1989, in Brasília and Davi Kopenawa's speech was reprinted in the *O Estado de São Paulo* newspaper dated February 14, 1989.

24. The Shamans' Death

1. As we have seen, these orphan spirits are referred to by the same term as posthumous children: *hapara pë* (Chapter 4).

2. Regarding these shamanic images of evil beings turned spirit helpers *(xapiri)*, see Chapter 7. Here, Davi Kopenawa also refers to the *Xuukari* (celestial seep-

age/dysentery), *Riori* (floods), *Rueri* (cloudy weather), *Xinarumari* (master of cotton), and *Krayari* (venomous caterpillar) evil beings.

3. This shaman and "great man" *(pata tʰë)* died in late 1989, during the most intense phase of the gold rush on Yanomami land. The *Hero u* River region was then a genuine "aerial wild west," and single-engine aircraft collisions and crashes were therefore frequent (see Chapter 15).

4. See the origin myth of peccaries (M 148) in which ancestors lost in the darkness and cold were plagued by a swarm of giant wasps (*xi wāri na kɨ*, "transformation wasps") until they turned into wild pigs.

5. Regarding flooding and "becoming other" in myth, see M 33 and Chapter 9 (origin of the outsiders). Also see Chapter 8 on the cosmological phenomena associated with the death of shamans killed at war or by enemy sorcerers.

6. See Chapter 8 regarding the shamans' work to prevent the sky from falling apart.

7. Davi Kopenawa, whose life was often threatened by the gold prospectors in the 1980s and 1990s, has more recently also been threatened by big cattle ranchers encroaching on the edges of Yanomami land.

8. See Chapter 20 and Davi Kopenawa's dream of the burning sky.

Words of *Omama*

1. This statement was recorded in the 1990s, when Davi Kopenawa was not yet forty.

2. It is considered necessary to undergo several initiation sessions over the years (two or three) to become an experienced shaman.

3. Regarding these spirits whom only the greatest shamans are able to own, see Chapter 7.

4. This elder's great shamanic reputation rested on his ability to regurgitate pathogenic objects (leaf packets containing sorcery plants, arrowheads and cotton of the *në wāri* evil beings, or hostile *xapiri* spirits).

5. *Ayokorari xapokori a* literally means "sterile *ayokora* cacique bird spirit." It is also called *Ayokorari haasipërima a*, the "*ayokora* cacique bird spirit of the left hand." It is opposed to *Ayokorari yai tʰaiwi a*, the "*ayokora* cacique bird spirit which really does" or *Ayokorari kateherima a*, the "*ayokora* cacique bird spirit of the right hand" (or the "beautiful *ayokora* cacique bird spirit"), or also *ayokora miamohamɨ a*, the "*ayokora* cacique bird spirit of the center."

6. The term used here, *oraka*, refers to the tube-shaped entrance of a bees' nest or the neck of a gourd.

7. The Yanomami compare the scent of onion and garlic to the odor of genitals.

8. Reference to the first phase of Kopenawa's initiation, described in Chapter 5.

9. Through this implicit comparison, Davi Kopenawa refers to the Bible, which he was taught in his youth was a book containing the "drawing of the words of *Teosi*" (*Teosi tʰë ã oni*) collected long ago to be remembered.

10. Shamans are referred to here as *noamatima t^{h}ë pë*, the "people who protect."

11. See the beginning of Chapter 4.

12. Another allusion to the proselytism of the New Tribes Mission pastors among the Yanomami; see Chapter 11.

How This Book Was Written

1. See Caratini, 2004.

2. Lévi-Strauss, 1983.

3. In France, historians have been bolder than anthropologists on this point: see Agulhon et al., 1987, on "ego-history." However, see Descola, 1994; Agier, 1997; Ghasarian, 2004; Dhoquois, 2008; Fassin and Bensa, 2008; Leservoisier and Vidal, 2008.

4. Gheerbrant, 1952.

5. The Guayabero are a Guahibo-speaking group living in the gallery forest along the Rio Guaviare, upstream from San José, which was then (in 1972) a small village. Their population is approximately 1,100. They were also visited in 1948 by the Amazon-Orinoco Expedition (Gheerbrant, 1952: 38–39). San José del Guaviare was founded in 1938 by rubber tappers. Today, it is a base for Colombian Army operations against the Revolutionary Armed Forces of Colombia (FARC). The Rio Guaviare is a tributary on the left bank of the Middle Orinoco, which originates in the Colombian Cordillera Oriental mountain range.

6. I echo here the famous statement from *Tristes Tropiques* about Tupi-Kawahib: "As close to me as a reflection in the mirror, I could touch them, but I could not understand them" (Lévi-Strauss, 1955: 397).

7. This term had just entered Americanist discourse in France after the publication of *La paix blanche: Introduction à l'ethnocide,* by Robert Jaulin (1970).

8. "The Third Bank of the River" (*A terceira margem do rio*) is the title of a story by the famous Brazilian writer João Guimarães Rosa (2001).

9. Paradoxically, despite the Americanist focus of the *Mythologiques,* Lévi-Strauss's theoretical contributions to Amazon studies didn't really take hold until the 1980s. See A. C. Taylor, 2004.

10. The Ikpeng are a Carib-speaking group numbering approximately 340 people today, located in the Parque Indígena do Xingu (PIX) in the northeast of Mato Grosso state. See Menget, 2001.

11. See Bloch, 2004: 353; Maybury-Lewis, 1967; and Rivière, 1969.

12. See Ribeiro, 1970, and Cardoso de Oliveira, 1964.

13. The invitation, which I answered in April 1974, initially raised the possibility for research among the northern Yanomami, the *Sanima* (Ramos and Taylor, 1973). Alcida Ramos and Kenneth Taylor were the first anthropologists who had conducted long-term fieldwork among the Yanomami in Brazil. They defended their doctoral thesis on the *Sanima* at the University of Wisconsin in 1972. A few months after receiving this invitation, the context of the work had changed and become the

Perimetral Yanoama project, organized by these two anthropologists at the University of Brasília under the auspices of FUNAI. See K. I. Taylor, 1975a, and Ramos and Taylor, eds., 1979.

14. Onchocerciasis (river blindness) is a parasitic disease caused by a filarial nematode. Letters from Kenneth Taylor to Bruce Albert, November 6, 1974: "I fully realize that onchocerciasis is an hideous disease and will entirely understand if you prefer to avoid the risk involved"; and December 1, 1974: "I very much hope you will decide to join us, the disease is horrible, but the prospect of a truly important job of work for the benefit of the Indians is very exciting" (Bruce Albert archives).

15. See Ramos, 1992.

16. Regarding these two faces of the "Wild Man," which has symbolized man's natural state since the Middle Ages, see White, 1978: chapter 7.

17. Literature about the Yanomami in Brazil was very recent and not yet widely available. Kenneth Taylor's doctoral thesis on animal classifications and food prohibitions among the *Sanima* in Brazil had just been published in Venezuela (K. I. Taylor, 1974). The doctoral thesis of Alcida Ramos on the *Sanima* (see Ramos, 1995) and of Judith Shapiro on the *Yanomae/Yanomama* of the highlands (1972), which both deal with social organization, as well as the doctoral thesis of the Protestant missionary John Peters (1973) about social change among the *Yanam/Ninam*, had not been published. I was only able to consult these works after I arrived in Brazil.

18. See note 14 above. Regarding the tasks assigned to the team I worked with at the Catrimani Mission (language training, study of the social and economic organization of local Yanomami communities, study of the relations between the mission and the Indians, monitoring the road works impacts), see K. I. Taylor, 1975d.

19. Our Lady of the Consolata is the patroness of the city of Turin, in northern Italy. The Catholic missionary congregation which bears her name was founded in 1901. This priest later studied anthropology in the United States and defended a master's thesis on the impact of the highway in the area of his former mission. See Saffirio, 1980.

20. See Saffirio, 1976.

21. Teams of topographers from the company responsible for building the highway arrived at the Catrimani Mission in January 1974. Just a few months later, highway workers were already a lot more numerous than the approximately 300 Yanomami divided among eight local groups in the area of the mission (Saffirio, 1976). A letter by Kenneth Taylor dated February 27, 1976, described the conditions there to me just before I arrived on site:

> The situation in the south of Roraima, in the area of the highway construction, is extremely grave as far as the interest and welfare of the Indians is concerned. They are in a state of constant difficulty with health problems, resulting from contact with the construction workers, and of considerable disruption of their economic life both because of these health problems and because of their fascination with the highway and the misguided tendency

of the "good-hearted" construction workers to give them food and second-hand clothing. At the Catrimani Mission, for example, the present dry season is almost over without the local villagers having done a thing towards preparing new fields for the next year's food production. (Bruce Albert archives)

22. The head of the FUNAI Ajarani Outpost described their situation in May 1975: "it would be very difficult to keep the Indians in the areas where they live with nothing having been offered to them by FUNAI as a symbol of fraternization, trust, and friendship . . . As soon as we depart to take up other work, they go to the canteens run by the roadwork companies to beg for everything—clothing, cooking-pots, machetes, etc." (Castro, 1975). Without communal house or gardens, these Indians were reduced to wandering along the highway, begging, prostituting themselves, or working in nearby sawmills (see Ramos, 1979).

23. I was given several nicknames in the course of my early fieldwork. My first name (Bruce) became *purusi* (which means "savanna" in Yanomami, the place where white people live), *purunama usi* (a type of thin bamboo, *Olyra latifolia*), or *prosi siki* (a very long Colubrid snake, *Pseutes sulphureus*), the two latter names probably referring to the fact that I am tall and was skinny. It is highly likely that other far less charitable nicknames were also used, as is customary, and never revealed to me. As I got older, my friends in the *Watoriki* community dubbed me *Horepë tʰëri a,* "inhabitant of the highlands," probably as a good-natured joke and allusion to my proclivity to insist so much along the years on the necessity of maintaining traditions. (The "people of the highlands" are frequently those with the least contact with white people, the more "traditional" Yanomami.)

24. I have borrowed the term "field baptism" from Caratini, 2004: 25.

25. See Ramos, 1975.

26. This project was proposed in June 1974 and officially approved in December 1974, but only really operated between October 1975 and January 1976 under the new name "Plano Yanoama."

27. *Makuta asihi* is the name of a big tree *(Bombacopsis* cf. *quinata)* covered with large thorns. Yanomami women like to wear its white filamentary flowers as earrings.

28. For a detailed explanation of the Plano Yanoama, see Bigio, 2007: chapter 4.

29. My reflections on the uncomfortable ambiguities of the "ethnographic situation," which were so striking in my first fieldwork, were later influenced by an article by Zempléni, 1984, from which I borrowed the above expression (110).

30. My "informants" often punctuated their explanations or demands by asking me to convey their point "to the leaders of the white people" *(napë pata pëha).*

31. "France Antarctique" was a French Protestant colony in the Guanabara Bay, Rio de Janeiro, Brazil, founded in 1555. It was ultimately destroyed by the Portuguese in 1567. See Navet, 1994–1995.

32. See Albert, 1997a, on "post-Malinowskian" fieldwork.

33. Some of this early material on kinship was presented to the International

Congress of Americanists meeting in Paris in September 1976 (see Ramos and Albert, 1977).

34. See Chapter 13. Prior to my trips there in 1975, the groups living along the upper Catrimani had only been visited once before, to my knowledge, by the founder of the Catrimani Mission, Father Calleri, in the late 1960s. When I still had not received any information about the epidemic, on May 23, 1977, I contacted the nongovernmental organization Survival International to seek funding for a healthcare project in the Rio Catrimani area. Then, on July 5, I wrote to the priest of the Catrimani Mission to secure his agreement. This letter received no response. On August 8, I wrote to the photographer Claudia Andujar (with whom I would found CCPY the following year) to ask for her help in convincing the Catrimani Mission to carry out the vaccinations. Although the Brazilian National Security Council had banned her from staying in the area at that time, she did everything she possibly could to push through my proposal (letters dated September 14, October 14, and November 11), but once again it remained unanswered.

35. See CCPY, 1979, and Bigio, 2007: chapter 5.

36. *Hewë nahi* refers to the *Centrolobium paraense*, a tree with very hard wood, prized by the Yanomami for construction of their communal houses.

37. I finally wrote this thesis after several new stays between 1979 and 1985 along the Rio Catrimani and other areas in Yanomami territory (see Albert, 1985).

38. An inflammatory disorder or infection of the inner ear that affects balance.

39. See Andujar, 2007: 168.

40. See Albert, 1997b: 187.

41. In a document dated October 1975, the Plano Yanoama coordinator indicates Davi Kopenawa's presence at the FUNAI outpost in Iauaretê, along the upper Rio Negro (see K. I. Taylor, 1975b).

42. Head of the Ajarani Outpost at kilometer 50 on the Perimetral Norte, he had been relieved of his responsibilities in October 1975 by the coordinator of the Plano Yanoama, who found his work as *sertanista* (field agent) in Yanomami territory to be useless (K. I. Taylor, 1975b).

43. In 1978, he wrote in one of his reports: "The Yanomami of the Catrimani Mission are subjugated under the yoke of an oppressor who leads them further towards primitivism every step of the way; they do not have the right to choose their own fate." See Costa, 1978.

44. See Costa, 1977, and Andujar, 2007: 166–167.

45. My first two fieldworks in Brazil took place during General Ernesto Geisel's presidency (1974–1979). FUNAI was run by General Ismarth de Araújo Oliveira, in direct contact with the dictatorship's notorious National Information Service (SNI). See "FUNAI espionou missionários na didadura," *Folha de São Paulo*, February 24, 2009.

46. Named in January 1975 to replace another *sertanista* shot with arrows by Indians resisting the encroachment of the Manaus–Boa Vista highway through their territory, he immediately declared to the press: "The Waimiri-Atroari deserve a lesson. We must teach them that they have committed a crime. I will wield an iron fist.

Their chiefs will be punished and, if possible, deported far away from their territory and their people. This is how they will learn that it is unacceptable to massacre civilized people. I'll go to an Indian village with an army patrol and, there, in front of the entire population, I'll give them a good example of our power. We will fire machine guns in the trees and explode grenades, making as loud a scene as possible, without wounding anyone, until the Waimiri-Atroari are convinced that we are stronger than they are" (*O Globo,* January 5, 1975). After making these statements, he was precipitously transferred out as head of the Ajarani Outpost in Yanomami territory.

47. A pro-Indian journalist, for example, rushed words into print about Davi Kopenawa after a brief visit to the Demini Outpost in early 1978:

> This base inculcates the deculturation of Yanoama groups living in the region. A typical example is one of the Indians contacted by FUNAI named Davi. He currently works as an interpreter for the Army's geographic service ... Davi is already ashamed of his indigenous identity. His presence at the Catrimani Mission, where the military camp is based, has demonstrated the miracles of the world of white people to some Indians from the mission: shirts made of synthetic fabrics ... printed bathing suit and comb ... Davi, has acquired enormous importance among the Yanoama because of his new status." (*Jornal de Brasília,* April 2, 1978)

48. The Rio Toototobi shamans, heavily influenced by the western Yanomami, have a much more exuberant shamanic style than those of the eastern Yanomami, with whom I worked more frequently along the Rio Catrimani and Rio Mucajaí.

49. I spent six months there as an ethnographic consultant for the production of a book of photographs issued by Time-Life in 1982.

50. See Andujar, 2007: 167, wherein Davi Kopenawa briefly relates how he made contact with members of the CCPY (Claudia Andujar, Carlo Zacquini, and Bruce Albert) and how this led to lasting friendships.

51. They had been affected by two successive epidemics of infectious disease in 1973 and in 1976 (Chapter 13).

52. Regarding this healthcare program conceived by Davi Kopenawa, see Albert, 1991; Turner and Kopenawa, 1991: 61; and Kopenawa, 1992. The missionaries finally left the Rio Toototobi in 1991 to head back downstream on the Rio Demini, establishing a new outpost there, "Novo Demini." In the end, only two Yanomami communities joined them.

53. Regarding this massacre and our role in bringing it to light, see Albert, 2005; Rocha, 1999: chapter 3; and Appendix D.

54. Turner and Kopenawa, 1991: 60. Regarding the AAA Special Commission, see http://www.aaanet.org/cmtes/cfhr/Report-of-the-Special-Commission-to-Investigate-the-Situation-of-the-Brazilian-Yanomami.cfm.

55. See the excellent analysis of this gold rush in Roraima by MacMillan, 1995.

56. This ban was also aimed at all members of the CCPY and the Catrimani missionaries. See Albert, 1990a: 125.

57. See Albert and Menegola, 1992.

58. See Albert and Kopenawa, 1990: 11–14. The interview was filmed by Beto Ricardo (Cedi/Instituto Socioambiental–ISA). The APC was a movement of parliamentarians, clergy, scientific associations, and NGOs that actively protested the decimation of the Yanomami in Brazil in 1989 and 1990. See APC, 1989 and 1990.

59. Commenting on Kopenawa's account, Claude Lévi-Strauss (1993: 5–7) wrote: "This conception of human solidarity and diversity, and their mutual implication, is striking in its grandeur. There is something like a symbol here. For it falls to one of the last spokesmen of one of so many societies on the path to extinction through our actions to state the principles of a wisdom which we are still too few to understand is also crucial to our own survival."

60. Turner and Kopenawa, 1991.

61. Ibid., 62.

62. Albert, 1993.

63. In 1995 and 1996, I was involved in setting up a pilot project for treating river blindness in the Rio Toototobi region for the Brazilian health ministry (Albert et al., 1995); then, from 1996 to 1999, in a project for a bilingual education system in Watoriki and Toototobi for the CCPY and the education ministry (Albert, 1997c). In 1997, I also began research on indigenous organizations and their plans for sustainable development throughout the Brazilian Amazon with the Instituto Socioambiental of São Paolo (Albert, 1997b, 2001, 2004). I helped a group of doctors to create a new NGO in 1999 for healthcare assistance to the Yanomami (Urihi Saúde Yanomami), and I stayed on its advisory board while becoming vice president of the CCPY the following year.

64. Lévi-Strauss, 1962: 290. See commentary by Wiseman, 2005: 406. Regarding the exhibition, see articles by G. Breerette (2003) and E. de Roux (2003) in Le Monde as well as the article by D. Thomas (2003) in Newsweek.

65. Albert and Kopenawa, 2003.

66. Whale Alley is also the title of a short book by Jean Malaurie in which a friendly dedication, to mark that occasion, signaled a new beginning for my manuscript (Malaurie, 2003).

67. The last pages of a short, recently published text dedicated "to the imaginary of the Inuit Nation" are a paradigm in this regard and resonate deeply with me (Malaurie, 2008).

68. Borges, 1987: 240, in reference, precisely, to translation.

69. See Basso, 1995; Hendricks, 1993; and Oakdale, 2005.

70. Brumble, 1988: chapter 3, and Duthil, 2006: chapter 2. (See also Krupat, 1994, and Wong, 1992, also on North American Amerindian autobiographies.)

71. Brumble, 1988: 75–76.

72. Under the pretext of a more faithful rendering, all of these ethnobiographies emphasize the role of their editors (as opposed to the classic textual model), and tend to make use of an interpretive apparatus that breaks up and surrounds the words of their "subject" to the point of phagocytosis. For ethnobiographies centered

on discourse analysis, see Hendricks, 1993. For critical essays, see Crapanzano, 1972. On the Amazon, see Muratorio, 1991, and Rubenstein, 2002. For more traditional ethnographic studies, see Keesing, 1978, or Shostak, 1981.

73. With regard to collaborative ethnobiographies, this is what Lejeune calls the "ethnological divide." See Lejeune, 1980: 271.

74. See Zempléni, 1984: 115.

75. Davi Kopenawa was schooled in basic literacy in his language by missionaries from the New Tribes Mission in Toototobi in the 1960s, but he received no further formal education.

76. Lejeune, 1980: 230.

77. An expression borrowed from ibid., 240, n. 1.

78. On the multiplicity of autobiographical "Is," see ibid., 235–236; Duthil, 2006: 159–160; and Aurégan, 2001: 51 and 428.

79. Balzac, 1977: 1020, after Aurégan, 2001: 397.

80. Lejeune, 1980: 239.

81. I chose to avoid mentioning these two elders by name in the text in accordance with Yanomami practice (see Chapter 1 and Chapter 10). The first died in 1997, the second is now about eighty years old.

82. See Albert, 1993: 244–246. In public remarks and in this book, Davi Kopenawa often cites his father-in-law as the person who inspired his shamanic prophecies. He made the same point in his interview with the representative of the AAA (see Turner and Kopenawa, 1991: 62). "I learned about this from Lourival, who is the headman of my village and my teacher; he is a shaman and also my father-in-law."

83. Viveiros de Castro, 2007: 47–48.

84. These sixty-three myths were published in English in the compilation of Yanomami oral literature edited by Wilbert and Simoneau, 1990. The expression "thick translation" was obviously inspired by Geertz's (1973) "thick description" as applied to ethnographic interpretation.

85. See Appendix A.

86. For a discussion about "closer" and "midway" methods of translation, as opposed to a more distant literary elaboration, see Lejeune, 1980: 290–300.

87. See Barthes, 1973: 71, about writing and the "pleasure of a text" as opposed to *"écrivance."*

88. An expression by Todorov, 1971: 77, cited by Duthil, 2006: 132.

89. See Lejeune, 1980: 304–307.

Appendix A. Ethnonym, Language, and Orthography

1. See Perri Ferreira, 2009: 17–18. This author views this mutual intelligibility as more dependent on (and proportional to) the frequency of contact between neighboring communities that speak different languages than on an absence of phonological or morphosyntactic differences.

2. Migliazza, 1972: 4c. The western Yanomami *(Yanomamɨ)*, the majority of

whom are located in Venezuela, constitute 59 percent of the group's population, followed by the eastern Yanomami *(Yanomam)*, who constitute approximately 21 percent and are located mostly in Brazil. The northern Yanomami *(Sanima)*, who are primarily located in Venezuela, represent almost 17 percent of the ethnic group, whereas the *Ninam/Yanam* of Brazil constitute barely 3 percent.

3. Ramirez, 1994: 25. A more recent survey of Amazonian languages, published by Dixon and Aikhenvald, 1999, considers Yanomami to be a unique language formed by a continuum of dialects rather than a language family (chapter 13). This hypothesis, however, is very far from being established, and more comparative field studies are needed. Moreover, it has been recently placed in doubt by Perri Ferreira, 2009: 17. To date, the most consistent linguistic studies in Brazil deal with the *Sanima* (Borgman, 1990), *Ninam/Yanam* (Gomez, 1990), *Yanomami* (Ramirez, 1994), and most recently, *Yanomam* (Perri Ferreira, 2009).

4. Migliazza, 1972: 35. Migliazza had raised instead the possibility of an unknown dialect of *Yanomam* in the Rio Ajarani area, which, at that time, was isolated.

5. This distinct phoneme, accepted in studies such as Borgman, 1990, and Lizot, 1996, is questioned by Ramirez, 1994: 61–62.

6. According to Ramirez, 1994: 35–36, this is a residual phoneme of limited distribution (never accompanying the vowels i, o, or u) and corresponds to the f of the upper Rio Parima area.

7. Ramirez, 1994: 236–237.

8. The written forms used by the missionaries, for example, used the letter l to transcribe the sound [r], and the symbols e and y with grave accent marks to represent central vowels, instead of $ë$ and i, which are used in the transcriptions in this book.

Appendix B. The Yanomami in Brazil

1. See Zerries, 1964.

2. See Becher, 1960.

3. Borofsky, 2005: 8, 39. This 1968 monograph is based on Chagnon's 1966 doctoral thesis.

4. This image was undoubtedly reinforced by the publication in English of Ettore Biocca's book (1970). This book, originally published in Italian in 1965, excerpted Helena Valero's exceptional account of captivity among the Yanomami, retaining only the most violent and spectacular scenes.

5. Lizot, 1985: xiv. The most recent paperback editions of this book in English (published by Canto in 1991 and 1997) feature a photograph of Davi Kopenawa on the cover, which, to the best of my knowledge, is unauthorized.

6. See Dorfman and Maier, 1990; Vanhecke, 1990.

7. See Chagnon, 1988, and Kamm, 1990.

8. See the *New York Times*, 1993, and Guiraut Denis, 1993.

9. See Tierney, 2000, as well as Borofsky, ed., 2005. For press coverage, see, e.g., Wilford and Romero, 2000; Roosevelt, 2000; or Birnbaum, 2000.

10. Of the dozens of ethnic groups that surrounded the Yanomami up until the end of the nineteenth century, only the Ye'kuana have survived.

11. Regarding this hypothesis and the data on which it is based, see Neel et al., 1972; Spielman et al., 1979; Migliazza, 1982; and Holmes, 1995.

12. On the history of Yanomami territorial expansion and its causes, see Albert, 1985 and 1990b; Chagnon, 1966 and 1974; Colchester, 1984; Good, 1995; Hames, 1983; Kunstadter, 1979; Lizot, 1984 and 1988; Ramirez, 1994; and Smole, 1976.

13. Capobianco, ed., 2001: 398–399.

14. On the history of the Yanomami relationships with their indigenous neighbors in Brazil, see Albert, 1985; Ramirez, 1994; and Le Tourneau, 2010.

15. See Le Tourneau, 2010: chapter 1, on the Brazilian Boundary Commission's work to define the border between Brazil and Venezuela.

16. This archipelago-like spatial clustering produced by early contacts was used in the 1970s by the Brazilian military as a reference to plan their attempt to dismember the Yanomami territory.

17. Regarding the situation of the Yanomami in Brazil at the end of the 1970s, see Ramos and Taylor, 1979.

18. This project was notable for its production of the first thematic map of the Brazilian Amazon, indicating geology, geomorphology, vegetation, and agricultural potentials.

19. See MacMillan, 1995, and Albert and Le Tourneau, 2005.

20. Again, see Albert and Le Tourneau, 2005.

21. See Appendix D for details of the Haximu Massacre in 1993. From 1991 to 1998, the Brazilian National Health Foundation (then known as the FNS) recorded the deaths of 1,211 Yanomami, the majority due to malaria and pneumonia. By May 2006, FUNAI estimated that there were still 700–800 gold prospectors entrenched in Yanomami territory, and their number has only grown since then.

22. See Ricardo and Rolla, 2005: 50.

23. See Le Tourneau, 2003, and Albert and Le Tourneau, 2004. In some cases, settlers and cattle ranchers have crossed over into Yanomami territory. According to a FUNAI study dating from October 2001, several cattle farms as well as thirty small settlements had already encroached on the Rio Ajarani area at that time.

24. See Elvidge et al., 2001, and Barbosa, 2003.

Appendix C. *Watoriki*

1. Other dwelling variations are, however, possible: a group of small communal houses; a main house and several small satellite houses; and sometimes a group of small rectangular houses.

2. A Dravidian kinship system classifies the social universe of any individual into two main groups ("kin" and "in-laws") based on distinctions of gender and generation, and implies a rule of marriage between people considered to be cross-cousins (children of opposite-sexed siblings).

3. Regarding kinship and multicommunity networks among the Eastern Yano-

mami see, for the lowlands, Albert, 1985, and, for the highlands, Duarte do Pateo, 2005.

4. See Albert, 2009, about this expression.

5. The Yanomami in *Watoriki* regularly consume the fruit of at least a dozen different species of palm (Albert and Milliken, 2009; 47–57).

6. Regarding the communal house in *Watoriki*, see Milliken and Albert, 1997b; Albert and Kopenawa, 2003; Albert and Le Tourneau, 2007; and Albert and Milliken, 2009: 73–88.

7. The *Hapakara hi* site is also known as *Yãri pora*, the place of the "Thunder Waterfall." *Hapakara hi* refers to the tatajuba tree *(Bagassa guianensis)*.

8. *Werihi sihipi u* is the "River of the *werihi sihi* trees" *(Pradosia surinamensis)*.

9. This situation created even greater dependence for Davi Kopenawa, as he was a young son-in-law living in his wife's father's house (uxorilocal residence). The Achuar headmen (Ecuador) use the same strategy to transform young bilingual teachers into dependent sons-in-law (A. C. Taylor, 1981: 661).

Appendix D. The Haximu Massacre

1. "Haximu" is the Brazilian version of the Yanomami toponym $H^w axima\ u$, the "Great Tinamou River." Prior to the massacre, the inhabitants of the $H^w axima\ u$ River, a local group from the Rio Orinoco headwaters, consisted of eighty-five people divided between two communal houses.

Ethnobiological Glossary

Plant Species Cited in Portuguese

açaí   *Euterpe precatoria*, a palm *(mai masi)*
bacaba   *Oenocarpus bacaba*, a palm *(hoko si)*
buriti   *Mauritia flexuosa*, a palm *(rio kosi)*

Fish Species Cited in Portuguese

curimatã   *Prochilodus* spp. *(maxaka watima a)*
jaraqui   *Semaprochilodus* spp. *(kohipëma a)*
surubim   *Pseudoplatystoma fasciatum (kurito a)*
tambaqui   *Colossoma macropomum* (no Yanomami name)
tucunaré   *Cichla ocellaris (kahiki rapema a)*

Animal Species Cited in English

ACOUCHI
waxoro a   *Myoprocta acouchy*, the red acouchi

AGOUTI
t^homi a   *Dasyprocta* spp., several species of the rodent genus *Dasyprocta*

ARAÇARI TOUCAN
miremire koxi   *Pteroglossus aracari*, the black-necked araçari, and *P. pluricinctus*, the many-banded araçari

ARMADILLO
opo a   *Dasypus novemcinctus*, the nine-banded long-nosed armadillo

BAT
hewë pë   generic name

BEE
puu na ki   generic name

BLACK FLY
ukuxi pë   *Simulium* spp., small hematophagous gnats (river blindness vector)

BOA

hetu kiki   *Constrictor constrictor*, the boa constrictor

BUTTERFLY

xia axi pë   generic name

CAIMAN

iwa a   *Caiman sclerops*, the spectacled caiman

iwa aurima a (or *kõekõe a*)   *Paleosuchus trigonatus*, Schneider's dwarf caiman or smooth-fronted caiman

poapoa a   *Melanosuchus niger*, the black caiman (largest of the neotropical crocodilians)

CATERPILLAR

oxeoxea pë   generic name

CICADA

rõrõ kona pë   unidentified; a common species of large cicada

COATI

yarixi a   *Nasua nasua*, the South American coati

COCK-OF-THE-ROCK

ehama ona   *Rupicola rupicola*, the Guianan cock-of-the-rock bird

CURASSOW

paari a   *Crax alector*, the black curassow bird

DEER

haya a   *Mazama americana*, the red brocket deer

DOLPHIN

ehuma a   *Inia geoffrensis*, the pink river dolphin

DOVE

horeto a   *Leptotila verreauxi*, the white-tipped dove

EARTHWORM

horema kiki   generic name

ELECTRIC EEL

kawahi kiki   *Electrophorus electricus*

FLY

prõõ pë   generic name

GIANT ANTEATER

tëpë a   *Myrmecophaga tridactyla*

GIANT ARMADILLO

waka a   *Priodontes giganteus* or *Priodontes maximus*

GIANT OTTER

kana a   *Pteronura basiliensis*

GREAT TINAMOU

h^waxima a   *Tinamus major*, the great tinamou bird

GUAN

maraxi a   *Pipile pipile*, the Trinidad piping guan bird

kurema a   *Penelope jacquacu*, Spix's guan, and *Penelope marail*, the marail guan

HORSEFLY

potoma pë   unidentified; large yellow-colored horseflies of the Tabanidae family

JAGUAR

tihi a   *Panthera onca*

ira a   *Panthera onca* among the western Yanomami (or *Xamat^hari*)

KINKAJOU

hera a   *Potos flavus*

LIZARD

waima aka   unidentified; a small common lizard

MACAW PARROT

ara wakërima a   *Ara macao*, the scarlet macaw

ara hana   *Ara ararauna*, the blue-and-yellow macaw

MONKEY

iro a   *Alouatta seniculus*, the red howler monkey

kusi si   *Saimiri sciureus*, the common squirrel monkey

kuukuu moxi   *Aotus trivirgatus*, the northern owl monkey or three-striped night monkey

paxo a   *Ateles belzebuth*, the white-bellied spider monkey

wixa a   *Chiropotes satanas*, the black saki or brown-bearded saki

yarima a   *Cebus albifrons*, the white-fronted capuchin

yōkoxi a   *Callicebus torquatus*, the collared titi monkey or yellow-handed titi

OCELOT

yao si   *Felis pardalis*

PACA

amot^ha a   *Agouti paca*

PARROT

werehe a   *Amazona farinosa*, the mealy Amazon

kurukae si   *Amazona amazonica*, the orange-winged Amazon

kuatoma a   *Amazona ochrocephala*, the yellow-crowned Amazon

PECCARY

poxe a   *Tayassu tajacu*, the collared peccary

warë a   *Tayassu pecari*, the white-lipped peccary

PIRANHA

taki pë   generic name

PUMA

tihi wakërima a   *Puma concolor*, the cougar or puma

hōō a   a variety of large puma

SCORPION

sihi a   *Tityus bahiensis*, the Brazilian black scorpion

SLOTH

yawere a (or *ximi a*)   *Bradypus tridactylus*, the pale-throated three-toed sloth

yawere si   *Bradypus variegatus*, the brown-throated three-toed sloth

SNAKE

oru kiki   generic name

TADPOLE
piokōma uxi pë   generic name

TAPIR
xama a   *Tapirus terrestris*

TORTOISE
totori a   *Geochelone denticulata*, the yellow-footed tortoise or Brazilian giant tortoise

TOUCANS
mayōpa a   *Ramphastos tucanus*, the red-billed or white-throated toucan
kreōmari a   *Ramphastos vitellinus*, the channel-billed toucan

TRUMPETER BIRD
yāpi a   *Psophia crepitans*, the gray-winged trumpeter

TURTLE
pisa a   *Podocnemis unifilis*, the yellow-spotted river turtle or yellow-headed sideneck turtle

VULTURE
watupa a   *Coragyps atratus*, the black vulture
watupa aurima a   *Sarcoramphus papa*, the king vulture

WASP
kopena na ki   generic name

Plant and Animal Species Cited in Yanomami

The names of the *xapiri* spirit helpers corresponding to animals and plants mentioned below are generally formed by the addition of the suffix *-ri a* (plur. *-ri pë*), which denotes excess, monstrosity, or the supernatural. For example: the tapir, a game animal, is designated as *xama a*, whereas the image of its mythical ancestor is called *Xamari a* ("Tapir"). It is this mythical image *(utupë)*, infinitely multiplied, that the shamans will "call," "bring down," and "make dance" under the form of innumerable "tapir spirit helpers" *(xamari pë)*.

ahōrōma asi   unidentified; a very large black ant equipped with strong mandibles
ama hi   *Elizabetha leiogyne*, a tree whose bark, when reduced to ashes, is used as an ingredient in *yakōana* hallucinogenic snuff
amatʰa hi   *Duguetia lepidota*, a tree whose bark, when reduced to ashes, is used as an ingredient in *yakōana* hallucinogenic snuff
apia hi   *Micropholis melinoniana*, the white balata, a tree with edible fruits
apuru uhi   *Cedrelinga catenaeformis*, the cedrorana (false cedar), a large tree whose trunk is hollowed out to make ceremonial containers *(huu tihika)* and whose bark is used as an ingredient in fish poison
aputuma upë   unidentified; large, brown-colored edible caterpillars
ara usihi   *Croton matourensis*, a softwood tree whose strong, fibrous bark is used to make temporary containers or hammocks
aria si   *Xanthosoma* sp., taro or cocoyam, a plant grown for its starchy tubers
aro kohi   *Hymenaea parvifolia*, jatoba, a large tree with edible fruits and resin that has medicinal properties
aroari kiki   *Cyperus* sp., a cultivated sorcery plant supposed to cause high fever and loss of consciousness; also generic name of sorcery plants

aroaroma koxi   *Selenidera piperivora,* the Guianan toucanet bird

ata hi   *Sanchezia* sp., a small tree whose red flowers are used as ornaments by the women; also used in the past by the Yanomami to prepare a vegetable salt

atari hi   *Mouriri* sp., a shrub whose straight branches are used to fabricate the shafts of harpoon arrow points for hunting large birds

ayokora a   *Cacicus cela,* the yellow-rumped cacique bird

ërama tʰotʰo   *Uncaria guianensis,* the cat's claw vine, a large woody liana with hooklike thorns that possesses medicinal properties

ëri si   *Astrocaryum aculeatum,* the tucumã, a spiny palm tree with edible fruits

ëxama a   *Campephilus rubricollis,* the red-necked woodpecker

hai hi   *Pseudolmedia laevigata,* a tree with small red edible fruits

hapakara hi   *Bagassa guianensis,* the tatajuba, a large tree with edible fruits

hātākua mo   *Ortalis motmot,* the little chachalaca or variable chachalaca bird

hawari hi   *Bertholletia excelsa,* the Brazil nut tree, whose edible fruits have high oil content and very valuable proteins

hayakoari hana ki   *Justicia pectoralis,* a cultivated variety of sorcery plant to which an effect of shamanic revelation is attributed. Its victims are supposed to lose consciousness and run into the forest until exhausted while their "image" is carried away by the tapirlike supernatural being *(Hayakoari)* that gives its name to this plant.

hëima si   *Cotinga cayana,* the spangled cotinga bird

herama a   *Daptrius ater,* the black caracara falcon

himara amohi   *Theobroma bicolor,* the cacau-do-Perú, a wild relative of the cocoa tree with edible fruits

hōahōama a (female) or *wakoxo a* (male)   *Speothos venaticus venaticus,* the bush dog, a small Amazonian canine

hoari a   *Eira barbara,* the tayra, a large omnivorous gray-headed marten, very fond of honey

hoko mahi   *Licaria aurea,* a tree in the laurel family whose trunk is hollowed out in order to make large ceremonial containers *(huu tihika)*

hoko si (or *hoko masi*)   *Oenocarpus bacaba,* the bacaba palm, whose fruit pulp, rich in vegetable fat, is used to prepare a greatly appreciated juice

hokoto uhi   *Eschweilera coriacea,* a tree in the Brazil nut family whose bark is used in the making of hammocks

hopë a   *Coendou prehensilis,* the Brazilian porcupine, a small, semi-arboreal porcupine with nocturnal habits

hōra a   unidentified; a large beetle

hore kiki   *Maranta arundinacea L.,* the arrowroot, a sorcery plant supposed to turn one's enemies into cowards

horoma a   *Iriartella setigera,* the paxiubinha, a small palm tree whose stem is used to make tubes for inhaling *yãkoana* snuff

horomana hi   *Pouteria cladantha,* the abiu, a tree with edible fruits

hotorea kosihi   *Couratari guianesis,* the tauari, a very large tree in the Brazil nut family that was used by the Yanomami in the past to prepare a vegetable salt

hrāehrāema a   *Otophryne robusta,* a small orange-bellied frog endemic to the forest highlands

hutuma a   *Momotus momota,* the blue-crowned motmot bird

hutureama nakasi   *Capito niger,* the black-spotted barbet bird *hʷāihʷāiyama a*   *Lipaugus vociferans*, the screaming piha bird

hʷatʰupa a   *Bufo* sp., a species of very large toad

hʷëri kiki   generic name of sorcery plants

irokoma si   *Heliconia bihai*, the wild plantain or macaw flower, a large herbaceous plant whose leaves are used for wrapping or temporary thatch

ixaro a   *Cacicus haemorrhous*, the red-rumped cacique bird

ixoa hi   *Osteophloem platyspermum*, the ucuubarana, a tree endowed with medicinal properties

kahu usihi   *Cecropia sciadophylla*, a species of trumpet tree, with a hollow trunk and edible fruits

kana a   *Pteronura brasiliensis*, the giant otter, a powerful carnivore that can reach 2.2 meters in length

kāokāoma a   *Micrastur ruficollis*, the barred forest falcon, a small solitary bird of prey that perches in the entangled vegetation and whose loud cry resembles a barking sound

karihirima kiki   *Bothrops jararaca*, the jararaca, a venomous pit viper responsible for the majority of the serious snakebite accidents among the Yanomami

kaxa pë   edible caterpillars of a butterfly from the Brassolinae subfamily

kaxi pë   *Solenopsis* sp., minuscule stinging red ants known as "fire ants." Their sting contains a moderately toxic alkaloid, *piperidine* (the same as that found in pepper), which is especially irritating to the mucous membranes, but which also attacks the temperature receptors of the skin, resulting in a burning sensation

koa axihana ki   *Clibadium sylvestre*, leaves of a cultivated plant used as a fish poison

kōanari si   *Oenocarpus bataua*, the patawa palm, whose fruit pulp, rich in vegetable fat, is used to prepare a greatly appreciated juice

koikoiyoma a   *Herpetotheres cachinnans*, the laughing falcon

komatima hi   *Peltogyne gracilipes*, the purple heart, a large tree with very hard wood

kona pë   unidentified; small black ants

kōōkata mo (or *rākohi a*)   *Aramides cajanea*, the gray-necked wood rail bird

kopari a   *Ibycter americanus*, the red-throated caracara falcon

kopena na ki   generic name for wasps

kori a   *Psarocolius decumanus*, the crested oropendola bird

kōromari a   *Mesembrinibis cayennensis*, the green ibis bird

kotopori usihi   *Croton palanostigma*, a softwood tree whose resistant bark is used to make temporary containers

koxoro na ki   *Trigona* cf. *dallatorreana*, orange bees known for the height of their flight

koyo pë   *Atta sexdens*, a species of leafcutter ant

kraya kiki   unidentified; large, yellow, edible caterpillars with red spots and venomous hairs. They generally colonize the *Vochysia ferruginea* trees that bear their name, *kraya nahi*

kree mo pë or *krëëkrëëma pë*   unidentified; a species of cicada

krepu uhi   *Inga edulis*, a species of ice-cream bean tree, a tree with greatly appreciated sweet fruits

krouma a   *Hyla boans*, the gladiator frog, and *H. geographica*, the map tree frog, species of small arboreal frog

kumi tʰotʰo   *Securidaca diversifolia*, a large woody vine with pink or purple flowers

kurira na ki   unidentified; small gray paper wasps, considered very aggressive

kurito a   *Pseudoplatystoma fasciatum*, the tiger shovelnose catfish or barred sorubim

kusārā si   unidentified; a black sparrow that feeds in noisy flocks in the tree tops. The Yanomami say that these birds tear the food frenetically from one another, "as dogs do"

kute mo pë   unidentified; a species of cicada

kuxixima a   unidentified; a very small brown bird that can generally be seen near the riverbanks

mahekoma hi   *Piper aduncum*, a wild species of pepper plant with medicinal properties, also used in the making of curare

maihiteriama a   *Colonia colonus*, the long-tailed tyrant bird

maika a   unidentified; a species of beetle

mai kohi   *Symphonia globulifera*, the mani, a tree whose resin is used to coat bow strings

mai masi   *Euterpe precatoria*, the tall assai palm, whose fruit pulp, rich in vegetable oil and antioxidants, is used to prepare a greatly appreciated juice

maka watixima a   *Thamnomanes caesius*, the cinereous antshrike bird

makina hi   *Endopleura uchi*, the uxi, a tree with edible fruits

makoa hi   *Talisia chartacea*, a tree with edible fruits

makoa hu   *Myrmornis torquata*, the wing-banded antbird

manaka ki   *Alstroemeria* sp., a type of cultivated lily used in sorcery; supposed to make women sterile

manaka si   *Socratea exhorriza*, the walking palm or paxiuba, a palm tree with high, thorny stilt roots. Its trunk is cut into small boards to make, among other things, the external walls of collective houses or platforms for diverse uses

maraxi a   *Pipile cumanensis*, the blue-throated piping guan bird

mani hi   *Guarea guidonia*, the American muskwood, a tree whose white flowers are popular among women as ear ornaments

marokoaxirioma a   *Ramphocelus carbo*, the silver-beaked tanager that often nests in cultivated *Bactris gasipaes* palms

masi kiki   *Heteropsis flexuosa*, the titica vines, used for basketwork and for binding the beams of houses

masihanari kohi   *Tabebuia capitata*, a very large ipê tree with very hard wood

maxara hana ki   *Justicia pectoralis* var. *stenophylla*, a cultivated plant whose fragrant leaves, when dried and ground, are used as an ingredient in *yākoana* hallucinogenic snuff

maxopoma na ki   *Melipona* sp., gray bees

maya kiki   unidentified; large, orange, edible caterpillars with stinging black hairs

mohuma a   *Harpia harpyja*, the harpy eagle, a very large forest eagle with a wingspan of two meters

momo hi   *Micrandra rossiana*, a tree of the highlands whose toxic fruits, in order to be made edible, must be roasted (to open their shell), dried, and soaked at length in the river

mora mahi   *Dacryodes peruviana*, the copalwood, a tree with edible fruits

moxa pë   generic term for maggots, small white Diptera larvae

nāi hi   *Micropholis* sp., a tree of the highlands with sweet, edible flowers

napore a   *Psarocolius viridis*, the green oropendola bird, and *P. yuracares*, the olive oropendola *nara xihi*   *Bixa orellana*, the annatto shrub, whose fruits contain a red pulp used as a dye

õema ahi   *Pourouma bicolor* spp. *digitata*, a species of trumpet tree, with edible fruits and leaves that serve as abrasives

õi na ki   *Trigona* sp., aggressive black bees

okarasi si   *Attalea maripa*, the inaja palm or American oil palm, whose fruits, rich in vegetable oils, are greatly appreciated

okoraxi hi   *Rinorea lindeniana*, a small hardwood tree used to make arrow nocks

oko xi ki   *Cyperus* sp., a cultivated plant used in women's sorcery, supposed to cause an intense fever associated with a state of jaundice in its male victims

õkraheama a   *Piaya cayana*, the squirrel cuckoo bird

operema axihi   *Couma macrocarpa*, the milk tree or sorva, a tree with edible fruits and a sticky sap

oruxi hi   *Anacardium giganteum*, the wild cashew tree

paa hana ki   *Geonoma baculifera*, a miniature palm of the humid undergrowth whose leaves are preferred for thatch

paara hi   *Anadenanthera peregrina*, the yopo tree, whose seeds serve to make a powerful hallucinogenic snuff *(paara a)* containing tryptamines and bufotenin

paari a   *Crax alector*, the black curassow bird

pahi hi   *Inga acreana*, a species of ice-cream bean tree, whose sweet fruits are greatly appreciated

paho a   unidentified; a small arboreal and nocturnal rodent of the Cricetidae family often seen in the roofs of communal houses

parapara hi   *Phyllanthus brasiliensis*, a cultivated plant whose leaves are used as fish poison

pari na ki   *Trigona* sp.; aggressive yellow bees

piri kona pë   *Azteca chartifex*, ants whose colonies occur in clusters of large, pendant cartonlike nests

pirima ahu t^{h}ot^{h}o   *Spondias mombin*, the mombin or hog plum tree, which has edible fruits

pirima ãrixi pë   unidentified; minuscule small red ticks

pirima hi ki   *Andropogon bicornis*, the West Indian foxtail, a kind of tall grass whose stems are used to make the small sticks that women wear at the corners of their lips and under their lower lips

piomari namo   *Cyanocorax violaceus*, the violaceous jay bird.

pokara a   *Odontophorus gujanensis*, the marbled wood quail bird

pooko hi   *Inga sarmentosa*, a species of ice-cream bean tree, whose sweet fruits are greatly appreciated

poopoma a   *Myrmothera campanisona*, the thrushlike antpitta, a bird frequently found in the tangled secondary vegetation

pora axi t^{h}ot^{h}o   *Posadaea sphaerocarpa*, a vine in the cucumber family from whose fruits are made small egg-shaped gourds

pore hi   *Eugenia flavescens*, a hardwood tree with edible fruits whose peeling bark constantly renews itself

poroa unahi   *Theobroma cacao*, the wild cocoa tree

poxe a   *Tayassu tajacu*, the collared peccary

prooroma koko   unidentified; a species of toad

proro a   *Lontra longicaudis*, the neotropical otter

purunama usi   *Olyra latifolia*, a thin, herbaceous bamboo used to make small flutes

purupuru namo   unidentified; a species of small monkey, sometimes confused with
   yõkoxi a, *Callicebus torquatus*; may be another species of the genus *Callicebus*

puu axi na ki   *Scaptotrigona* sp., small bees

puu hana ki   *Justicia* sp., a wild herb whose fragant leaves (literally "honey leaves") are
   used as a very popular ornament for women, who wear them in bouquets fixed in
   their cotton armbands

puu xapiri na ki   unidentified; black bees

raema si ki   unidentified; large brown edible caterpillars

rahaka pë   *Guadua* spp., a genus of bamboo used to make the lance-shaped arrow
   points of the same name for hunting large game

rai natihi   *Anaxagorea acuminata*, a tree whose fibrous bark is used to make head
   carrying straps

rapa hi   *Martiodendron* sp., an imposing tree with a resistant wood and smooth bark

rasa si   *Bactris gasipaes*, the peach palm, a cultivated palm with highly appreciated
   fruits, but whose harvest is rendered difficult by its very thorny trunk

remoremo moxi   *Centris* sp., large solitary bees frequently seen on the sand of
   riverbanks

repoma na ki   *Trigona* sp., yellow bees that nest underground

rio kosi   *Mauritia flexuosa*, the buriti palm, whose fruits, rich in vegetable fat, are
   greatly appreciated

roha a   *Gonatodes humeralis*, the bridled forest gecko, a species of small arboreal lizard

rõrõ kona pë   unidentified; a species of large cicada

ruapa hi   *Caryocar villosum*, the Piquiá, a large tree whose fruits possess a thick and
   very fatty edible mesocarp that is particularly appreciated

ruru asi   *Phenakospermum guianense*, the false banana, a tall plant whose fruits contain
   edible seeds and whose broad, fanlike leaves serve to cover improvised forest
   shelters

ruru hi   *Rhodostemonodaphne grandis*, a tree in the laurel family, whose trunk is
   hollowed out in order to make large ceremonial containers *(huu tihika)*

sei si ki   generic name of a set of small, brightly colored birds of different genera whose
   feathers are very popular in male ornaments: *Atlapetes* sp., *Cyanocompsa* sp.,
   Ammodramus sp., *Hylophilus* sp., *Cyanerpes* sp., *Dacnis* sp., *Diglossa* sp., *Tangara* sp.,
   Thraupis sp.

siekekema a   *Hypocnemis cantator*, the Guianan warbling antbird

simotori a   *Titanus giganteus*, the titan beetle, a xylophagous beetle with large
   mandibles that can reach over fifteen centimeters in length

sitipari si   *Saltator maximus*, the buff-throated saltator bird

tãĩtãima pë   unidentified; a species of cicada

tãrakoma a   *Pipra* sp., a species of manakin bird

taritari axi   *Euphonia xanthogaster*, the orange-bellied euphonia bird

teateama a   *Gampsonyx swainsonii*, the pearl kite

tʰooroma asi   another name for the red-necked woodpecker, *ëxama a*, *Campephilus
   rubricollis*

tʰora a   *Guadua latifolia*, a species of large bamboo from which Yanomami hunters
   make the quivers to keep their arrow points

tima na ki   *Trigona* sp., black bees that nest in the roots of trees

tokori a   *Cecropia peltata* and *C. obtusa*, species of trumpet tree abundant in natural clearings (treefalls) and ancient garden locations

tooro a   male of *Bufo marinus*, the cane or giant toad, a species of large terrestrial toad whose name comes from the onomatopeia of its loud nocturnal call, noticeable during the rainy season. Its parotid glands produce a toxic milky secretion. See *yoyo a*

ukuxi pë   black flies, small hematophagous gnats (river blindness vector)

wāha aki   *Dioscorea trifida*, edible tubers of the cush-cush yam

waima aka   unidentified; a small lizard commonly found in gardens

waka moxi ki   *Cyperus* sp., a cultivated plant used in sorcery, supposed to provoke a violent convulsive state

wakoa a   *Leucopternis melanops*, the black-faced hawk

wakopo na ki   *Trigona* sp., very aggressive yellow bees

wapo kohi   *Clathrotropis macrocarpa*, the aromata or cabari tree, whose toxic fruits are rendered edible by a long succession of boiling and soaking in the river

warama aka   *Megalobulimus oblongus*, a large forest snail

warapa kohi   *Protium* spp., a genus of tree whose flammable resin, also endowed with medicinal properties, is used for various purposes

warea koxiki   *Lycosa* sp., a kind of large, venomous tarantula

wari mahi   *Ceiba pentandra*, the kapok tree, a very tall tree with huge buttress roots whose soft wood trunk is hollowed out to make large ceremonial containers *(huu tihika)*

waroma kiki   *Corallus caninus*, the emerald tree boa, a large, nonvenomous, nocturnal and arboreal boa species with highly developed front teeth

wāsikara a   *Tupinambis teguixin*, the black or golden tegu, a very large terrestrial lizard (eighty centimeters to one meter in length)

watupa aurima a   *Sarcoramphus papa*, the king vulture

wayapaxi a   *Sciurus igniventris*, the northern Amazon red squirrel

wayawaya apë   unidentified; edible dark brown caterpillars

wayohoma a   *Nyctibius griseus*, the common potoo, a kind of nightjar bird

werehe a   *Amazona farinosa*, the mealy Amazon parrot

weri nahi   *Posoqueria latifolia*, the tree jasmine, whose white flowers are very popular among women as ear ornaments

weto mo   *Primolius maracana*, the blue-winged macaw parrot

wisawisama si   *Tangara chilensis*, the paradise tanager bird

witiwitima namo   *Elanoides forficatus*, the swallow-tailed kite bird

xaki na ki   *Trigona amalthea*, very aggressive black bees whose honey is not considered edible

xapo kiki   *Cyperus* sp., a cultivated plant used in sorcery, supposed to make women sterile

xaraka ahi   *Manilkara huberi*, the massaranduba or balata, a tree with edible fruits and very resistant wood

xaraka si   *Gynerium sagittatum*, the wild cane or bitter cane, a cultivated tall grass up to five meters in length from which Yanomami hunters make their arrows

xiho a   *Paraponera clavata*, the bullet ant, a large venomous black ant with an extremely painful sting

xiri na pë   unidentified; orange ants that travel in very large groups *xiroxiro a*   *Cypseloides* sp., a species of swift

xitopari hi   *Jacaranda copaia*, the jacaranda or caroba, a large tree with mauve flowers and resistant wood. The smoke of its burnt leaves is used to ward off mosquitoes in the forest

xōa a   *Caladium bicolor*, a cultivated tuberous aroid used by women for love magic

xoapema a   *Cymbilaimus lineatus*, the fasciated antshrike bird

xopa hi   *Helicostylis tomentosa*, the amora preta, a tree with edible fruits

xotokoma a   *Trogon melanurus*, the black-tailed trogon bird, and *T. collaris*, the collared trogon

xotʰetʰema a   *Piculus chrysochloros*, the golden-green woodpecker bird

xoo mosi   *Astrocaryum gynacanthum*, the mumbaca palm, with a thorny trunk and edible orange fruits

xuwāri na ki   *Stelopolybia* sp., large black wasps

yākoana hi or *yākoana a*   *Virola elongata*, a tree whose resin is used by shamans to prepare their hallucinogenic snuff *(yākoana a)* and whose principal active ingredient is dimethyltryptamine

yāma asi ki   *Ananas* sp., a cultivated relative of the pineapple from whose leaves is drawn a strong fiber to make strings

yamanama na ki   *Scaptotrigona* sp., small black bees

yamara aka   *Potamotrygon* sp., a freshwater stingray

yaraka asi pë   generic name for various species of small freshwater tropical fish of the Characidae family

yaremaxi hi   *Brosimum utile*, the cow tree or amapá-doce, a tree whose bark is beaten to make baby-carrying slings

yaro xi ki   generic name for hunting charms made from the small tubers of cultivated plants of the *Cyperus* genus

yawara hi   *Micropholis* sp., the curupixá, a tree with edible fruits

yipi hi   *Sorocea muriculata* ssp. *uaupensis*, literally the "menarche tree," a shrub with crimson peduncles

yoi si   probably *Attalea excelsa*, the urucuri palm, which can reach a height of thirty meters and whose fruits contain edible seeds

yōkihima usi   *Dendrocincla fuliginosa*, the plain-brown woodcreeper bird

yopo una ki   *Asplundia* sp., an epiphytic plant from whose ashes the Yanomami used to prepare a vegetable salt

yōriama a   *Crypturellus soui*, the little tinamou bird

yōrixiama a   *Turdus fumigatus*, the cocoa thrush bird

yoropori a   *Maduca sexta*, the tobacco hornworm; large green caterpillar of the sphinx moth; a defoliator of tobacco leaves and a nicotine-resistant insect

yoyo a   female of *Bufo marinus*, the cane or giant toad, significantly larger than males (see *tooro a*)

Geographic Glossary

Toponyms Cited in Portuguese

Ajarani, Rio   tributary on the right bank of the Rio Branco, whose upper course spreads out in the proximity of the right bank of the Rio Catrimani (in the region of the Catrimani Catholic mission)

Ajuricaba Outpost   FUNAI outpost situated on the middle Rio Demini. This outpost was opened by the Indian Protection Service (SPI) between 1941 and 1942 on the Rio Demini (near the Auatsinaua rapids and Genipapo village) on the occasion of the first visit by the Brazilian Boundary Commission to the region. It then had other locations: at the mouth of the Rio Toototobi (1943) and on the Rio Mapulaú (1947), before being practically abandoned and then reestablished on the Rio Demini in 1957.

Ananaliú, Rio   tributary on the left bank of the upper Rio Demini that borders the northern Serra do Demini range

Apiaú, Rio   tributary on the right bank of the lower Rio Mucajaí

Aracá, Rio   tributary on the right bank of the lower Rio Demini

Barcelos   small town situated on the right bank of the Rio Negro upstream of Manaus, facing the mouth of the Rio Demini. In 1972, Barcelos had about 15,000 inhabitants. It now has 25,715 (2010).

Boa Vista   capital of the state of Roraima, situated in the extreme north of Brazil. The town had approximately 5,200 inhabitants in the 1950s and 290,741 inhabitants in 2011.

Cachoeira dos Índios   waterfall of the lower Rio Aracá

Caracaraí   small town situated about 155 kilometers south of Boa Vista, along the road leading to Manaus, whose population increased from 2,200 inhabitants in 1976 to 18,384 people in 2010

Casiquiare, Rio   a natural navigable channel of 200 kilometers linking the Rio Orinoco to the basin of the Rio Negro

Castanho (Igarapé)   small tributary on the left bank of the lower Rio Catrimani (downstream of the Piranteira waterfall)

Catrimani, Rio   large tributary of the lower Rio Branco

Catrimani Mission   Catholic missionary outpost opened on the river of the same name by Italian Fathers of the Order of the Consolata (Turin) in 1965

Cauaboris, Rio   tributary on the left bank of the upper Rio Negro, which has its source
    in the Serra Imeri range
Couto de Magalhães, Rio   tributary on the right bank of the upper Rio Mucajai
Cutaíba, Rio   tributary on the right bank of the upper Rio Uraricoera, running south of
    the Serra Uafaranda range
Demini Outpost   FUNAI outpost created in 1977 on a former construction site of the
    Perimetral Norte highway, which had been abandoned the previous year. The
    Yanomami led by Davi Kopenawa's father-in-law began contact with this outpost in
    1978.
Demini, Rio   large tributary on the left bank of the Rio Negro
Iauaretê   village established around a Catholic mission founded in 1929 by the
    Salesian Fathers at the confluence of the Rio Uaupés and Papuri, in the region of the
    upper Rio Negro, near the border with Colombia. A FUNAI outpost was opened
    there in 1974.
Jundiá, Rio   tributary on the right bank of the upper Rio Catrimani
Jutaí, Rio   tributary on the right bank of the Rio Demini; downstream of the mouth of
    the Rio Toototobi
Lobo d'Almada, Rio   tributary on the right bank of the upper Rio Catrimani
Maiá, Rio   tributary on the left bank of the Rio Cauaboris
Manaus   capital of the state of Amazonas, a city of 1,802,525 inhabitants (2010)
    situated 758 kilometers south of Boa Vista, capital of the state of Roraima. The
    population of Manaus was 279,151 inhabitants in 1950 and 343,038 in 1960.
Mapulaú, Rio   tributary on the left bank of the upper Rio Demini; upstream of the Rio
    Ananaliú
Mapulaú Outpost   temporary FUNAI outpost opened in 1974 downstream of the river
    of the same name; abandoned in 1976, and then burned down by the Yanomami in
    1977.
Mucajaí, Rio   large tributary on the right bank of the Rio Branco
Novo, Rio   tributary of the upper Rio Apiaú
Paapiú Outpost   FUNAI outpost established in 1981 on the Rio Couto de Magalhães
Padauiri, Rio   tributary on the left bank of the Rio Negro
Parima, Rio   one of the two major tributaries of the upper Rio Uraricoera
São Gabriel da Cachoeira   town of 37,300 inhabitants (2010) situated in the region of
    the upper Rio Negro, 860 kilometers from Manaus, in the northwest of the state of
    Amazonas. The population of São Gabriel was 13,420 people in 1970.
Serra do Melo Nunes   mountain range situated between the upper Mucajaí, Parima,
    and Uraricoera rivers
Serra Parima   mountain range that constitutes the watershed between the upper Rio
    Orinoco and the upper Rio Parima
Serra dos Porcos   mountain range situated south of the village of Iauaretê, in the basin
    of the upper Rio Negro
Siapa, Rio   tributary of the Casiquiare channel in Venezuela
Surucucus Outpost   FUNAI outpost founded in 1976 in the region of the upper Rio
    Parima on a plateau of about a thousand meters in altitude (the Serra das
    Surucucus)
Taraú, Rio   tributary on the right bank of the upper Rio Demini
Toototobi, Rio   tributary on the left bank of the upper Rio Demini

Toototobi Mission   former missionary outpost of the American evangelical organization the New Tribes Mission (NTM), founded between 1960 and 1963 on the upper
Rio Toototobi, then abandoned in 1991. The site is currently occupied by a health
outpost of the Brazilian Secretary for Indigenous Health (SESAI).

Uraricaá, Rio   tributary on the left bank of the middle Rio Uraricoera

Uraricoera, Rio   one of the two major tributaries of the Rio Branco

Ethnonyms Cited in Portuguese

Apurinã   Arawak-speaking group of over 4,000 people who live in the basin of the Rio
Purus, one of the major tributaries on the right bank of the upper Amazon (Rio
Solimões). The Apurinã began to have intense contact with white people (rubber
tappers) at the end of the nineteenth century.

Bahuana   early Arawak-speaking group of the Rio Demini, now extinct

Kayapó   Gê-speaking group whose vast territory (with a surface area equivalent to that
of Austria) is situated on the plateau of central Brazil, in the basin of the Rio Xingu.
The various present-day Kayapó subgroups total more than 6,000 people. They
began to accept peaceful contact with the white people in the 1950s.

Krenak   Macro-Gê-speaking group of just over 200 people situated on the Rio Doce, in
the state of Minas Gerais, descendants of the "Botocudos," who have been
systematically exterminated since the Portuguese colonial period

Maku   hunter-gatherers who speak an isolated language and live in the interfluvial
forest between Rio Tiquié and Rio Papuri (tributaries of the Uaupés, upper Rio
Negro). Their population is about 2,600 in Brazil and approximately 700 in
Colombia. The Maku's inaccessible habitat somewhat protected them from the
historic contacts that affected the riverine peoples of the upper Rio Negro, like the
Tukano.

Makuxi   Carib-speaking group of more than 23,400 people in Brazil and 9,500 in
Guyana who occupy the savannas of the eastern part of the state of Roraima. The
Makuxi have been in contact with white people (slave hunters and colonists) since
the mid-eighteenth century during the Portuguese colonial penetration of the Rio
Branco basin.

Munduruku   Tupi-speaking group of about 10,000 people who today occupy
discontinuous territories found mainly in the basin of the Rio Tapajós, one of the
major tributaries on the right bank of the Amazon. After a long warring period, the
Munduruku came into peaceful contact with white people at the end of the eighteenth century.

Pauxiana   early Carib-speaking group, extinct since the beginning of the twentieth
century, whose territory extended from the mid Rio Catrimani to the lower Rio
Mucajaí

Sateré-Mawé   Tupi-speaking group with a population of almost 9,200 people living in
the middle Amazon river region, at the border of the states of Amazonas and Pará.
The Sateré-Mawé had their first contacts with Jesuit missionaries in the late
seventeenth century.

Suruí (Paiter)   Tupi-Mondé-speaking group of about 1,000 people situated in the basin
of the Rio Branco (tributary of the Rio Roosevelt, basin of the Rio Madeira), in the
state of Rondônia. After a history of violent contacts with the regional economic

frontier since the end of the nineteenth century, the first official contacts of the
Suruí with the FUNAI "pacification" teams date from 1969.

Taurepang (Pemon)   Carib-speaking group of almost 600 people in Brazil and over
20,600 in Venezuela. Inhabitants of the savannas of the Rio Branco, like the
Makuxi and the Wapixana, they had their first contacts with white people in the mid-
eighteenth century.

Tariano   group of about 1,900 people in Brazil (and 200 in Colombia) who originally
spoke Arawak but later adopted the eastern Tukano language. The Tariano live in the
basin of the upper Rio Negro (middle Uaupés, lower Papuri, and upper Iauiari). The
colonial Portuguese slave hunters penetrated this region during the mid-eighteenth
century. Then it was dominated by the Franciscan and Salesian missionaries from
the end of the nineteenth century.

Tikuna   largest indigenous group of the Brazilian Amazon, whose tonal language is
considered an isolate. There are 35,000 Tikuna in Brazil, where they occupy various
tributaries on the left bank of the upper Solimões. (There are 4,200 in Peru and just
over 4,500 in Colombia.) The group had its first contacts with white people through
Spanish Jesuit missionaries at the end of the seventeenth century.

Tukano   most important group of the eastern Tukano linguistic family, with about
6,240 people in Brazil (and 6,330 in Colombia). They occupy mainly the Rio Tiquié,
Papuri, and Uaupés (upper Rio Negro). Like the Tariano, the early Tukano dealt with
Portuguese slave hunters from the mid-eighteenth century, then with Catholic
missionaries from the end of the nineteenth century.

Waiãpi   Tupi-Guarani–speaking group that numbers about 900 people in Brazil (and
just over 400 in French Guiana), where it is concentrated in the basin of the Rio
Amapari (tributary of the Rio Araguari, state of Amapá). The Waiãpi had their first
official contacts with white people through FUNAI teams in 1973 during the
opening of the Perimetral Norte highway.

Waimiri-Atroari   Carib-speaking group situated on the left bank of the lower Rio Negro
(basin of the Rio Jauaperi and Rio Camanaú) with about 1,120 people. After a long
history of violent conflicts with the regional population in search of forest products,
the territory of the Waimiri-Atroari was cut, between 1972 and 1977, by the road
linking Manaus to Boa Vista. After a dramatic period of "pacification" (that almost
decimated them) by FUNAI and the military in charge of opening the road, this
group has, since the 1980s, experienced a rapid growth in population.

Wapixana   Arawak-speaking group of 7,000 people in Brazil (and 6,000 in Guyana)
living in the savannas of the eastern part of the state of Roraima, notably the region
of the Serra da Lua, between the Rio Branco and Rio Tacutu, as well as on the lower
Rio Uraricoera. The Wapixana, like the Makuxi, their Carib neighbors, have been in
contact with white people since the mid-eighteenth century.

Warekena   In contact with white people since the beginning of the eighteenth century,
this originally Arawak-speaking group now mainly speaks *Nheengatu*, a lingua
franca based on Tupi-Guarani spread by Carmelite missionaries in the colonial
period. The Warekena number around 800 people in Brazil (and almost 500 in
Venezuela). They live on the Rio Xié, a tributary of the upper Rio Negro.

Xikrin   Kayapó subgroup (Gê linguistic family) of just over 1,340 people who occupy
the basin of the Rio Itacaiúnas (tributary on the left bank of the lower Tocantins) and
Rio Bacajá (tributary on the right bank of the Xingu) in the state of Pará. The first
contacts of the Xikrin with the Indian Protection Service date from the 1950s.

Ye'kuana   Carib-speaking group, northern neighbors of the Yanomami, with almost
450 people in Brazil (and more than 4,800 in Venezuela), situated primarily on the
Rio Auaris (a main tributary of the Rio Uraricoera). The history of Ye'kuana contacts
with white people began with the Spanish colonization in the eighteenth century,
followed by the brutal intrusion of the rubber tappers at the beginning of the
twentieth century.

Sources: *Online Encyclopedia Povos Indígenas no Brasil*—Instituto Socioambiental,
São Paulo: http://pib.socioambiental.org/en

Toponyms and Ethnonyms Cited in Yanomami

Amat^ha u   small tributary on the right bank of the headwaters of the Rio Orinoco; site
occupied by the ancestors of the Rio Toototobi Yanomami

Amikoapë t^hëri   group situated on the upper Rio Mucajaí in the first decades of the
twentieth century

Arahai t^hëri   ancestral group of the present inhabitants of the Rio Catrimani that
occupied a site at the headwaters of the Rio Mucajaí at the beginning of the
twentieth century

Ariwaa t^hëri   Xamat^hari (western Yanomami) group who lived in the region of the
headwaters of the Rio Demini in the 1960s and known, later on, by the name
H^wayasiki t^hëri (region of the *Parawa u* River)

Hapakara hi   site occupied in the early 1970s by the group led by Davi Kopenawa's
father-in-law on the upper Rio Lobo d'Almada

Haranari u   Rio Ananaliú

Hayowa t^hëri   Xamat^hari group situated between the upper Rio Siapa and the upper Rio
Orinoco at the beginning of the twentieth century

Hayowari   hill situated at the headwaters of the Rio Orinoco and the Rio Parima.
Mythical location of the transformation of a group of Yanomami ancestors, the
Hayowari t^hëri, who gave origin to white people

Hero u   Rio Couto de Magalhães, tributary on the right bank of the upper Rio Mucajaí

Hewë nahipi   site opened (after the abandonment of *Makuta asihipi*) in 1976 on the
lower Rio Jundiá, tributary on the right bank of the Rio Catrimani

H^wara u   upper course of the Rio Orinoco

H^waxi t^hëri   enemy group of the people of the upper Rio Toototobi at the beginning of
the 1950s, situated in the highlands of the Rio Orinoco and Rio Parima headwaters

H^waxima u   small tributary on the right bank of the headwaters of the Rio Orinoco

H^waya u   site occupied during the the the early 1970s on the river of the same name,
tributary on the right bank of the middle Rio Lobo d'Almada

Iwahikaropë t^hëri   Xamat^hari group of the upper Rio Padauiri, tributary on the left bank
of the Rio Negro in Brazil

Kapirota u   Rio Jutaí

Kaxipi u   Rio Jundiá

Kõana u   small tributary on the left bank of the headwaters of the Rio Orinoco,
previously occupied by the ancestors of the Rio Toototobi Yanomami

Kokoi u   Rio Demini

Konapuma t^hëri   Xamat^hari group of the upper Rio Siapa in Venezuela

Maharu u   Rio Mapulaú

Mai koxi (or *Mai koko*)   designation attributed in the past to the inhabitants of the Rio Catrimani by their enemies living on the Rio Toototobi

Maima siki u   small tributary on the left bank of the Rio Mapulaú (upstream of the *Werihi sihipi u*)

Mait^ha   extinct indigenous group with which the early Yanomami were in contact until the first decades of the twentieth century in the region of the *Takaimaki* mountain (see below) and from whom they obtained metal tools

Makuta asihipi   site occupied during the first half of the 1970s on the Rio Jundiá, tributary of the Rio Catrimani

Mani hipi   site occupied during the first half of the 1970s and located between the upper Rio Jundiá and the middle Rio Lobo d'Almada, another tributary of the upper Rio Catrimani

Manito u   small tributary on the left bank of the headwaters of the Rio Orinoco occupied by the ancestors of the Rio Toototobi Yanomami

Marakana   site inhabited in the 1950s by the group of Davi Kopenawa's relatives. The collective house of *Marakana* was visited by an expedition of SPI and the missionaries of the New Tribes Mission in June 1958. One of these missionaries, J. McKnight, describes it as an oval structure of 58 meters by 41 meters, inhabited by some 200 people (McKnight, 1958: 10).

Moxi hatëtëma   isolated Yanomami of the Rio Apiaú headwaters. See also *Yawari*.

Mõra mahi araopë   site located on the upper course of the Rio Toototobi and occupied by the forefathers of Davi Kopenawa between 1930 and 1940

Opiki t^hëri   group inhabiting the Rio Catrimani region in the 1970s

Parahori   term used by the eastern Yanomami for a group of highland Yanomami living on the left bank of the upper Rio Parima

Parawa u   upper course of the Rio Demini

Puu t^ha u   Rio Cutaíba

Sina t^ha   site located on the upper course of the Rio Toototobi and occupied by the close allies of the group of Davi Kopenawa's forefathers (see *Mõra mahi araopë* and *Yoyo roopë*) at the end of the 1940s

Takai maki   the Serra do Melo Nunes range

Tëpëxina hiopë t^hëri   group formerly situated fifteen kilometers north of the FUNAI outpost of the Serra das Surucucus (upper Rio Parima)

T^hoot^hot^hopi   site on the upper Rio Toototobi occupied in the 1930s by a group that has since disappeared (the *Xihopi t^hëri*) and reoccupied in the early 1960s by Davi Kopenawa's parents' community. This site was then chosen by the New Tribes Mission for one of their missionary outposts (1963–1991). This name was transformed by the white people to "Toototobi" (Toototobi Mission and Rio Toototobi).

Uxi u   Rio Lobo d'Almada and name of a site midstream that was occupied during the 1970s

Waika   designation for the eastern Yanomami by the western Yanomami *(Xamat^hari)*

Waka t^ha u   Rio Catrimani

Waka t^ha u t^hëri   group of the upper Rio Catrimani situated next to the Catrimani Catholic mission

Wanapi u   tributary on the left bank of the upper Demini

Warëpi u   tributary on the left bank of the Rio Cunha Vilar *(Paxoto u)*, whose upper course, parallel to that of the upper Rio Toototobi, was occupied in the 1950s and 1960s by a group allied to the forefathers of Davi Kopenawa

Wari mahi   site on the upper Rio Toototobi, occupied by the relatives of Davi Kopenawa at the beginning of the 1960s (following that of *Marakana*)

Watata si   extinct indigenous group with which the early Yanomami were in contact until the first decades of the twentieth century in the region of the upper Rio Parima and from whom they obtained metal tools

Watoriki   the Serra do Demini range situated on the left bank of the upper Rio Demini, between the Rio Ananaliú and Rio Filafilaú. The Demini FUNAI outpost was created in 1977 at the foot of this rock mountain.

Wawanawë tʰëri   *Xamatʰari* group situated on the Rio Cauaboris and Rio Maiá (the latter being a tributary of the former)

Weerei kiki   rocky peak belonging to the Serra do Demini range, situated between the upper courses of the Rio Ananaliú and the Xeriana stream

Werihi sihipi   small tributary on the right bank of the Rio Mapulaú (downstream of the *Maima siki u*)

Weyahana u   Rio Toototobi

Weyuku tʰëri   *Xamatʰari* group which, in the 1960s, lived on a tributary of the Rio Taraú (itself a tributary on the right bank of the upper course of the Rio Demini)

Xamatʰari   designation for the western Yanomami by the eastern Yanomami *(Waika)*

Xama xi pora   large waterfall of the upper Rio Parima

Xiriana   designation for the *Ninam (Yanam)*-speaking Yanomami of Rio Mucajai and Rio Uraricaá by the eastern Yanomami

Yawari   designation for the Yanomami of Rio Ajarani and Rio Apiaú by the eastern Yanomami

Yoyo roopë   site located at the headwaters of the Rio Toototobi and occupied by the forefathers of Davi Kopenawa in the years 1930–1940

References

Ação pela Cidadania (APC). 1989. *Roraima, o aviso de morte. Relatório sobre a viagem da Comissão da Ação pela Cidadania ao Estado de Roraima entre 9 e 12 de junho de 1989.* São Paulo: Comissão pela Criação do Parque Yanomami.

———. 1990. *Yanomami: A todos os povos da terra. Segundo Relatório da Ação pela Cidadania sobre o caso Yanomami, referente à acontecimentos do período junho de 1989 a maio de 1990.* São Paulo: Comissão pela Criação do Parque Yanomami.

Agier, M., ed. 1997. *Anthropologues en danger: L'engagement sur le terrain.* Paris: Jean-Michel Place.

Aguiar, B. D. de. 1946. *Relatório—Comissão Brasileira Demarcadora de Limites* (1a *Divisão*). Belém: Comissão Brasileira Demarcadora de Limites.

Agulhon, M., P. Chaunu, G. Duby, R. Girardet, J. le Goff, M. Perrot, and R. Remond. 1987. *Essais d'ego-histoire.* Paris: Gallimard (Bibliothèque des Histoires).

Albert, B. 1985. "Temps du sang, temps des cendres: Représentation de la maladie, espace politique et système rituel chez les Yanomami du sud-est (Amazonie brésilienne)." Ph.D. diss., Université de Paris X Nanterre.

———. 1988. "La fumée du métal: Histoire et représentations du contact chez les Yanomami du Brésil." *L'Homme* 106–107: 87–119.

———. 1989. "Yanomami 'Violence': Inclusive Fitness or Ethnographer's Representation." *Current Anthropology* 30: 637–640.

———. 1990a. "Développement amazonien et sécurité nationale: Les Indiens Yanomami face au projet 'Calha Norte.'" *Ethnies* 11–12: 116–127.

———. 1990b. "On Yanomami Warfare: A Rejoinder." *Current Anthropology* 31: 558–562.

———. 1991. "Situação do garimpo na bacia do rio Demini (Amazonas)" and "Garimpo e malaria na área do alto Toototobi (Amazonas)." *Boletim URIHI* 14. São Paulo: Comissão Pró-Yanomami.

———. 1992. "Indian Lands, Environmental Policy, and Military Geopolitics in the Development of the Brazilian Amazon: The Case of the Yanomami." *Development and Change* 23 (1): 35–70.

———. 1993. "L'or cannibale et la chute du ciel: Une critique chamanique de l'économie politique de la nature." *L'Homme* 126–128: 353–382.

———. 1997a. "'Ethnographic situation' and ethnic movements: Notes on post-Malinowskian fieldwork." *Critique of Anthropology* 17 (1): 53–65.

———. 1997b. "Territorialité, ethnopolitique et développement: À propos du mouvement indien en Amazonie brésilienne." *Cahiers des Amériques latines* 23: 177–210. Also published as: "Territoriality, Ethnopolitics, and Development: The Indian Movement in the Brazilian Amazon." In *The Land Within: Indigenous Territory and the Perception of Environment,* ed. A. Surrallès and P. G. Hierro, 200–233. Copenhagen: International Work Group for Indigenous Affairs, 2005.

———. 1997c. *Palavras escritas para nos curar: Escola dos* Watorikɨ tʰëri pë. São Paulo: Comissão Pró-Yanomami / Ministério da Educação / Programa das Nações Unidas para o Desenvolvimento.

———. 2001. "Associations amérindiennes et développement durable en Amazonie brésilienne." *Recherches amérindiennes au Québec* 31 (3): 49–58.

———. 2004. "Les Indiens et l'État au Brésil." *Problèmes d'Amérique latine* 52: 63–83.

———. 2005. "Human Rights and Research Ethics among Indigenous People: Final Comments." In *Yanomami: The Fierce Controversy and What We Can Learn from It,* ed. R. Borofsky, 210–233. Berkeley: University of California Press.

———. 2009. "Native Land: Perspectives from Other Places." In *Native Land: Stop Eject,* ed. P. Virilio and R. Depardon, 37–58. Paris: Fondation Cartier pour l'art contemporain.

Albert, B., C. Esteves de Oliveira, D. Alves Francisco, G. Evelim Coelho, J.-B. Vieira, M. Filgueira da Villa, and V. Py-Daniel. 1995. *Projeto piloto de assistência às áreas endêmicas de oncocercose nos polos base de Toototobi e Balawa ú: Relatório final.* Brasília: Ministério da Saúde / Programa das Nações Unidas para o Desenvolvimento.

Albert, B., and G. Gomez. 1997. *Saúde Yanomami: Um manual etnolingüístico.* Belém: Museu Paraense Emílio Goeldi.

Albert, B., and D. Kopenawa. 1990. "Xawara: O ouro canibal e a queda do céu. Depoimento de Davi Kopenawa." In *Yanomami: A todos os povos da terra,* Ação pela Cidadania, 11–14. São Paulo: Comissão pela Criação do Parque Yanomami. French translation: 1993. "Fièvres de l'or." Special issue *Chroniques d'une conquête, Ethnies* 14: 39–44.

———. 2003. *Yanomami: L'esprit de la forêt.* Paris: Actes Sud / Fondation Cartier pour l'art contemporain.

Albert, B., and F.-M. Le Tourneau. 2004. "Florestas Nacionais na Terra Indígena Yanomami—Um cavalo de Tróia ambiental?" In *Terras Indígenas and Unidades de Conservação da Natureza,* 372–383. São Paulo: Instituto Socioambiental.

———. 2005. "Homoxi: Ruée vers l'or chez les Indiens Yanomami du Haut Mucajaí, Brésil." *Autrepart* 34: 3–28.

———. 2007. "Ethnogeography and Resource Use among the Yanomami Indians: Towards a Model of Reticular Space." *Current Anthropology* 48 (4): 584–592.

Albert, B., and I. Menegola. 1992. "O impacto sanitário dos garimpos em áreas indígenas: O caso Yanomami." Rio de Janeiro: Proceedings of the symposium "Forest '90" (Manaus, October 7–13, 1990), 12–16.

Albert, B., and W. Milliken. 2009. *Urihi a. A terra-floresta yanomami.* São Paulo: Instituto Socioambiental / IRD.

Albert, B., and M. W. Oliveira. 2011. "Yanomami: Novos 'isolados' ou antigos resistentes?" In *Povos Indígenas no Brasil 2006–2010,* ed. B. Ricardo and F. Ricardo, 279–283. São Paulo: Instituto Socioambiental.

Alès, C. 2003. "La horticultura yanomami y la problemática de los medios de sabanas en la Amazonía venezolana." In *Caminos cruzados: Ensayos en antropología social, etnoecología y etnoeducación,* ed. C. Alès and J. Chiappino, 389–421. Caracas: IRD / Universidades de los Andes (GIAL).

Andrade Gomes, A. 1958. *Relatório,* November 20, 1958. Manaus: Serviço de Proteção aos Índios / Posto Ajuricaba (1a *Inspetoria Regional*).

———. 1959. *Ofício* 18, May 24, 1959. Manaus: Serviço de Proteção aos Índios / Posto Ajuricaba (1a *Inspetoria Regional*).

Andujar, C. 2007. *Yanomami: La danse des images.* Paris: Marval Éditions.

Arantes, J. B. 1974. *Relatório,* October 1, 1974. Manaus: Fundação Nacional do Índio.

Arvello-Jiménez, N. 1971. *Political Relations in a Tribal Society: A Study of the Ye'kwana Indians of Venezuela.* Ithaca, NY: Cornell University Press.

Aurégan, P. 2001. *Des récits et des hommes: Un autre regard sur les sciences de l'homme.* Paris: Nathan / Plon (collection Terre Humaine).

Avila, J. B. de, and J. de S. Campos. 1959. "Observações de um acampamento de Índios Padauari e Paquidari." *Boletim da Sociedade de Geografia de Lisboa,* série 77, 7–9: 259–272.

Baines, S. G. 1991. *É a FUNAI que sabe: A frente de atração Waimiri-Atroari.* Belém: Museu Paraense Emílio Goeldi.

———. 1994. *Epidemics, the Waimiri-Atroari Indians and the Politics of Demography.* Série Antropologia 162. Brasília: Universidade de Brasília.

Balzac, H. de. 1977. *Facino Cane. La Comédie humaine,* vol. 6: *Études de mœurs: Scènes de la vie parisienne.* Paris: Gallimard (Bibliothèque de la Pléiade).

Barbosa, R. I. 2003. "Incêndios florestais em Roraima: Implicações ecológicas e lições para o desenvolvimento sustentado." In *Fronteira agro-pecuária e Terra Indígena Yanomami em Roraima,* Documentos Yanomami 3, ed. B. Albert, 43–54. Brasília: Comissão Pró-Yanomami.

Barthes, R. 1973. *Le Plaisir du texte.* Paris: Éditions du Seuil (collection Points).

Basso, E. B. 1995. *The Last Cannibals: A South American Oral History.* Austin: University of Texas Press.

Becher, H. n.d. *Relatório sobre uma viagem de pesquisas no norte do Brasil na região compreendida entre os Rios Demini e Aracá.* Rio de Janeiro: Serviço de Proteção aos Índios / Ministério da Agricultura.

———. 1957. "Die Yanonami: Ein beitrag zur frage der völkergruppierung zwischen Rio Branco, Uraricuéra, Serra Parima und Rio Negro." *Wiener Völkerkundliche Mitteilungen* 5 (1): 13–20.

————. 1960. *Die Surara und Pakidai: Zwei Yanonami-Stämme in Nordwest Brasilien*. Hamburg: Mitteilungen aus dem Museum für Volkerkunde, vol. 26.

Bezerra de Lima, F. 1974. *Relatório,* March 1, 1974. Manaus: Fundação Nacional do Índio (1a *Delegacia Regional*).

Bigio, E. dos Santos. 2007. "Programa(s) de Índio(s): Falas, contradições, ações interinstitucionais e representações sobre Índios no Brasil e na Venezuela (1960–1992)." Ph.D. diss., Universidade de Brasília.

Biocca, E. 1965. *Yanoama: Dal raconto di una dona rapita degli Indi*. Bari: Leonardo da Vinci.

————. 1968. *Yanoama: Récit d'une femme brésilienne enlevée par les Indiens*. Paris: Plon (collection Terre Humaine).

————. 1970. *Yanoama: The Narrative of a White Girl Kidnapped by Amazonian Indians*. New York: Dutton.

Birnbaum, J. 2000. "Les Indiens Yanomami ont-ils été victimes d'expériences eugéniques?" *Le Monde,* October 1, 2000.

Bloch, M. 2004. "Lévi-Strauss chez les Britanniques." In *L'Herne. Lévi-Strauss,* ed. M. Izard, 349–356. Paris: Éditions de l'Herne.

Borges, J.-L. 1987. *Livre de préfaces*. Paris: Gallimard (collection Folio).

Borgman, D. 1990. "Sanumá." In *Handbook of Amazonian Languages,* ed. D. C. Derbyshire and G. K. Pullum, 2:17–248. The Hague: Mouton.

Borofsky, R., ed. 2005. *Yanomami: The Fierce Controversy and What We Can Learn from It*. Berkeley: University of California Press.

Breerette, G. 2003. "Produire des œuvres avec les Indiens Yanomami." *Le Monde,* May 30, 2003.

Brooks, E., R. Fuerst, J. Hemming, and F. Huxley. 1973. *Tribes of the Amazon Basin 1972*. London: C. Knight.

Brumble, H. D. 1988. *American Indian Autobiography*. Berkeley: University of California Press.

Campbell, A. T. 1989. *To Square with Genesis: Causal Statements and Shamanic Ideas in Wayãpi*. Edinburgh: Edinburgh University Press.

Capobianco, J.-P. R., ed. 2001. *Biodiversidade na Amazônia Brasileira*. São Paulo: Editora Estação Liberdade / Instituto Socioambiental.

Caratini, S. 2004. *Les non-dits de l'anthropologie*. Paris: Presses Universitaires de France (collection Libelles).

Cardoso de Oliveira, R. 1964. *O Índio e o mundo dos Brancos*. São Paulo: Pioneira.

Carrera Rubio, J. 2004. "The Fertility of Words: Aspects of Language and Sociality among the Yanomami People of Venezuela." Ph.D. diss., University of Saint Andrews.

Carvalho, J. P. F. de. 1982. *Waimiri Atroari: A história que ainda não foi contada*. Brasília: by the author.

Castro, M. B., B. Albert, and W. C. Pfeiffer. 1991. "Mercury Levels in Yanomami Indians' Hair from Roraima-Brazil." In *Heavy Metals in the Environment,* ed. J. G. Farmer, 367–370. Edinburgh: CEP Consultants.

Castro, O. de. 1975. *Relatório,* May 4, 1975. Boa Vista: Fundação Nacional do Índio.

Chagnon, N. A. 1966. "Yanomamö Warfare, Social Organization and Marriage Alliances." Ph.D. diss., University of Michigan.

———. 1968. *Yanomamö: The Fierce People.* New York: Holt, Rinehart and Winston.

———. 1974. *Studying the Yanomamö.* New York: Holt, Rinehart and Winston.

———. 1988. "Life Histories, Blood Revenge, and Warfare in a Tribal Population." *Science* 239: 985–992.

Chateaubriand, F.-R. de. 1969. "Voyage en Amérique." In *Œuvres romanesques et voyages.* Paris: Gallimard (Bibliothèque de la Pléiade).

Cocco, L. 1987. *Iyëwei-teri: Quince años entre los Yanomamos.* 2nd ed. Caracas: Escuela Técnica Don Bosco.

Colchester, M. 1984. "Rethinking Stone Age Economics: Some Speculations Concerning the Pre-Colombian Yanoama Economy." *Human Ecology* 12 (3): 291–314.

Comissão Pró-Yanomami (CCPY). 1979. "Yanomami Indian Park, Proposal and Justification." Report prepared by B. Albert and C. Zacquini, under the direction of C. Andujar. In *The Yanoama in Brazil 1979,* ed. A. R. Ramos and K. I. Taylor. IWGIA Document 37. Copenhagen: International Work Group for Indigenous Affairs.

———. 1989a. *Boletim Urihi* 10. São Paulo: Comissão Pró-Yanomami.

———. 1989b. "Mineração: O esbulho das terras Yanomami. Histórico das invasões 1975–1989." *Boletim Urihi* 11. São Paulo: Comissão Pró-Yanomami.

Costa, S. A. da. 1976a. *Relatório,* July 14, 1976. Boa Vista: Fundação Nacional do Índio.

———. 1976b. *Relatório,* no. 3/FAY/1976. Boa Vista: Fundação Nacional do Índio.

———. 1977. *Relatório,* December 5, 1977. Boa Vista: Fundação Nacional do Índio.

———. 1978. *Relatório,* April 27, 1978. Boa Vista: Fundação Nacional do Índio.

Crapanzano, V. 1972. *The Fifth World of Forster Bennet: A Portrait of a Navaho.* New York: Viking.

Crutzen, P., and E. F. Stoermer. 2000. "The 'Anthropocene.'" *Global Change* (IGBP Newsletter) 41: 17–18.

Descola, P. 1994. "Rétrospections." *Gradhiva* 16: 15–27.

Dhoquois, A. 2008. *Comment je suis devenu ethnologue.* Paris: Le Cavalier Bleu Éditions.

Dixon, R. M. W., and A. Y. Aikhenvald, eds. 1999. *The Amazonian Languages.* Cambridge: Cambridge University Press.

Dorfman, A., and J. Maier. 1990. "Assault in the Amazon." *Time Magazine,* November 5, 1990.

Dorst, J. 1996. "Les oiseaux ne sont pas tombés du ciel." In *Comme un oiseau,* ed. H. Chandès, 47–74. Paris: Gallimard / Électa: Fondation Cartier pour l'art contemporain.

Duarte do Pateo, R. 2005. "Niyayu: Relações de antagonismo e aliança entre os Yanomam da Serra das Surucucus." Ph.D. diss., Universidade de São Paulo.

Duthil, F. 2006. *Histoire de femmes aborigènes*. Paris: Presses Universitaires de France/Le Monde.

Elvidge, C. D., V. R. Hobson, K. E. Baugh, J. B. Dietz, Y. E. Shimabukuro, T. Krug, E. M. L. Novo, and F. R. Echavarria. 2001. "DMSP-OLS Estimation of Tropical Forest Area Impacted by Surface Fires in Roraima, Brazil: 1995 versus 1998." *International Journal of Remote Sensing* (22) 14: 2661–2673.

Fassin, D., and A. Bensa, eds. 2008. *Les Politiques de l'enquête: Épreuves ethnographiques*. Paris: La Découverte.

Figueiredo Costa, G. P. 1967. *Relatório*, August 22, 1967. Manaus: Serviço de Proteção aos Índios.

Fragoso, J. M. V. 1997. "Desapariciones locales del baquiro labiado *(Tayassu pecari)* en la Amazonia: Migración, sobre-cosecha, o epidemia." In *Manejo de fauna silvestre en la Amazonia*, ed. T. G. Fang et al. Lima: UNAP-University of Florida-UNDP/GEF-Universidad Mayor de San Andrés.

Fuerst, R. 1967. "Die Gemeinschaftswohnung der Xiriana am Rio Toototobi: Beitrag zur kenntnis der Yanomami-Indianer in Brasilien." *Zeitschrift für ethnologie* 92 (1): 103–113.

Fundação Nacional do Índio (FUNAI). 1975. *Relatório*, September 3, 1975: *Projeto de Emergência Roraima*. Brasília: Fundação Nacional do Índio.

Geertz, C. 1973. *The Interpretation of Cultures*. New York: Basic Books.

Geffray, C. 1995. *Chroniques de la servitude en Amazonie brésilienne*. Paris: Karthala.

———. 2001. *Trésors: Anthropologie analytique de la valeur*. Strasbourg: Éditions Arcanes.

Ghasarian, C., ed. 2004. *De l'ethnographie à l'anthropologie réflexive: Nouveaux terrains, nouvelles pratiques, nouveaux enjeux*. Paris: Armand Colin (collection U).

Gheerbrant, A. 1952. *L'Expedition Orénoque-Amazone*. Paris: Gallimard.

Golden, T. 1991. "Talk about Culture Shock: Ant People in Sky-High Huts." *New York Times*, April 17, 1991, B1, 4.

Gomez, G. G. 1990. "The Shiriana Dialect of Yanam (Northern Brazil)." Ph.D. diss., Columbia University.

Good, K. 1989. "Yanomami Hunting Patterns: Trekking and Garden Relocation as an Adaptation to Game Availability in Amazonia, Venezuela." Ph.D. diss., University of Florida.

———. 1995. "Yanomami of Venezuela: Foragers or Farmers—Which Came First?" In *Indigenous Peoples and the Future of Amazonia: An Ecological Anthropology of an Endangered World*, ed. L. E. Sponsel, 113–120. Tucson: University of Arizona Press.

Gourou, P. 1982. *Terre de bonne espérance: Le monde tropical*. Paris: Plon (collection Terre Humaine).

Guimarães, S. M. F. 2005. "Cosmologia Sanumá: O xamã e a constituição do ser." Ph.D. diss., University of Brasília.

Guimarães Rosa, J. 2001. *Primeiras Estórias*. Rio de Janeiro: Nova Fronteira.

Guiraut Denis, H. 1993. "Brésil: Après le massacre de plusieurs dizaines d'Indiens, les Yanomami exigent le départ des chercheurs d'or de leur territoire." *Le Monde*, August 24, 1993.

Hames, R. 1983. "The Settlement Pattern of a Yanomamö Population Block: A Behavioural Ecological Interpretation." In *Adaptive Responses of Native Amazonians*, ed. R. B. Hames and W. T. Vickers, 393–427. New York: Academic Press.

Hartman, Mrs. B. 1968. "On Another Planet." *Brown Gold* 25 (9).

Heinen, D. H. 1983–1984. "Themes in Political Organization: The Caribs and Their Neighbours. Introduction." *Antropológica* 59–62: 1–8.

Hendricks, J. W. 1993. *To Drink of Death: The Narrative of a Shuar Warrior*. Tucson: University of Arizona Press.

Holmes, R. 1995. "Small Is Adaptive: Nutritional Anthropometry of Native Amazonians." In *Indigenous Peoples and the Future of Amazonia: An Ecological Anthropology of an Endangered World*, ed. L. E. Sponsel, 121–148. Tucson: University of Arizona Press.

Horst, C. 1977. *Relatório*, July 1977. Brasília: Fundação Nacional do Índio.

Huber, O., J. A. Steyermark, G. T. Prance, and C. Alès. 1984. "The Vegetation of the Sierra Parima, Venezuela-Brazil: Some Results of Recent Exploration." *Brittonia* 36 (2): 104–139.

Jaulin, R. 1970. *La Paix blanche: Introduction à l'ethnocide*. Paris: Éditions du Seuil (collection Combats).

Jovita, M. de L. 1948. *Roteiro Etnográfico, Comissão Brasileira Demarcadora de Limites* (1a *Divisão*) 1941–1943. Belém: Comissão Brasileira Demarcadora de Limites.

Kamm, T., 1990. "Amazon Tragedy: White Man's Malaria and Pollution Imperil Remote Tribe in Brazil." *Wall Street Journal*, March 22, 1990.

Keesing, R. M. 1978. *Elota's Story: The Life and Times of a Solomon Islands Big Man*. St. Lucia, Australia: University of Queensland Press.

Koch-Grünberg, T. 1982. *Del Roraima al Orinoco*, vol. 3. Caracas: Ediciones del Banco Central de Venezuela. (Orig. pub. 1924.)

Kopenawa, D. 1992. "O projeto de saúde Demini: Mensagem para Bruce Albert gravada por Lucimara Montejane." *Boletim URIHI* 16. São Paulo: Comissão Pró-Yanomami.

Krupat, A. 1994. *Native American Autobiography*. Madison: University of Wisconsin Press.

Kunstadter, P. 1979. "Démographie." In *Écosystèmes forestiers tropicaux*, 345–380. Paris: UNESCO.

Lacorte, J. G., and R. Veronesi. 1976. "Influenza (grippe)." In *Doenças infeciosas e parasitárias*, ed. R. Veronesi. Rio de Janeiro: Guanabara Koogan.

Lazarin, M., and L. Vessani. 1987. *Xiriana, Índios que garimpam. Relatório de pesquisa na área Yanomami (Roraima)*. Universidade federal de Goiás-CNPq, mimeo.

Lejeune, P. 1980. *Je est un autre: L'autobiographie, de la littérature aux medias*. Paris: Éditions du Seuil (collection Poétique).

Leservoisier, O., and D. Vidal, eds. 2008. *L'Anthropologie face à ses objets: Nouveaux contextes ethnographiques*. Paris: Éditions des archives contemporaines.

Le Tourneau, F.-M. 2003. "Colonização agrícola e áreas protegidas no oeste de Roraima." In *Fronteira agro-pecuária e Terra Indígena Yanomami em Roraima*, Documentos Yanomami 3, ed. B. Albert, 11–42. Brasília: Comissão Pró-Yanomami.

———. 2010. *Les Yanomami du Brésil, géographie d'un territoire amérindien*. Paris: Belin (collection Mappemonde).

Le Tourneau, F. M., and B. Albert. 2010. "Homoxi (1989–2004): O impacto ambiental das atividades garimpeiras na Terra Indígena Yanomami (Roraima)." In *Roraima: Homem, ambiente e ecologia*, ed. R. I. Barbosa and V. F. Melo, 155–170. Boa Vista: FEMACT.

Lévi-Strauss, C. 1955. *Tristes Tropiques*. Paris: Plon (collection Terre Humaine; édition Pocket, 2001).

———. 1962. *La Pensée sauvage*. Paris: Plon.

———. 1983. *Structural Anthropology*, vol. 2. Chicago: University of Chicago Press.

———. 1993. "Présentation." *Chroniques d'une Conquête, Ethnies* 14: 5–7.

———. 1996. "L'origine de la couleur des oiseaux." In *Comme un oiseau*, ed. H. Chandès, 23–41. Paris: Gallimard / Électa: Fondation Cartier pour l'art contemporain.

Lizot, J. 1974. "Histoires indiennes d'amour." *Les Temps modernes* 339: 1–34.

———. 1976. *Le cercle des feux: Faits et dits des Indiens Yanomami*. Paris: Éditions du Seuil.

———. 1984. *Les Yanõmami centraux*. Paris: Éditions de L'EHESS (Cahiers de l'Homme).

———. 1985. *Tales of the Yanomami: Daily Life in the Venezuelan Forest*. Cambridge: Cambridge University Press; Paris: Maison des Sciences de l'Homme.

———. 1986. "La Recolección y las Causas de su Fluctuación." *Extracta* 5: 35–40.

———. 1987. "Compte rendu de Valero, H., 1984, *Yo soy Napëyoma: Relato de una mujer raptada por los indígenas Yanomami*." *L'Homme* 27 (101): 176–178.

———. 1988. "Los Yanõmami." In *Los aborígenes de Venezuela*, vol. 3, ed. J. Lizot, 479–583. *Etnologia contemporanea*. Caracas: Fundación La Salle de Ciencias Naturales / Monte Ávila.

———. 1994. "Words in the Night: The Ceremonial Dialogue—One Expression of Peaceful Relationships among the Yanomami." In *The Anthropology of Peace and Nonviolence*, ed. L. Sponsel and T. Gregor. London: Lynne Rienner.

———. 1996. *Introducción a la lengua yãnomami: Morfología*. Caracas: Vicariato Apostólico de Puerto Ayacucho.

———. 2004. *Diccionario enciclopédico de la lengua yãnomãmi*. Puerto Ayacucho: Vicariato Apostólico.

Lopes de Araujo, F. X. 1884. *Relatório da Comissão Brasileira Demarcadora de Limites Brasil-Venezuela (1879–1884)*. Rio de Janeiro: Itamaraty.

MacMillan, G. 1995. *At the End of the Rainbow? Gold, Land and People in the Brazilian Amazon*. London: Earthscan.

Magalhães, D. de. 1943. *Normas para atração e pacificação.* Manaus: Serviço de Proteção aos Índios (1a IR).

Malaurie, J. 2003. *L'Allée des baleines.* Paris: Éditions Mille et une nuits.

———. 2008. *Terre Mère.* Paris: CNRS Éditions.

Marcos, J. J. 1976. *Relatório,* July 2, 1976. Brasília: Fundação Nacional do Índio.

Matsumoto, Y., K. Fujimoto, and R. Tsuda. 1974. *Yanomamutachi.* Osaka: Tankenbu Kansai Daigaku.

Mattei-Müller, M.-C. 2007. *Lengua y cultura Yanomami: Diccionario ilustrado Yanomami-Español/Español-Yanomami.* Caracas: Epsilon Libros.

Maybury-Lewis, D. 1967. *Akwe-Shavante Society.* Oxford: Clarendon Press.

McKnight, J. 1958. "Tototobi." *Brown Gold* 16 (6).

Melo, M. G. de. 1982. *Relatório,* January 1982. Brasília: Fundação Nacional do Índio.

———. 1985. *Relatório,* June 5, 1985. Boa Vista: Fundação Nacional do Índio.

Mendes, C. 1990. *Mon combat pour la forêt.* Paris: Éditions du Seuil.

Menget, P. 2001. *Em nome dos outros: Classificação das relações sociais entre os Txicão do Alto Xingu.* Lisboa: Assírio e Alvim.

Migliazza, E. 1972. "Yanomama grammar and intelligibility." Ph.D. diss., Indiana University.

———. 1982. "Linguistic Prehistory and the Refuge Model in Amazonia." In *Biological Diversification in the Tropics,* G. T. Prance, 497–519. New York: Columbia University Press.

Milliken, W., and B. Albert. 1996. "The Use of Medicinal Plants by the Yanomami Indians of Brazil." *Economic Botany* 50 (1): 10–25.

———. 1997a. "The Use of Medicinal Plants by the Yanomami Indians of Brazil. Part II." *Economic Botany* 51 (3): 264–278.

———. 1997b. "The Construction of a New Yanomami Round-house." *Journal of Ethnobiology* 17 (2): 215–233.

Milliken, W., and B. Albert, eds. 2002. *Homoxi: Degraded Areas in the Yanomami Territory, Roraima, Brazil.* Available at www.proyanomami.org.br/frame1/noticia.asp?id=1388

Mont'alverne Pires, F. 1974. *Relatório,* November 27, 1974. Manaus: Fundação Nacional do Índio (1a *Delegacia Regional*).

Monteiro Caltaneão, A. C. 1977. *Relatório,* April 22, 1977. Boa Vista: Instituto Brasileiro de Desenvolvimento Florestal/Roraima.

Moore, B. 1973. "Three-hour War at Surucucus." *Brown Gold* 31(6).

Muratorio, B. 1991. *The Life and Times of Grandfather Alonso: Culture and History in the Upper Amazon.* New Brunswick, NJ: Rutgers University Press.

Navet, E. 1994–1995. "Le rôle des truchements dans les relations franco-amérindiennes sur la côte du Brésil au XVIe siècle. Documents coloniaux," *Amerindia* 19–20: 39–49.

Neel, J. V., T. Arends, C. Brewer, N. Chagnon, H. Gershowitz, M. Layrisse, J. MacCluer, E. Migliazza, W. Oliver, F. Salzano, R. Spielman, R. Ward, and L. Weitkamp. 1972. "Studies on the Yanomama Indians." *Proceedings of the*

Fourth International Congress of Human Genetics, 96–111. Amsterdam: Excerpta Medica.

Neel, J. V., W. R. Centerwall, and N. A. Chagnon. 1970. "Notes on the Effects of Measles and Measles Vaccine in a Virgin Soil Population." *American Journal of Epidemiology* 91 (4): 418–429.

New York Times. 1993. "Death in the Rain Forest." August 27, 1993, A28.

Oakdale, S. 2005. *I Foresee My Life: The Ritual Performance of Autobiography in an Amazonian Community*. Lincoln: University of Nebraska Press.

O'Connor, G. 1997. *Amazon Journal: Dispatches from a Vanishing Frontier*. New York: Dutton.

Oliveira, D. de. 1977. Memo no. 202/COAMA/77C, June 14, 1977. Brasília: Fundação Nacional do Índio.

Oliveira, L. P. de. 1959. *Relatório. Campanha 1958/1959. Rio Toototobi*. Belém: 1a Comissão Brasileira Demarcadora de Limites.

Pacheco Rogedo, I. M. 1976. *Relatório*, July 5, 1976. Brasília: Fundação Nacional do Índio.

Paixão, A. M. da. 1977. *Relatório*. Brasília, FUNAI.

Pearce, F. 2009. "Rainforests May Pump Winds Worldwide." *New Scientist* 2072: 6–7.

Perri Ferreira, H. 2009. "Los clasificadores nominales en el Yanomama de Papiu (Brazil)." Master's thesis, Centro de Investigaciones y Estudios Superiores en Antropologia Social (Ciesas), Mexico.

Peters, J. H. 1973. "The Effects of Western Material Goods on the Social Structure of the Family among the Shiriana." Ph.D. diss., University of Michigan.

Poulson, T. and M. 1968. "Great Things God Hath Done . . ." *Brown Gold* 26 (2).

Ramirez, H. 1992. "Le Bahuana: Une nouvelle langue de la famille arawak." *Chantiers Amerindia* 17.

———. 1994. "Le Parler yanomamɨ des Xamatauteri." Ph.D. diss., Université d'Aix-en-Provence.

Ramos, A. R. 1975. *Manual para treinamento na lingua Yanomam*. Brasília: Fundação Universidade de Brasília (Trabalhos de Ciencias Sociais, Série Antropologia).

———. 1979. "Yanoama Indians in Northern Brazil Threatened by Highway." In *The Yanoama in Brazil 1979*, ed. A. R. Ramos and K. I. Taylor, 1–41. IWGIA Document 37. Copenhagen: International Work Group for Indigenous Affairs.

———. 1992. "Reflecting on the Yanomami: Ethnographic Images and the Pursuit of the Exotic." In *Rereading Cultural Anthropology*, ed. G. E. Marcus, 48–68. Durham, NC: Duke University Press.

———. 1995. *Sanumá Memories: Yanomami Ethnography in Times of Crisis*. Madison: University of Wisconsin Press.

Ramos, A. R., and B. Albert. 1977. "Yanoama Descent and Affinity: The Sanumá / Yanomam Contrast." In *Actes du XLIIe Congrès international des Américanistes*, 2:71–90. Paris: Société des Américanistes.

Ramos, A. R., M. A. Lazarin, and G. Goodwin Gomez. 1986. "Yanomami em tempo de ouro. Relatório de pesquisa." In *Culturas Indígenas de la Amazonia*. Madrid: Biblioteca Quinto Centenario, 73–83.

Ramos, A. R., and K. I. Taylor. 1973. *Research Opportunity in North Brazil*. Brasília: mimeo.

———, eds. 1979. *The Yanoama in Brazil 1979*. IWGIA Document 37. Copenhagen: International Work Group for Indigenous Affairs.

Ribeiro, D. 1970. *Os Índios e a Civilização: A integração das populações indígenas no Brasil moderno*. Rio de Janeiro: Civilização Brasileira.

Ricardo, F., and A. Rolla. 2005. *Mineração em Terras Indígenas na Amazônia brasileira*. São Paulo: Instituto Socioambiental.

Rivière, P. 1969. *Marriage among the Trio: A Principle of Social Organization*. Oxford: Clarendon Press.

Rocha, J. 1999. *Murder in the Rainforest: The Yanomami, the Gold Miners and the Amazon*. London: Latin American Bureau. Updated edition in Portuguese: 2007. *Haximu. O massacre dos Yanomami e as suas conseqüências*. São Paulo: Casa Amarela.

Roosevelt, M. 2000. "Yanomami: What Have We Done to Them?" *Time Magazine*, October 2, 2000.

Roux, E. de. 2003. "Comment artistes et chamans se sont rencontrés." *Le Monde*, May 30, 2003.

Rubenstein, S. 2002. *Alejandro Tsakimp: A Shuar Healer in the Margins of History*. Lincoln: University of Nebraska Press.

Saffirio, J.-B. 1976. "Relatório Missão Catrimani, outubro 1965– outubro 1975," *Boletim do CIMI (Conselho Indigenista Misssionário)* 25: 15–18.

———. 1980. "Some Social and Economic Changes among the Yanomama of Northern Brazil (Roraima): A Comparison of 'Forest' and 'Highway' Villages." Master's thesis, University of Pittsburgh.

Schweickardt, J. C. 2002. *Magia e Religião na Modernidade: Os rezadores em Manaus*. Manaus: Editora da Universidade do Amazonas.

Shapiro, J. 1972. "Sex Roles and Social Structure among the Yanomama Indians of Northern Brazil." Ph.D. diss., Columbia University.

Shelley, A. J. 1976. "Observações preliminares sobre a transmissão da oncocercose no rio Toototobi, Amazonas, Brasil." *Acta Amazonica* 6 (3): 327–334.

Shostak, M. 1981. *Nisa: The Life and Words of a !Kung Woman*. Cambridge, MA: Harvard University Press.

Smiljanic, M. I. 1999. "O Corpo Cósmico: O Xamanismo entre os Yanomae do Alto Toototobi." Ph.D. diss., Universidade de Brasília.

———. 2002. "Os enviados de Dom Bosco entre os Masiripiwëiteri: O impacto missionário sobre o sistema social e cultural dos Yanomami ocidentais (Amazonas, Brasil)." *Journal de la Société des Américanistes* 88: 137–158.

———. 2003. *Cristãos Conversos, Xamãs Professos: Infantício, cristianismo e contato interétnico entre os Yanomae do alto Toototobi*, ms.

Smole, W. J. 1976. *The Yanoama Indians: A Cultural Geography*. Austin: University of Texas Press.

Spielman, R. S., E. C. Migliazza, J. V. Neel, H. Gershowitz, and R. T. de Araúz. 1979. "The Evolutionary Relationships of Two Populations: A Study of the Guaymí and the Yanomama." *Current Anthropology* 20 (2): 377–388.

Sponsel, L. 1986. "Amazon Ecology and Adaptation." *Annual Review of Anthropology* 15: 67–97.

Taylor, A. C. 1981. "God-Wealth: The Achuar and the Missions." In *Transformations and Ethnicity in Modern Ecuador*, ed. N. E. Whitten, 647–667. Urbana: University of Illinois Press.

———. 2004. "Don Quichotte en Amérique: Claude Lévi-Strauss et l'anthropologie américaniste." In *L'Herne. Lévi-Strauss*, ed. M. Izard, 92–98. Paris: Éditions de l'Herne.

Taylor, K. I. 1974. *Sanumá Fauna: Prohibitions and Classifications*. Caracas: Fundación La Salle de Ciencias Naturales.

———. 1975a. "Descrição sumária do Projeto Yanoama." In *Política e ação indigenista brasileira*. Brasília: Fundação Nacional do Índio.

———. 1975b. *Memo ao chefe da COAMA*, October 31, 1975. Brasília: Fundação Nacional do Índio.

———. 1975c. *Memo ao DGO*, October 6, 1975. Brasília: Fundação Nacional do Índio.

———. 1975d. *Viagem ao rio Ajarani e à Missão Catrimani (TF de Roraima), janeiro–maio de 1975. Memo ao DEP*, January 5, 1975. Brasília: Fundação Nacional do Índio.

———. 1979. "Development against the Yanoama: The Case of Mining and Agriculture." In *The Yanoama in Brazil 1979*, ed. A. R. Ramos and K. I. Taylor. IWGIA Document 37. Copenhagen: International Work Group for Indigenous Affairs.

Thomas, D. 2003. "Portrait of a Tribe." *Newsweek*, July 14, 2003.

Tierney, P. 2000. *Darkness in El Dorado: How Scientists and Journalists Devastated the Amazon*. New York: W. W. Norton.

Time-Life. 1982. *Aborigines of the Amazon Rain Forest: The Yanomami*, photographs: V. Englebert; text: R. Hanbury-Tenison and editors of Time-Life; anthropological consultant: B. Albert. Amsterdam: Time-Life Books (Peoples of the Wild Series).

Todorov, T. 1971. *Poétique de la prose*. Paris: Éditions du Seuil.

Toototobi Gang, The. 1970a. "Satan's Counter-Attack." *Brown Gold* 27 (10).

———. 1970b. "Latest from Toototobi." *Brown Gold* 28 (2).

Turner, T., and D. Kopenawa. 1991. "'I Fight Because I Am Alive': An Interview with Davi Kopenawa Yanomami." *Cultural Survival Quarterly* 91: 59–64.

Valero, H. 1984. *Yo soy Napëyoma: Relato de una mujer raptada por los indígenas Yanomami*. Caracas: Fundación La Salle de Ciencias Naturales.

Vanhecke, C. 1990. "La détresse des Indiens Yanomami: Malgré les promesses du nouveau gouvernement brésilien, le grand pillage de l'Amazonie continue." *Le Monde*, August 2, 1990.

Viveiros de Castro, E. 2007. "La forêt des miroirs: Quelques notes sur l'ontologie des esprits amazoniens." In *La Nature des esprits dans les cosmologies autochtones*, ed. F. B. Laugrand and J. G. Oosten, 45–74. Quebec: Les Presses de l'Université Laval.

Wardlaw, K. 1964. "Uaica News." *Brown Gold* 22 (4).

———. 1968. "A Far Greater Tragedy." *Brown Gold* 25 (10).

———. 1970a. "Change of Heart." *Brown Gold* 27 (11): 5.

———. 1970b. "Trouble at Toototobi." *Brown Gold* 27 (11): 6–8.

Wardlaw, Mrs. K. 1965a. "A Little Progress." *Brown Gold* 22 (12).

———. 1965b. "Report on Air Drop February 23." *Brown Gold* 22 (12).

Wardlaw, K. and M. 1968a. "Uaica Breakthrough." *Brown Gold* 25 (12).

———. 1968b. "Among the Uaica." *Brown Gold* 26 (2).

White, H. 1978. *Tropics of Discourse: Essays in Cultural Criticism*. Baltimore: Johns Hopkins University Press.

Wilbert, J., and K. Simoneau, eds. 1990. *Folk Literature of the Yanomami Indians*. Los Angeles: UCLA Latin American Center Publications.

Wilford, J. N., and S. Romero. 2000. "Book Seeks to Indict Anthropologists Who Studied Brazil Indians." *New York Times*, September 28, 2000.

Wiseman, B. 2005. "La réconciliation." *L'Homme* 175–176: 397–418.

Wong, H. D. 1992. *Sending My Heart Back across the Years: Tradition and Innovation in Native American Autobiography*. New York: Oxford University Press.

Yssao, K. 1975. *Relatório de viagem à Rodovia Perimetral Norte BR-210—Tr. Caracaraí—Rio Padauarí 14 a 15/12/74*, February 25, 1975. Brasília: Centro de Documentação Memória Camargo Corrêa (CDMCC).

Zempléni, A. 1984. "Secret et sujétion: Pourquoi ses 'informateurs' parlent-ils à l'ethnologue?" *Traverses* 30–31: 102–115.

Zerries, O. 1964. *Waika: Die Kulturgeschichtliche Stellung der Waika-Indianer des Oberen Orinoco im Rahmen der Völkerkunde Südamerikas*. Munich: Klauss Renner Verlag.

Zimmerman, P. 1960. "Visit with the Xirianos." *Brown Gold* 18 (3).

Acknowledgments

My deepest gratitude goes first to those who, at various times, made the writing of this book possible, both directly and indirectly, through their friendship, support, advice, suggestions, and encouragement: Patrick Menget, Alcida Ramos, Hervé Chandès, and Jean Malaurie. Their crucial role in the origin and publication of this book is described in the last chapter, "How This Book Was Written." I am also especially grateful to my wife, Gabriela Levy, for her constant patience and sagacious help over the many days and intense work that went into this writing.

I am much obliged to Professor Claude Lévi-Strauss (†) and my colleagues and friends Eduardo Viveiros de Castro, Manuela Carneiro da Cunha, and François-Michel Le Tourneau, who have been kind enough to read and comment on all or parts of the manuscript at various stages. I am indebted to Gale Goodwin Gomez and Helder Ferreira Perri for kindly providing helpful comments on linguistic aspects of the text, and to the latter, along with Maurice Tomioka Nilsson, for many useful ethnobiological observations. I owe special thanks for their generous patience to François-Michel Le Tourneau in producing the original versions of the maps included in the book, and to William Milliken for putting at my disposal so many botanical indentifications and so much information. My thanks go as well to the following friends and institutions for kindly opening their photo archives: Raymond Depardon, Dafran Gomes Macário, William Milliken, Fiona Watson (Survival International), and Marcos Wesley de Oliveira (Instituto Socioambiental).

Last but not least, I am very grateful to Rebecca Byers and Gale Goodwin Gomez for their support in bringing about the publication of this book in English, as well as to Elizabeth Knoll, a wonderful editor, and Nicholas Elliott and Alison Dundy, very talented translators, who made this book possible with dedication and heroic effort. My gratitude also goes to the editorial and production staff at Harvard University Press for their invaluable dedication to this project.

—B. A.

Index